VEGETATION DESCRIPTION
AND DATA ANALYSIS

VEGETATION DESCRIPTION AND DATA ANALYSIS

A PRACTICAL APPROACH

Second Edition

Martin Kent

University of Plymouth

A John Wiley & Sons, Ltd., Publication

Library of Congress Cataloging-in-Publication Data

Kent, M., 1950–
　Vegetation description and data analysis : a practical approach / Martin Kent. – 2nd ed.
　　p. cm.
　Includes bibliographical references and index.
　ISBN 978-0-471-49092-0 (cloth) – ISBN 978-0-471-49093-7 (paper)
　1. Plant ecology. 2. Plant communities. 3. Vegetation surveys. 4. Plant ecology–Data processing.
5. Plant communities–Data processing. 6. Vegetation surveys–Data processing. I. Kent, M., 1950–
Vegetation description and analysis. II. Title.
　QK901.K37 2011
　581.7–dc23

　　　　　　　　　　　　　　　　　　　　　　　　　　　　　　　2011030218

A catalogue record for this book is available from the British Library.

This book is published in the following electronic format: ePDF 9781119944782; ePub 9781119962397; Mobi 9781119962403

Set in 10/12pt Times by Aptara Inc., New Delhi, India

Printed in Singapore by Markono Print Media Pte Ltd

First Impression 2012

This second edition is dedicated to my dear friend, academic colleague and former co-author, Paddy Coker, who sadly died in July 2005. His enthusiasm for and enjoyment of the subjects of plant ecology, vegetation science and computing are greatly missed.

Contents

Preface to the second edition

The success of the first edition of this text was as much of a surprise to the authors as it was to both the publishers and its academic and scientific audience. The enthusiastic response demonstrated that there was a very clear need for a book that sought to simplify the complexities of vegetation description and multivariate analysis in the context of vegetation data and plant ecology. The first edition went through eight reprints and only went out of print in 2005. A second edition has been at the proposal stage for over ten years and it is only now, following retirement from mainstream academic life, that the author has found the time to develop the project further. Nevertheless, the need for such a text appears to be as great as ever.

In the 19 years since the publication of the first edition, a great deal has changed in the world of vegetation science and plant ecology. What is now one of the key journals in the subject, *Journal of Vegetation Science,* was only founded in 1990, two years before the publication of the first edition of this book in 1992. The sister journal *Applied Vegetation Science* only appeared in 1997. Numerous other journals relevant to the subject have evolved during the intervening years and the whole field of multivariate analysis has extended its application across the full range of ecological sciences. While actual methods and techniques have evolved only relatively slowly over this time, far more significant changes have occurred in the world of computer hardware and software. As the review of software in Chapter 9 of this new edition demonstrates, there are now numerous quite sophisticated packages available for vegetation data analysis, and a whole new approach has emerged at the research frontier using the R language and related packages.

Over the past 20 years, the use of methods of vegetation description and analysis has extended globally. Originally the province of a relatively small group of academics in the United Kingdom, Europe, North America and Australia, examination of the range of research locations of scientific papers published in *Journal of Vegetation Science* and *Applied Vegetation Science* clearly demonstrates that the scope and application of vegetation description and analysis by scientists and academics is now truly worldwide. One of the paradoxes of this is that during that same period, both plant ecology and vegetation science can only be described as having taken a back seat in the author's home country of the UK. This is partly because many of the most exciting avenues for research and exploration in vegetation science lie elsewhere on the globe, particularly in the tropics. A key theme of the new edition is to demonstrate and foster this worldwide perspective of the subject. This is a perspective that is all the more important because of the universal threats to biodiversity and the limited success of the numerous valiant efforts at biological conservation across the globe in the face of human exploitation and so-called development.

Lastly, the author would like to thank all those students, particularly those who have completed his Masters course in ecology and multivariate analysis at the University of Plymouth over the past 12 years, for their enthusiasm for and commitment to this subject. While the greater range and complexity of methods and of computer software today means that this text cannot possibly cover every aspect of the subject of vegetation science at the research level, ultimately, the purpose of this book is to introduce, simplify and explain quite complex things for the improvement of understanding and to assist with learning. I have never forgotten the comment of a very well-known vegetation research scientist, who came up to me at a conference after the publication of the first edition and told me in no uncertain terms that *'you should never have published that book – it makes things too easy for students and removes the mystique'*!! I knew from that moment onwards that the book had every potential to achieve its objectives and it is my fervent hope that this second edition manages to build even more successfully upon that achievement.

Martin Kent
Plymouth, UK
March 2011

Acknowledgements

Many people assisted in writing the first edition of this book. In particular Martin Kent would once again like to thank Dr Tom Dargie (private ecological consultant, Boreas Ecology – http://www.boreasecology.com/), who has always shared his enthusiasm and interest for the subject, Professor David Gilbertson (School of Geography, Earth and Environmental Sciences, University of Plymouth), Dr Ken Thompson (formerly of the Department of Biological Sciences, University of Sheffield) and Dr Peter Wathern (formerly of the Department of Biological Sciences, University of Aberystwyth, Wales), all of whom gave their support and encouragement. A good number of years ago, Dr Nicholas J. Cox of the Department of Geography at the University of Durham read the first edition of the text with an exceptionally trained and critical eye, noting the many small errors therein, and his corrections have helped greatly with this revision. I thank him for his efforts.

Martin Kent wishes to offer very special thanks to Professor Robin Pakeman of The James Hutton Research Institute (formerly the Macaulay Land Use Research Institute), Craigiebuckler, Aberdeen, Scotland, for his kindness and patience in carefully reading and correcting the manuscript of this new edition and making a number of very helpful suggestions for its improvement. Very particular thanks also go to Jamie Quinn of the Cartographic Resources Unit in the School of Geography, Earth and Environmental Sciences at the University of Plymouth, who redrew most of the original figures and diagrams, as well as many new ones. The volume of work became far more than either of us originally realised but, as the quality of the diagrams demonstrate, he has succeeded admirably. Dr Rana Moyeed of the Department of Statistics, Computing and Mathematics at the University of Plymouth deserves special mention for his tolerance of my many questions concerning statistical analysis relevant to the writing of the revision.

My very good friends, Professor Liquan Zhang of the State Key Laboratory for Estuarine and Coastal Research (SKLEC) and Dr Xihua Wang of the Department of Environmental Science, both from East China Normal University, Shanghai, kindly funded a study visit to China in September–October 2010, which acted as a catalyst for embarking on this second edition. I thank them both very warmly indeed. The University of Plymouth and particularly the School of Geography, Earth and Environmental Science have also provided invaluable support throughout the revision and in particular I would wish to thank my colleague Dr Ruth Weaver.

In truth, this edition would never have appeared at all, were it not for the dogged perseverance of a succession of editors at John Wiley and Sons (now Wiley-Blackwell), Keily Larkins, Rachael Ballard and most especially Fiona Woods. Fiona was the one who finally succeeded, and I offer very grateful thanks to you, Izzy Canning, Sarah Karim and your colleagues for all your hard work in seeing this second edition through to publication.

Finally, I owe an enormous debt of gratitude to my wife Gay for her infinite patience and support during the writing and revision of the book. It is also dedicated to our children Jonathan, Joseph, Holly and Kitty and her husband Ben, and to our grandchildren Sam and Tom.

Copyright and authorship of all figures and tables are acknowledged in the appropriate captions. The authors are grateful to Routledge publishers for permission to include diagrams from P. Gould and R. White (1986) *Mental Maps* in Chapter 6.

The author also wishes to thank Professors Bruce McCune and James Grace and MjM Software Design for permission to include material from McCune, B. and Grace, J.B. (2002) *Analysis of Ecological Communities*, MjM Software Design in Figures 4.1 and 6.23 and the 'Landscape Analogy' text for non-metric multidimensional scaling in Chapter 6. Dr. Peter Henderson of Pisces Conservation Ltd. is thanked for permission to present the data in Table 6.5. Dr. Jane Robbins (School of Animal, Rural and Environmental Sciences, Nottingham Trent University, UK) and Professor John Matthews (Department of Environment and Society, Swansea University, UK) kindly gave permission to use their research published in *Journal of Vegetation Science* and in *Arctic, Antarctic and Alpine Research* as a case study and Plate 1.2.

Every effort has been made to try to trace the copyright holders of material reproduced here. In a small number of cases, this has proved impossible and the author and publisher would be grateful for any further information that would enable them to do so.

Martin Kent
School of Geography, Earth and Environmental Sciences
University of Plymouth, UK
March 2011

Safety in the field

All fieldwork is potentially dangerous, even when carried out in local, well-known areas. Precautions should always be taken and local safety codes adhered to. The following recommendations are important:

(1) Always obtain an up-to-date weather forecast.
(2) Take advice from local experts if in doubt.
(3) Be aware of potential health problems of any members of the party.
(4) Collect the addresses and telephone numbers of family or friends of every member of the party.
(5) Leave this information and details of the route to be followed with a responsible person at the base and an expected time of return.
(6) If possible, carry a mobile phone and a geographical positioning system (GPS) device but bear in mind that signals and reception may be weak or non-existent in remote areas.
(7) Never, ever, carry out fieldwork alone: a group of three or four leaves one or two people free to go for help, while a second person can stay with an injured or ill colleague.
(8) All members of the party should have had a tetanus injection. Always take note of travel advice relating to preventative inoculations and medicines in the fieldwork locality.
(9) Be extra careful in certain habitats such as wetlands, bogs and swamps. Working in the tropics carries special potential dangers.
(10) Be prepared for the worst that can happen in terms of bad weather or an accident. Responsible members of the party must be familiar with basic first aid and safety procedures. The following equipment is essential, depending on environment:

> suitable footwear (usually stout boots), appropriate clothing, waterproofs with hood, over-trousers, warm hat and gloves, sunhat and sunscreen, water, first aid kit, insect repellent, torch with batteries, whistle, emergency rations including glucose sweets, spare warm clothing and socks, survival blanket or lightweight tent, map and compass.

(11) The standard SOS signal for torches or whistles is three short signals, three long and three short.

In the United Kingdom, all those responsible for organising fieldwork and research overseas should be aware of the British Standard 8848 (2007) + Amendment 1 (2009) which provides clear guidelines for good practice. Similar documents and information exist in many other countries.

ACCESS

Always obtain permission from landowners, farmers and other relevant agencies before carrying out fieldwork on their land. By far the majority will gladly give permission provided it is requested before going onto their land.

DISCLAIMER

While every reasonable care has been taken, neither the author nor the publisher accept any liability for any injury, accident, loss or consequent damage, however caused, arising from this book or any information contained therein.

Chapter 1
The nature of quantitative plant ecology and vegetation science

THE NATURE OF VEGETATION

Dictionary definitions usually describe vegetation as 'plants collectively' or 'plant growth in the mass'. To the plant ecologist and vegetation scientist, this definition is completely inadequate and perhaps conforms to the view of many students (and teachers and lecturers!), who see it as 'a frightening and unknown mass of green, shrouded in technical terms and Latin Names' (Randall, 1978: p. 3). This book is concerned with the techniques for both collecting and analysing data on vegetation with the primary aim of making sense of the 'frightening and unknown mass of green'. As such, it is a text on quantitative plant ecology, which is a clearly recognisable subdiscipline of ecology and biogeography. The field of quantitative plant ecology is also related to an area of research known as vegetation science, which in addition to vegetation description and analysis, also includes plant population biology, species strategies and vegetation dynamics (successional processes and vegetation change) (van der Maarel, 1984a, 2005a,b). Most researchers and students take the phrase 'vegetation description and data analysis' to mean the collection of vegetation data, followed by analysis, usually using complex mathematical methods. However, in the 1980s and 1990s, there was a distinct tendency for the processes of analysis to become an end in themselves. An important aim of the previous edition of this book was to show that quantitative plant ecology and vegetation description and data analysis can and perhaps should be primarily ecological rather than mathematical in emphasis. The only way that variations in vegetation and plant species distributions can be properly understood and explained is within an ecological framework. This introduces the fundamental point that vegetation is always an integral part of an ecosystem (Tansley, 1935; Waring, 1989; Willis, 1997; Dickinson and Murphy, 1998; Leuschner, 2005) and can only be studied by fully exploring its role within that ecosystem. Vegetation cannot be isolated as a separate entity from the ecosystem within which it exists.

The building blocks of vegetation are individual plants. Each plant is classified according to a hierarchical system of identification and nomenclature using carefully selected criteria of physiognomy and growth form. The individuals of a species, taken together, form a species

Vegetation Description and Data Analysis: A Practical Approach, Second Edition. Martin Kent.
© 2012 John Wiley & Sons, Ltd. Published 2012 by John Wiley & Sons, Ltd.

population, and within the local area of a few square metres to perhaps as much as a square kilometre, groups of plant species populations that are found growing together are known as plant communities or plant species assemblages. Much more will be said of plant communities and species assemblages later in Chapter 2, but within plant communities, the presence or absence of particular species is of primary importance. After this, the amount or abundance of each species present is of interest. Although most vegetation data are still collected at the species level, one of the more interesting developments of the past 20 years has been in alternative methods of describing vegetation, such as plant functional types and taxonomic, morphological and structural surrogates (Ramsay *et al.*, 2006). This book is concerned with reasons and methods for collecting data of these kinds, and with techniques for their analysis.

The importance of vegetation within ecology is three-fold. Firstly, in most terrestrial parts of the world, with the exception of the hot and cold deserts, vegetation is the most obvious physical representation of an ecosystem. When ecologists talk about different ecosystem types, they usually equate these with different vegetation types and the dominant species life-forms within them. Secondly, most vegetation is the result of primary production, where solar energy is transformed through photosynthesis by different plant species into green plant tissue. The net primary production, which is the amount of green plant tissue accumulated within the area of a particular vegetation type over a given period of time, represents the base of the trophic pyramid. All other organisms in both the grazing and detrital food webs are ultimately dependent upon that base for their food supply. Thirdly, vegetation also acts as the habitat within which the organisms live, grow, reproduce and die. In the case of the grazing food web, it is among the above-ground parts of plants. With the detrital web, it is on the surface and below ground among the roots. Taken together, these three points show the central importance of vegetation to ecology and demonstrate the need for methods to assist with both description and data analysis (Anderson and Kikkawa, 1986; Cherrett, 1989; Barbour *et al.*, 1999; van der Maarel, 2005a,b).

WHY STUDY VEGETATION?

There are many situations where vegetation merits study. The commonest examples of the use of vegetation description are in the recognition and definition of different vegetation types and plant communities, which is known as the science of phytosociology, the mapping of vegetation communities and types, the study of relationships between plant species distributions, environmental controls and their interactions with humans and animals, and the study of vegetation as a habitat for animals, birds and insects. Change in vegetation over time may also need to be described using concepts of succession and climax.

Information on vegetation may be required to help to solve an ecological problem, for biological conservation and management purposes, as an input to environmental impact statements, to monitor management practices, or to provide the basis for prediction of possible future changes in plant species distributions and linked to both human impacts on habitats via land use practices and also climate.

A useful distinction is into aspects of study that are academic, as opposed to those which can be termed applied. In the academic case, vegetation may be described and data analysed largely for their own sake. Applied studies are where vegetation data are collected and analysed with the aim of providing information of relevance to some ecological problem, often to do with environmental conservation and ecosystem management or the prediction of future environmental and ecological change. Many examples of research include elements of both.

CASE STUDIES

Throughout this book, many different examples of the application of methods for the description and analysis of vegetation will be presented. A brief introduction to four contrasted case studies serves to demonstrate the diversity of situations where vegetation may need to be surveyed and data collected and analysed.

Case Study 1: Evergreen broad-leaved forest in Eastern China: its ecology and conservation and the importance of resprouting in forest restoration (Wang et al., 2007)

Evergreen broad-leaved forest (EBLF) is now recognised as an important global vegetation formation type that contributes to both the biodiversity and the sustainable development of the subtropical regions of China. Discussion of the forests is omitted in Archibold (1995), although they are mapped as EBLF in the more recent overview of world vegetation types by Box and Fujiwara (2005). While its biogeographical status in China still remains a matter of debate, unfortunately, the extent of the EBLF has decreased very significantly due to long-term anthropogenic disturbance, including deforestation, logging and fire, and much of the forest is now degraded to plantation, secondary forests, shrub and grassland communities. Song (1988, 1995) provided the most valuable review in English of both the position of the Chinese EBLF within the world vegetation formation types and the overall characteristics of the forest. In China, it occurs between 24°–32°N and 99°–123°E and formerly covered around 25% of the area of the country (Figure 1.1). It lies within areas dominated by a subtropical monsoon climate and the forests occupy mountainous and hilly areas across the south and east of China (Plate 1.1). The forests are extremely diverse, particularly in terms of tree and shrub species (phanerophytes 50–80% of species – see Chapter 3), ranging from over 100 vascular plant species/400 m^2 in the south, to 30–45 species in the north of its distribution (Song, 1988). The dominant species of the EBLF come from only a few genera, together with some ancient coniferous species, many of which have 'broad leaves'.

To inform forest conservation and management, Wang *et al.* (2007) described research into major plant community types and underlying environmental gradients (see Chapter 2) of degraded EBLF around Tiantong and Dongqian Lake near Ningbo in Eastern China (Figure 1.1; Plate 1.1), and examined the importance of vegetative resprouting as a key mechanism in secondary succession following forest clearance. Species composition was described from 199 10 m × 10 m plots (Chapter 3) and analysed using various methods presented later in this book (Two-Way Indicator Species Analysis [TWINSPAN] – Chapter 8; and canonical correspondence analysis [CCA] ordination – Chapter 6). Some 22 degraded and mature forest community types were identified, while CCA indicated that a primary vegetation gradient was related to the distance of sample plot from mature forest, which was closely linked to altitude and slope. The secondary gradient corresponded to successional stage and disturbance. The roles of resprouting and reseeding characteristics in forest regeneration were researched firstly by 10 m × 10 m plots taken from selected TWINSPAN groups, and secondly by 20 m × 20 m plots in representative areas of forest at different ages – 1, 20, 43 and 60 years, and in an area of mature forest – 100+ years.

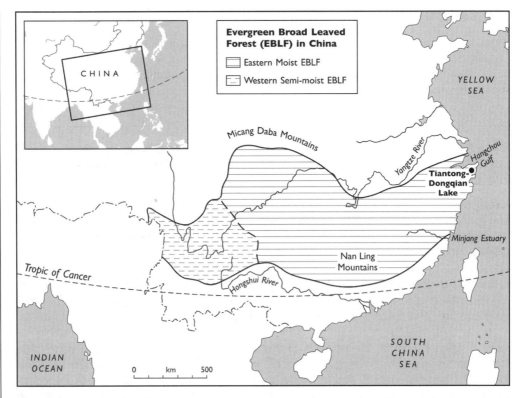

Figure 1.1 The distribution of evergreen broad-leaved forest (EBLF) in China and the separation of forest into the Western semi-moist and Eastern moist forest types (after Song, 1988; Wang *et al.*, 2007: reproduced with kind permission of Elsevier).

The importance of resprouting in the regeneration of many EBLF tree and shrub species was demonstrated, a process linked to ideas of the persistence niche – resprouting from stumps is an important means of persistence for many species. Existing remnant forests should be conserved, but forest restoration is also essential and will benefit from understanding of the importance of tree/shrub resprouting, as well as seedling recruitment in forest regeneration. Further work is in progress on seedbanks, germination success and both inter- and intra-specific competition within Chinese EBLF to assist with successful conservation and management of this rare forest type (Chapter 6 – Case studies).

Case Study 2: Pioneer vegetation on glacier forelands in southern Norway (Robbins and Matthews, 2009, 2010)

Climate change and its ecological impact is one of the most important and also controversial environmental topics at the present time. A widely observed phenomenon in Europe and Scandinavia over the past 100 years has been the retreat of mountain glaciers in response to increased summer temperatures and relatively low winter precipitation, and Nesje *et al.* (2008) have predicted that up to 98% of Norwegian glaciers may have disappeared by 2100. Robbins and Matthews (2009) saw the retreat of such glaciers as a valuable opportunity to study the earliest stages of vegetation colonisation (primary succession) and the manner in which plant species are responding the availability of new terrain on glacier forelands (Plate 1.2). They were concerned firstly to examine the species composition of early pioneer stages, and secondly to see whether these highly disturbed sites are colonised by consistent sets of species or whether species composition is more dependent on chance (stochastic) factors that tend to produce more random and variable collections of species. McCook (1994), Walker and del Moral (2003), Pickett and Cadenasso (2005) and Pickett *et al.* (2008) present the most recent reviews of processes of primary succession.

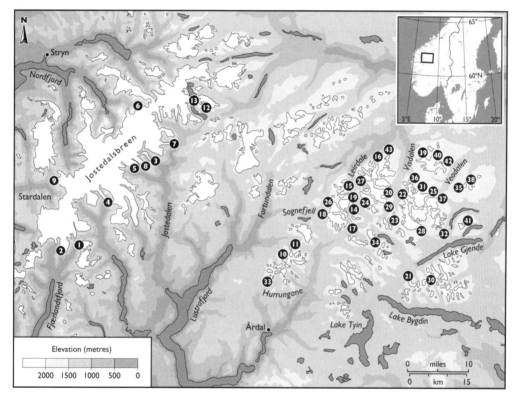

Figure 1.2   The locations of the 43 glacier forelands in southern Norway studied by Robbins and Matthews (2009, 2010). Redrawn and reproduced with kind permission from Wiley-Blackwell.

A total of 43 glacier forelands in the Jotunheim and Jostedalsbreen regions of southern Norway, with an altitudinal range of 80–1860 m, were sampled (Figure 1.2). The vegetation data were collected in a particularly interesting manner, using rectangular quadrats for contiguous (adjacent) sampling along transects away from the glacier snout in each case. The detail of this is presented in a case study at the end of Chapter 3. In addition to the vegetation data, two regional explanatory environmental variables were collected at each quadrat, altitude and distance eastwards from a fixed reference point, representing a continentality index. Earlier surveys had also examined more local habitat factors, such as snow distribution, microsite conditions, and measures of the randomness of points of initial colonisation (stochasticity) (Whittaker, 1989, 1991).

As with the Chinese case study above, data analysis involved techniques of multivariate analysis – numerical classification using a form of similarity analysis known as flexible-β followed by use of a multi-response permutation procedure (MRPP) (see Chapter 8) and ordination using non-metric multidimensional scaling (NMS) (McCune and Grace, 2002) (see Chapter 6). An example NMS ordination plot, displaying the results for 42 sites, with four classification groups superimposed is shown in Figure 1.3. This type of plot is fully explained in Chapter 6.

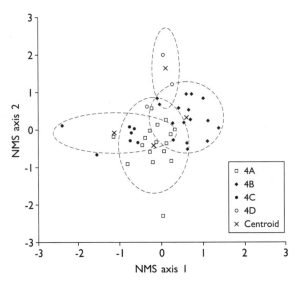

Figure 1.3 Non-metric multidimensional scaling (NMS) ordination plot of the 42 glacier foreland sample sites with four numerical classification groups superimposed. One site was removed due to a complete absence of vegetation in the pioneer zone. Redrawn and reproduced with kind permission from Wiley-Blackwell.

The key result of the research was that the vegetation could be seen as one broad vegetation type characterised by *Poa alpina* (Alpine meadow grass) and *Oxyria digyna* (Mountain sorrel), indicated in general terms by the overlapping of the classification group circles on Figure 1.3 (see Chapter 8). Nevertheless, within this overall group, there was a high degree of variability, but two emerging subcommunities could be identified and these were shown to be linked to both site altitude and continentality. Whereas the very earliest stages of colonisation could be said to be highly stochastically determined (i.e. by randomness and chance), the beginnings of the organisation of changes in species composition into two successional pathways could be observed. There are many interesting features to this work, including the regional scale of vegetation description, the underlying model of vegetation change and dynamics, the sampling design and its relevance to the topic of vegetation response to climate change.

Case Study 3: Vegetation description and data analysis to inform the conservation of a rare plant species – Lobelia urens L. (the heath lobelia) in southern England (Dinsdale et al., 1997, 2000)

The successful conservation of rare plant species usually requires a range of information and scientific research linked to both the autecology (the study of a single species in relation to its ecology and environment) and synecology (the study of a community of species in relation to their ecology and environment) of the species involved in order to assist with management practice. *Lobelia urens* L. (the heath lobelia) (Plate 1.3) is a perennial rhizomatous herb that shows a Lusitanian distribution in Europe and North Africa extending from Morocco, Madeira and the Azores in the south, along the Atlantic coast through Portugal, Spain and France, as far north as Belgium. However, in the UK, at the extreme north of its range, *L. urens* is today limited to six locations in the southern coastal counties of England, although historical records indicate that it may have been found on 19 sites altogether. Dinsdale *et al.* (1997) surveyed the historical and documentary evidence for its distribution and completed extensive botanical surveys to try to understand the plant communities and species assemblages within which it grows (its phytosociology) and to assess both environmental controls and limitations on the species.

At the six remaining sites in southern England, a total of 95 0.5 m × 0.5 m quadrats, containing 122 plant species, were recorded, using a Domin abundance scale and a carefully devised sampling strategy (see Chapter 3). Data on 16 environmental variables were also collected (Table 1.1) in order to summarise the environmental variability and the factors possibly limiting its distribution. Analysis of the data for phytosociological purposes involved using the numerical classification method called Two-Way Indicator Species Analysis (TWINSPAN: Hill, 1979b) (see Chapter 8), while the floristic variation and correlations with the environmental variables in the 95 quadrats were assessed using the method of ordination known as canonical correspondence analysis (CCA) (ter Braak, 1986a, 1987, 1988a,b) (see Chapter 6).

Table 1.1 The 16 environmental variables measured in the survey of *Lobelia urens* L. (the heath lobelia) at six sites in southern England (Dinsdale *et al.*, 1997). Reproduced with kind permission of Wiley-Blackwell.

Microclimate and habitat	Soil variables
1. Height of dominant vegetation (cm)	8. Soil texture (6-point ordinal scale)
2. Slope (degrees)	9. Soil structure (4-point ordinal scale)
3. Bare ground (% cover; 5-point Domin scale)	10. Soil calcium (mg/cm^3)
4. Litter cover (% cover; 5-point Domin scale)	11. Soil sodium (mg/cm^3)
5. Litter type (5-point ordinal scale)	12. Soil magnesium (mg/cm^3)
6. Microtopography (5-point ordinal scale)	13. Soil phosphorus (mg/cm^3)
7. Exposure (5-point ordinal scale)	14. Soil pH
	15. Organic matter (g/cm^3)
	16. Moisture content (g/cm^3)

This research, as with the previous two case studies, demonstrates a common feature of vegetation and ecological survey, in that very large quantities of data on both vegetation and associated environmental variables are often collected. Such data are described as *multivariate*, because they contain information on many species and many environmental factors or variables. Most of the methods of data analysis described later in this book are designed to look for order and pattern within these types of data.

The first part of the data analysis involved numerical classification of the 95 quadrats using TWINSPAN and resulted in six groups of quadrats (A–F) that were each characterised in terms of their within-group species composition and were taken as representing the range of plant community types within which *L. urens* occurs (Table 1.2). The important next step, however, was to enter the quadrats into another computer program, TABLEFIT (Hill, 1989, 1991), which compares the species composition of the quadrats within each of the groups with a large database of sample quadrats from recognised key vegetation types in the British Isles, as defined in the National Vegetation Classification (NVC) (Rodwell, 1991, 1992a,b, 1995a, 2000, 2006). The NVC is described in more detail in the case studies of Chapters 3 and 8. Use of the NVC enabled each of the six groups A–F to be matched with their nearest key vegetation types in Britain, each of which is characterised by a code, such as M25, W25 or H3.

Table 1.2 The six Two-Way Indicator Species Analysis (TWINSPAN) groups (A–F) resulting from numerical classification of the 95 quadrats in the *L. urens* survey matched with their nearest vegetation types in the British National Vegetation Classification (NVC) (Dinsdale *et al.*, 1997). Reproduced with kind permission of Wiley-Blackwell.

TWINSPAN group	NVC group
A	M25 – *Molinia caerulea–Potentilla erecta* mire
B	M25c – *Molinia caerulea–Potentilla erecta* mire
C	W23 – *Ulex europaeus–Rubus fruticosus* scrub
	W25 – *Pteridium aquilinum–Rubus fruticosus* under scrub
D	M25 – *Molinia caerulea–Potentilla erecta* mire
	H3 – *Ulex minor–Agrostis curtisii* heath
E	W23 – *Ulex europaeus–Rubus fruticosus* scrub
	W25 – *Pteridium aquilinum–Rubus fruticosus* under scrub
F	W23 – *Ulex europaeus–Rubus fruticosus* scrub

In the second part of the data analysis, the results of the ordination using canonical correspondence analysis (CCA) are shown in Figure 1.4. The way to interpret this plot is explained fully in Chapter 6. However, the scatter of points on the ordination plot shows that the six sites where *L. urens* occurs are environmentally heterogeneous and there was no obvious limiting response to many of the environmental variables. However, soil texture, microtopography, soil moisture and soil structure were all shown to be extremely important because they are represented by the longest arrows on the plot. Taken together, these were seen as indicators of the importance of microhabitat linked to disturbance, particularly related to the availability of suitable sites for germination of *L. urens* seed.

Thus a new set of hypotheses related to a new research project were established, whereby using both field survey methods and experimental seed beds, the germination and survival rates of *L. urens* seeds and seedlings were measured in great detail and linked to microhabitat conditions (Dinsdale *et al.*, 2000). Depressions in the soil and seedbed created by grazers were shown to encourage the most successful germination which was assisted further by the presence of moss, although also probably impaired

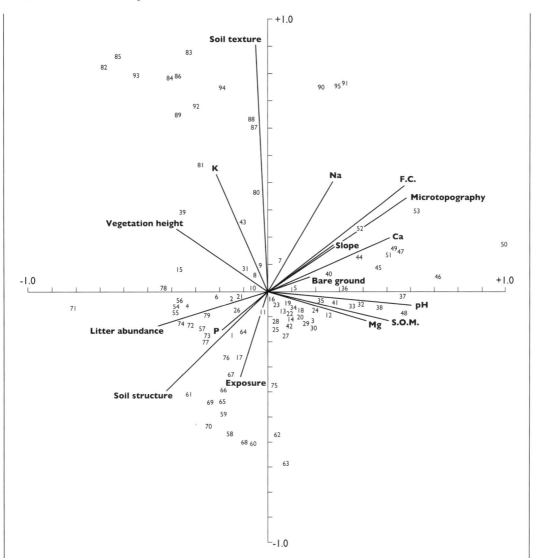

Figure 1.4 A CCA sample–environment biplot of the 95 quadrats collected from six sites in southern England (Dinsdale *et al.*, 1997). Redrawn and reproduced with kind permission of Wiley-Blackwell.

by any significant accumulations of litter. These results had considerable significance in that producing the right microhabitat conditions for germination in conservation management is vital for the future survival of the species.

There are numerous interesting aspects to this work. The focus on a single rare species, the sampling design and the matching of community types to a national vegetation classification are important. However, perhaps most important of all is how the use of an initial descriptive survey of vegetation and environmental factors followed by multivariate data analysis led to the generation of new hypotheses and further research on the germination characteristics of the species, linked to microhabitat conditions. The importance of large-scale vegetation description as a precursor to much more focused and detailed research is emphasised later in this chapter.

Case Study 4: The grazing ecology of the Serengeti grasslands (Anderson et al., 2007; Augustine and McNaughton, 2004; McNaughton et al., 1998; McNaughton, 1983, 1985)

This case study is the only one that has been retained from the first chapter of the first edition of this book. The reasons are two-fold. Firstly, it concerns the tropical grasslands or savannas of the world, which comprise one of the most important and extensive world vegetation formation types and a world biome. Such grasslands are predominantly composed of grass and herb species but often have a significant tree component. The factors determining the status of savanna vegetation have long been a matter of debate. Seasonality of climate, fire, grazing, edaphic or soil factors and geomorphology have all been described as of importance in the origin and maintenance of the vegetation (Huntley and Walker, 1982; Boulière, 1983; Boulière and Hadley, 1983; Cole, 1986, 1987; Archibold, 1995; Mistry, 2000).

Secondly, the work of McNaughton on savanna environments represents an excellent example of the manner in which a research agenda in vegetation ecology develops over time. The following main section of this case study is taken from the original summary of McNaughton's 1983 and 1985 papers.

The Serengeti takes its name from the Masai word for a broad open plain and is situated on the border between Kenya and Tanzania, stretching from Lake Victoria in the west to the volcanic Crater Highlands in the east (Figure 1.5). The Serengeti is famous for its fauna, both the large herds of ungulates such as the wildebeest and zebra (Plate 1.4) which roam the open plains, as well as its carnivores, including lions, cheetahs and jackals. According to McNaughton, three distinct subregions are identified: (a) the open Serengeti Plains of the southeast; (b) the Western Corridor which is used by grazers during seasonal changes and whenever rain falls on the Serengeti Plains during the wet season; and (c) the north, which is used by migrants at the height of the dry season. The greater part of the plains area is included within two national parks which have been established largely to protect the fauna: the Serengeti National Park in Tanzania and the Masai Mara Game Reserve in Kenya.

McNaughton made the point that prior to his work in the Serengeti, most studies of tropical grassland had been based on descriptive observation and physiognomic classification – description by growth or life form (see Chapter 3). Such studies had often been made in relation to soils and to the distributions of the large herds of ungulates that are characteristic of the biome (Archibold, 1995; Mistry, 2000). Detailed quantitative studies of plant community structure (species composition and association or groupings of species growing together at the same location) were virtually absent. At the start of his original article, McNaughton posed several questions concerning community structure in the vegetation:

(1) Are there repeating combinations of species and community types that occur more or less frequently?
(2) Are there consistent patterns of species abundance and diversity that provide insight into community organisation?

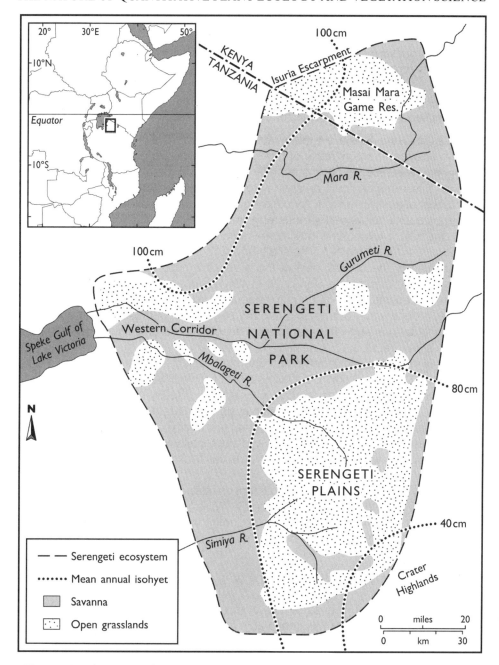

Figure 1.5 The Serengeti Plains of Kenya and Tanzania (McNaughton, 1983). Modified and redrawn with kind permission of the Ecological Society of America.

(3) How important is spatial heterogeneity (local variation of plant communities in space) and what is its role in community organisation?
(4) What environmental factors influence species abundance, spatial distribution and community organisation?
(5) If consistent patterns of community properties and environmental relationships exist, why have they developed and what can be inferred from them about the mechanisms and functional consequences of community organisation?

These questions are typical of those asked at the start of a great deal of work in vegetation science; in many parts of the world, rigorous description of vegetation has only comparatively recently been completed or has yet to be attempted.

Although this research could be seen as primarily academic in emphasis, there were also a number of applied reasons for carrying out the survey. Following control of rinderpest, an acute viral disease of various ungulates and cattle, the population of grazing animals, particularly wildebeest, had increased dramatically from c. 220,000 wildebeest in 1961 to c. 1.4 million in 1977. The scale and extent of burning had declined following the designation of the greater part of the area as a national park. Elephants, which were only reintroduced to the Serengeti in 1951, had undergone a population explosion and were causing much damage to vegetation by uprooting trees and laying waste large areas around vital waterholes in the dry season. All of these factors undoubtedly were exerting serious and often detrimental effects on the grassland ecosystems and community structure.

For all these reasons, McNaughton proceeded to select 105 sites within both national parks to sample the vegetation. Detailed aspects of sampling are described in the papers and the data were analysed once again by methods of ordination and classification as described in Chapters 6 and 8. In the final analysis, 17 different community types could be recognised, largely characterised by perennial grasses. Examination of environmental controls demonstrated that, while rainfall and seasonality were significant in separating different community types, grazing intensity was critical, with a secondary gradient being attributable to soil texture.

At the end of the paper, there is an excellent discussion of where vegetation research should go on to from the results of the survey. McNaughton argued that formulation of hypotheses and design of experiments to quantify the exact effects of grazers within each of the vegetation types is necessary. However, there are many other questions that might be asked:

(a) What is the species composition of the grassland and particularly its local spatial variation?
(b) What are the phenological strategies of the grasses (how does each species time the major events of its life (existence as a seed or vegetative rhizome, germination, growth and leafing, flowering, fruiting, death)?
(c) Is there an overlying tree canopy? How dense is it and what is its impact on the underlying vegetation?
(d) What are the grazing species and what are their densities? Do they preferentially graze certain species at certain times of the year?
(e) Has the vegetation been burned lately? If so, how often and how extensively?
(f) Is the soil wet or dry? What are the detailed chemical and physical properties of the soils and how are they related to the grazing regime?

Since the mid-1980s and the publication of the original papers, McNaughton and his various research teams have published over 100 papers related to research on savanna environments, many of which are directly related to the questions and hypotheses outlined above. A search of the Google Scholar database or the Web of Science or Scopus databases available in many university libraries will reveal the range of papers – good examples are Anderson *et al.* (2007), Augustine and McNaughton (2004) and McNaughton *et al.* (1998). Careful reading of these papers will show that many of these questions have been answered but many more remain and require careful thought, hypothesis generation and experimentation (Figures 1.6 and 1.7) to be answered. A further problem is that many of these questions overlap; for example, are the effects of grazing or burning the same on different soil types?

In summary, McNaughton's work demonstrates how a relatively thorough and complete survey of the vegetation of an area in response to one set of aims and objectives generates even more questions which may require many more years of detailed experimental ecology in order to be answered. Having completed this primary survey of vegetation community types and environmental factors in the 1980s, over the next 20–30 years, McNaughton also went on to examine many aspects of primary productivity of vegetation and soil nutrients in relation to grazing, and this work in turn demonstrated the importance of vegetation as both food supply for higher trophic levels and as a habitat for both herbivores and carnivores.

THE SCIENTIFIC APPROACH

Induction and deduction

As with all aspects of science, vegetation description and analysis must be approached in a logical and systematic fashion. All vegetation description and analysis must have a purpose. Nevertheless, there are many different approaches to vegetation study. The collection and analysis of vegetation data provides the principal form of ecosystem description and classification. However, simply to describe and observe patterns of variation in vegetation data over space is not the sole aim of vegetation science. A further very important concept is the idea of explanation. Explanation is the attempt to answer the question 'Why?'. 'Why is one area of vegetation different to another?' 'Why are certain plant species found in some locations but not in others?' 'Why do certain vegetation types appear to be undergoing change – due either to natural processes of succession or to human-induced effects?' 'Why is the vegetation of a particular area under stress and showing signs of damage or disease?'

To attempt to answer these kinds of questions and find explanations, we have to have some view or theory of the manner in which the world functions. For any part of science, there is always an existing body of knowledge and theory, which is generally accepted at the time. Often this initial body of knowledge has been collected by a process known as induction (Figure 1.6a). With induction, the data are collected without formulation of prior hypotheses (descriptions of existing notions of how the ecological world functions) or any preconceived ideas, and explanations are then derived from the data collected. Ideas of induction date from Frances Bacon back in the early seventeenth century and John Stuart Mill in the nineteenth century. They argued that one can only start to make valid explanations once all the necessary facts and information relevant to the situation have been collected. Such data were thought to be particularly secure, because the

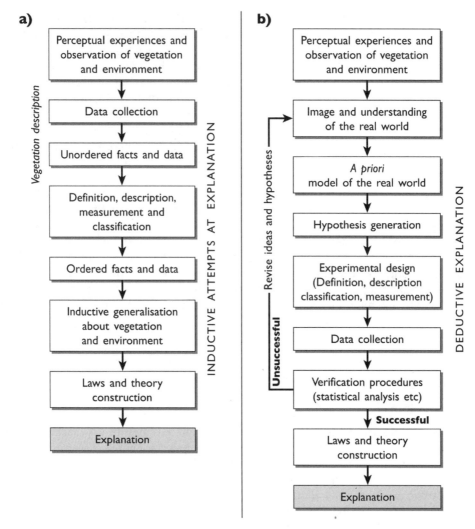

Figure 1.6 (a) Inductive and (b) deductive approaches to scientific enquiry. Reproduced with permission from John Wiley & Sons, Inc.

collection and ordering of facts for their own sake are not based on any biased or selective guesses or hypotheses. The problems of the inductive approach are many and particularly centre on the inefficient process of collecting a large amount of data which may not be immediately (or even eventually) relevant or useful and which cannot ever prove convincingly a theory or hypothesis.

On the basis of this existing knowledge and theory, a scientific approach may often involve the generation and testing of hypotheses concerning observed variations in vegetation cover and their causes. This is usually known as the deductive approach (Figure 1.6b). The sequence of activities as shown in Figure 1.6b is commonly described as 'scientific method'. The method involves setting up both null and alternative hypotheses (described fully in Chapter 5) and is often also described as the null hypothesis testing (NHT) approach. It follows a rational and logical sequence of thought processes and actions. Most scientists would probably subscribe to that view and are thus known as rationalists. An important point is that hypotheses are generated from the existing body of theory, and data collection and analysis are then based on the notion of accepting or rejecting the hypothesis. The basic principles of deductive thinking are explained in much greater detail in many texts, for example, Haines-Young and Petch (1986), Ford (2000), Scheiner (2001), Scheiner and Gurevitch (2001), Quinn and Keough (2002) and Wilson (2003).

The testing of hypotheses and proving them true (verification) or false (falsification) is central to the process of deduction and was discussed at length by Popper (1972a,b; 1976) Haines-Young and Petch (1986) and Schrader-Frechette and McCoy (1993). The emphasis of a lot of scientific publication on proving hypotheses true (verification), rather than disproving them (falsification), has led to the notion of logical positivism. Other views are those of Kuhn (1962, 1970) and Feyerabend (1975, 1978). They are known as relativists. Kuhn argued that the history of science demonstrates that usually most scientific work is not concerned with trying to overthrow existing theories and develop new ones. Rather, scientists tend to accept existing established theories and ideas and use them to solve 'puzzles' within different parts of their subject. (He deliberately chose the word 'puzzles'.) Well-established theories and concepts that are believed by most to be tried and tested are known as paradigms. The ecosystem concept is a good example of such a paradigm. Large numbers of scientists will normally be working within different areas of a subject, largely accepting their paradigms. This is known as 'normal science'. Only every now and then are a large number of results taken together or a particularly radical piece of research is completed, and existing theories and paradigms drastically revised. The time when this occurs is known as a period of scientific revolution. Certain paradigms in plant ecology, such as the concept of the plant community and successional processes (see Chapter 2), although readily accepted by many ecologists, are still matters of considerable debate, and the nature of these paradigms is almost certain to change over the next century in the way described by Kuhn.

Still more controversial views come from Feyerabend, who attacked the whole of scientific method (hypothesis generation, verification, falsification) and argued that there are many different routes to scientific results, many of which are far less precise than most scientists would admit. There is thus no one scientific method, but a series of alternative approaches which on some occasions may partly overlap and which work with varying degrees of objectivity. He states that the only principle that can be applied to all circumstances is that 'anything goes' (Feyerabend, 1978: p. 39). While this in itself may seem an extreme viewpoint, it is of considerable relevance to quantitative plant ecology and vegetation science.

In many parts of the world, it is not possible to devise hypotheses immediately, since basic descriptions of the vegetation are completely absent or at best extremely generalised. Nowhere is this more true than in the tropics. As a result, a considerable amount of vegetation research is still purely descriptive or observational in nature and inductive in approach (Figure 1.6a).

Many research programmes involve elements of both induction and deduction. The vegetation of an area may be described and analysed without any prior hypotheses, resulting in an initial fuller understanding of the major vegetation types, their environmental controls and human impacts. Study of the results should then lead through the process of hypothesis generation, testing and deduction to more detailed work on specific questions generated by the original inductive research. A large amount of vegetation work still remains purely descriptive and ends with a list of the major vegetation types and their controlling factors, often with a minimal attempt at explanation. Ideally, however, this should only be a starting point for more detailed research on specific aspects and problems, involving the deductive approach and hypothesis generation and testing. The links between inductive and deductive research in vegetation science are shown in Figure 1.7. Hairston (1989), Wiegleb (1989), Eberhardt and Thomas (1991), Inchausti (1994), Ford (2000), Quinn

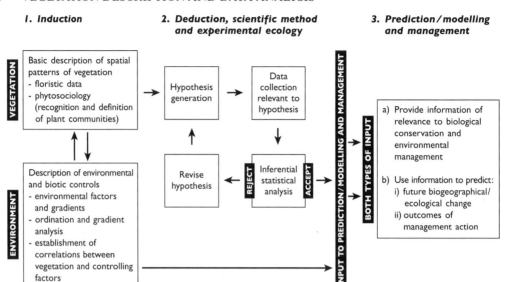

Figure 1.7 A model for inductive and deductive research in vegetation science. Original work of Martin Kent.

and Keough (2002) and particularly MacNeil (2008) discuss these issues further and provide an introduction to experimental ecology in relation to hypothesis generation.

Multiple working hypotheses

To make explanations, many ecologists now use Chamberlin's method of multiple working hypotheses (Chamberlin, 1890, reprinted 1965). This involves the consideration of as many likely explanations of the distribution of vegetation in terms of environmental controls and the effects of humans and animals as possible, followed by a series of experimental studies designed to reject the hypotheses until the most likely ones remain as reasonable explanations. Often it may be possible to assess the likelihood of particular explanations being correct. Scientists adopting that approach have often been termed probabilists. A related idea is Occam's Razor, named after William of Occam (c. 1300–1349), who argued that the simplest and most straightforward explanation should be sufficient until there is good reason to believe that a more complex one is required.

Bayesian statistical approaches to hypothesis testing

In recent years, the traditional deductive (NHT) approach to hypothesis testing, as described above, has been criticised extensively. Deduction is also known as the 'frequentist approach' because it relies on the expectation that, if a set of data were to be collected repeatedly, the expected frequency that the patterns in the data would be observed can be determined. At the same time, the potential value and relevance of alternative Bayesian approaches to hypothesis testing in ecology have been introduced (Dennis, 1996; Hilborn and Mangel, 1997; Anderson *et al.*, 2000; Anderson and Burnham, 2002; Burnham and Anderson, 2002; Brooks, 2003; Clark, 2005; Stephens *et al.*, 2005, 2007; McCarthy and Masters, 2005; McCarthy, 2007; Lukacs *et al.*, 2007; Kéry, 2010; Link and Barker, 2010). Bayes was a seventeenth-century mathematician who introduced a set of rules for what is known as 'conditional probability'. There is insufficient space to present the full arguments here, although some further discussion is included in Chapter 5. Nevertheless, this is an important issue in contemporary ecology. Major criticisms of classical (NHT) hypothesis testing centre on: (a) the dependence of test results on sample size and how large a sample should be collected to increase the chances of a statistically significant result; (b) the fact that the null hypothesis is always false; (c) failure to reject the null hypothesis is not the same as evidence supporting the null hypothesis; (d) the arbitrary 'traditional' p-values for significance (typically $p = 0.01$ and $p = 0.05$); and (e) the tendency to resort to 'data dredging' (collecting and analysing data on more and more variables to try to come up with adequate explanations) when attempting to revise rejected hypotheses (Johnson, 1999, 2002; Eberhardt, 2003).

The Bayesian approach sees population parameters such as means, standard deviations and regression coefficients as random or unknown variables. Rather than erecting hypotheses to be tested and accepted or rejected, as in classical hypothesis testing, the approach evaluates several hypotheses and assesses the chance or probability of each hypothesis being the most valid. In this way several hypotheses may be compared and it may thus be seen as an approach that is closer to Chamberlin's Multiple Working Hypothesis idea.

An important aspect of Bayesian approaches is that for each hypothesis, they initially incorporate prior information through what are termed 'prior probabilities' based on existing knowledge and previous research. Also, as McCarthy (2007: p. 8) states "…. instead of asking 'What is the probability of observing data given that various hypotheses are true?' Bayesian methods ask: *What is the probability of the hypotheses being true given the observed data?*" The Bayesian method then takes the prior probabilities for a given hypothesis and integrates them with the observed data to give the 'posterior probability' for each hypothesis. These are then compared, using various statistics. The approach can be integrated with many forms of classical statistics, for example, measures of central tendency and dispersion, correlation and regression and ANOVA (Analysis of Variance) (see Chapter 5). Bayesian ideas are not without their critics either. Considerable debate centres on the difficulties in defining and incorporating the prior information to derive probabilities for a given hypothesis. Others have argued that the prior probabilities exert comparatively little influence on the posterior probabilities compared with the observed data (Quinn and Keough, 2002; Stephens *et al*., 2005, 2007). Thus far, Bayesian approaches have had very limited use in vegetation science. Part of the reason for this is that much work is still inductive and involves the search for pattern and order in data, with the idea of hypothesis generation rather than hypothesis testing. In the context of vegetation science and plant ecology, probably the best way to see Bayesian methods is as a potential parallel approach to hypothesis testing in ecology, rather than as a universal replacement for the classical scientific method.

Prediction

A final stage of scientific method is prediction. Once hypotheses have been generated, tested and accepted, or in the case of Bayesian analysis, evaluated, it becomes possible to use the results to predict future outcomes. Construction of mathematical and statistical models is particularly important in prediction and is related to description and explanation of relationships in the real world, resulting from the operation of processes, rather than just those processes themselves. They are often known as empirical models (Hilborn and Mangel, 1997). In vegetation science and quantitative plant ecology, this can be particularly important for understanding in environmental management and biological conservation.

Understanding the relationships between theory and practice in vegetation science is extremely important and is discussed at length in Chapter 2. Valuable further discussion is provided in Podani (1994, 2000), Austin (1999a,b, 2005), Noy-Meir and van der Maarel (1987), van der Maarel (1989, 2005a,b) and Waite (2000).

PROBLEMS IN QUANTITATIVE PLANT ECOLOGY

There are a number of difficulties facing the student taking a course or carrying out research involving the use of vegetation description and data analysis and quantitative plant ecology.

The importance of an overview of the whole subject of ecology

Although this text is focused on vegetation description and data analysis, it is nevertheless vital that all students and researchers have a broad appreciation of the whole subject of ecology. There are many excellent ecology textbooks available. The following is a list of useful texts: McIntosh (1986); Chapman and Reiss (1992); Schrader-Frechette and McCoy (1993); Putman (1994); Crawley (1997); Odum (1997); Bradbury (1998); Morin (1999); Gotelli (2001); Krebs (2001); Silvertown and Charlesworth (2001); Gurevitch *et al.* (2002); Townsend and Begon (2002); Begon *et al.* (2006); and Keddy (2007).

Temperate and tropical ecology

Most ecological theory and method has evolved through the activities of academics and researchers working in the temperate regions of the world (northern Europe and North America). While many ideas and techniques transfer readily from these areas to other biomes and vegetation formation types, some do not. This problem is most acute in the tropical regions of the world, and in particular in the tropical rain and moist forests and the tropical grasslands or savannas. Research over the past decades has shown that concepts of the plant community, succession and climax, diversity and species richness all require careful application and often drastic revision when dealing with tropical forest and savanna environments. It is also increasingly argued that many of the conceptual and practical problems in the ecology of non-tropical areas of the world may be solved by research carried out in the tropics. Describing tropical ecology, Deshmukh (1986) stated (p. iv): 'Many notions that seemed well-founded a few years ago are actually based upon flimsy evidence (sometimes none at all) and many recent ideas contradict rather than confirm what seemed like established ideas'. This view has been reinforced more recently by various authors, notably Richards (1996), Whitmore (1998) and Osborne (2000). The same problems apply to methods of vegetation description and data analysis. Standard methods such as the use of quadrats (Chapter 3) and methods of classification and ordination (chapters 6–9) may be totally inappropriate or require considerable modification to work in rainforest and savanna areas. Further introductory discussion of these issues is found in Mabberley (1991) and Whitmore (1998), while information on the rate of destruction of tropical forests in particular areas is given in Laurance *et al.* (2002), Sodhi and Brook (2006) and Primack and Corlett (2011).

Species identification

Problems of plant identification present enormous difficulties to both students and more advanced researchers alike. For most developed regions, published floras exist, although even these require careful training in use. However, if students are just beginning their studies, it is best to keep to relatively simple local situations, where the flora may be well known. Where reliable identification is important, samples may be taken to the local expert or museum for identification. However, usually, and particularly where rarities are known to occur, plant specimens should never be collected; instead, good-quality colour photographs should be taken. Advances in molecular ecology and ecological genetics have also resulted in some interesting questions over the accuracy of some species identifications (Winston, 1999; Isaac *et al.*, 2004; Ramsay *et al.*, 2006). The development of DNA barcoding (Janzen and CBL Plant Working Group, 2009) represents a particularly important area for future research.

Sampling designs

Sampling is discussed at length in Chapter 3. However, in general, this subject has received less attention from vegetation scientists than it should. For example, important differences exist in approaches to sampling where vegetation is simply being described and classified perhaps for phytosociological purposes (inductive approaches – Figure 1.6a), as opposed to research where plant species response to environmental factors is of interest and is being examined by experimental methods involving hypothesis generation and testing which may require random sampling (deductive approaches – Figure 1.6b).

The apparent complexity of techniques for analysing vegetation data

Plant community data are multivariate in nature. Data are usually collected for samples and species giving a raw data matrix (Tables 4.1–4.4). To make sense of these data requires methods for data reduction – methods that create order out of chaos and which will search for and summarise patterns of variation within the data. All the methods of classification and ordination described in Chapters 6 and 8 are of this type and are based on methods of matrix analysis. In detail, these are very difficult techniques for the average student to understand. The approach taken in this book is that it is what the methods can show ecologically and in their application that matters. While it is important to eventually try to understand how the more complex methods work mathematically, for the student beginning studies in plant ecology, the real understanding lies in the demonstration of ecological applications and in the ability to interpret results in the context of the ecological problems for which the data were collected in the first place. That is not to deny that the methods do not have their problems and limitations. Almost without exception, they do. However, once again, these can be appreciated most effectively at the stage of interpretation of results.

The choice of methods for classification and ordination

The mathematical emphasis that underlies the analysis of the site/species matrices which are typical of vegetation data has provided a large number of different methods for analysis, each with particular advantages and disadvantages. For both numerical classification and ordination, there are at least three or four methods in fairly common use and a choice of 15–20 methods overall in each category. This array of methods is bewildering to beginners and even to practised researchers. There is still no absolute agreement on which methods are 'best' in any given situation. For example, at present, both similarity analysis and Two-Way Indicator Species Analysis (TWINSPAN) are widely used for community classification, while non-metric multi-dimensional scaling (NMS) and detrended (DCA) or canonical (CCA) correspondence analysis are the most widely applied methods of ordination, despite many criticisms (e.g. McCune and Grace, 2002). Although comparative evaluations of different techniques have been applied on many occasions, objective tests of different methods on real-world data have never clearly been able to suggest one optimal method. Thus a 'pluralistic approach' now dominates, as discussed in Kent

(2006) and von Wehrden *et al.* (2009). This problem will be discussed again several times later in the book.

Computer software

Application of multivariate analysis requires the use of computers and more importantly, computer software. Most of the techniques for analysis described in this book are only made possible by using appropriate computer programs.

A wide range of computer programs is now available to perform rapid data analyses using the techniques presented in this book. These are reviewed in Chapter 9. A particularly important aspect of this is the notion of 'user-friendliness'. For the average undergraduate, probably the best packages to use are *PC-ORD* written by McCune and Mefford (1999, 2010), which is admirably supported by their accompanying texts – McCune and Grace (2002) and Peck (2010) – and those available from Pisces Conservation Ltd., again supported by a very useful handbook (Henderson and Seaby, 2008). A much wider range of possibilities are available for postgraduates and experienced researchers and these are also introduced in Chapter 9, notably those packages that are linked to the R software.

Interpretation of results

Most textbooks on vegetation description and multivariate analysis are poor at describing and demonstrating the value of the results of community analyses. Students work extremely hard at project planning, quadrat description, species identification, data preparation and computer analysis, only to find that they can make little sense of their results. Yet data collection and analysis are only means to an end. If the methods are taught in the context of a useful and valid ecological problem, then this difficulty over interpretation is usually resolved. For this reason, this book places a strong emphasis on the case study approach to demonstrate ecological applications.

Table 1.3 The 'Ten Suggestions to Strengthen the Science of Ecology' of Belovsky *et al.* (2004). With kind permission of the American Institute of Biological Sciences.

---

1. Issues come in and out of fashion in ecology, like the latest *haute couture*, without scientific resolution.
2. There is a lack of appreciation of past literature; this in part, leads to ecology's fickleness towards central issues.
3. There is inadequate integration of empirical and theoretical ecology.
4. There is inadequate integration of natural history and experimentation.
5. There often is an implicit belief that ecological patterns are the result of single causes, but ecology's complex nature may be due to multiple causation (Hilborn and Mangel, 1997; Shurin *et al.*, 2001).
6. Applications of equilibrial and disequilibrial perspectives are often misleading in terms of explaining ecological patterns.
7. There is inadequate replication over time and space in ecological studies.
8. Data incompatibility and lack of rigour in obtaining data, and especially reliance on qualitative measures of ecological dynamics, often hinder the comparison of existing long-term and multi-location data.
9. Methodology and statistics should not be driving forces in ecology.
10. Ecology as a fundamental science is sometimes seen as distinct from the application of ecology to solve environmental problems.

Common errors in ecological research

Perhaps the best way to end this chapter is to recommend the paper by Belovsky *et al.* (2004), who cite '*Ten suggestions to strengthen the science of ecology*' (Table 1.3). Although written for the wider subject of ecology as a whole, they are equally appropriate for the subfield of vegetation science and plant ecology, and their paper must be seen as essential reading for all ecologists. Reference back to these points will be made throughout the book.

Chapter 2
Environmental gradients, plant communities and vegetation dynamics

This chapter examines the underlying concepts and theories of how plant species and vegetation are organised and how they change in relation to both space and time. Critical to both space and time is the notion of scale.

SCALES OF STUDY

Ecosystems and vegetation may be studied at a range of scales from an individual leaf up to the level of the whole biosphere. The various levels of ecosystem recognition are summarised in Figure 2.1. The concept of nesting of systems is important. All levels of ecosystems may be seen as subsystems of the biosphere (Rowe, 1961; Leuschner, 2005). Each subsystem occupies a progressively smaller area in space. As vegetation is the most obvious external feature on the basis of which ecosystems tend to be defined and classified, it follows that smaller and smaller units of vegetation can be recognised, from the biosphere through vegetation formations down to the individual plant and leaf (Figure 2.1).

Much early work in both plant ecology and biogeography concentrated on the world biomes and vegetation formations, which represented the first major subdivision of the biosphere (Eyre, 1968; Archibold, 1995; Box and Fujiwara, 2005). Inevitably, most of these studies were extremely generalised. In the latter half of the last century, studies became focused at the lower, more local scale of the plant community, firstly because this is the scale at which populations and individuals of a plant species could be identified and grouped together to characterise the vegetation of an area of a few square metres to several square kilometres. Secondly, the community scale was seen as important because it is at this scale that humans can make best sense of the nature and variation of the vegetation cover of the Earth. Thirdly, it is at this community scale that human activity in changing vegetation cover takes place, and thus conservation and environmental management practices and policies may be applied.

In Figure 2.2, the community or patch scale (the word 'patch' applies to one patch in the landscape matrix, corresponding to a particular plant community or land-use type) lies below the regional/landscape scale on the one hand and above the individual species/plant scale on

Vegetation Description and Data Analysis: A Practical Approach, Second Edition. Martin Kent.
© 2012 John Wiley & Sons, Ltd. Published 2012 by John Wiley & Sons, Ltd.

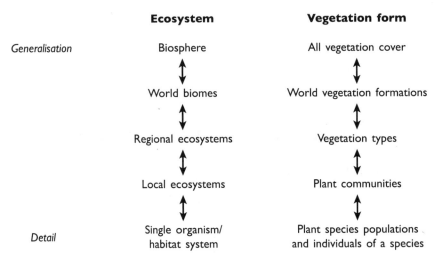

Figure 2.1   The various scales of study of ecosystems and vegetation. After Rowe (1961), with kind permission of the Ecological Society of America.

the other (Kent, 2009a,b). All research in vegetation science must be placed into context through recognition of links with these two scales above and below the community/patch scale. A nested approach is required, in which the relationships between these scales are clearly identified. These spatial scales also overlap with organisational scales, and organised communities obviously exist at a range of spatial scales (Rowe, 1961).

This problem of scale in ecology was widely discussed during the 1980s and 1990s and many papers were published on the subject (Allen and Wileyto, 1983; Giller and Gee, 1987; Meentenmeer and Box, 1987; May, 1989; O'Neill, 1989; Wiens, 1989; Allen and Hoekstra, 1990; Rahel, 1990; Hoekstra *et al.*, 1991; Levin, 1992; Jonsson and Moen, 1998; Wilson *et al.*, 1998). More recent contributions come from Parker (2001), Schneider (2001), Harte *et al.* (2005), Beever *et al.* (2006) and Sandel and Smith (2009). A major problem has always been the diverse and conflicting use of terminology, but also more recently, with the advent of macroecology and the description and analysis of species distributions at wider spatial scales (Brown, 1995; Gaston, 2003; Blackburn and Gaston, 2004, 2006; Kent, 2005, 2007b), discussion of scale has focused on single, rather than composite or community species distributions (e.g. Rahbek, 2005) and the debate has inevitably moved away from a primary emphasis on plants and vegetation to a much wider range of organisms. Plant community ecology occupies an interesting position between reductionist ecology, which dominated ecology in the 1980s and 1990s, and macroecology which has become a significant focus since 2000 (Figure 2.3).

THE CONCEPT OF THE PLANT COMMUNITY

Introduction

When an ecologist stands on a hilltop and surveys a landscape dominated by semi-natural vegetation in any part of the world, the major areas of difference in space that can be identified will be those of plant communities or plant species assemblages. Major distinctions will be made on the basis of physiognomy or the growth form of the vegetation, for example, woodland as opposed

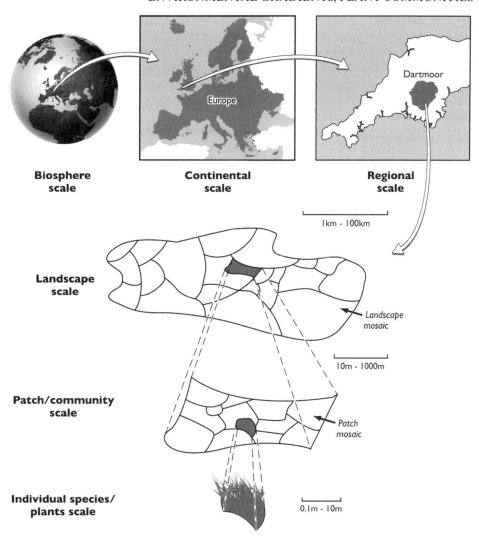

Figure 2.2   Scales in landscape ecology and biogeography (adapted from Forman and Godron, 1986, and Kent, 2009a).

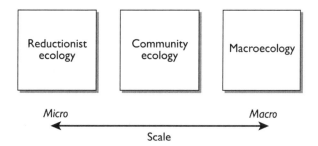

Figure 2.3   Community ecology in relation to reductionist ecology and macroecology. Original work of Martin Kent.

to scrub or grassland. These units will also represent the major subdivisions of the landscape in functional terms as ecosystems. More subtle changes in the landscape will also be evident in variations in colour between different areas of vegetation with perhaps the same physiognomy. These colour variations will be a reflection of differences in plant species composition. A considerable part of plant ecology and vegetation science is concerned with methods for actually characterising and defining these areas or patches as different plant communities. Thus the concept of the plant community is absolutely fundamental to the whole discipline.

The plant community can be defined as the collection of plant species growing together in a particular location that show a definite association or affinity with each other. The idea of association is very important and implies that certain species are found growing together in certain locations and environments more frequently than would be expected by chance. Similarly, different combinations of species will be found together in other environments more frequently than would have been expected by chance. Most environments of the world support certain associated species which can therefore be characterised as a plant community.

In the past decade, in the eyes of some ecologists, the concept of the plant community has become equated with the idea of species assemblage (Lockwood *et al.*, 1997; Weiher *et al.*, 1998; Belyea and Lancaster, 1999; Weiher and Keddy, 1999a,b; Wilson, 1999). The origins of the idea appear to lie primarily in organism groups other than plants, and the notion of assembly seems to imply that species are separate, independent and non-interacting entities or components that can be 'assembled' into some greater object, equivalent to a building or a product such as a car. However, one of the most important features of a plant community is that species interact and show interdependence, and the resulting community composition is the result of that species interaction through time (Callaway, 1997; Brooker and Callaghan, 1998; Lortie *et al.*, 2004). This idea is well summarised in the definition of the plant community provided by Shimwell (1971: p.1):

> ... not merely a random aggregation of plants but an organised complex with a typical species composition and morphological structure which have resulted from the interaction of species populations through time.

Environmental limiting factors

The reason why certain species choose to grow together in a particular environment will usually be because they have similar requirements for existence in terms of abiotic (non-living) environmental factors such as light, temperature, water, drainage and soil nutrients. They may also share the ability to tolerate the activities of living animals and humans, such as grazing, burning, cutting or trampling, which collectively are known as biotic factors.

If one environmental factor is taken, for example, temperature or soil moisture, and the abundance of a species is plotted across its range on variation the result may approximate to a normal or Gaussian curve (Figure 2.4). This variation of species abundance in response to an environmental factor is known as an environmental gradient. If several species are associated in a community, it is often assumed that their abundance curves in relation to environmental factors will be broadly similar. However, studies of species response to environmental gradients suggest that in practice, species curves vary enormously (Figure 2.5). The width and height of the curve for each species will be very different, indicating differences of tolerance range. Also the form of the curve will rarely be of the idealised perfect bell shape of Figure 2.4. Instead, curves will be skewed, bimodal, polymodal or have a 'plateau' shape (Austin and Austin, 1980; Austin *et al.*, 1984; Austin, 1987, 1999a,b, 2005; Austin and Smith, 1989) (Figure 2.5).

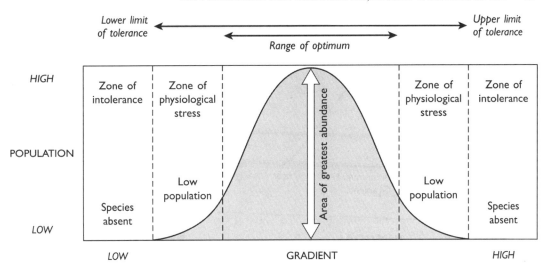

Figure 2.4   The Gaussian or normal distribution of abundance of one species along a single environmental gradient. After Cox and Moore (2005), redrawn with kind permission of Wiley-Blackwell.

Another feature of all Gaussian species response curves is the 'zero truncation problem'. Most species only occupy part of an environmental gradient. Points that lie beyond the ends of a species response curve all have a zero value for abundance, with the result that for those parts of the gradient where the species is absent, there is no information on how unfavourable that part of the environmental gradient is for the species. It is clearly not possible to derive negative values for a species' abundance once it is no longer present on an environmental gradient!

A further feature of species response curves is that the curve is usually fitted around the highest abundance values for each point on the environmental gradient. However, many other lower abundance values usually occur for the species at each point to 'fill' the area under the curve. Zero abundance values for some samples will often occur in the region of the species optimum in the centre of the curve, for example. The explanation for this lies in the fact that species are responding not just to the one single environmental factor but to multiple factors and gradients, and also to interactions between species, their life cycle strategies and to random (stochastic) processes.

This introduces the further complication that a species found growing at a point on the Earth's surface will virtually always be responding to more than one environmental factor. Thus each species will have a different environmental response curve for every environmental factor and each curve will differ in form. The ultimate favourableness or unfavourableness of a site for growth of a certain species will be represented by the collective positions of the site on each of the response curves for the species on each environmental gradient. For some factors, the site will be near the central, most abundant part of the curve and those environmental conditions will be near optimal. For others, it may be at or near the extremes. The points that are near to the ends of a response curve for a species are of great significance, because if one of the points is at or beyond the limits of the curve, the conditions will be too unfavourable for the species to grow. The factor for which this occurs is known as the master limiting factor. However, since several environmental factors determine the growth of a plant, unfavourable conditions for a species in terms of one factor may sometimes be compensated for by another. This is known as factor compensation.

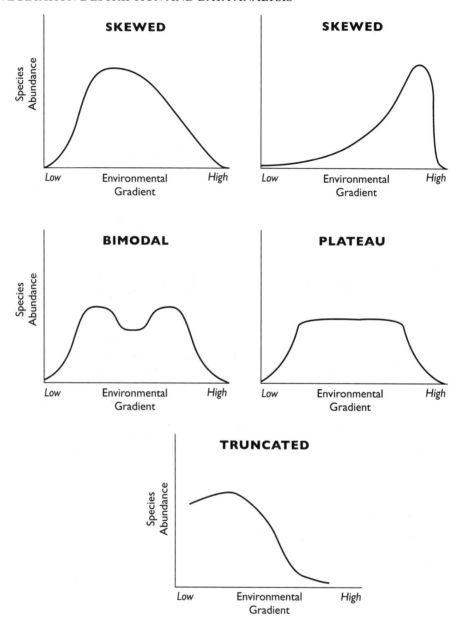

Figure 2.5   Typical forms of species abundance distribution curves found in the real world. Original work of Martin Kent.

It is because species are responding to multiple environmental/biotic gradients that multivariate analysis is so important to vegetation science (McGarigal *et al*., 2000). This links back to suggestion 5 of Belovsky's ten suggestions to strengthen the science of ecology (Belovsky *et al*., 2004; Table 1.3), which stresses the importance of multiple interacting environmental and biotic factors in determining species response.

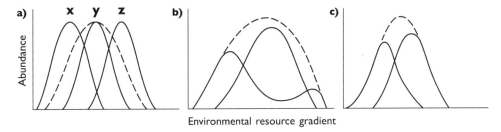

Figure 2.6 Relationships between the fundamental and realised niches of species. The fundamental niche of species y is shown by the dashed line, while the solid lines and shaded areas represent the realised niches of species x, y and z – see text for explanation (adapted from Mueller-Dombois and Ellenberg, 1974; Austin, 1999a, 2005).

Austin and Smith (1989) and Austin (2005) described environmental factors or variables represented as gradients as falling into three categories: (a) direct – variables or factors that have a direct influence on plant growth, for example, temperature and soil pH; (b) indirect – also sometimes called proxy variables, because they act indirectly through other direct variables; the best example is altitude, which can only exert influence through other direct variables, such as precipitation, temperature or wind exposure (Lookingbill and Urban, 2005); (c) resource variables – those used by plants in the course of growth, for example soil nutrients such as nitrogen or phosphorus. Some variables may be both resource and direct factors, for example water, which is a resource variable but becomes a direct variable when it is surplus to requirements and causes anaerobic (waterlogged) conditions.

A related concept is that of the niche. This is defined as '... the limits, for all important environmental features, within which individuals of a species can survive, grow and reproduce' (Begon *et al.*, 1990: p.963). The concept of the niche has been widely debated and discussed, for example in Whittaker *et al.* (1973), Schoener (1989), Leibold (1995), Silvertown (2004), Begon *et al.* (2006) and Godsoe (2010). A distinction is made between the fundamental niche, sometimes also called the physiological niche, which represents that part of the environmental gradient that a species response curve should be able to occupy in the absence of any other species, and the realised niche, which is the position that the curve may occupy in the presence of other species (Hutchinson, 1957). In Figure 2.6a, the fundamental niche of species y (dashed line) is reduced to its realised niche (solid line) by competition and the realised niches of species x and z (solid lines). In Figure 2.6b, species y is displaced from the optimal part of its fundamental niche by species x, and in Figure 2.6c, species x and y have shared out the fundamental niche space between them.

An appreciation of these ideas is important in the understanding and conceptualisation of the plant community. If species do grow in association with each other, many of them should have similar although rarely identical response curves and hence tolerance to the prevailing environmental conditions.

THE DEBATE ON THE EXISTENCE OF PLANT COMMUNITIES

The idea of the plant community was hotly debated by early plant ecologists and is still a matter of contention today. Two American ecologists, F. E. Clements and H.A. Gleason expressed the most extreme viewpoints.

Clements' view of the plant community

Clements (1916, 1928) saw plant communities as clearly recognisable and definable entities which repeated themselves with great regularity over a given region of the Earth's surface. Clements' view of the plant community is known as the 'organismic' concept, in which the various species comprising the vegetation at a point on the Earth's surface were likened to the organs and parts of the body of an animal or human. Putting all the parts together made a kind of super-organism that comprised the plant community, and the organism (plant community) could not function without all its organs present.

Within North America, Clements defined three major groups of vegetation, which he called climaxes: forests, scrub and grasslands. Each of these three climax types was then divided into a number of formations; for example, the forest climaxes were woodland (*Pinus* and *Juniperus*), montane forest (*Pinus* and *Pseudotsuga*), coast forest (*Thuja* and *Tsuga*), sub-alpine forest (*Abies* and *Picea*), boreal forest (*Picea* and *Larix*), lake forest (*Pinus* and *Tsuga*), deciduous forest (*Quercus* and *Fagus*), isthmian forest and insular forest. Each of these associations was further divided into 'associations', which were recognised on the basis of one or more characteristic dominant species. Clements also saw each of these associations as a 'climax community' which, given sufficient time and relative long-term stability, would come into equilibrium with climate. For this reason, his theory was known as the climatic climax or monoclimax theory. The Clementsian view has also been described as the 'community-unit' idea (Whittaker, 1951, 1953).

Clements's ideas are presented as an 'abstract' model of species response to a single environmental gradient in Figure 2.7a. Species growing together to form an association are assumed to have similar overlapping response curves. The questionmarks in Figure 2.7a correspond to the boundary areas between two community types. Unfortunately, Clements failed to explain how such areas would fit into his organismic model.

Gleason's view of the plant community

Gleason (1917, 1926, 1939) saw all plant species distributed as a continuum. He argued that plants respond to variations in environmental factors, and those factors vary continuously in both space and time. As a result, the combination of plant species found at any given point on the Earth's surface is unique. Every species has a different distribution or tolerance range and abundance over that range – a different size and shape of curve as in Figure 2.7b. The assemblage of plants growing in an area is not only the result of environmental selection but also species migration. Any area is continuously receiving propagules of species. The success of these species would depend upon the combination of environmental factors at that site and the tolerance ranges of the invading species. Gleason argued that the range of permutations of combinations of environmental factors, together with the different tolerance ranges of the species, would always give a different combination and abundance of species. Sampling along those gradients would always produce a different mix of species composition and abundance and samples could thus never be generalised into clearly defined plant communities. Taken to its extreme, Gleason's view was that plant communities, although they exist in the sense of a group of species at one point in space, cannot be identified as combinations of associated species repeating over space. He was thus fundamentally at odds with Clements (Nicholson and McIntosh, 2002).

Interesting discussions of these early viewpoints on the nature of the plant community are found in Tansley (1920, 1935), Cain (1934a) and Whittaker (1951). Sobolev and Utekhin (1978) and Robotnov (1979) introduced the work of L.G. Ramenskii, a Russian ecologist who had been

a)

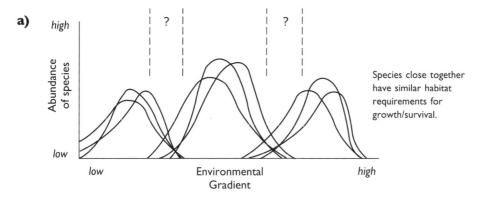

Species close together have similar habitat requirements for growth/survival.

b)

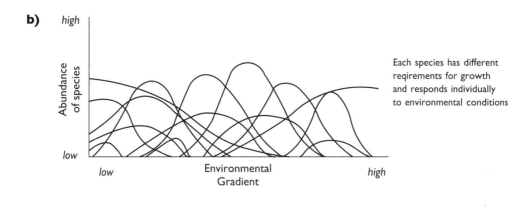

Each species has different reqirements for growth and responds individually to environmental conditions

c)

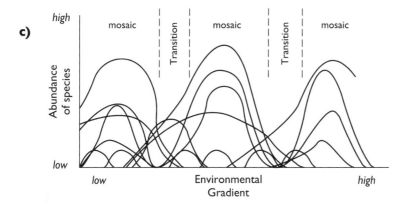

Figure 2.7 (a) The Clementsian view of plant communities expressed as species response curves along an environmental gradient; (b) the Gleasonian view of plant communities expressed as species response curves along an environmental gradient; (c) Whittaker's climax pattern hypothesis expressed as species response curves along an environmental gradient (Kent *et al.*, 1997). Reproduced with permission from Sage Publications.

largely ignored in the Western literature. They made the point that Ramenskii published ideas on the continuum concept and the plant community in the early decades of the last century and that, independently, he evolved theories that were very similar to those of Gleason in America (Ramenskii, 1930, 1938).

The 'climax pattern hypothesis' (Whittaker, 1953)

Whittaker (1953) and Whittaker and Levin (1977) provided a partial compromise to the highly contrasted views of Clements and Gleason in the 'climax pattern hypothesis'. This envisaged a mosaic of different communities with repeating patterns at the regional scale, correlated with particular combinations of controlling environmental factors. Whittaker argued that in any region, such as the Great Smoky Mountains of Tennessee and North Carolina, broadly similar conditions in terms of environmental factors and biotic pressures will occur over considerable areas. Where these combinations repeat themselves, the vegetation is also repeated, like similar fragments within a mosaic (for example, ROC and OCF in Figure 2.8). However, not all areas could be placed within one or other of these forest types, nor were the boundaries as clear as shown in the idealised diagram of Figure 2.8. Often, one forest type would grade into another across an area of transition or boundary between any two forest types, so that, while perhaps 60–80% of the vegetation could be put into one definite forest type, 20–40% could not, because they were transitional between types. Until recently, such transitional areas were largely neglected in plant ecology and most research concentrated on plant communities themselves. However, over the past decade, such transitions or boundaries have received more attention through research into boundary recognition and definition (Kent *et al.*, 1997, 2006; Fortin *et al.*, 2000; Fortin and Dale, 2005).

Using the same types of graphs for a single environmental gradient, as in Figures 2.7a and b, an approximation to the real world can be made as in Figure 2.7c. Over considerable areas, groups

BG	Beech gap	HB	Heath bald
CF	Cove forest	OCF	Chestnut oak-chestnut forest
F	Fraser fir forest	OCH	Chestnut oak-chestnut heath
GB	Grassy bald	OH	Oak-hickory forest
H	Hemlock forest		

P	Pine forest and pine heath
ROC	Red oak-chestnut forest
S	Spruce forest
SF	Spruce-fir forest

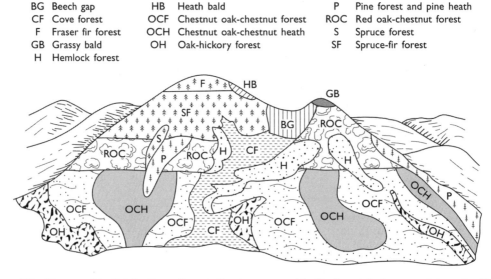

Figure 2.8 Topographic distributions of vegetation types on an idealised west-facing mountain and valley in the Great Smoky Mountains, USA. Redrawn from Whittaker (1956) with kind permission of the Ecological Society of America.

of species may repeat themselves in space in the manner of Clements, but there are other areas in between that contain a mix of species from both adjacent communities and that do not correspond to recognised major community types for the region. These latter areas are closer to the concepts of Gleason in that they are more 'individual' in character. More importantly, conceptually, they also represent the transitional or boundary areas between plant community types. However, in all three graphs of Figure 2.7, it is important to realise that the species curves and distributions are plotted along a single environmental gradient in conceptual or abstract environmental space, and the distances along the x axis in each case are not direct spatial distances.

The hierarchical continuum concept (Collins *et al.*, 1993)

Further advances in the conceptualisation of plant communities came from Kolasa (1989), Hanski (1982, 1998, 1999), Collins *et al.* (1993), Hanski *et al.* (1993) and Hanski and Gyllenberg (1993, 1997). Collins *et al.* (1993) and Hoagland and Collins (1997) described the 'hierarchical continuum concept', derived from a combination of the individualistic distribution of species, hierarchical assemblage structure and the core–satellite hypothesis of Hanski (1982). The 'core–satellite' model states that within any region, a positive relationship exists between the number of sites at which a plant species occurs and its average abundance in a region (Figure 2.9a). Four distinct types of species distribution can be recognised (Figure 2.9b):

(1) *Core species* – species that are widely distributed across the region with high abundance.
(2) *Satellite species* – those with limited distribution in the region and low abundance.
(3) *Urban species* – species with limited distribution but high abundance when they occur.
(4) *Rural species* – species with widespread distribution but low abundance when they occur.

The use of the terms 'urban' and 'rural' was unfortunate. They do not imply that such species are exclusively found in those actual environments and it is important to emphasise that the terminology instead refers to the types of species distribution and abundance properties. Hanski (1982) also suggested that most species tend to have abundance/distribution relationships that lie along the continuum from satellite to core types, represented by the line of positive correlation in Figure 2.9a. Thus 'urban' and 'rural' abundance/distribution types are rarer. Further discussion by Nee *et al.* (1991) covered the possibility that the distribution of satellite and core species is not a continuous spectrum, as shown in Figure 2.9a, but instead is bimodal, corresponding to the well-known 'J-shaped' curve of species abundances proposed by Raunkaier (1934) (Chapter 3; Figure 3.11), with most species having either high or low levels of abundance when they occur and comparatively few species showing levels of abundance between the two extremes. However, Nee *et al.* (1991) speculated as to whether such a bimodal distribution of abundances is an artefact of sampling rather than being present in the real world. This issue remains unresolved.

These four types of species abundances and distributions are envisaged as a hierarchical continuum and may be combined to show the composition of a typical community responding to a single environmental gradient, as shown in Figure 2.10. This model also incorporates the ideas of Kolasa (1989) that species distribution within communities corresponds to a non-nested hierarchy with varying levels of abundance. In Figure 2.10, a three-level hierarchy is evident with 'core'/regional species widely distributed at level 1 plus the occasional 'urban' species, 'urban' species at level 2, and a mixture of 'rural' and 'satellite' species at level 3.

The hierarchical continuum model may also be linked to that of the climax pattern idea presented in Figure 2.7c. Combining the basic single environmental gradient model of Figure 2.7c with the hierarchical structure of Figure 2.10 gives an even more developed conceptual model of plant

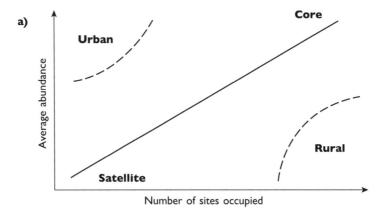

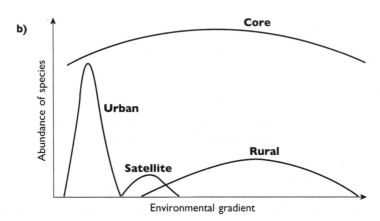

Figure 2.9 (a) The relationship between the number of sites occupied by a species and the average abundance of a species in a given region, showing the position of core, satellite, urban and rural species; (b) the distributions of core, satellite, urban and rural species along a hypothetical gradient. Redrawn from Hanski (1982, 1991) and Collins *et al*. (1993), with kind permission of the Wiley-Blackwell.

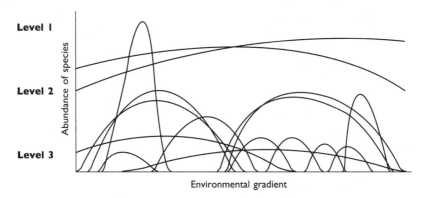

Figure 2.10 The hierarchical continuum model expressed as species response curves along an environmental gradient. Redrawn from Collins *et al*. (1993), with kind permission Wiley-Blackwell.

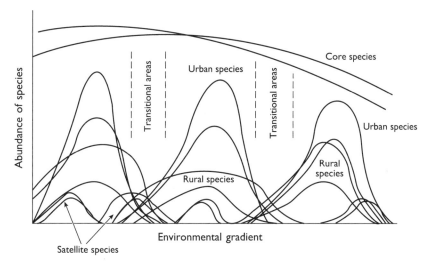

Figure 2.11   The hierarchical continuum model combined with the climax pattern hypothesis expressed as species response curves along an environmental gradient (Kent *et al*., 1997). Reproduced with permission from Sage Publications and the author.

communities and transitional areas (Figure 2.11). However, even this model is seriously limited by showing the responses of multiple species to only a single environmental gradient. In the real world, multiple species are responding to multiple environmental gradients. It is this added complexity that makes conceptualisation difficult.

Nevertheless, this model still suggests that communities, in the sense of Clements, definitely exist in most natural and semi-natural vegetation, and a set of 'community types', repeating in space, can be recognised in most, if not all, instances. However, in any region, considerable areas of vegetation occur between these major plant community types. These contain a more 'individualistic' distribution of species in the sense of Gleason and varying combinations of species from the recognised major vegetation types and, as a result, as described above, correspond to transitional or boundary zones.

A further important idea is that variability exists in the degrees of 'sharpness' with which Clements-type communities may be identified in any region. Most natural and semi-natural communities tend to have diffuse or 'fuzzy' boundaries, whereas those modified or created by human activity are usually more sharply or crisply defined (Fortin, 1999; Fortin *et al*., 2000). However, even in natural communities, there may be a spectrum of different degrees of distinctiveness of Clements-type communities and the transitional areas between them.

Yet further difficulties with the conceptual models of Figures 2.7 and 2.11 come from the work of Austin (1980, 1987), Austin and Smith (1989) and Austin and Gaywood (1994), which throws open to question the theoretical model of Gaussian species response curves along environmental gradients and thus the validity of the curves in Figures 2.7 and 2.11. Austin's empirical testing of the existence of bell-shaped response curves suggested that the form of the curve is never the perfect or near-perfect bell shape of the theoretical model. Instead, curves are often highly skewed, possibly bimodal or have a 'plateau' shape (Figure 2.5). Austin and Gaywood (1994) indicated that skewed response curves for Eucalypt species in Australia show a pattern in the direction of the skew, with the tail of the species response curve nearly always depending on the position of the species along the environmental gradient.

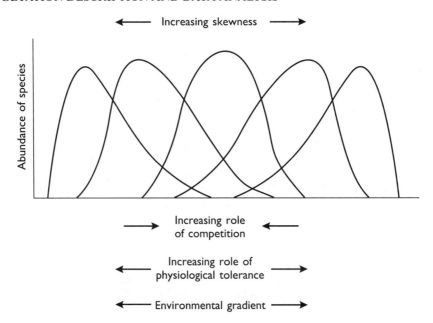

Figure 2.12 A model of the changing form of species response curves along an environmental gradient. Redrawn and adapted from Austin (1990).

As shown in Figure 2.12, Austin (1990) predicted that there will be an increasing degree of skewness as the mode of a species response curve moves towards either end of an environmental gradient. However, Heikkinen and Mäkipää (2010) questioned these findings, demonstrating that further research is still required to assist with the conceptualisation and understanding of plant and vegetation response to environmental/biotic factors and gradients.

The intrinsic properties of plants themselves and plant species strategies

The species that grow together to form a community have also usually proved that they can coexist with each other. Much early plant ecology was very deterministic in nature and tended to assume that the combination of environmental and biotic factors occurring at a particular point in space were the only controls of the plant species growing there. Now it is generally agreed that the properties of the plants themselves are also very important. These properties are related to plant physiognomy and physiology and include the idea of plant species strategies and plant population biology (Grime, 1974, 1977, 1993, 2001; Harper, 1977; Tilman, 1982, 1988; van der Maarel, 1984b, 2005b; Silvertown and Charlesworth, 2001). Through evolution, each species has evolved a set of characteristics of physiognomy and physiology that improve its chances of survival in certain environments. Numerous components of the life history of a plant can be varied to produce the different strategies. Reproduction mechanisms (seed versus clonal growth or vegetative reproduction), growth form and rate, timing of germination, growth, flowering and fruiting, reproduction and death (phenology) are all examples of aspects of physiognomy and physiology which are different for each species. An example of differences of strategy among five British grasses when comparing just the one aspect of timing of seed production shows how varied species strategies may be (Figure 2.13). The grasses have varied their strategies so that in each case

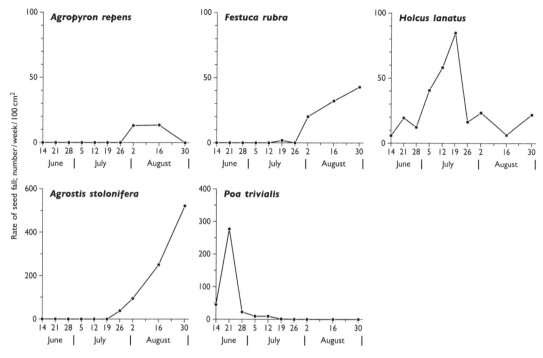

Figure 2.13 The time count of release of seeds from five grass species in a meadow. Redrawn with kind permission from Mortimer (1974).

their seed has an optimal chance of germination success. No two species have evolved identical strategies for timing of seed dispersal. In other examples, however, similar strategies may enable some species to coexist in a certain location. Alternatively, they may cause them to come into direct competition with each other.

Grime and Tilman's models of species strategies

Grime's (1979, 2001) CSR model of species strategies argued that plants are exposed to stress (resource limitation, particularly soil nutrients) and disturbance (removal of biomass) to varying extents, and this results in three plant species strategies. The competitive strategy (C) characterises plants that grow in resource-rich environments with low or no stress and that are undisturbed. Stress-tolerators (S) are species that can cope with high stress due to limitation of resources and low disturbance, while ruderals (R) are species that prosper in environments with high availability of resources but are also subject to high levels of disturbance. The combination of high stress and high disturbance represents an environmental space that cannot be occupied by species. A range of combinations and intermediates also occur, for example, an indifferent CSR strategy. Tilman (1982, 1988), on the other hand, modelled plant growth as a function of one or more resources and considered plant strategies to be adaptations to different combinations of the availability of resources. Tilman (1982) introduced the idea of minimum resource availability R^*, which represents the minimal needs of a plant species, depending on the quantities necessary for successful growth and the ability of the environment to provide the resources. Species with the lowest R^* are seen as being the most competitive.

It is important to be aware that some plant ecologists do not like the use of the species strategy concept and that it represents a controversial idea. Craine (2005, 2007) has presented a detailed review of the contrasting theories of Grime and Tilman and the ongoing debate on the roles of environmental resources and ideas of competition (see below) in determining the structuring of plant communities (Grime, 1974, 1977, 1993, 2001; Grime *et al.*, 1997; Tilman, 1982, 1988). Grime (2007) and Tilman (2007) provided replies to Craine's initial article. Their discussion provides an excellent example of academic debate being carried on through the literature that must be seen as essential reading for all students of vegetation science and plant ecology. It also links to the suggestions to strengthen the science of ecology of Belovsky *et al.* (2004) (Table 1.3), particularly suggestion 3. Grime and Tilman's work clearly demonstrates the ongoing search for links between empirical evidence and theoretical ecology.

INCORPORATION OF IDEAS OF SPECIES INTERACTION INTO CONCEPTS OF THE PLANT COMMUNITY

Competition, facilitation and coexistence

Given the combination of environmental controls at a certain location, plant species will also compete to occupy a position and coexist with other species that may already be there. Mechanisms of competition are many and varied. Growth form and physiognomy, growth rate, shading effects, deposition of litter, release of toxic substances from roots and in litter (allelopathy) and differences in reproductive strategy are all examples of such mechanisms. Competition can also occur between individuals of the same species (intraspecific competition) or between individuals of different species (interspecific competition). Often species may survive within a community just by being there first and pre-empting space. Walker *et al.* (1989) developed this idea with their 'ecological field theory'. Good introductions and reviews of competition and coexistence are provided in Fitter (1987), Law and Watkinson (1989), Zobel (1992), Keddy (2001, 2007), Silvertown and Charlesworth (2001), van Andel (2005) and Begon *et al.* (2006).

The importance of competition, interaction and coexistence is shown in Figure 2.14. Rorison (1969) carried out experiments on the environmental tolerance of four plant species grown under both field and laboratory conditions along an environmental gradient of pH. The resulting abundance curves for the four species are very different both one from another and between field and laboratory. The abundance distribution along the gradient for the field situation (the ecological curve) is in all cases different and more limited than for the laboratory situation (the physiological curve). Rorison argued that one of the reasons for this difference was the removal of the interspecific competition found in the real world (field) from the laboratory experiment. In the absence of competitors, species were able to grow across a much wider range of pH. This empirical research links to the ideas of fundamental (physiological) and realised niches and of niche partitioning and displacement illustrated in Figure 2.6. The wider aspects of the interactions between resources and competition are again highlighted in the debate between Grime and Tilman introduced in the previous section of this chapter.

Positive interaction and facilitation are equally, if not even more, important than negative interaction (Bertness and Callaway, 1994; Callaway, 1995, 1998; Callaway and Walker, 1997; Brooker and Callaghan, 1998; Brooker and Callaway, 2009). Adjacent plants compete with each other (negative) but also provide benefits for each other (positive) in the form of factors such as shade

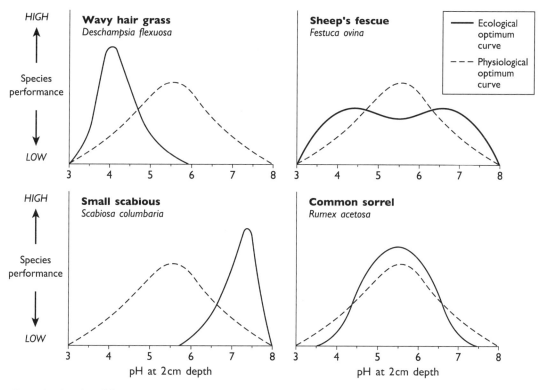

Figure 2.14   The difference between the behaviour of four plant species growing in the field (ecological optimum curve) and under non-competitive conditions in controlled laboratory conditions (physiological optimum curve). Redrawn from Rorison (1969), with kind permission of Wiley-Blackwell.

and shelter, raising of nutrient levels, conservation of moisture, soil oxygenation, protection from grazers, improved soil microflora and fauna, attraction of pollinators and transfer of resources. All of these are forms of facilitation, encouraging neighbouring plants to survive. The result is to encourage species coexistence and hence diversity and species richness within a particular community (Cavieres and Badano, 2009; Navas and Violle, 2009).

The continuing debate over the nature of the plant community

Even today, ecologists still differ strongly over their conceptualisation of plant communities (Austin, 1986, 1999a, 2005; Barkman, 1990). Nevertheless, a majority of ecologists would agree about the existence of plant communities as sets of species that repeat themselves over space within the landscape mosaic, and that individual pieces of the mosaic have boundaries or edges of varying sharpness (Kent *et al.*, 2006). Thus their viewpoint lies somewhere between the two extremes of Clements and Gleason and ties in with Whittaker's 'climax pattern hypothesis'.

Most quantitative plant ecologists who use classification methods (Chapters 7 and 8) would tend towards the views of Clements because, by definition, classification assumes that samples of vegetation composition (species and their abundances) can be grouped into types. Other workers, however, particularly in the US, would argue that classification of vegetation samples into groups is impossible. Instead, vegetation samples can only be arranged along environmental gradients as continua, using ordination techniques as described in Chapter 6. They thus tend to agree more with the views of Gleason. However, particularly with the flexibility of most computer packages for numerical classification and ordination methods, most vegetation scientists routinely use both groups of techniques in a complementary fashion to explore the patterns within their vegetation data, which lends support to the views of Whittaker and the climax pattern hypothesis.

Austin (2005) provided an excellent review of the 'continuum versus community controversy', as he described it. The continuum (Gleasonian) view has predominated in North America, while the community (or community-unit) view (Clementsian) has been more favoured in Britain and Europe and particularly by the group on mainland Europe known as the continental phytosociologists, formerly led by J. Braun-Blanquet (see Chapter 7). Austin (2005: p.81) made a very important point, that:

> ... Whether vegetation is discontinuous depends on the perspective of the viewer. Viewed from a landscape perspective it [vegetation] is often discontinuous. In environmental space it is usually thought to be continuous.

In other words, when we look at the landscape in the real world, we often see plant communities as mosaics (Clementsian), while underlying conceptual models of environmental gradients and species response curves makes us think of abstract continua (Gleasonian) – the viewed real world (Clementsian) is different from our conceptual environmental gradient model (Gleasonian).

The integrated community concept of Lortie *et al.* (2004)

Lortie *et al*. (2004) have presented the most recent advances in the conceptualisation of the plant community. They argue that the 'continuum versus community' debate has caused plant ecologists to neglect the significance of interactions within plant communities and how those interactions affect theories of plant community structure and vegetation–environment relationships. Interactions between plant species are both negative and positive. Negative interactions occur though inter- and intraspecific competition, leading to niche partitioning, as outlined above (Figure 2.6). However, positive interactions are equally significant. They stress that hundreds of research studies have now demonstrated positive interaction and, in particular, facilitation between plant species (e.g. Bertness and Callaway, 1994; Callaway, 1995, 1998; Brooker and Callaghan, 1998; Brooker and Callaway, 2009). Complex webs of species interactions, both positive and negative, occur in all communities. Lortie *et al*. (2004) predict that community structure is ultimately determined by

> ... synergistic non-linear interactions involving: (i) stochastic processes; (ii) the specific tolerances of species to the suite of local abiotic conditions; (iii) positive and negative direct and indirect interactions among plants and (iv) direct interactions from other organisms. (Lortie *et al*., 2004: p.434)

Their model, which they term the 'integrated community concept' (IC), is presented in Figure 2.15. The key aspect of the model is that each of the four processes can be of greater or lesser importance in the structuring of a plant community at a given point in space and time. Communities show different degrees of dependencies or integration between species controlled by the relative importance of each of the filters in Figure 2.15. Species within every community will exhibit a spectrum of dependence that is related to stochastic or chance processes. Strong interdependence and positive interaction is related to non-randomness at one extreme, while independence is related to high levels of randomness at the other. The position of any one species on that spectrum will vary in both space and time, and the degree of independence or interdependence

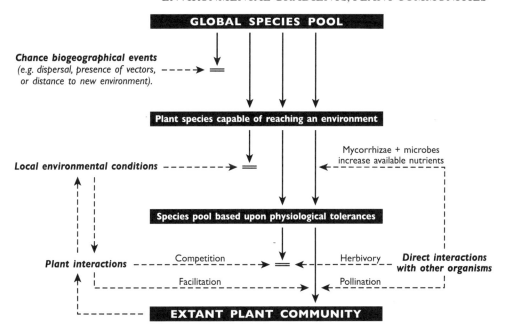

Figure 2.15   The main processes or filters that structure a plant community. The IC concept proposes that all four processes can be important in determining the extant plant community at a given site, but that the relative importance of each process will vary in space and time. Each process/filter is represented by a pair of horizontal lines and the corresponding description is in bold italics adjacent to the symbol (subsets of a process such as herbivory or competition are labelled in plain text). Solid arrows depict the movement of species through the filters, and dashed lines illustrate where each process might influence the plant community. Redrawn from Lortie *et al*. (2004: p.434) and reproduced with kind permission of Wiley-Blackwell.

of a whole community at one point in space and time will reflect the cumulative degrees of independence or interdependence of its constituent species.

Their key argument is that the fact that many communities demonstrate high degrees of positive interaction and interdependence does not support the idea of the universal occurrence of the Gleasonian individualistic view of community structure that has been prevalent in many parts of the world over the past 30 years. Instead, their model suggests that plant communities show different degrees of interaction at different points in time and space and may correspond more or less to either the Clementsian or Gleasonian models accordingly. Communities showing strong positive interactions are much more likely to be Clementsian in nature.

The other important feature of the model is the idea that significant positive and negative interactions not only occur among plants themselves, but also occur between plants and higher trophic levels within the whole ecosystem of which the vegetation is but a part. Thus grazing or even biotic impacts of any kind can be seen as positive or negative interactions that determine the overall status of species within a particular plant community. In addition, there may be even further interactions between species at any trophic level and characteristics of the environment, such as burrowing effects of animals, wind exposure reduction by trees, or physical and chemical effects of litter on weathering and erosion (Jones and Callaway, 2007).

In conclusion to this whole section on the nature of the plant community, the critical point is that vegetation scientists (and ecologists generally) still do not have a fully validated model of how species respond to their environment and to biotic impacts, and appear to have consistently neglected the importance of species interactions. This is a statement that should concern all plant ecologists and vegetation scientists, since a majority of ecologists appear to believe that the nature of species–environment relationships and species interactions are clearly understood and all questions have been answered (Pickett and Kolasa, 1989). Since species/environment together with species interaction response models still require verification, then there must still be many aspects of the nature of the plant community that have yet to be clarified. Much further research and analysis remains to be completed. Another quote from Austin (2005: p.81) seems the best way to conclude this section:

> At the present time there are many unanswered questions in vegetation science but there are also too many unquestioned answers.

HUMAN ACTIVITY, PLANT COMMUNITIES AND LAND USE – THE DISCIPLINE OF LANDSCAPE ECOLOGY

Over wide parts of the globe, human populations have modified plant communities extensively. Where natural and semi-natural vegetation remains, one of the most obvious effects has been to sharpen community boundaries, so that species and community distributions are often much more like those in Figure 2.7a.

Where human activity has completely removed the former vegetation cover, it has normally been replaced with some form of land use, such as agriculture, industry or urbanisation. With some land uses, such as arable cultivation, for example, vegetation becomes almost completely absent. However, in others, highly modified vegetation types (for example parks and gardens) may thrive. Often, between different land-use types, there are small or even quite extensive areas where remnants of the previous natural or semi-natural vegetation, such as road verges and hedgerows, may flourish, albeit in a highly modified form.

In response to these changes, over the past 25 years the subdiscipline of landscape ecology has evolved (Forman and Godron, 1986; Naveh and Lieberman, 1994; Forman, 1995; Kupfer, 1995; Kent et al., 1997; Sanderson and Harris, 2000; Bastian, 2001; Turner et al., 2001; de Blois et al., 2002; Burel and Baudry, 2003; Turner, 2005a,b, 2010; Wiens and Moss, 2005; Kent, 2007a, 2009b). Landscape ecology is concerned with the description, analysis and explanation of the spatial patterns of plant community and land-use types within a given landscape or region. As described above (Figures 2.2 and 2.8), landscapes are composed of 'patches' that are distributed as a mosaic within any local area. The mosaic of patches at the patch scale within a landscape can be aggregated to give the landscape scale, and both higher and lower scales may be identified above the landscape scale and below the patch scale (Figure 2.2). These individual patches have varying degrees of 'naturalness', and a spectrum of patch types with varying intensity of human impact and modification, known as land-cover types, may be identified. Linear patches and features, such as river courses and hedgerows, are known as corridors, and again boundaries between patches or corridors of different types are of considerable interest. The links between the landscape patch idea and the concept of environmental gradients is important and is reviewed in Urban et al. (2002) and McGarigal and Cushman (2005).

The methodology of landscape ecology is closely linked to recent developments in both remote sensing and geographical information systems (GIS) (Haines-Young et al., 1993a,b; Bissonette, 1997; Klopatek and Gardner, 1999; Farina, 2000, 2006; Turner et al., 2001; Wiens and Moss, 2005;

Kent, 2007a, 2009a,b). The key point is that vegetation science is increasingly becoming linked to and integrated with landscape ecology.

THE TIME FACTOR IN THE STUDY OF VEGETATION

Vegetation dynamics – concepts of succession and climax

Plant communities are dynamic in nature. In all parts of the world with seasonality of climate, the community changes with the seasons in terms of both presence of species and in their relative abundance. Over longer periods of time, community composition will often change according to the principles of succession and climax. Succession involves the immigration and extinction of species together with changes in their relative abundances. Primary succession occurs on bare ground where vegetation has not previously been found. Such sites are relatively small in extent on a world scale and are represented by landslips, new surfaces created by human activity or new volcanic islands emerging out of the ocean as in the case of Krakatoa in the Western Pacific in 1883 and Surtsey in the Atlantic, south of Iceland, in 1963. Several successive groups of species may invade as seral stages. Secondary successions occur when an established vegetation cover at a particular seral stage climax is removed or modified to an earlier seral stage, usually by some form of disturbance (Sousa, 1984; Pickett and White, 1985; Laska, 2001). The end product of succession is the climax community.

Whereas succession implies changes in species composition through time, the climax concept is based on the idea of relative stability. The nature of climax communities has always been controversial. Clements (1916, 1936) was responsible for the original ideas of succession and climax, and it was he who first applied the terms sere, seral stage and climax to successional processes and plant communities. He also introduced the concept of the climatic climax, arguing that, ultimately, all communities in a given region come into equilibrium with the regional climate. For this reason, his ideas were known as the monoclimax theory. Tansley (1935) argued that observation of communities showed that, although in theory most vegetation might reach an equilibrium with climate, this was unrealistic because of the time periods required were so long. Also, factors other than climate could hold a community relatively stable for considerable periods of time, such as soil-related (edaphic), topographic (geomorphic) and biotic (human and animal) factors. He thus proposed the polyclimax theory whereby, within a given region, communities could be in relatively stable equilibrium with any one or a combination of the above factors.

The notion of climatic climax has also been questioned by the very large amount of research completed over the past 40 years into Quaternary and Holocene ecology and palaeoecology (Birks and Birks, 1980; Lowe and Walker, 1997; Roberts, 1998; Anderson et al., 2007). World climates are increasingly shown to have been more and more dynamic over the Quaternary period, with the result that the idea of stable climax communities existing for long periods of time must be called into question. Wholesale movements of world vegetation formation types are known to have occurred throughout the Quaternary at all latitudes, with many smaller but still significant fluctuations superimposed upon them.

Although concepts of long-term stable climaxes are now questioned, there is no doubt that prior to the arrival of humans, extensive regional vegetation types existed over large tracts of the globe. The problem for present-day community ecologists is that in many parts of the globe, only a comparatively small proportion of this former natural vegetation remains. Thus in the majority of situations, study of plant communities involves the examination of communities that are subject to some form of successional change, or which are held as plagio- or biotic climaxes by human

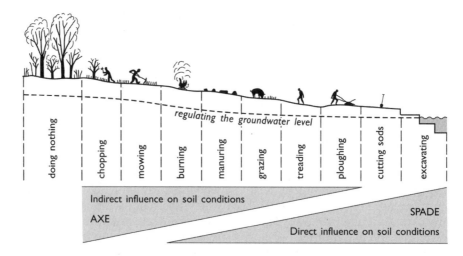

Figure 2.16   The spectrum of biotic controls on plant communities. Redrawn from Bakker (1979), with kind permission of Junk, The Hague.

activities (burning, cutting, mowing, treading) or animals (grazing, treading, dunging) and are subject to disturbance of some kind. The nature of these controls is summarised in Figure 2.16, where the spectrum of biotic controls of vegetation is illustrated. Maintenance of many communities requires the continuance of the land management activities that created the communities in the first place. This is one of the most important reasons why active management of vegetation and communities for conservation is necessary. However, the application of management practice usually requires detailed research to provide information to form the basis of management planning. The importance of methods for vegetation description and data analysis in this process has already been stressed in the heath lobelia case study presented in Chapter 1.

A considerable number of valuable summaries and critical reviews of successional and climax theory have been published over the past 40 years: Drury and Nisbet (1973), Connell and Slatyer (1977), Noble and Slatyer (1980), Crawley (1997), Gibson and Brown (1986), Gray *et al.* (1987), Huston and Smith (1987), van der Maarel (1988), Smith and Huston (1989), Burrows (1990), Luken (1990), Glenn-Lewin *et al.* (1992), Glenn-Lewin and van der Maarel (1992), van Andel *et al.* (1993), McCook (1994), Pickett and Cadenasso (2005) and Pickett *et al.* (2008).

The major debate still centres on whether it is valid to see succession as a community process where seral stages replace each other, as Clements originally envisaged, or whether it is more realistic to concentrate on the level of the individual plant and plant species in terms of immigration, interaction and extinction within a probabilistic framework linked more to the original ideas of Gleason (1917, 1927). Connell and Slatyer (1977) produced some of the most formative ideas on succession, developing the ideas of Egler (1954) and Drury and Nisbet (1973) to produce their three 'alternative' models of succession: facilitation, tolerance and inhibition (Figure 2.17). The three approaches are essentially descriptive and were not intended to provide a complete explanation for successional processes, but they have engendered healthy debate in the successional literature (see references above) ever since.

Pickett *et al.* (2008) presented a revised Clementsian-based framework based on the ideas of MacMahon (1981). Figure 2.18 shows a hierarchical causal framework for succession or

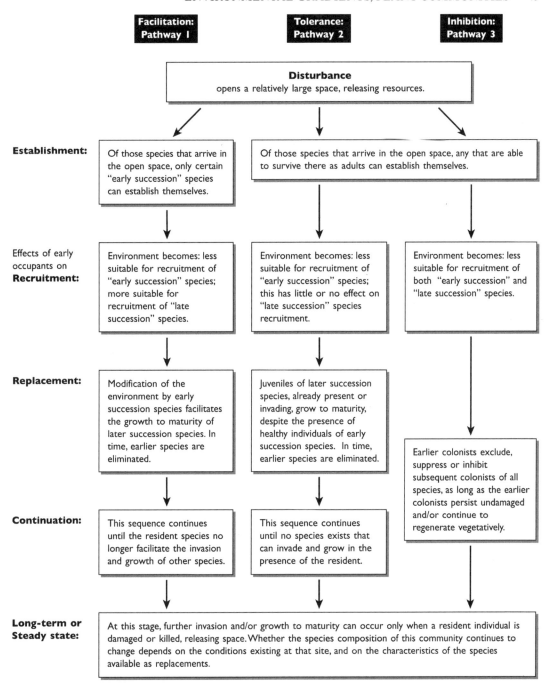

Figure 2.17   The three pathways of succession according to Connell and Slatyer (1977). In Pathway 1, early colonisers facilitate later species. Pathway 2 involves tolerance, and in Pathway 3, early colonisers inhibit later species. At each stage, further disturbance may return the system to the starting point. Redrawn from Connell and Slatyer (1977), with kind permission from the American Society of Naturalists.

I. Process   **II. General Causes**   **III. Specific Mechanisms**

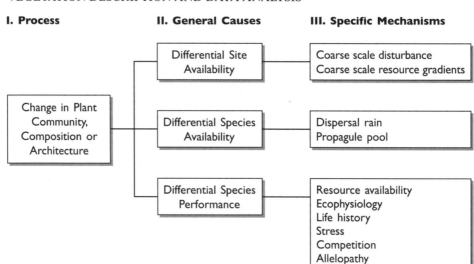

Figure 2.18   A hierarchical framework for succession or vegetation dynamics. Level I specifies the most inclusive process to be considered. Level II shows the three major component processes that result in vegetation dynamics. Level III shows more specific mechanisms that contribute to each of the three differentials that generate succession. The entry for consumers in Level III includes herbivory, predation and disease. The mechanisms on Level III may themselves be disaggregated into still more specific mechanisms, depending on the needs of particular models (cf. Pickett and Cadenasso, 2005). These models result in the net effects of facilitation, tolerance and inhibition. Redrawn from Pickett and Cadenasso (2005) and Pickett *et al.* (2008), with kind permission of Wiley-Blackwell.

vegetation dynamics with three major differentials as drivers of vegetation dynamics: (a) differential site availability; (b) differential species availability; and (c) differential species performance. Figure 2.19a presents the original five mechanistic causes of Clements, and Figure 2.19b the revised framework.

Two further aspects of succession are worthy of discussion. The first is the idea of chronosequences. Because of the problems of long-term time-based research into succession, a common approach has been to adopt the ergodic hypothesis, which involves the substitution of space for time. Thus within a given area, examples of different stages of succession are assumed to be represented by different communities in space at one point in time. Johnson and Miyanishi (2008) tested the assumptions of chronosequences in succession and examined four classic situations where they have been used: dune, hydrarch, glacial till and old-field successions. Comparison of chronosequence data with data from full time-based research in each case gave serious cause for concern, and they suggested that the attractive simplicity of such chronosequences (they can be found in most basic ecology textbooks) was invalidated entirely by empirical evidence.

The second is the concept of patch dynamics (Pickett and White, 1985; Pickett and Cadenasso, 1995). As outlined above, the patch concept is fundamental to landscape ecology (Forman and Godron, 1986; Turner *et al.*, 2001; Turner, 2005a,b). A patch is an area within the landscape, of any scale, that differs from its surrounding area in terms of species composition and structure (Figure 2.2). It may also differ in its abiotic character (Pickett and Cadenasso, 1995). Patches are often created by disturbance, by microclimate or by edaphic (soil) conditions (Pickett and

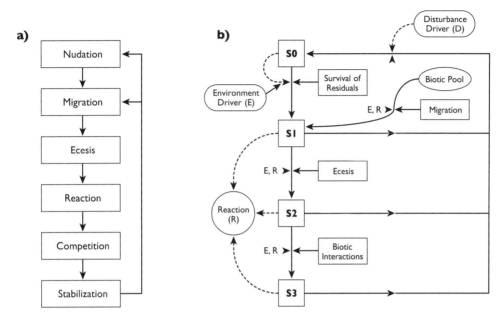

Figure 2.19 (a) The six causes of succession as described by Clements (1916). The middle four are better described as sequential phases of each successional stage, while nudation only occurs at the outset, and stabilisation is not a cause but an outcome. (b) A flow model of successional processes incorporating the five mechanistic causes of Clements (1916) and updating competition to all relevant biotic interactions. Clements' concept of nudation is updated in the processes included in the disturbance driver here. Connections between reaction (R) and the environment (E) are indicated by those letters. Figure 2.18 suggests the richness of major causes and some of the specific mechanisms that may be incorporated in models based on this template. Adapted and redrawn from Pickett *et al.* (2008), with kind permission of Wiley-Blackwell.

White, 1985). Patch dynamics correspond to changes within patches due to successional processes and interactions amongst adjacent patches and the surrounding landscape matrix. The result is a shifting mosaic of patches through time at the landscape scale. This shifting mosaic of patches may or may not be in equilibrium with prevailing conditions.

Peters *et al.* (2006) presented a conceptual framework for the study of patch dynamics at multiple scales from a root or leaf patch through an individual plant patch to the assemblage patch, and argued that ultimately a landscape mosaic is made up of patches from these multiple scales. They examined patch boundaries or transitions and defined three types of transition between patches: (a) directional transitions – for example, the encroachment of woody shrubs and trees into grassland (successional change within a patch); (b) stationary transitions – boundaries that change little over time, often related to sharp environmental changes, such as sudden changes in topography, slope angle or aspect; (c) shifting transitions – boundaries that remain relatively constant over time but that show movement back and forth over shorter periods of time due to short-term fluctuations in environmental factors, meso/micro climate conditions and biotic controls such as grazing intensity and fire.

Finally, mention must be made of the links between successional processes and landscape or ecological restoration (Walker and del Moral, 2003, 2008; Walker *et al.*, 2007). Much restoration and reclamation work relies upon the functioning of successional processes within a programme of restoration management. Understanding the importance of vegetation dynamics and species interactions is critical to ecosystem restoration success.

CONCLUSION

The aim of this chapter has been to introduce the key underlying concepts and models that underpin the description and analysis of vegetation. The development of such understanding is essential before any research or project work in vegetation science is commenced. All data collection and analysis should be performed only once the student or researcher has a good appreciation of the concepts of the plant community and vegetation dynamics that have been presented. However, a recurring theme has been that existing theory and models are still incomplete and require further testing and development. It is therefore very important that researchers and students realise their limitations.

Also, the literature on the concept of the plant community, environmental gradients and on succession and vegetation dynamics shows very clearly the importance of suggestion 2 in Belovsky *et al.*'s (2004) ten suggestions for ecologists (Table 1.3). Students can only fully appreciate the ongoing debates on these fundamental concepts in vegetation science if they take the trouble to go back and read the original papers and books on these topics that stretch back over a century.

Chapter 3

The description of vegetation in the field

INITIAL CONSIDERATIONS

The methods of vegetation description employed in a particular project will depend upon a number of factors:

(a) The research objectives and the purpose of the survey – the features and characteristics of the vegetation to be described will vary according to the overall aims and objectives.
(b) The scale of the study – very different description methods will be required for a survey covering many thousands of square kilometres compared with very detailed studies of a small area of perhaps a few hundred square metres.
(c) The overall habitat type – different techniques are necessary for different habitat types and growth forms. Thus a method suited to forests in the southeastern US will be totally inappropriate for a tropical grassland or savanna in West Africa or tropical rainforest in Brazil.
(d) Resources available – finance, equipment, human resources and time. Vegetation description of any sizeable area will require sufficient resources of all four to be available.

Several other questions will also be important:

(1) Is it necessary or relevant to identify all of the plant species present?
(2) If identification is required, do published floras already exist?
(3) Are environmental and biotic data to be collected, and if so, which variables should be measured and are the appropriate equipment and resources available to do so?
(4) What methods of vegetation data analysis are going to be used and how do these relate to the research objectives? Certain types of data analysis impose conditions and constraints on the form of data collected in the field.
(5) Has the dynamic nature of the vegetation been taken into account? The time of year when the recording is to take place is critical, and information on the successional and climax status of the vegetation may be extremely valuable (Kirby *et al.*, 1986).
(6) Where more than one recorder is being used, have the problems of ensuring consistency between different workers been addressed?
(7) The balance between speed and accuracy of the description method should be assessed. Often the time available for survey and the amount of data that can be collected will be limited by

Vegetation Description and Data Analysis: A Practical Approach, Second Edition. Martin Kent.
© 2012 John Wiley & Sons, Ltd. Published 2012 by John Wiley & Sons, Ltd.

seasonality of climate or weather conditions and human resources. The maxim that the best and most suitable quality of data must be collected in the minimum or available time is a useful one to apply.

Sutherland (2006a,b) describes the process of 'reverse planning' of projects. Work out the ultimate aims and objectives first and then work backwards, bearing in mind all the points above. Not infrequently, the original ideas may emerge as being over-ambitious and thus the project will have to be redesigned or scaled down to realistic proportions. This is often a hard lesson for students, but it is much better to revise and review the feasibility of a project completely before commencing any actual work in the field. Adaptability and flexibility are key factors, but any reappraisal of a project should always be made bearing in mind the whole project proposal. Cumulative small changes to the project should be avoided.

Finally, it is important to see field data collection as only part of the whole project. The collection of data should never be isolated from the research aims and objectives or from methods of data analysis (Austin, 1991a,b; Jongman *et al.*, 1995; Bullock, 2006; Sutherland, 2006b).

PHYSIOGNOMIC AND FLORISTIC DATA

Methods of vegetation description fall into two categories:

(a) Physiognomic or structural – where description is based upon external morphology, life-form, stratification and size of the species present.
(b) Floristic – where the species present in the study are identified and their presence/absence or abundance is recorded.

Physiognomic and structural methods have been used primarily for the classification of vegetation at small scales (over large areas), such as world vegetation formations. Floristic analyses have usually been applied at the large scale (over small areas) particularly at the level of the plant community. However, physiognomic and structural methods have their uses at the community scale, for example in habitat classification (see below). Floristic data can only be collected in a very generalised form at scales higher than the community level.

TECHNIQUES OF VEGETATION DESCRIPTION BASED ON PHYSIOGNOMY AND STRUCTURE

Raunkaier's life-form classification

One of the most famous physiognomic techniques of vegetation description is the life-form method of Raunkaier (1934, 1937). He devised a biological spectrum based on the height above ground of the perennating buds of each species, which are the parts of the plant from which growth commences in the next favourable growing season. His life-form classification was based upon the assumption that species morphology is closely related to climatic controls, the humid tropics representing the most favourable conditions for species in terms of solar radiation, temperature and precipitation, while species in other environments, deficient in moisture, solar radiation or temperature, show varying degrees of adaptation and response in the positioning of their perennating buds. The main groups of his classification are summarised in Table 3.1. Five main categories are recognised, each of which may be subdivided further. The phanerophytes, for example, are divided on the basis of height, into Nanophanerophytes (less than 2 m in height), Microphanerophytes (2–8 m), Mesophanerophytes (8–30 m) and Megaphanerophytes (over 30 m). Cryptophytes are

Table 3.1   The major categories of the Raunkaier life-form classification system (1937). Reproduced with permission from John Wiley & Sons, Inc.

Group 1 PHANEROPHYTES
Species with perennating buds emerging from aerial parts of the plant:
(a) evergreen phanerophytes without bud scales
(b) evergreen phanerophytes with bud scales
(c) deciduous phanerophytes with bud scales

Each of these types may be classified according to height:

Megaphanerophytes (>30 m)
Mesophanerophytes (8–30 m)
Microphanerophytes (2–8 m)
Nanophanerophytes (<2 m)

Group 2 CHAMAEPHYTES
Species with perennating buds borne on aerial parts close to the ground (below 2 m)
They may be woody or herbaceous:
(a) Suffruticose chamaephytes – after the main growth period, upper shoots die so that only the lower parts of the plant remain in the unfavourable period
(b) Passive chamaephytes – at the onset of adverse conditions, shoots weaken and fall to the ground, becoming procumbent. They get some protection from environmental stress
(c) Active chamaephytes – shoots are only produced along the ground and remain so in the unfavourable season
(d) Cushion chamaephytes – a modification of passive types, where shoots are arranged so close together that they cannot fall over and the close packing of all shoots forms a cushion

Group 3 HEMICRYPTOPHYTES
All above-ground parts of the plant die back in unfavourable conditions and buds are borne at ground level:
(a) Protohemicryptophytes – leaves become better developed up the stem of the plant. Poorly developed leaves protect the bud in early stages of growth
(b) Partial rosette plants – the developed leaves form a rosette at the base of the plant in the first year of growth. The following year an elongated aerial shoot may form
(c) Rosette plants – leaves are restricted to a basal rosette with an elongated aerial shoot, which is exclusively flower-bearing

Group 4 CRYPTOPHYTES
Plant species with buds or shoot apices that survive the unfavourable period below ground or under water
(a) Geophytes – plants with subterranean organs such as bulbs, rhizomes and tubers from which shoots emerge in the next growing season
(b) Helophytes – plants with their perennating buds in soil or mud below water and which produce shoots reaching above water
(c) Hydrophytes – species with buds that lie under water and survive the unfavourable season by budding from rhizomes under water or from detached vegetative buds which sink to the bottom

Group 5 THEROPHYTES
Plants that survive the unfavourable period as seeds. Species are thus annuals and complete their life cycle from seed to seed in the favourable summer months.

---

subdivided into geophytes (plants with rhizomes, bulbs or tubers), helophytes (plants with perennating organs in soil or mud under water and with aerial shoots above water) and hydrophytes (plants with perennating buds under water and with floating or submerged leaves). More detailed categories may be recognised, based upon the degree of protection of the buds and whether or

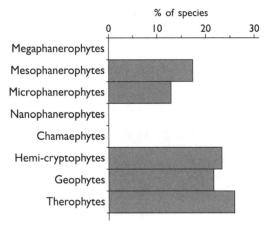

Figure 3.1   Savanna Raunkaier life-form spectrum, Olokomeji Forest Reserve, Nigeria (Hopkins, 1965; redrawn with kind permission of Heinemann Publishers).

not the species are deciduous. A full outline of the classification with some minor modifications is given in Mueller-Dombois and Ellenberg (1974), and a new category of aerophytes was suggested by Galán de Mera *et al.* (1999).

Application of life-form classification involves categorising all the plants in a study area into Raunkaier groups. The results are usually then presented as a bar graph showing the percentages of species in each. Good examples of the use of life-form spectra come from the work of Hopkins (1965) and Richards (1996). Hopkins drew a spectrum for savanna vegetation in Nigeria (Figure 3.1), while Richards constructed one for tropical rainforest in British Guiana (now Guyana) in South America (Figure 3.2). The structure and life-forms of the two different environments are clearly shown, with the phanerophytes dominant in the rainforest where the climate is at an optimum for growth, whereas a much greater diversity of types and adaptations is found in Hopkins' savanna example, where there is a marked dry season. Hemicryptophytes, geophytes and therophytes constitute around 70% of the species present in the savanna. More recent examples, demonstrating close relationships between Raunkaier life-form spectra and climatic gradients, are provided by Danin and Orshan (1990) in Israel, and Batalha and Martins (2002) in Brazil.

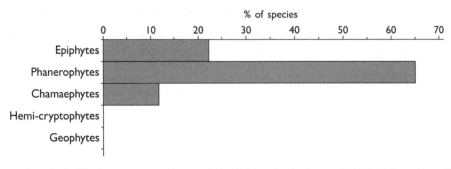

Figure 3.2   Raunkaier life-form spectrum for tropical rainforest in the former British Guiana (now Guyana), South America (Richards, 1996; with kind permission of Cambridge University Press).

Although life-form spectra have traditionally been used to describe world vegetation formation types, they can profitably be used in certain situations at the community level. If locally severe environments are to be compared, life-form data can provide more information than floristics. An example is in the study of colonisation of derelict land and spoil heaps in industrial and urban areas which have been highly modified by human activity. Here the severity of the micro-environment for plant growth is reflected in the wide range of different life-form types that occur (see Down (1973) – case studies).

The structural–physiognomic classification schemes of Danserau, Küchler and Fosberg

These three systems are grouped together because of similarities in their concept. However, they each have differences in approach.

Danserau's method (1951, 1957)

Danserau devised a description method based on six sets of criteria: (a) plant life form; (b) plant size; (c) cover; (d) plant function (deciduous or evergreen); (e) leaf size and shape; and (f) leaf texture (Figure 3.3). No detailed information on the species present other than the dominants is required, allowing the technique to be used rapidly by inexperienced workers over large areas. Each of the dominant plants found within an area is described using firstly the symbols, shapes and shading shown in Figure 3.3 to produce a symbolic profile diagram, and secondly the lettering system associated with the subdivisions in each category. The Danserau method, although logical and easy to use, has not been widely applied, despite its worldwide scope and potential. Part of the reason for this may be the rather abstract symbols of the profile diagrams, plus evidence that when used, the method often requires modification to suit particular circumstances, as demonstrated in the case study at the end of this chapter.

Küchler's method (1967)

Küchler's method is hierarchical in nature, beginning with a subdivision of vegetation into two broad categories – woody and herbaceous. Within the first category, seven woody types are distinguished, while in the second, three classes are recognised (Table 3.2). Within these ten physiognomic groups, further differentiation of vegetation may be achieved by the recording of whether or not they show certain specialised life-forms and leaf characteristics, and on the basis of height and cover. Results are presented as formulae, using letters and numbers. The system is also intended as a basis for vegetation mapping.

Fosberg's method (1961)

The technique of Fosberg is important in that he presented his method as a means of describing vegetation within the International Biological Program (IBP) which was established during the 1960s with the aim of quantifying the primary production and energy budgets of all major world ecosystem types (Newbould, 1965; Peterken, 1967). The purpose of the Fosberg approach is to provide a classification of vegetation at the world scale. The criteria used had to be structural, because any floristic data would have been far too detailed and difficult to obtain. Also, comparisons between differing parts of the world based on floristics would have been impossible and of little relevance.

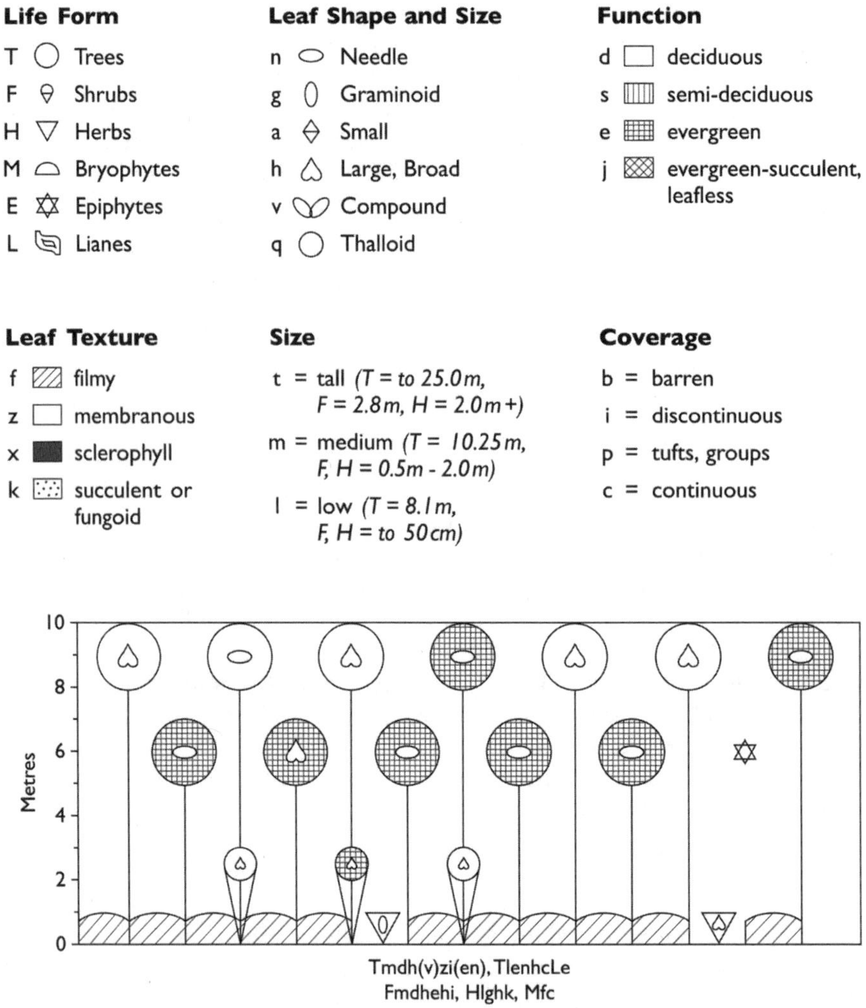

Life Form

T ◯ Trees

F ♀ Shrubs

H ▽ Herbs

M ◠ Bryophytes

E ✡ Epiphytes

L 🖑 Lianes

Leaf Shape and Size

n ◠ Needle

g ◔ Graminoid

a ◇ Small

h △ Large, Broad

v ♡ Compound

q ◯ Thalloid

Function

d ▢ deciduous

s ▥ semi-deciduous

e ▦ evergreen

j ▩ evergreen-succulent, leafless

Leaf Texture

f ▨ filmy

z ▢ membranous

x ▮ sclerophyll

k ▦ succulent or fungoid

Size

t = tall *(T = to 25.0m, F = 2.8m, H = 2.0m+)*

m = medium *(T = 10.25m, F, H = 0.5m - 2.0m)*

l = low *(T = 8.1m, F, H = to 50cm)*

Coverage

b = barren

i = discontinuous

p = tufts, groups

c = continuous

Tmdh(v)zi(en), TlenhcLe
Fmdhehi, Hlghk, Mfc

Figure 3.3 The symbols used in the construction of profile diagrams to denote the six major classes of the Danserau structural vegetation description method (Danserau, 1951), and an example of a Danserau profile diagram from the Killarney *Quercus-Taxus* woodlands at Muckross (Shimwell, 1971; redrawn with kind permission of Sidgwick and Jackson).

The method is summarised in Figure 3.4. The classification is hierarchical and starts with a subdivision of vegetation into categories of open, closed or sparse. Within each of these, a series of 31 formation classes are recognised at a second level (Figure 3.4). Height and continuity of the vegetation are the main criteria used. Further subdivision within each of the 31 classes at a third level is possible using plant function (whether the dominant foliage is evergreen or deciduous). Beyond that, a fourth level of division is made using the properties of leaf texture, leaf size and growth form of the dominant species. At this fourth level, mapping of units in the field can be carried out, with each category being called a formation group.

Table 3.2   A summary of Küchler's method for structural description of vegetation (after Küchler, 1967). Reproduced with permission from John Wiley & Sons, Inc.

LIFE-FORM CATEGORIES

BASIC LIFE FORMS		SPECIAL LIFE FORMS	
Woody plants		Climbers (lianas)	C
Broadleaf evergreen	B	Stem succulents	K
Broadleaf deciduous	D	Tuft plants	T
Needleleaf evergreen	E	Bamboos	V
Needleleaf deciduous	N	Epiphytes	X
Aphyllous	O		
Semi-deciduous (B + D)	S	**LEAF CHARACTERISTICS**	
Mixed (D + E)	M	Hard (sclerophyll)	h
Herbaceous plants		Soft	w
Graminoids	G	Succulent	k
Forbs	H	Large (400 cm^2)	l
Lichens, mosses	L	Small (<400 cm^2)	s

STRUCTURAL CATEGORIES

Height (stratification)	Coverage
8 = >35.0 m	c = continuous (>75%)
7 = 20.0–30.0 m	I = interrupted (50–75%)
6 = 10.0–20.0 m	p = parklike, in patches (25–50%)
5 = 5.0–10.0 m	r = rare (6–25%)
4 = 2.0–5.0 m	b = barely present, sporadic (1–5%)
3 = 0.5–2.0 m	a = almost present, extremely scarce (<1.0%)
2 = 0.1–0.5 m	
1 = <0.1 m	

Sampling for physiognomic and structural description

The basis of sampling for physiognomic and structural description is rarely discussed and this is a major criticism of the whole approach. A physiognomic or structural description of a vegetation type is usually made in an area that the researcher considers to be typical or representative. While this is probably the only realistic way in which these methods may be applied, it raises some interesting questions about what is representative. Choosing a representative area implies a degree of familiarity with the vegetation of the region under study, and students, in particular, may not possess sufficient breadth of knowledge or experience. The element of bias and subjectivity of sample selection is thus always there. Nevertheless, all physiognomic and structural description methods, with their slightly varying approaches, serve as a practical means of collecting and organising vegetation data for general purposes at the regional, continental or world scale.

As an example, Goldsmith (1974) carried out an assessment of the Fosberg method as part of a rapid survey of vegetation for conservation purposes on the Spanish island of Mallorca in the Mediterranean. He concluded that Fosberg's method offered a quick and repeatable approach to vegetation description, providing a useful framework for an ecological inventory of the island. However, he also found that attempts at vegetation mapping using Fosberg's classification groups resulted in some difficulties, largely because mapping on an extensive scale demanded the use of aerial photography or a suitable range of hilltop vantage points in order to show accurately the extent of the Fosberg vegetation types.

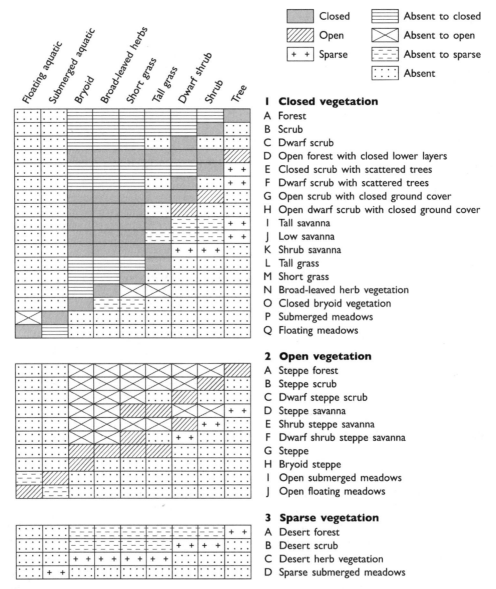

Figure 3.4   A summary of the Fosberg structural system (Peterken, 1967; redrawn with kind permission of Wiley-Blackwell).

Structural and physiognomic methods have been shown to be of considerable value in the tropics and particularly in tropical forest environments where floristic data are difficult both to obtain and to analyse. The value of structural data in Australian rainforest environments was shown by Webb *et al.* (1970, 1976) and Webb (1978); while Werger and Sprangers (1982) compared structural and floristic descriptions in dry tropical communities in India. More recently, the value of physiognomic methods for extensive vegetation survey in Australia has been demonstrated by Sun *et al.* (1997) and the Australian National Vegetation Information System (2003).

Habitat classification (Elton and Miller, 1954; Elton, 1966)

The structural approach to vegetation description is also well illustrated by the habitat description and classification system originally devised by Elton and Miller (1954) and also described in Elton (1966). The method was originally proposed as a rapid method of habitat survey for zoologists, but has subsequently been used for more general surveys of ecosystems and habitats for rural planning purposes and as part of techniques for ecological evaluation. The main assumption of the method is that structural complexity in the vegetation, as represented by the degree of layering or stratification present, can be equated with habitat diversity, which will in turn encourage animal and to a lesser extent plant species diversity. Three major habitat systems were originally recognised: terrestrial, aquatic, and the aquatic–terrestrial transition. The terrestrial habitat system is of greatest interest and is divided into four categories: open ground, field layer, scrub and woodland, on the basis of the height of the dominant species (Figure 3.5).

When combined with the measurement of the area of each habitat type recognised, the technique provides a rapid method for assessing ecosystem structure and diversity. Where all four layers are present, as for example at the edge of a temperate deciduous woodland or in open woodland, habitat and species diversity will be high. Areas of open grassland will have lower habitat diversity and thus less animal and sometimes plant species diversity.

A further refinement of the method involves counting the number of plant species within a sample quadrat of each habitat layer recognised in an area. The information on habitat structure, area of each type and plant species diversity may then be combined in various ways to enable comparison of differing vegetation types. Examples of the application of this approach are demonstrated in the case study at the end of this chapter.

Linear features, such as hedgerows and road verges, may be counted as 'edge' types and their length between blocks of vegetation or their total length within a convenient size of grid can be measured. Habitats adjacent to water bodies are assigned to the separate aquatic–terrestrial transition category.

Habitat classification has the advantage of being very simple to apply and can be very rapid in its application. Despite its simplicity, it is a very efficient means of collecting information on ecosystem structure and function using vegetation as a general index. It is also important in describing vegetation as a habitat for animals, insects and birds (Kusch *et al.*, 2004; Gunnarsson *et al.*, 2006; Tabeni *et al.*, 2007). The technique is also easily adapted, and although originally devised for use in temperate forest regions, is capable of modification and refinement for use with many different world vegetation types, depending on the purpose for which it is being used and the environment under study. Habitat mapping was widely used in the UK Phase I Habitat Mapping in the 1980s and 1990s, but despite its apparent simplicity, problems were found in the consistent application of the method by lone and poorly qualified surveyors in the field (Cherrill and McClean, 1999).

METHODS OF VEGETATION DESCRIPTION BASED ON FLORISTICS

The nature and problems of floristic data

The description of vegetation on the basis of floristics usually means that individual species within the community being studied have to be identified. Floristic description immediately raises four problems. The first is the identification of plant species. The second is whether or not to collect data on the abundance of each species and if so, how to measure abundance. The third is whether other forms of description than species identification are the best way to describe the vegetation,

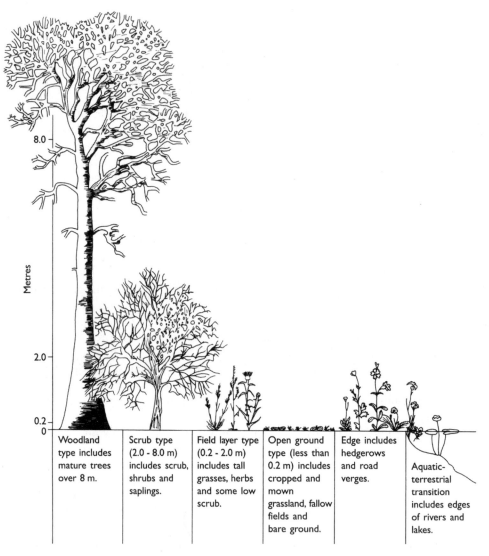

Figure 3.5   The habitat description approach of Elton and Miller (1954).

given the research project aims and objectives, and the fourth is the problem of sampling. Bullock (2006) provides a useful overview of the field description of plants and vegetation.

Species identification

Most species have both Latin and English, local or common names. Often the common names are easiest to remember, particularly if they relate to some obvious characteristic of the plant (Bebbington, 2005). Unfortunately, however, common names vary from region to region and some plants have several such names. For example, one of the commonest shrubs of British heathlands,

Vaccinium myrtillus, is known by at least four English names, bilberry, blueberry, whortleberry and whim/winberry. Similarly, *Arum maculatum*, a frequent plant of hedges and woodlands in England, is known as both 'lords and ladies' and 'cuckoo-pint'. If vegetation description is to be truly scientific, then the use of the Latin binomial is essential. In the case of *Arum maculatum*, *Arum* refers to the genus and *maculatum* to the species. Where plants are unknown to the recorder, the correct approach is to use a flora.

For Britain, the recommended floras for higher plants are Stace (2010) and Clapham, Tutin and Moore (1987), which is also produced in a smaller excursion edition (Clapham, Tutin and Warburg, 1981). The original four-volume flora is now available with illustrations (Clapham, Tutin, Warburg and Roles, 2010). For Europe, the *Flora Europaea* (Tutin *et al.*, 1964–1993) extends to five volumes. A computerised version of the *Flora* was released in 2001 and it is now available online from the Royal Botanic Gardens at Kew in London. Recently a very useful vegetative key to the British flora has been produced (Poland and Clement, 2009). A more recent and ongoing computerised flora for Europe is the Euro+Med Plant Base project, which provides an online database and information system. Initially funded under EU Framework V, the project is still in its early phases and information is available at http://www.emplantbase.ord/home.html.

Frodin (2001) is an excellent bibliographic source for floras and species checklists of the world, arranged geographically and with detailed comment on the history of floras in each world region as well as maps. The lists are not exhaustive, as was attempted in the earlier volumes by Blake and Atwood (1942) and Blake (1961). However, more detailed floras exist in many parts of the world and it is best to research these from the Internet or from within each country.

In other parts of the world, no systematic description of plant species may be available or only small local floras may be available for selected areas. In such situations, it may become necessary to compile a flora prior to or during work on vegetation description. Local help is usually invaluable in such circumstances. Sometimes, however, the time-consuming nature of such work may be sufficient to cause a researcher to abandon a project based on floristics altogether, or to seriously consider the use of structural and physiognomic methods, with identification of just dominant species. As always, the decisions depend on the aims and objectives of the project and the resources and expertise available.

Floras

The classification systems used in botanical floras employ a hierarchical dichotomous key for plant identification, and use of a flora usually takes much practice to perfect. Use of the key is based on the presence and absence of distinctive morphological and physiological properties of plants, and the hierarchy of taxonomic nomenclature is shown in Figure 3.6. There is an International Code for Botanical Nomenclature that applies the following rules for plant identification (Gilbertson *et al.*, 1985; Weber *et al.*, 2000):

(1) One taxon – a single taxonomic unit of any scale (e.g. a variety, a species, a genus) may have only one valid name.
(2) More than one taxon may not have the same name.
(3) In cases not in agreement with the first two points, the decision on which name is to be used is based upon priority of publication.
(4) The identity of the plant to which the name has been given is according to the type specimen designated by the author as representative.

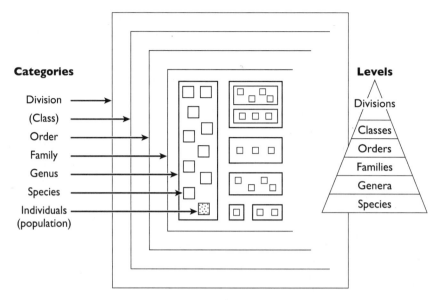

Figure 3.6 The hierarchy of botanical nomenclature (after Tivy, 1982; with kind permission of Longman Publishing).

The latest edition of the code published in 2006 (known as the Vienna Code) is available online at http://ibot.sav.sk/icbn/main.htm. In practice, the application of taxonomic rules is complex and is best left to the experts. Names are, for example, changed from time to time. This occurred for a number of grass species in Britain when the revised Excursion Flora was published in 1981 (Clapham *et al.*, 1981). One widely known example of such a change is the common bent, a grass which was formerly known as *Agrostis tenuis* but which was changed to *Agrostis capillaris* for reasons that probably lie in (3) above, with the name *capillaris* having been given to the species elsewhere at an earlier date, but the relationship between the two having only more recently been appreciated.

Separate floras are often compiled for different groups of plants, for example, grasses and sedges and for bryophytes (mosses and liverworts) and for lichens. In many countries, pictorial floras now exist, which make identification for the student very much easier. Such floras are usually structured by families with their own keys, often based on flower structure and colour. Beginners, however, often use them with a random search procedure at first, and this quickly enables them to build up an understanding of families and genera and their relationships, so that the working of the key becomes clear.

Interestingly, evidence suggests that there is both an 'accessibility' factor (Todd and White, 2009) and a 'proximity to professional botanist effect' (Moerman and Estabrook, 2006) that means that more species are likely to have been fully described in the most accessible locations, especially if they are near to ecological research institutes or universities with ecology/botany departments!

Quadrats – the sampling frame for recording plant species

Quadrats

The usual means of sampling vegetation for floristic description is the quadrat. Traditionally, quadrats are square, although rectangular and even circular quadrats have been used. The purpose of a quadrat is to establish a standard area for examining the vegetation. Quadrat size is very important and will vary from one type of vegetation to another. Methods have been devised to estimate the optimum size of quadrat for a particular community type and are based on the concepts of minimal area and species–area curves. The most frequently used approach is derived from the methodology of the Braun-Blanquet school of vegetation classification and phytosociology (Chapter 7) and involves starting with the area that is considered to be the smallest feasible quadrat size, usually just containing one or two species, then doubling the size of the quadrat, counting the number of species present again, doubling the size of quadrat again, counting the number of species and so on, until no new species are recorded at a doubling of quadrat size. The resulting graph of species numbers against quadrat size is known as the minimal area or species–area curve (Figure 3.7). If a homogeneous area of vegetation is taken, then the curve of species numbers levels off and the point at which this occurs is taken as the minimal area for sampling that community. The recommended quadrat size should then be a little larger than the minimal area (Cain, 1938).

In practice, minimal area curves are often less easy to define and much confusion surrounds their use (Barkman, 1989). The reason is that the method is really only suitable as part of the overall Braun-Blanquet approach to subjective vegetation classification (Chapter 7), where a vegetation sample or relevé, as it is known, is deliberately chosen as being a uniform and representative sample of the plant community being described. The method only works well if the vegetation being sampled is truly homogeneous and is not an edge or transition/boundary between two vegetation types. If non-uniform areas are taken, then the minimal area curve may level off within one locally homogeneous area but then start to rise again as the doubling of quadrat size starts to sample a different community type or transition between community types. The selection of homogeneous areas also presupposes a certain amount of knowledge about the vegetation being studied, and students, in particular, are unlikely to have this knowledge. It also assumes that community types can be clearly defined and does not make it clear how to treat transitional zones between communities. Despite these comments, minimal area curves may still help in the establishment of quadrat size, provided that the above points are realised.

As an alternative to minimal area, general guidelines may be laid down for the optimum size of quadrat for selected vegetation types. These are presented in Table 3.3, but they should not be taken as universally appropriate for every situation and should be treated with particular caution in tropical environments. Generally, larger quadrats are better than smaller ones (Archaux *et al.*, 2007) and they should always be significantly larger than the largest growth forms, as is the case with students throwing quadrats of 5 m × 5 m in upland *Calluna vulgaris* (heather or ling) in Plate 3.1. Larger quadrats are also more likely to reduce pattern effects (see below).

Table 3.3 Suggested quadrat sizes for certain vegetation types. Reproduced with permission from John Wiley & Sons, Inc.

Vegetation type	Quadrat size
Bryophyte and lichen communities	0.5 m × 0.5 m
Grasslands, dwarf heaths	1 × 1 m–2 × 2 m
Shrubby heaths, tall herbs and grassland communities	2 × 2 m–4 × 4 m
Scrub, woodland shrubs, small trees	10 m × 10 m
Woodland canopies	20 m × 20 m–50 × 50 m (or use plotless sampling)

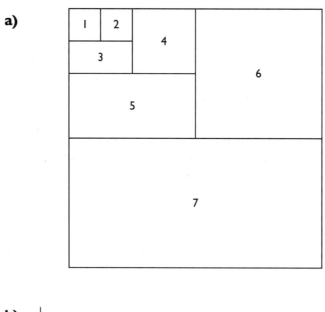

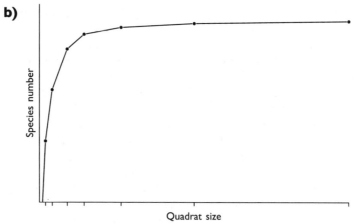

Figure 3.7   Progressive doubling of quadrat size for minimal area and derivation of species area curves. Reproduced with permission from Longman.

Between 1991 and 2000, the National Vegetation Classification (NVC) survey of the UK was published (Rodwell, 1991, 1992a,b, 1995a,b, 2000). The purpose was to describe the whole of the vegetation of Britain (see case study). In the methodology, 2 m × 2 m quadrats were used for herbaceous vegetation, 4 m × 4 m quadrats for taller or more open field layers, woodland field layers and scrub, and 50 m × 50 m for sparse scrub and woodland canopies. In the tropics and particularly in tropical and subtropical forests, even larger quadrats may be required. Quadrats or plots of up to 1000 m × 500 m are used by researchers at the Centre for Tropical Forest Studies at the Smithsonian Tropical Research Institute (http://www.ctfs.si.edu/). The sheer scale required to carry out vegetation description in tropical forests is daunting, even for the most committed of researchers.

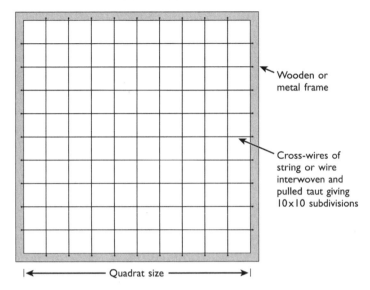

Wooden or
metal frame

Cross-wires of
string or wire
interwoven and
pulled taut giving
10×10 subdivisions

|◀——— Quadrat size ———▶|

Figure 3.8 A conventional square quadrat with subdivisions. Reproduced with permission from John Wiley & Sons, Inc.

The conventional square quadrat has a wooden or metal frame and is often subdivided with lengths of wire or string (Figure 3.8; Plate 3.2). This subdivision increases the accuracy of recording when using certain measures of abundance, since each subunit of the quadrat may be examined separately. The most common subdivisions are into tenths, fifths or quarters of each side of the quadrat, giving 100, 25 or 16 subunits respectively. A common alternative to a rigid quadrat frame is to simply lay out four coloured pegs or knitting needles, linked with brightly coloured string or cord at the appropriate quadrat size (Plate 3.3). In communities with large growth forms (forest and scrub), use of quadrat frames is obviously impossible and quadrats are usually laid out using tape measures and with ranging poles at the corners.

Vegetation pattern and quadrat size

The concept of pattern in vegetation refers to the manner in which the individuals of a given species are distributed within a plant community. It is also dependent upon the size of the species in relation to the size of the quadrat. Species can exhibit clustered, random or regular patterns (Figure 3.9). Depending upon quadrat size and whether the species is distributed in a clustered, random or regular fashion, different results will be obtained by throwing the small quadrat in Figure 3.9 into each of the larger square areas, even though those square areas all have the same numbers of individual plants. If the quadrat size was changed to the size of the larger square areas, there would, however, be no observable differences in the numbers of individuals and the effects of pattern would be masked. Using the small quadrat in Figure 3.9, the regularly distributed species (a) has a much greater chance of being sampled than the clustered species (b), while the randomly distributed species (c) would be somewhere in between. Kershaw and Looney (1985) and Causton (1988) made the point that plant species tend towards a clustered, rather than a random or regular pattern.

A useful laboratory exercise to demonstrate the effects of different quadrat sizes in relation to variations in plant size and the effects of pattern was devised by Williams *et al.* (1979). Although

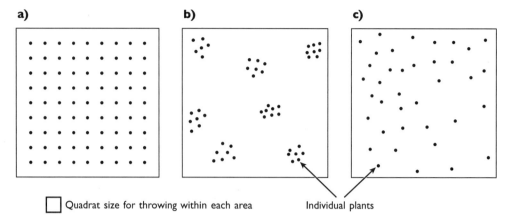

Figure 3.9 Pattern in vegetation: (a) regular (b) clustered (c) random. Reproduced with permission from John Wiley & Sons, Inc.

the problem of pattern is well known to ecologists, the recognition and solution of the problem that every species of plant in a community has a different pattern and size is virtually impossible, since theoretically, one would need a different size of quadrat for every species. The problem is reduced, however, if the largest size of quadrat to suit all species in a particular community or survey is chosen. The other alternative is to use nested quadrats, with different sizes of quadrats for the different sized components of the vegetation. For example, in a wood, 1 m × 1 m could be used for ground flora, 5 m × 5 m for shrubs, saplings and scrub, and 20 m × 20 m for trees. If such different-sized quadrats are used, the data from each smaller quadrat size and component of the vegetation at each level will have to be multiplied up to match the largest quadrat size.

Permanent quadrats and vegetation monitoring

Studies of vegetation dynamics, succession, or of the effects of management practices on vegetation, as well as the potential ongoing effects of climate change, may require the establishment of permanent quadrats, where the recording of species at the same location may be made a number of times over a period of years (Herben, 1996). Such methods form part of vegetation monitoring programmes (Goldsmith, 1991; Spellerberg, 1991; Bakker *et al.*, 1996). Corners of quadrats are usually marked out with coloured pegs, and accurate survey, including plane table methods and use of surveying equipment (Gilbertson *et al.*, 1985), may be useful if individuals of a species are being studied. Individuals may also be tagged with a marker within the permanent quadrat if, for example, competition within and between species or species population dynamics are being studied. Very useful reviews of permanent quadrats and the philosophy behind their use are provided in Austin (1981), Bakker *et al.* (1996), Herben (1996) and Barker (2001). Barker (2001) provides detailed discussion of many aspects of vegetation survey and permanent quadrats within the context of environmental monitoring. There are many problems in ensuring consistency in and standardisation of repeated measurements within permanent quadrats for the assessment of vegetation dynamics, as is stressed in suggestion 8 of Belovsky *et al.* (2004) (Table 1.3). Precise relocation of quadrats is often more difficult than might initially be imagined (Bonham and Reich, 2009), the method of recording should always be kept the same at different points in time (see the case study of Ball, 1974), and secure storage of long-term data is essential.

Bekker *et al.* (2007) edited a volume of *Journal of Vegetation Science* that presented results from various long-term vegetation monitoring projects. As examples, Smits *et al.* (2002) described 70 years of permanent plots in the Netherlands, and del Moral (2007, 2009) showed how 20 permanent plots along a 1 km transect characterised early successional changes following the volcanic eruption of Mount St. Helens, Washington, USA in 1980. Other papers stressed various problems attached to long-term monitoring. The need for consistency and continuity in recording species and their abundance, particularly when the personnel involved change, was emphasised by Vittoz and Guisan (2007) and Milberg *et al.* (2008), as well as the increased accuracy resulting from using multiple observers. Le Duc *et al.* (2007) stressed the problems and value of using databases for long-term data storage, in addition to the need for error checking and the need to make stored data accessible following technological changes in both computer hardware and software.

Measurement of species abundance

A key problem in the assessment of species abundance is what constitutes an individual plant. Many species of flowering plant are clearly identifiable as individuals (genets), while others, such as grasses and bryophytes are not. Also some species exhibit clonal growth patterns and vegetative reproduction, growing with a network of connected shoots or ramets (e.g. grass tillers) under or on the surface of the ground. Variation in plant size (trees as opposed to herbs and grasses) means that counting individuals and then comparing results for different-sized species can become meaningless.

Presence/absence or qualitative data

In the measurement of abundance, a very important distinction is made between presence/absence and abundance data. As the name suggests, with presence/absence data, only the occurrence of a species within a quadrat is noted and there is no measurement of the amount of each species. Hence the data are qualitative. The decision as to whether or not to record abundance values depends upon the aims of the project and the time and resources available for recording. Presence/absence methods are extremely rapid to use and the results represent the simplest form of vegetation data. All the following measures of abundance implicitly include the assessment of presence/absence and can be reduced to this form if necessary. The abundance data contain additional information. However, in terms of information content, the presence of a species provides a significant proportion of the total information on species composition and the additional information provided by the assessment of abundance is more limited (Smartt *et al.*, 1974, 1976; Greig-Smith, 1983; Ramsay *et al.*, 2006).

Abundance measures or quantitative data

Abundance measures can be categorised into two types:

(1) *Subjective* – these are estimated by eye and thus values will vary from one recorder to another.
(2) *Objective* – where more accurate and precise measures are taken which should not vary from one recorder to another.

Table 3.4 The Braun-Blanquet and Domin cover scales. Reproduced with permission from John Wiley & Sons, Inc.

Value	Braun Blanquet	Domin
+	<1% cover	A single individual – no measurable cover
1	1–5% cover	1–2 individuals – no measurable cover– individuals with normal vigour
2	6–25% cover	Several individuals but less than 1% cover
3	26–50% cover	1–4% cover – many individuals
4	51–75% cover	5–10% cover
5	76–100% cover	11–25% cover
6		26–33% cover
7		34–50% cover
8		51–75% cover
9		76–90% cover
10		91–100% cover

Subjective measures include:

Frequency symbols   The most subjective and descriptive method of vegetation description is that of frequency symbols. The following 'DAFOR' scale is used to characterise species: dominant (d), abundant (a), frequent (f), occasional (o) and rare (r), with the prefixes 'locally' and 'very' where appropriate. This scale has been used more often where a whole site such as an area of grassland or a small wood was being studied rather than with quadrats.

Cover estimated by eye   Cover is defined as the area of ground within a quadrat that is occupied by the above-ground parts of each species when viewed from above. Cover is usually estimated visually as a percentage, but stratification or multiple layering of vegetation will often result in total cover values of well over 100%. A number of recording scales are available. Some workers use values in the range 0–100% at 5% or 10% intervals, giving even-sized classes. Others use the Domin scale and the Braun-Blanquet scale, where the range 0–100% is partitioned into five or ten classes with smaller graduations nearer to the bottom of the scale (Table 3.4). Since estimation is done by eye, there is certain to be a degree of error in recording. A recorder will tend to overestimate species that are in flower, attractive and conspicuous and which he or she knows, and to underestimate others. Nevertheless, the method is rapid to use and the problems of subjectivity may have been over-emphasised (Fenner, 1997) (see below). Use of cover estimates is essential when describing grassland species where individuals of a species cannot be identified. Variability and sources of bias in cover estimates are discussed by Kennedy and Addison (1987) and Bergstedt *et al.* (2009), while Chiarucci *et al.* (1999) compared the conclusions derived from cover, as opposed to biomass estimates.

Objective measures include:

Density   This is a count of the numbers of individuals of a species within the quadrat. Subdivided quadrats are useful to increase accuracy in counting and the method presupposes that individuals of a species can be recognised, which is a serious problem in grassland and with multiple stemmed shrubs. Most monocarpic species (annuals and biennials) grow as identifiable individuals, while herbaceous polycarpic species (perennials) possess a more complex growth form, the best examples being many species of grass. Species with rhizomes can cause further problems since they will often be connected underground. Density is most frequently measured in studies of herb species or tree saplings and is rarely used in the description of whole communities. Good examples are population studies of rare orchids in chalk grassland nature reserves, or censuses of tree saplings in woodlands. The method is time-consuming and often where there are large numbers of individuals, individual plants will have to be tagged as they are counted in order to avoid double counting or miscounting. Density counts are entirely dependent on quadrat size, which must be kept constant because the number of individuals counted is entirely a function of the size of area examined. Density is also affected by pattern of species in relation to quadrat size, as discussed above. Lastly, comparison of densities for different-sized species using the same quadrat size is nonsensical because many more individuals of a small species can grow within the same area compared with a large species.

Frequency   This is defined as the probability or chance of finding a species in a given sample area or quadrat. Recording of frequency involves either throwing a series of quadrats within a local area and recording the presence and absence of species in each (local frequency), or using subunits of a quadrat and recording presence/absence of a species in each subunit. If a relatively small area of vegetation is taken and 100 quadrats are thrown within it, then the proportion of quadrats that contain a species is the local frequency. Thus if a species occurs in 63 of those quadrats, it has a frequency of 63%. The most effective use of frequency is, however, where a quadrat is subdivided into 10 × 10 units, as in Figure 3.8 and Plate 3.2, and presence/absence is recorded in each of the 100 contiguous subunits to give a percentage score. This technique probably provides the most accurate data of the various measures discussed so far, but is time-consuming to use. Frequency is dependent upon quadrat size, plant size and patterning in the vegetation. There tends to be a bias towards abundant small species, as opposed to patchy dominants.

Cover estimation using computerised analysis   Several researchers have investigated the potential for measuring cover by using computerised analysis of digital photographs of quadrats. Dietz and Steinlein (1996), Hansjörg and Steinlein (1996), Vanha-Majamaa *et al.* (2000), Olmstead *et al.* (2004), Seefeldt and Booth (2006) and Gallegos Torell and Glimskär (2009) compared measurements of vegetation cover in the field with those for digital photographic images of the same quadrats. While these can be successful in low-growing vegetation, the image analysis software (e.g. ERDAS Imagine) needs to be 'trained' and calibrated and can be prone to errors when compared with visual estimates, particularly if the vegetation is species-rich. Nevertheless, digital image analysis can prove more accurate and sometimes faster, once calibration is complete. Visual estimates are also generally quicker, if less accurate in more varied vegetation, and digital analysis becomes much more difficult and eventually impossible in tall herb, scrub and forest communities.

Cover estimation using a cover pin frame   Objective measurement of cover can be achieved by use of a cover pin frame (Figure 3.10), which consists of a row of ten pins in a wooden frame, with the length of the frame equal to one side of the quadrat. The pins are lowered vertically on to the ground and the species that they touch are recorded. This procedure is repeated ten times at ten equal intervals across the quadrat to give a total of 100 pin samples. The pin diameter should be as small as possible, since tip size has been shown to correlate with exaggeration of cover values. If more than one species is touched as the pin is lowered, more than one 'hit' is recorded which can result once again in a total cover value for a quadrat of over 100%. The method is lengthy to use and is affected by patterning in the vegetation. It is also difficult to use in tall-growing or shrubby

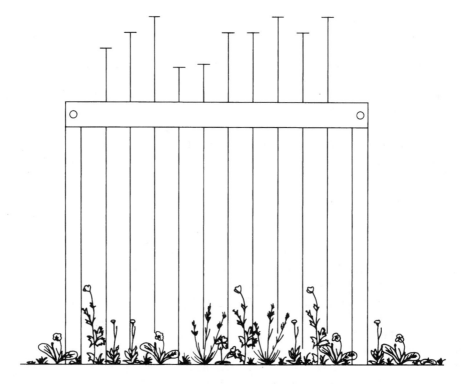

Figure 3.10   A cover pin frame. Reproduced with permission from John Wiley & Sons, Inc.

vegetation and is best suited to grassland. Thus, although point-frequency estimation of cover may appear to be more accurate, it is a much more time-consuming approach, is only suitable for low-growing vegetation, and has problems in that it subsamples the vegetation (Bråkenhielm and Liu, 1995). Some researchers have argued for the merits of inclined point quadrats, with pins brought down on the vegetation at an angle, as it allows for the different architecture of grasses, forbs and small shrubs (Grant, 1993).

Biomass, yield and performance   Experiments to determine primary productivity and studies of competition between plant species often require information on the quantities of plant tissue present in a quadrat. Biomass or standing crop is the amount of plant material present in a quadrat at one point in time. The figure is obtained by clipping the above-ground vegetation using shears or secateurs and sorting the resulting crop into species. The dry weight of each species is then found by drying them for 24 hours at 100°C and weighing the dried plant material. The result is expressed in g/unit area. If samples of vegetation are taken from adjacent plots at different points in time, any increase in biomass between the first and second harvest is known as the yield or productivity of the species and the community. Care must be taken in the sampling of such adjacent plots where the assumption is made that they are representative of the same community type. With both yield and biomass estimates, the point of clipping at the ground surface must be carefully chosen and standardised. Ideally, root biomass and yield should also be measured but this can only be done with great difficulty. No completely satisfactory method of assessing root biomass exists, although methods such as coring followed by estimation of the volume of roots in the core and techniques involving the washing of roots have been tried. All methods are extremely time-consuming.

Biomass data are only occasionally used directly as a measure of abundance because of the time and effort involved in gathering the data and the destructive nature of the measurements. Also, in order to eliminate the possible overriding effect of between-quadrat biomass differences, it may be necessary to transform data by expressing the weight of each species as a percentage of the total biomass in the quadrat. This measure is known as percentage biomass.

Assessment of performance involves the measurement of some relevant part of a plant to provide an index of growth rate or vigour. Typical measurements are leaf size, length and shape, plant height and flower or fruit characteristics, the latter providing information of relevance to reproductive effort.

Root and shoot frequency and presence/absence data

In any survey, it must be clearly stated whether species are being recorded as present and having abundance only if they are rooted in the quadrat or if any of their above ground parts extend into the quadrat, even though they may be rooted outside. This problem has traditionally been described for frequency measures, but clearly is equally applicable to most other measures and particularly presence/absence data. Fehmi (2009) has reviewed three common definitions of cover: (a) aerial cover – the proportion of each species at the uppermost surface of the vegetation (the aerial view); (b) species cover – the cover of the upper layers of each plant species independent of any overhanging cover of other species; and (c) leaf cover – all the layers of each species from the uppermost surface to the surface of the soil, which is roughly equivalent to biomass. Fehmi (2009) has stressed the importance of clearly stating which has been applied, since estimates will vary depending on definition.

Unbounded, bounded and partially bounded data

Smartt et al. (1974, 1976) made a useful division of abundance measures into:

(1) Unbounded measures, or those with no fixed upper limit – e.g. biomass, density.
(2) Partially bounded measures, where there is a limit in that the quadrat area is equated with 100%, but multiple layers can cause abundance values to exceed this figure – e.g. percentage cover.
(3) Bounded measures, where the upper limit is fixed – e.g. frequency recorded in a subdivided quadrat.

Comparisons of the effectiveness of presence/absence and abundance measures

Smartt et al. (1974, 1976) investigated the properties of different measures of species abundance by recording the same vegetation using a range of different methods. They concluded that, although the measure chosen for a given project should depend primarily on the purpose for which the data are being collected, measures of cover, even assessed subjectively, were a good approximation to superficially more accurate measures such as frequency and biomass. Another reason why subject cover scales often provide rapid, yet relatively accurate description lies in the evidence of Raunkaier's 'J' curve (Figure 3.11) and the Law of Frequencies (Raunkaier, 1928). This law states

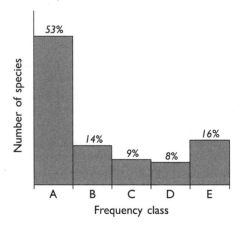

Figure 3.11   The Raunkaier 'J' curve. Redrawn from Kershaw and Looney (1985), with kind permission of Edward Arnold Publishers/Hodder and Stoughton.

that the numbers of species of a community in the five percentage frequency classes 0–20, 21–40, 41–60, 61–80, 81–100 are distributed:

$$A > B > C \geq\leq D < E$$

The graph shows that, as might be expected, the greatest number of species occur with low frequency (A: 0–20% – 53%) but the next highest percentage of species is for species with high frequency (E: 81–100% – 16%). These results have been interpreted in many ways and have to be treated with caution because of the usual factors of quadrat size, plant size and pattern. However, what they also suggest is that a majority of species occur with either very low or relatively high frequency. The same principle probably applies to cover, in that many (but not all) communities contain several dominant species with high frequency and cover, together with a number of associated species with low frequency and cover. Thus, usually, minor variations in recording of percentage frequency and cover do not greatly affect results of subsequent community analyses. An interesting perspective on these ideas is provided by Olff and Bakker (1998).

Reliability of recording species presence–absence and abundance

Numerous researchers have completed experiments into the accuracy of recording vegetation presence–absence and abundance. In relation to presence–absence and thus measures of species richness in quadrats, it has often been demonstrated that the observed number of species is considerably smaller than the actual richness, when a thorough survey is completed after an initial attempt. According to various surveys, on average, a typical recorder could miss between 10% to 30% of species (Sykes *et al.*, 1983; Nilsson and Nilsson, 1985; Lepš and Hadincova, 1992; Scott and Hallam, 2003; Kercher *et al.*, 2003; Klímeš, 2003; Archaux *et al.*, 2006; Vittoz and Guisan, 2007; Elphick, 2008; Milberg *et al.*, 2008). Many factors affect the accuracy of recording, including the experience and botanical knowledge of the recorder, fatigue, weather conditions, time of year, overall species richness, the proportion of plants with low abundance and the degree of consistency between workers, where multiple recorders are used (Rich and Woodruff, 1992; Bråkenhielm and Liu, 1995).

These problems have led various botanists to suggest that multiple observers (or at least two) should record quadrat abundances simultaneously (Floyd and Anderson, 1987; Nilsson, 1992; Klímeš et al., 2001; Archaux et al., 2006; Archaux, 2009). Clearly, however, there are implications in terms of the increase in time and resources required, as well as the inevitable trade-off in that fewer samples are likely to be collected. Where different observers are involved, or for long-term monitoring in permanent quadrats, simply collecting presence–absence data, rather than abundance, has been proposed as a more consistent measure for vegetation survey (Critchley and Poulton, 1998; Ringvall et al., 2005). Also Vellend et al. (2008) have indicated that rare species, and/or those that are difficult to identify, may be omitted from a data-set with comparatively minimal loss of overall information content. Thus the statistical power of many vegetation surveys may be maximised by removing difficult species during the survey, with the result that an increased number of sites or samples may be recorded.

Line-intercept methods

In environments where the vegetation is sparse, such as in some Mediterranean, semi-arid, and hot and cold desert environments, the normal square quadrat may not be suitable for sampling. Instead line-intercept methods may be used. Typically a 10 m length of tape is laid out and all species intercepting or touching the line are recorded (Plate 3.4). A form of percentage cover data can be collected by measuring the length of line touching each species present or over bare ground or rock. The method can also be extended, so that all species within a certain width on either side of the line are included, which is analogous to a very small belt transect (see below). This approach may also be valuable in woodland and scrubland, where shrubs are often difficult to measure quantitatively, and in studies of plant colonisation of derelict and degraded land, where plants will often be sparsely distributed. Line-intercept methods and their advantages and limitations are reviewed in Muttlak and Sabooghi-Alvandi (1993) and Barabesi and Fattorinin (1998).

RECORDING VEGETATION OTHER THAN BY SPECIES – SURROGATES AND PLANT FUNCTIONAL TYPES (PFTs)

In vegetation description, it is often assumed that it is essential to work at the species level, and also that estimates of abundance are vital for adequate description. Apart from the problems of defining some species consistently and appropriately (Winston, 1999; Isaac et al., 2004), there are many situations where use of species taxonomy is not required or relevant. Recording mere species lists (presence–absence data) with no information on relative abundance can make significant savings and may still result in adequate description of vegetation (Greig-Smith, 1983; Ramsay et al., 2006). The large number of species recorded in many vegetation surveys often generate undue complexity and noise that may mask the main trends in the data (Woodward et al., 1997; Duckworth et al., 2000; Villaseñor et al., 2005; Vellend et al., 2008). Amalgamating species into functionally similar groups that share ecological traits and play similar roles in the community (Boutin and Keddy, 1993) often results in a reduction of noise and a clearer signal. These non-phylogenetic groupings are known as plant functional types or PFTs (Smith et al., 1993; Box, 1996; Díaz and Cabido, 1997; Gitay and Noble, 1997; Pugnaire and Valladares, 1999; Weiher et al., 1999; Duckworth et al., 2000; Semenova and van der Maarel, 2000; Westoby et al., 2002).

Members of a PFT share similar morphological, physiological and/or life history traits, with the differences between members within a PFT being smaller than those between types (Halloy, 1990). PFTs can be based on single (e.g. Raunkiaer, 1934) or multiple (e.g. Box, 1996) characteristics

(Weiher *et al.*, 1999; Cornelissen *et al.*, 2003; Díaz *et al.*, 2004). Through analysis, using the multivariate techniques of numerical classification described in Chapter 8, combinations of traits can be determined that characterise vegetation types (Westoby and Leishman, 1997; Pillar, 1999; Westoby, 1999; Pillar and Sosinski, 2003) and thus enable the delimitation of PFTs from species sharing these trait combinations. Studies that derive PFTs from morphological traits are based on the assumption that morphology is the key to function, although the relationship between morphology and function is not always clear (Díaz and Cabido, 1997; Shugart, 1997). The use of PFTs does not, in theory, rely on species identification, and so this approach has potential value as a surrogate, particularly in regions where the flora is not well described.

An important aspect of the derivation of PFTs is the distinction between 'hard' and 'soft' traits for the description of species functions (Hodgson *et al.*, 1999). 'Soft' traits are those that are comparatively easy to measure, usually in the field with a minimum of equipment and expense. Typical are morphological characteristics, such as growth form, life form, specific leaf area, leaf hairiness/thorniness/succulence, root structures and adaptations, or phenological and reproductive characteristics, such as lifespan, deciduousness, shoot phenology, seed size and mass or number, pollination or dispersal mode or vegetative reproduction. 'Hard' traits, in contrast, require either or both laboratory analysis or experimentation, for example, relative growth rate, photosynthesis, leaf nutrient content, DNA content, nitrogen fixation ability or drought sensitivity or tolerance (Duckworth *et al.*, 2000). Researchers have now derived lists of key 'hard' and 'soft' traits that appear to apply widely in a range of world vegetation types (Weiher *et al.*, 1999; Westoby, 1999; Westoby *et al.*, 2002; Díaz *et al.*, 2004; Hunt *et al.*, 2004; Pakeman *et al.*, 2009). However, given the time and expense required to measure 'hard' and 'soft' traits, there is a difficult trade-off between deciding if all, or perhaps a subset, of species in a given community need to have their traits measured, which traits to measure, and whether a subset will adequately describe the whole assemblage (Pakeman and Quested, 2007).

Clearly, the decision as to whether to use PFTs as descriptors, as opposed to traditional species taxonomy, depends upon the aims and objectives of the research project. A project with the goal of phytosociology would employ species taxonomic data, while projects using vegetation description for ecosystem management and the restoration of degraded land could find PFTs or some other surrogate measure of vegetation a more suitable approach (Gondard *et al.*, 2003; Ramsay *et al.*, 2006).

MEASUREMENT OF VEGETATION QUALITY OR CONDITION

While most vegetation description is usually concerned with the presence and abundance of plant species, the assessment of vegetation quality or condition can be important, particularly in relation to the quality of grasslands and rangelands for grazing, the effects of repeated burning, or the extent of tree regeneration in woodlands and forests (Keith and Gorrod, 2006). This is a substantial topic, and for grass and rangelands there is a significant amount of literature in agricultural journals. Nevertheless, a few pointers can be given. Davies *et al.* (1993) summarised the literature on grass sward measurement, with a particularly useful contribution on the measurement of vegetation quality from Grant (1993), while Jerram and Drewitt (1998) produced a set of guidelines for the assessment of vegetation conditions in English upland grasslands, heathlands and wetlands. Gibbons and Freudenberger (2006) completed an overall review of field methods for the assessment of vegetation condition. A key area is the use of subjective (visual) estimates of grass and rangeland condition compared with more time-consuming but objective measurements based on biomass and productivity (Gorrod and Keith, 2009; Cook *et al.*, 2010).

SAMPLING DESIGN FOR VEGETATION DESCRIPTION AND ANALYSIS

The sampling of vegetation for description and analysis is an area that has received insufficient attention in the literature. Even in published work, sampling designs are often insufficiently well explained or sometimes ignored altogether, yet choosing the right approach is a vital part of any vegetation survey work. The primary determinant of the sampling design should be the aims and objectives of the project. However, factors such as the time and resources available for study, the type of habitat and proposed methods of data analysis and presentation must also be considered (Maher *et al.*, 1994).

Spatial and temporal autocorrelation

All sampling of vegetation and associated environmental and biotic variables takes place in both space and time. Various implications result from this, the most important of which is the realisation that all vegetation and environmental data sampling is therefore characterised by both spatial and temporal autocorrelation. Positive spatial autocorrelation implies that phenomena or samples that are close together will be more similar to each other, while those further away will be increasingly dissimilar, a concept that was first stated in Tobler's Law that 'everything is related to everything else but near things are more related than distant things' (Tobler, 1970: p. 234). Ver Hoef *et al.* (2001: p. 218) put this more elegantly when they defined spatial autocorrelation as 'the tendency for random variables to covary as a function of their locations in space'.

In ecology, the importance of understanding and taking account of spatial autocorrelation has recently been re-emphasised by Gustafson (1998), Sokal *et al.* (1998a,b), Koenig (1999), Fortin (1999), Lennon (2000), Mistral *et al.* (2000), Fortin and Dale (2005) and Jetz *et al.* (2005), following earlier commentary by Sokal and Oden (1978a,b), Legendre and Fortin (1989) and Legendre (1993). These authors stress that, until recently, most ecologists have conveniently ignored the effects of spatial autocorrelation, and even today a majority of ecologists still prefer to downplay or disregard the impacts of spatial dependency on their analyses of ecological data. A useful simple introduction to the concept is provided in Cox (1989).

Spatially stratified sampling

A large amount of work in plant community ecology, either deliberately or implicitly, employs spatially stratified sampling. The principle of stratification is that the vegetation of the area under study is divided up before samples are chosen on the basis of major and usually very obvious variations within it. The samples or quadrats are then allocated to these different areas using the various types of sampling strategy discussed below. Stratification would normally be carried out by an initial reconnaissance of the survey area and/or examination of any aerial photography or satellite imagery that may be available. The major divisions will be made primarily on the basis of differences in growth form, physiognomy and structure of the vegetation (e.g. woodland, scrub, tall herb, short-cropped grassland, open ground) with secondary divisions made on variations in dominant species. Other criteria for stratification are obviously available, for example, areas of vegetation subject to different management regimes, important environmental differences, such as aspect, geology or slope form, or areas that are known to have had differing time periods since they were last disturbed and will have thus undergone different degrees of successional change.

There is much to recommend some form of spatially stratified sampling design because major sources of variation in the vegetation are recognised before sampling commences. Smartt (1978) took this idea even further in proposing his 'flexible' systematic model for sampling vegetation. The sampling design is based upon the assumption that the greater the diversity and the rate of change of the vegetation cover over a given distance, the more intensively it should be sampled. The approach involves laying down a network or skeleton of primary sampling points over the whole of the study area to define a framework. Analysis of the degree of variation in floristic composition between each of these primary points then enables secondary sampling points to be allocated to areas where a large change in species composition has been found between two primary points. Where there is little change, then further samples are not required. The result of this is that more samples are taken where variation in floristic composition is high and less where it is low. Smartt argued that since most studies of vegetation are looking for variations within and between plant communities, this is a very efficient approach and time is not wasted on repeated sampling of essentially similar vegetation. This method may be viewed as a special form of spatially stratified sampling, since samples are being allocated on the basis of some predetermined criterion, in this case, the extent to which floristic diversity varies locally over the survey area.

Replication, pseudoreplication and control samples

When taking samples for research that involves the deductive approach (Chapter 1) and what is usually termed experimental design, it is important that samples are replicated and that there is a control. The purpose of replication is to ensure that the entire range of variation in the population of the phenomenon being studied is being sampled. This will facilitate the assessment of sampling errors. As an example, if a project is examining the effects of burning frequency on a type of vegetation, different areas of the vegetation mosaic may be demarcated by spatially stratifying the sample into perhaps three levels of burning – frequent, occasional, and never burned. Samples need to be allocated to those three areas, but if the samples are concentrated in only a small proportion of the area of each burning category, the sample will be biased and a large part of each area will not have been adequately sampled. The result will be that the community that is subject to burning will appear to be more homogeneous than it is and sample errors will be small, giving the erroneous idea that a high level of precision has been achieved.

If ten samples were being allocated to each category, it might appear reasonable to take just one patch of each of the three burning types from the vegetation mosaic and throw ten apparently replicate quadrats within each. These would not be a true replicates, however, and would be an example of what is termed 'pseudoreplication' (Hurlbert, 1984), because the ten samples in each burning type are really just repeated measures of the same sample site. An unbiased sample would take ten samples from ten separate patches of each burning type in the vegetation mosaic, ideally selected at random (see next section), and spread as widely across the whole area as possible. The subject of pseudoreplication has been a matter of considerable debate (Oksanen, 2001; Hurlbert, 2004). Experimental designs involving 'treatments', such as the burning in this example, also need control samples. In this case the control samples would be those taken from the parts of the vegetation that have not been subject to burning at all.

Random sampling

Most descriptive vegetational work, which is inductive in approach (Chapter 1), does not require random sampling. Indeed, a great deal of descriptive vegetation survey is biased in its sampling and uses some form of spatially stratified sampling design as discussed in the previous section. Nevertheless, random sampling still remains a possibility for use in descriptive studies, perhaps combined with the spatially stratified approach.

However, if the deductive approach using hypothesis generation and testing is being applied, including the application of inferential statistics (statistical methods for hypothesis testing based on probability), then random sampling is essential, because most statistical methods for hypothesis testing based on probability assume that all observations or samples are independent of each other. A common situation where this may be necessary is in the exploration and testing of relationships between plant species and environmental factors, using correlation and regression techniques (Chapter 5).

Strict application of random sampling means that every point within the survey area should have an equal chance of being chosen on each sampling occasion. To take a random sample, a grid of coordinates is set up over the survey area and pairs of random numbers are taken to locate each quadrat. Tables of random numbers are available in many elementary statistical textbooks (e.g. Gilbertson *et al.*, 1985) and can be generated on some hand calculators or on computers. Other sources are playing cards, diaries or telephone directories. Where aerial photographs or satellite images are available, random points can be placed on the photographs or images, although exact location of those points subsequently in the field can prove difficult in rough terrain, thick scrub or woodland. An extended discussion of the many difficulties involved in doing this is presented in Causton (1988) and Greenwood and Robinson (2006). Precise location of random points in the field is actually quite difficult.

Where a grid is difficult to set up, for example in scrub or woodland, an alternative approach that has been suggested is to use so-called 'random walk procedures', whereby a sample point is located taking a random number between 0° and 360° to give a compass bearing, followed by another random number for a number of paces. Several of these can be strung together to increase the degree of randomisation between points. However, since the location of each next point is still to some extent dependent on the previous one, so-called random walks are not strictly random. To overcome these problems, a global positioning system (GPS) device may be used, but these are generally accurate to only around 20 m. Once a random point is located approximately by GPS, bias can be removed by then taking a random walk with a random number of paces and a random compass bearing.

The possible implications of non-random sampling in vegetation survey for inferential statistical analysis was the subject of a theme volume of the journal *Folia Geobotanica* in 2007. An initial article by Lájer (2007) pointed out that most vegetation data-sets collected for phytosociology, particularly using the Braun-Blanquet continental school of phytosociology system and its derivatives (Chapter 7, where so-called 'preferential sampling' is employed), are inherently biased and non-random. They thus violate the requirement for samples to be independent, for individual variables to be normally distributed and for the requirement of homogeneity of variances (see Chapter 5). A series of researchers were invited to reply and their responses were in general agreement with the above points (Chiarucci, 2007; Diekmann *et al.*, 2007; Hédl, 2007; Økland, 2007; Wilson, 2007), although they stressed that some inferential techniques are quite 'robust' when these requirements are not met and a possibly even greater problem is that of spatial autocorrelation. However, the practicalities of achieving a truly random spatial sample were also emphasised (Botta-Dukát *et al.*, 2007; Chiarucci, 2007; Lepš and Šmilauer, 2007; Ricotta, 2007; Roleček *et al.*, 2007), as well as the fact that very large amounts of data have been collected by European phytosociologists over the past 100 years, albeit in a biased fashion, and these data nevertheless still represent a valuable resource.

In conclusion, it is worth stressing that spatially biased or non-random data collection in inductive or observational research is entirely valid. However, such data should not then be 're-used' within a hypothesis testing/deductive framework (see Chapter 1). Instead, new research data collected in a truly random manner should be used to answer the research question and test any newly generated hypotheses.

Systematic sampling

Systematic sampling involves the location of sampling points at regular or systematic intervals. The size of the sampling interval is extremely important and may be a fixed interval, such as 100 m, or a regular number of paces. If quadrats are taken at fixed intervals, then care has to be taken that the sampling interval does not coincide with any pattern in the vegetation due either to the properties of plants themselves or to some regularly distributed environmental control. An example is where the floristics of an old meadow in England are being examined and past agricultural practice has left some form of ridge and furrow system, with damper conditions in the furrows giving rise to one set of species, and drier conditions on the ridges resulting in another. If the phase length of the ridge and furrow coincides with the sampling interval, only ridges or furrows could be selected.

Transects

The transect approach is very popular in vegetation work. A transect is a line along which samples of vegetation are taken and are usually set up deliberately across areas where there are rapid changes in vegetation and marked spatial environmental gradients. Most transects are thus biased in their location, although it is clearly possible to locate the start and end of a transect at random and then take samples along the line connecting the two points. Classic examples of laying-out of transects across gradients are up hillsides, where slope angle, drainage and altitude combine, across major changes in geology or ecotones such as terrestrial/aquatic/marine transition zones such as river and lake edges, salt marshes and sand dunes. The main purpose in using transects in these situations is to describe maximum variation over the shortest distance in the minimum time.

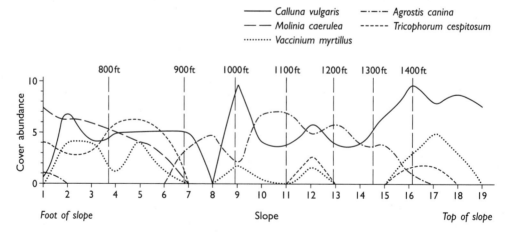

Figure 3.12   Results of a continuous belt transect up a hillslope near Creetown, Kircudbrightshire, Scotland, showing variation in cover-abundance of selected species (Tivy, 1982; redrawn with kind permission of Longman Publishing).

Where quadrats are laid out next to each other or contiguously along a transect line, the result is a belt transect. Figure 3.12 shows an example where the changes in species composition up a Scottish hillside have been described. Another good example comes from the study of recreation ecology and footpaths, where short belt transects are laid out across paths to illustrate the effects of walker pressure on vegetation (Figure 3.13). If abundance data have been collected, histograms

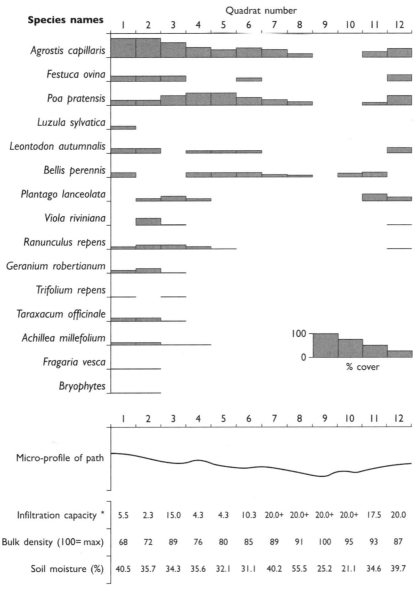

Figure 3.13 A belt transect across a heavily used footpath at Aysgarth Falls, Wensleydale, Yorkshire, England. Reproduced with permission from John Wiley & Sons, Inc.

may be drawn and environmental data plotted on the same diagram to enable visual comparison to be made of correlations between species distribution and environmental factors.

A sophisticated form of transect sampling across large areas was described by Austin and Heyligers (1989). Using climatic, topographical and lithological characteristics of a 20,000 km^2 forested area in New South Wales, Australia, a series of gradsects (transects incorporating significant environmental gradients) were selected as representative of the environmental variability in the area. The method used computerised databases for the environmental data and numerical classification (Chapter 8) to define sampling classes. The aim was to obtain a data-set that was representative of the major environmental gradients. Wessells *et al*. (1998) evaluated the approach and suggested further improvements to the method.

Grids and isonomes

For certain specialised and detailed studies, the transect approach can be broadened out into a grid, where a large number of quadrats are placed adjacent to each other in the grid and species abundances and environmental factors are recorded in each quadrat. Plotting of values for abundances or environmental factors for each quadrat or cell of the grid allows a contour diagram to be constructed, which is known as an isonome. Separate contour maps can be produced for each species or environmental factor and when overlaid, visual correlation of species distributions and their relationships with environmental factors can be explored. A good example of this approach is provided in Kershaw and Looney (1985), where the distributions of two species of *Carex* are shown to be clearly correlated with microtopography.

Plotless sampling

In forest and woodland ecosystems and where vegetation is sparsely distributed, as in certain maquis and garigue Mediterranean vegetation types, semi-desert or high alpine communities, the use of conventional quadrat analysis and sampling may be limited. In the case of woodlands, sampling of the tree cover would require very large quadrats and often the ground flora may be totally impenetrable. In sparse communities, the large amount of bare ground may cause problems in recording, and it is the presence and absence of individual plants and the distances between them which is important. Several sampling techniques have been devised to overcome these problems and are known collectively as plotless sampling. The simplest of these is the 'nearest individual method', which involves the location of random sampling points throughout the area. In woodlands, the distance to the nearest n individuals of each tree species is recorded as shown for species A, B and C in Figure 3.14. Successive distance measurements are taken and the whole procedure is then repeated for a series of random points. The density of each species is then derived from the formulae:

$$\text{Mean area} = (\text{mean distance to nearest } n \text{ individuals of a species})^2$$

$$\text{Tree density for a species} = \sqrt{\frac{\text{Mean area}}{2}}$$

A complete review of plotless sampling methods is given in Mueller-Dombois and Ellenberg (1974), the use of distance measurements to map trees is described by Boose *et al*. (1998), and an example of plotless sampling in forest community analysis is provided by Bryant *et al*. (2004).

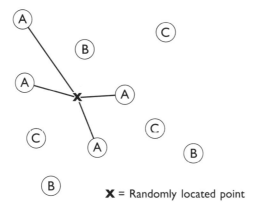

X = Randomly located point

Figure 3.14   The 'nearest individual' method of plotless sampling where $n = 4$. Reproduced with permission from John Wiley & Sons, Inc.

How many samples should be collected?

The answer to this question will depend primarily on the research objectives, but will also be determined by the resources available and the various factors outlined at the start of this chapter. One commonly suggested approach to sample number determination is the species accumulation curve (Barker, 2001). The numbers of species found in 20–30 quadrats within a particular vegetation type is recorded and the results plotted against the number of samples. Where the curve levels off indicates the number of samples needed to describe most of the species in that vegetation type. However, the area of the vegetation type being surveyed should be uniform, edge effects should be avoided, and the curves are likely to be different for differing vegetation types, even within a comparatively local area.

For inferential statistical analysis, a minimum sample size of around 30 has often been suggested, since this is a figure that is likely to provide reasonable estimates for significance testing and probability estimates (Eckblad, 1991; Pillar, 1998). However, in principle, this assumes independent random sampling and the actual figure will also depend on the various factors mentioned above. Greenwood and Robinson (2006) described how to calculate the means and confidence limits for a set of observations and stated that:

> ... generally speaking, the precision of the overall estimate depends on the square root of the number of replicate samples. Thus to halve the width of the confidence interval, one needs to quadruple the number of replicates. Notice that, unless the sampling fraction is large, it is the number of samples (observations) that determines the precision of one's estimate, not the sampling fraction. One does not need to take more samples from larger areas. (p.27)

CHECKSHEETS

Speed and accuracy in recording will always be helped by the drawing up of a standardised checksheet prior to a survey. Usually the checksheet will be designed so that all the information relating to one quadrat or vegetation sample is entered on a single A4 sheet (Figure 3.15).

Department of Geographical Sciences University of Plymouth	Department of Archaeology & Prehistory University of Sheffield

Vegetation Survey
South Uist - Outer Hebrides

RECORDERS: MK/BB

SITE CODE: F1   QUADRAT NUMBER: 15

GRID REF: 729282   ALTITUDE: 5m

SLOPE ANGLE: 1°   ASPECT: 290°

SOIL TIN NUMBER: 725   pH: 7.4

% SOIL MOISTURE: 23.4   ORGANIC MATTER: 8.4%

PEAT/SOIL DEPTHS: 25/25/25/25/25

Species:	% Cover
Plantago lanceolata	50
Trifolium repens	10
Festuca rubra	60
Anacamptis pyramidalis	1
Rhinanthus minor	5
Sinapis arvensis	5
Euphrasia officinalis (agg.)	3
Cerastium glomeratum	1
Equisetum arvense	25
Luzula multiflora	2

Site description and comments:

Edge and back slope of machair sand dunes towards Upper Loch Kildonan.

Figure 3.15 A checksheet for a vegetation survey across South Uist, Outer Hebrides, Scotland. Reproduced with permission from John Wiley & Sons, Inc.

Three sets of information are generally collected:

(a) Information on general site conditions – location, grid reference, global positioning system (GPS) information, transect number, quadrat number, surrounding terrain and land use, weather conditions and any other similar information.

(b) A species checklist or space for a species list to be written in by hand in the field. Ideally, these should use the Latin binomial.
(c) Environmental and biotic data concerning the quadrat and the vegetation being described.

The nature of these data will depend on the aims and objectives of the survey. Typically, in an initial descriptive survey, data enabling the primary gradients of variation in vegetation should be collected, such as soil moisture, aspect, soil depth, light conditions, soil nutrient status, information on grazing, burning and management. Many methods for the study of soil variables are given in Smith and Atkinson (1975), Briggs (1977), Ball (1986), Allen *et al.* (1986), Gerrard (2000), Ashman and Puri (2002), White (2006) and Brady and Weil (2004, 2007), while techniques for geomorphic and climatic factors are presented in Hanwell and Newson (1973), Briggs *et al.* (1997), Watts and Halliwell (1996) and Birkeland (1999). Ehrenfeld *et al.* (1997) warned of the dangers of recording 'easy' measures, such as soil moisture, which can be difficult to standardise if the survey is completed under varying weather conditions and in differing seasons. Similarly, soil pH may vary under different weather and soil moisture conditions. Further useful advice on procedures for quadrat survey and checksheet design are given in Bunce and Shaw (1973), Waite (2000), Henderson (2003) and Sutherland (2006b).

In addition to ensuring standardisation of recording, use of a checksheet makes loss of data during and following field survey less likely, particularly if several recorders are working together. The phenomenon of the 'data drain', whereby part or sometimes all of the data disappear or suddenly appear unintelligible is greatly reduced.

In terms of data storage, hand-held computers for use in the field are now available, but unfortunately, particularly in adverse weather conditions, can be notoriously unreliable, along with their batteries. It is often best still to use traditional checksheets (pen/pencil and paper) and to transfer data to computer spreadsheets (e.g. Microsoft EXCEL® or Lotus 1-2-3®) immediately upon return to home or laboratory. Considerable care is required in the layout of spreadsheets, and these should be planned in advance of fieldwork and with methods of data analysis and available computer software in mind. No transfer of data from checksheets or notebooks is ever error-free, and transferred data should be double- or triple-checked before any data analysis commences. The original data sheets should always be kept in a safe place as a final precaution against data loss.

ELLENBERG INDICATOR VALUES (IVs)

Ellenberg Indicator Values (IVs) (Ellenberg, 1979, 1988; Ellenberg *et al.*, 1992) were originally derived using expert judgement in placing higher plant species on a loosely defined 9-point ordinal scale for seven environmental variables: soil moisture, pH, nutrients, light, temperature, continentality and salinity. The values were based upon Ellenberg's observations on species in Germany and Central Europe and represent his estimation of the realised niche of each species along the seven environmental gradients. If species composition is recorded in a quadrat or sample, the mean IV scores for those species on any one of the seven environmental factors give an indication of the overall position of the sample along the gradient for that factor.

Although Ellenberg was an extremely astute ecologist, criticisms of the approach stem from its obvious subjectivity and also the fact that the values are really only valid in the regions where they were originally devised (Thompson *et al.*, 1993). Various attempts have been made to revise the scores for higher plants in other locations, notably for the United Kingdom by Hill *et al.* (1999, 2000) and the Faröe islands by Lawesson (2003). The values were also extended to cover bryophytes in Central Europe by Ellenberg *et al.* (1992) and Vevle (1999), working in Scandinavia.

IVs have been used for many ecological studies, including the prediction of yield in a long-term experiment (Hill and Carey, 1997), to identify habitats that have experienced eutrophication at the national scale (Smart *et al.*, 2003a,b), to interpret succession on sand dunes (van der Maarel *et al.*, 1985) and in fen vegetation (Southall *et al.*, 2003), and to describe eutrophication and desiccation in wetlands (Latour *et al.*, 1994). Various researchers have examined correlations between both IV scores derived from species found in quadrats on the one hand, and real world field data on the other, and species response curves for pH, soil moisture and nutrients, with mixed success (Ertsen *et al.*, 1998; Schaffers and Sýkora, 2000; Lawesson, 2003; Lawesson *et al.*, 2003; Wamelink *et al.*, 2002, 2005; Pakeman *et al.*, 2008). Pakeman *et al.* (2008) emphasised the problems of interactions between the environmental gradients, such as those for pH and soil moisture. An interesting case study using change in Ellenberg IVs to monitor long-term change in fertiliser applications in grassland is presented by Chytrý *et al.* (2009).

VEGETATION DESCRIPTION AT THE LANDSCAPE SCALE – AERIAL PHOTOGRAPHY, SATELLITE IMAGERY AND GEOGRAPHICAL INFORMATION SYSTEMS

Depending upon research aims and objectives, vegetation survey at the patch/quadrat scale is frequently supported by information gained at the landscape scale (Figure 2.2). Aerial photography and particularly remote sensing and satellite imagery are now available for all of the globe, notably Google Earth®, which can be very valuable at the initial reconnaissance stage of any vegetation survey.

Vegetation mapping is now routinely performed from such imagery as LANDSAT™ and the Normalized Difference Vegetation Index (NDVI) data derived from the Advanced Very High Resolution Radiometer (AVHRR) (Roughgarden *et al.*, 1991; Walsh *et al.*, 1994; Ravan *et al.*, 1995; Bredenkamp *et al.*, 1998; Alexander and Millington, 2000; Aplin, 2005; McDermid *et al.*, 2005; Wang *et al.*, 2009; Franklin, 2009; Horning *et al.*, 2010), and should always involve ground-truthing, whereby image characteristics are validated by quadrat survey on the ground (Treitz *et al.*, 1992). Such data have also proved useful in broad-scale phenological research (Ricotta and Avena, 2000). All the aspects of field description covered in this chapter are relevant to the planning of ground-truthing and survey.

Understanding the relationship between vegetation and its spectral characteristics is essential in the interpretation of remote sensing images. A number of studies of natural and semi-natural vegetation investigated different approaches to the quantification of the relationship between species composition or vegetation characteristics such as biomass or plant productivity, and spectral response using imaging spectroscopy (Curran, 1983; Trodd, 1996; Armitage *et al.*, 2000, 2004; Schmidtlein, 2005; Zhang *et al.*, 2008). Different vegetation types show varying patterns of spectral reflectance due to the differences in the tone, shape or texture of their components. However, Cochrane (2000), Armitage *et al.* (2000, 2004), Artigras and Yang (2004), Schmidtlein and Sassin (2004) and Zhang *et al.* (2008) demonstrated that vegetation types with the same overall physiognomy but which varied in floristic composition were often difficult to differentiate using both spectroscopy and remote sensing, resulting in misinterpretation and misclassification of remote sensing images. Nevertheless, the use of remote sensing and analysis of spectral characteristics represents an increasingly important aspect of vegetation description.

Case studies

Use of the Raunkaier life-form spectrum to describe vegetation communities on derelict colliery spoil heaps – Down (1973)

Most spoil heaps resulting from the activities of mining represent extremely harsh environments for plant colonisation and establishment. Spoil produced by deep coal mining activity presents some of the most severe problems of all. Colliery spoil is a mixture of shales, mudstones and sandstones, within which the coal seams were interbedded underground. By volume, the shales are dominant in the heaps and it is their physical and chemical characteristics that determine the severity of the surface of the heap for plant growth (Kent, 1982, 1987a; Harris *et al.*, 1996).

The shales are predominantly grey-black in colour, with a low albedo, resulting in very high summer temperatures (up to 60°C) and a high diurnal variation in temperature. Water availability is low in summer and yet waterlogging may occur in winter. The spoil is often tipped at very steep angles, so that slope wash and erosion are serious problems and the shales weather down to fine clays that are easily transported. Many essential plant nutrients, particularly nitrogen and phosphorus, are deficient or in chemical forms unavailable to plants. Toxicity problems occur due to the evolution of forms of sulphuric acid from the mineral pyrites (FeS_2) during weathering of shales and due to high concentrations of soluble salts. Soil organism populations are low or non-existent.

Reclamation is desirable for both landscape and safety reasons, but because of cost this usually means only the establishment of some form of grass sward. However, before reclamation of a site, information on problems of vegetation establishment may be obtained from a survey of the existing natural colonising vegetation, particularly if there appear to be large variations from one part of a tip to another. The life-form of the species present is also of interest, since those species that can colonise such a harsh environment are presumably well adapted to the prevailing conditions.

Description of vegetation using floristic methods is often of limited value, since these sites are at early successional stages and contain many ruderal species. Ruderals are colonisers of bare and disturbed ground and take advantage of the lack of competition from other species that enter later in succession. Also on most sites, the source of colonising species is the surrounding semi-natural vegetation and this may be highly variable from one site to another. The result is that there is a wide variety of early natural colonisers. Thus, using floristics as a means of assessing nutrient deficiency and toxicity problems on a proposed reclamation site is rarely very useful.

As a result of these problems, researchers have turned to life-form methods to describe the vegetation of spoil heaps. Down (1973) worked on five disused colliery spoil heaps in Somerset, England, where the age of each heap was known, ranging from 12 to 98 years. He recorded both life-form and floristic data with a view to assessing the severity of the microclimate and environment for plant growth.

The results of the floristic data showed that 73 vascular plant species were found, but only ten of these showed a definite trend with age and site condition. The life-form spectrum for each tip was also recorded and the results are shown in Table 3.5.

Table 3.5 Life-form spectra of plant species on colliery spoil heaps of five differing ages in Somerset, England (Down, 1973). Results as percentages of total number of species on each tip. Reproduced with kind permission of Academic Press.

Life form	Spoil age (years)				
	12	15	21	55	98
MM	0.0	7.1	0.0	7.9	0.0
M	4.5	0.0	7.1	10.5	2.9
Chh	4.5	7.2	3.6	5.3	5.9
H	9.2	7.2	2.6	2.6	5.9
Hp	4.5	10.7	3.6	21.1	32.4
Hs	22.7	32.1	35.7	21.1	29.4
Hr	31.8	25.0	33.1	23.6	11.8
Total H	68.2	75.0	75.0	68.4	79.5
Gr	0.0	3.6	3.6	0.0	0.0
Grh	0.0	0.0	0.0	0.0	2.9
Th	22.8	7.1	10.7	7.9	8.8

Key to life forms:
MM: Phanerophytes, woody plants with buds more than 25 cm above soil level, plants higher than 8 m.
M: Phanerophytes, plants 2–8 m high.
Chh: Herbaceous chamaephytes with buds above soil level but below 25 cm.
H: Hemicryptophytes, herbs with buds at soil level.
    H – undivided hemicryptophytes.
    Hp – protohemicryptophytes with basal leaves smaller than stem leaves.
    Hs – partial rosette hemicryptophytes (semi-rosette), basal leaves larger than stem leaves.
    Hr – rosette hemicryptophytes, leafless flowering stems with basal leaf rosette.
Gr: Geophytes with buds or roots.
Grh: Geophytes with rhizomes.
Th: Therophytes, passing the unfavourable season as seeds.

The life-form data demonstrate two important points. Firstly, the percentages of rosette species (Hr) declined from 31.8% to 11.8% as the tips aged. This was accompanied by an increase in protohemicryptophytes (species with partially formed basal leaves which protect the perennating buds and with better developed leaves higher up the stem) with age. Secondly, therophytes (species that spend the unfavourable period of the year as seeds) represented 22.7% of the original flora but declined to 9.8% over 98 years.

The colonising flora was thus dominated by hemicryptophytes and therophytes. The importance of the rosette hemicryptophytes as colonising species had not been noticed before, but appears to be a function of their life-form. They have long tap roots which ensure that they are firmly anchored in unstable soils and which enable them to extract water from depth when the soil is subject to drought in the summer months. The rosette form of the basal leaves also serves to reduce temperature variability around the sensitive perennating buds of the plant and cuts down evapotranspiration around the stem/spoil interface.

The decline in therophytes is due to the dominance of ruderal species at early stages of colonisation and their inability to compete with perennial grasses and herbs at later stages of succession.

Down concluded with the comment that in most schemes for the reclamation of colliery waste, perennial grasses and legumes are planted. These are life-forms unlike most early natural colonisers of spoil heaps, and many examples exist where heaps that have been reclaimed using grass species have suffered serious die-back and regression. He suggested that a partial solution may lie in the sowing of more rosette species in reclamation seed mixes.

This case study shows how life-form description methods, originally devised for use at the world scale to show plant response to gross differences in world climate, may be applied equally successfully at the much more local and detailed scale of a colliery spoil heap to assist with reclamation studies. However, at both scales, it is plant response to limitations of macro- and microclimate and habitat that makes the use of Raunkaier's methods possible.

Application of Elton and Miller's habitat classification system within methods of ecological evaluation and habitat assessment (Kent, 1972; Kent and Smart, 1981; Goldsmith, 1975)

The conservation and management of semi-natural vegetation in Britain involves not just the establishment of a national network of nature reserves and protected sites, but also concern for all areas of semi-natural vegetation that remain, especially within the highly modified agricultural landscape of lowland Britain. For a conservation policy to be established and implemented for the remaining vegetation, some form of inventory and assessment of such sites is necessary, and furthermore, some estimation of their 'ecological value' may be attempted.

In the 1980s, a range of techniques were proposed for the ecological evaluation and assessment of the remaining 'islands' of semi-natural vegetation remaining within highly developed landscapes (Goldsmith, 1983; Usher, 1986; Spellerberg, 1992). In Britain, a substantial impetus came in the 1970s from the requirement to provide information for county structure plans. Methods devised by Kent (1972), Goldsmith (1975) and Kent and Smart (1981) included the use of Elton and Miller's habitat classification system, although with some modification in each case.

In the method of Kent (1972) and Kent and Smart (1981), all areas of remnant semi-natural vegetation within a small study region of 10 km × 10 km of predominantly arable agriculture were regarded as individual sites (Figure 3.16). Each site was first divided into Elton and Miller habitat formation types: open ground, field layer, scrub, woodland, edge and aquatic-terrestrial transition (Figure 3.5). Numbers of plant species were then enumerated for all layers in each habitat type recognised. One problem was the minimum scale at which a single habitat type could be identified. Some areas were a mosaic of two or even three habitat types, for example, a tall grass meadow invaded by intermittent patches of scrub. The identification of each patch of scrub as a separate formation type was impossible. Instead, a combination of habitat types was recognised: field layer/scrub. Some woodlands thus had three or four layers or habitat types present.

The area of each site was measured from maps using a planimeter, the edge length of each site was recorded, distances were measured to the nearest other sites of semi-natural vegetation, and various aspects of the existing management and use of sites were noted. Sites were then grouped together, using the ordination and classification methods described in Chapters 6 and 8, and a series of groups of different sites emerged, representing the main types of habitat within each 10 km × 10 km area. The number of

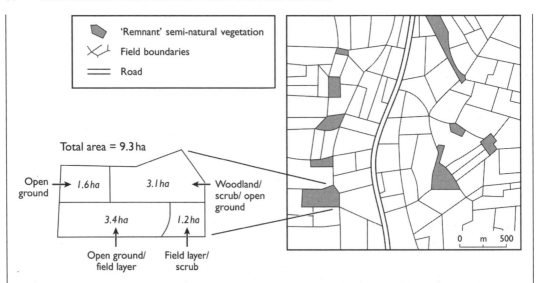

Figure 3.16 Using Elton and Miller habitat description of remnant semi-natural vegetation in the lowland agricultural landscape of Britain (Kent, 1972; Kent and Smart, 1981). Reproduced with permission from John Wiley & Sons, Inc.

individual sites within each group and their characteristics put a relative scarcity value on the amount of that habitat within the local region. Although numerical methods were used, in a small area the sites could easily be grouped subjectively on the basis of the habitat data collected.

This classification of habitats could then be used as an index for planning purposes. If development or modification of a site was proposed, its position in the classification could be looked up and the planner or decision-maker could then assess the habitat characterisation of that site in an informed manner.

In the second study (Goldsmith, 1975), Elton and Miller's habitat classification was used in a different manner. Rather than simply making an inventory of sites of semi-natural vegetation, the aim was to survey the whole agricultural landscape, including arable and grazing land, and an attempt was made to assess 'ecological value'. The basis of 'ecological value' is discussed at some length in Goldsmith's article, but the basic concept is that some habitats are commonly agreed to be more 'interesting' and 'valuable' than others by professional and academic ecologists and planners.

Following a transect along the Wye Valley to the Black Mountains on the English–Welsh border, 1 km × 1 km blocks were taken and each block or square was allocated to one of three land system types:

I. Unenclosed upland, mostly moorland over 300 m.
II. Enclosed cultivated land, mostly permanent pasture.
III. Enclosed flat land, most arable, in the valley bottoms.

Within each of these land systems, habitats were recorded as:

(1) Arable and ley
(2) Permanent pasture

(3) Rough grazing
(4) Woodland
    (a) deciduous/mixed
    (b) coniferous
    (c) scrub
    (d) orchard
(5) Hedges and hedgerows
(6) Streams.

For each habitat type within each grid square, four variables were measured:

(1) Extent (E) – the area of habitat types 1–4 in hectares and the lengths of linear habitats 5 and 6.
(2) Rarity (R) – recorded for each habitat type in each land system and calculated from R = 100% – the % area of each habitat type within each land system.
(3) Plant species diversity (S) – 20 m × 20 m quadrats were located in each habitat type within each land system.
(4) Animal species diversity (V) – the stratification of the vegetation using Elton and Miller's ideas, where the number of vertical layers was recorded from 1 for open ground to 4 for well-developed woodland.

Thus the habitats were recorded first, and then the structural nature of the vegetation within each habitat type was taken as an index of animal species diversity. The approach is clearly derived from the original ideas of Elton and Miller.

The index of ecological value for each grid square was finally derived by multiplying together the four variables for each habitat type and then adding the results for each of the six habitat types present within a grid square. The results were then rescaled in the range 1–20 to enable them to be mapped in convenient classes.

The main point to stress is the assumption that animal, and to a lesser extent plant, species diversity is correlated with stratification and vertical structure of vegetation. Both methods were applied to relatively small areas in the first instance and would require modification if used more widely or if an attempt was made to formulate a standardised approach that could be used by all planning authorities across the country. Interestingly, the whole subject of ecological evaluation and habitat assessment that was of great interest to many ecologists in the 1980s, culminating in Spellerberg's book on the subject in 1992, has subsequently gone into decline, partly due to criticisms over the subjectivity of many approaches and the difficulties of agreeing on any standardisation of methodology. In that respect it is a good example of suggestion 1 of Belovsky *et al.* (2004) (Table 1.3), that issues come into and out of ecology like *haute couture*, often without scientific resolution.

Use of the Danserau universal structural system for recording and mapping vegetation in South-East Queensland, Australia (Dale, 1979)

The aim of this survey was to make an inventory of the vegetation of South-East Queensland in Australia before the production of a vegetation map. As such, the aim was primarily academic. The categorisation of vegetation types and the map would then provide a framework for more detailed research in the area. Structural and physiognomic approaches to vegetation description were considered to be the most suitable because of the large area to be surveyed.

The problems of sampling for physiognomic survey are highlighted, in that five major vegetation types were recognised subjectively before applying the Danserau method for description. Diagrams constructed from a transect through each type are shown in Figure 3.17. No mention is made of how the transects were located, although it would be logical to assume that they were placed across what were believed to be the most typical or representative areas.

Problems were also encountered in the application of the standard Danserau method to the range of semi-open *Eucalypt* woodlands. The main difficulties came in the representation of all forms of lianes, determining seasonality for all species and in the over-representation of certain species, such as the bottle tree (*Brachychiton* sp.) on the diagrams. As a result, Dale produced his own set of modified symbols to suit his particular circumstances (Figure 3.18), which were partly adapted from the work of Danserau *et al.* (1966). They described modifications of the standard types with respect to crown shape, leaf size, stem diameter, angle of branching and height of principal branching. In Figure 3.18, crown shapes are drawn to scale and the line on the trunk section of each tree corresponds to the height of principal branching. The diamond shape below each trunk refers to the scrub nature of the *Eucalyptus* species.

Thus while the Dansereau method was shown to be valuable in distinguishing between major vegetation types, for more detailed surveys of subgroups of *Eucalyptus* types in Queensland, modifications and revisions were required. Such adaptation of existing methods to suit varying circumstances and environments is commonplace in vegetation description. This study also shows how physiognomic methods have their place in certain world vegetation formation types where floristic description can only be carried out with difficulty.

Domin, percentage cover and frequency scales and the study of floristic changes in grassland on the Isle of Rhum, Scotland, following the reduction or exclusion of grazing (Ball, 1974)

The island of Rhum in the Inner Hebrides of the west coast of Scotland is one of Britain's largest nature reserves, with an area of 10,620 ha. The greater part of the island consists of upland grassland communities dominated by the grasses *Agrostis capillaris* (common bent) and *Festuca rubra* (red fescue), and wet heathland and bog areas with *Calluna vulgaris* (ling or Scots heather), *Molinia caerulea* (purple moor grass) and *Trichophorum cespitosum* (deer sedge). The vegetation has been subjected to intensive grazing and burning practices, the main grazers being red deer, sheep, cattle, and to a lesser extent, ponies and goats.

In 1957, when the Nature Conservancy Council took over the whole island as a nature reserve, the sheep were removed and the deer population was reduced. Burning practices were also controlled and carried out on a more regularised basis. In order to examine changes in the vegetation resulting from the new pattern of management after 1957, thirteen 5 m × 5 m experimental exclosures or permanent plots were established on various community types in 1958. The vegetation of each area was described by a 10-point Domin scale using randomly located 4 m × 4 m quadrats inside and outside each exclosure in 1958, 1962 and 1970.

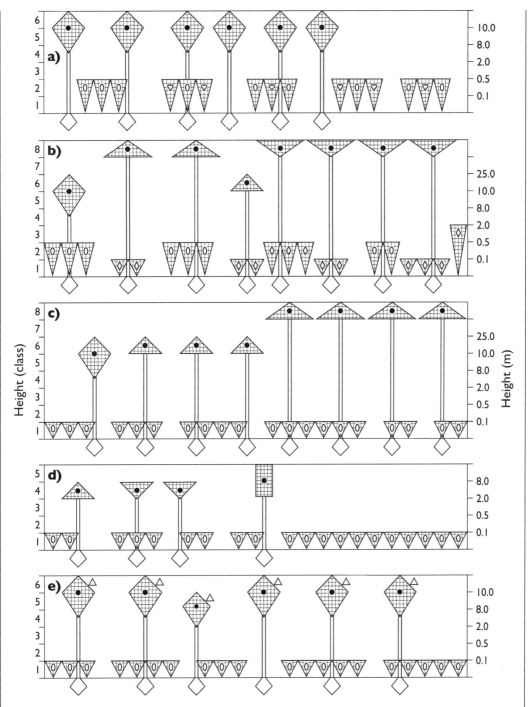

Figure 3.17 Standard Danserau diagrams of the vegetation of part of South-Eastern Queensland, Australia (Dale, 1979; redrawn with kind permission of Kluwer Academic Publishers).

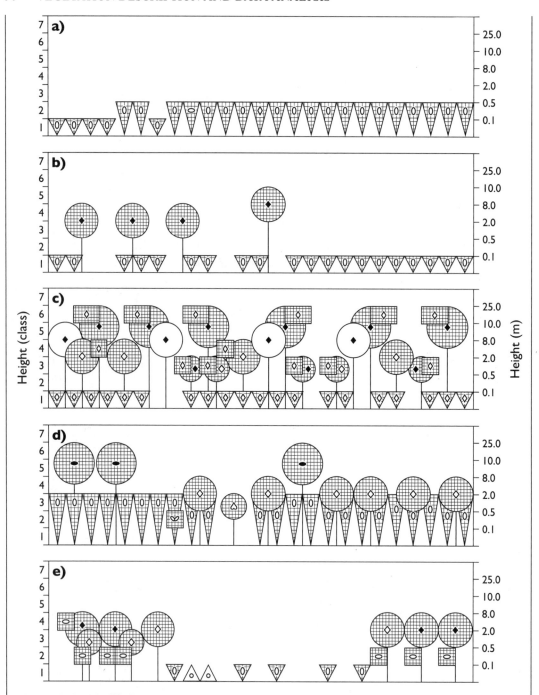

Figure 3.18 Modified Danserau diagrams for *Eucalypt* woodlands of South-East Queensland, Australia. (a) *E. maculata,* (b) *E. moluccana,* (c) *E. tereticornis,* (d) *E. melanophloia,* (e) *E. crebra* (Dale, 1979; redrawn with kind permission of Kluwer Academic Publishers).

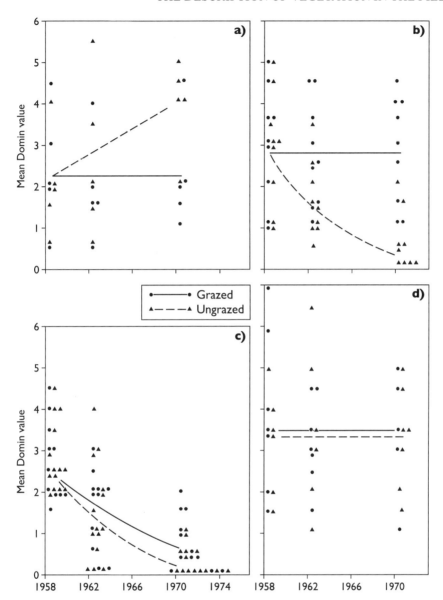

Figure 3.19 Changes in mean Domin value of four groups of species in grazed and ungrazed plots on species-poor *Agrostis-Festuca* grassland on the Isle of Rhum, Scotland, 1958–70. Means are taken from three replicate plots (Ball, 1974; redrawn with kind permission of Academic Press).

Figure 3.19 shows the changes observed within one community type, the species-poor Agrostis–Festuca grasslands. The Domin values for four groups of species (a–d) in the grassland sward have been plotted for the three sampling intervals over the 12-year period. The trend in the mean Domin values of the enclosed plots across the three sampling intervals is shown by a dashed line, while that in the grazed plots is shown by a solid line.

The ungrazed plots show a marked increase in the cover and dominance of group (a) species, notably *Poa pratensis* (meadow grass) and *Festuca rubra*, over the experimental period, and a marked fall in the Domin cover values of other grasses and herbs in groups (b) and (c), notably the grasses *Anthoxanthum odoratum* (sweet vernal grass) and *Cynosurus cristata* (crested dog's tail) and the herbs *Plantago lanceolata* (ribwort plantain), *Polygala serpyllifolia* (heath milkwort), *Euphrasia officinalis* (eyebright), *Achillea millefolium* (yarrow) and *Veronica officinalis* (common speedwell). Species in group (d) have remained constant after the exclusion of grazing and include *Agrostis capillaris*, *Holcus lanatus* (Yorkshire fog), *Koeleria gracilis* (crested hair grass), *Ranunculus acris* (common meadow buttercup), *Potentilla erecta* (tormentil) and *Rumex acetosella* (sheep's sorrel).

In contrast, the grazed plots showed that species composition and abundance has not changed at all in groups (a), (b) and (d), whereas there was a significant fall in group (c) species, although to a lesser extent than in the ungrazed plots. Thus the effect of a relaxation in grazing pressure was to reduce the abundance of many species quite dramatically to the advantage of the two main grass species, which due to their greater height and tussock structure could gain competitive advantage over the herb species that are of lower stature.

One problem with this study was that the experimental recorder changed between 1962 and 1970. Therefore in 1969, a check was made on the results by carrying out a percentage cover and frequency analysis along a transect located by random numbers, which traversed the boundary of each exclosure. Three replicate transects of 16 quadrats were thrown, eight inside each exclosure and eight outside. A 50 cm × 50 cm quadrat, subdivided by cross-wires into 25 10 cm × 10 cm squares, was used and the cross-wire intersections were used to provide a 16-point sampling frame in each quadrat for cover analysis.

The principle that underlay this experimental check was that the original study using Domin values was carried out through time by recording change at the same point in space. However, with the change of researcher, a different approach was adopted whereby at the same point in time, quadrats at different locations in space were examined both inside and outside the exclosures. Thus in the second analysis, variation in space was substituted for variation through time and it was assumed that the grazed grasslands around the enclosures in 1969–70 were the same as those grasslands inside when the exclosures were originally established in 1958. The other interesting aspect of this survey was that in 1969 the percentage cover and frequency were recorded, as opposed to Domin cover values alone. The results of this second survey, for the species-poor *Agrostis–Festuca* grasslands, are shown in Figure 3.20. Two points emerged:

(1) The two measures of abundance, frequency and percentage cover, showed significant differences in results. Most important was that three species, *Arrhenatherum elatius* (false oat-grass), *Luzula sylvatica* (great wood-rush) and *Plantago lanceolata* (ribwort plantain), were recorded only by frequency and were missed in the point cover estimates. Also the degree of difference between the two measures appeared to be much greater in the grazed than in the ungrazed plots.

(2) When compared with the first survey, some predictable similarities occurred but differences also existed. Following exclusion of grazing, *Poa pratensis* and *Festuca rubra* were increased in cover, although *Festuca rubra* did not increase in frequency and the species of groups (b) and (c) in the Domin analysis were shown to be much

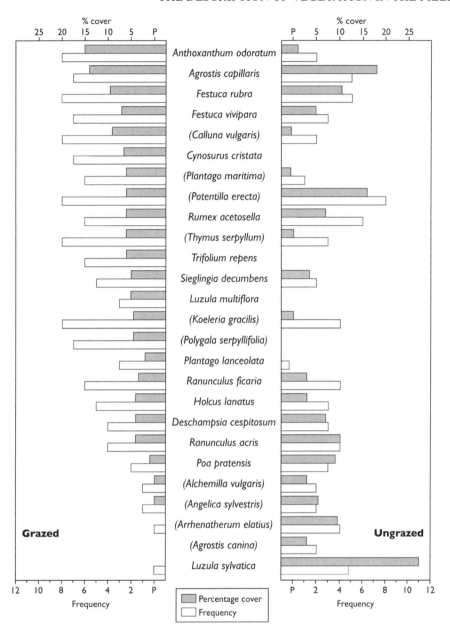

Figure 3.20 Floristic differences as shown by percentage cover and frequency on grazed and ungrazed plots in species-poor *Agrostis-Festuca* grasslands on the Isle of Rhum, Scotland, in 1969. P: present, species in parentheses in one plot only. Solid columns show the percentage cover, open columns the frequency. Percentage cover and frequency data are averaged from 24 plots in each case (3 replicates × 8 quadrats) (Ball, 1974; redrawn with kind permission of Academic Press).

reduced or eliminated. However, *Agrostis capillaris*, *Potentilla erecta* and *Ranunculus acris* were much increased in cover but not in frequency, and certain species emerged as being of importance in the enclosed plots, notably *Luzula sylvatica*, *Agrostis canina* (velvet bent) and *Arrhenatherum elatius*. Although overall trends were the same, possibilities existed for differences in detailed interpretation as a result of variations in the abundance measures used.

The implications of these results for the future management of the *Agrostis-Festuca* communities were interesting. The changes in the dominance of the grasses in the sward following the exclusion of grazers were probably not of great significance in terms of the food supply for the grazing animals. The general level of grazing on the island had been reduced, and from the point of view of range management, the only problem was that a reduction in grazing leads to an increase in plant litter and taller vegetation, resulting in an increased level of flammability of the vegetation. However, one of the main reasons for declaring the Isle of Rhum a nature reserve and a major goal of the management plan was to maintain high floristic diversity in all the vegetation types present. This demonstrated that, although the observed changes might not have been important for upland agriculture, species diversity and thus the aims of nature conservation were affected. The answer appeared to lie in increasing the grazing pressure on the swards again in order to keep the more aggressive pasture grasses in check. Towards this end, the annual cull of the red deer on Rhum was severely reduced in the mid- to late 1970s.

Fieldwork design for the analysis of pioneer vegetation on glacier forelands in southern Norway (Robbins and Matthews, 2009, 2010)

This case study was introduced in Chapter 1. The aims were, firstly, to examine the species composition of the early pioneer stages of vegetation development following glacier retreat, and secondly, to see whether these highly disturbed sites are colonised by consistent sets of species or whether species composition is more dependent on chance (stochastic) factors that tend to produce more random and variable collections of species. Vegetation data on pioneer vegetation growing on 43 glacier forelands in Norway were collected (Figure 1.2).

The sampling design for this research was particularly interesting. Distance away from the snout of each glacier was taken to represent a chronosequence (Matthews, 1992; Foster and Tilman, 2000). Recording of higher (vascular) species was carried out using transects of contiguous (adjacent) quadrats, 2 m × 0.5 m in size, commencing with the individual plant closest to the glacier snout, and then continuing until 100 occurrences of the species had been recorded at each of the 43 sites. A measure of local frequency was thus recorded by noting the presence of each species in a quadrat. This approach was particularly appropriate for this extensive survey, where many sites had to be recorded rapidly in terrain that was sparsely vegetated. The method was relatively objective and avoided the problem of having to separate individual plants where growth forms were contorted in the extreme environment. Transect lengths were highly variable, ranging from 6 m to 213 m with a mean of 55.2 m, and the transect starting distance from the glacier snout varied from 0.5 m to 164.5 m, with an average of 36.1 m. The distance from the glacier snout to the end of each transect varied from 6.5 m to 298 m. Longer transects correlated with relatively large glaciers, where the rates of retreat were greatest, large areas of uncolonisable bedrock, or where open water occurred or there were high levels of periglacial frost action. One site was dropped from the analyses described in Chapters 6 and 8 because of a complete absence of pioneer vegetation in the whole of the foreland zone, resulting in a total of 42 sites being included.

In terms of starting position, transects were located optimally where there was a flat, relatively even surface in front of a glacier, avoiding water features where possible. Glacial till as a substrate for plant growth, rather than bare rock, was obviously preferable. Each transect was characterised in terms of altitude and continentality, measured by distance east from a point 5 km to the west of the most easterly glacier foreland sampled.

The key features of this sampling strategy were the need for a rapid, yet accurate method of vegetation description across a large number of different sites, the adaptability required to record in sparsely vegetated terrain, and the incorporation of the underlying principle of chronosequence in front of each retreating glacier. This project clearly demonstrates the flexibility required in planning vegetation survey and the need for lateral thinking to ensure that the aims and objectives take account of local environmental conditions. The chronosequence concept is particularly important in underpinning this research, and it is thus interesting to assess this sampling strategy in the context of some of the critical comments on chronosequences made in the paper by Johnson and Miyanishi (2008) introduced in Chapter 2.

Sampling design for a vegetation survey of the Faröe Islands (Tomlinson, 1981; Milner, 1978)

The flora of the Faröe Islands, which lie midway between Shetland and Iceland in the North Atlantic, are relatively well documented. However, the nature of the plant communities and their similarity to the vegetation on other Scottish islands and mainland Britain had not been studied in such depth. Thus in June–July 1978, a team of ecologists visited the two islands of Vágar and Mykines with the aim of describing the vegetation and plant communities.

They were confronted with a typical problem in terms of vegetation sampling, which was to describe the maximum variation of the plant communities of the area but in a relatively short space of time. Tomlinson devised a method of stratified sampling which was based on information published from a similar survey of the Shetland Islands completed by the Institute of Terrestrial Ecology in 1975 (Milner, 1978). In the Shetland survey, 150 physical attributes on the 1-inch Ordnance Survey map were recorded for each 1 km grid square. These attributes included factors such as distance from the coast and aspect, and could be grouped into three categories:

(a) factors related to the coast and coastal features;
(b) factors related to high altitude;
(c) factors related to the size and type of water bodies.

The measurements for all grid squares on Shetland were then analysed using numerical classification (Chapter 8) and ordination (Chapter 6), from which 16 groups emerged. The result was effectively a broad-scale classification of habitats based on physiographical characteristics. Since the Faröes are similar to Shetland in their general physiography, Tomlinson decided to apply the 16-group classification for Shetland to the Faröes in

Table 3.6 The nine physiographical categories from the ecological survey of Shetland (Milner, 1978) used as a basis for stratified sampling in the vegetation of the Faröe Islands (Tomlinson, 1981; with kind permission of Elsevier).

Land classes	General physical appearance	Number of squares	Proportion surveyed (%)
A	West-facing coasts, typically with high cliffs	82	3.6
B	East-facing exposed coasts	39	7.6
C	West-facing sheltered coasts, typically with gently sloping shores	19	15.8
D	East-facing sheltered coasts	48	6.2
E	Uplands (altitude exceeding 250 m) with prominent	14	21.4
F	Upland plateau and summits lacking freshwater features	295	1.0
G	Lowlands with prominent freshwater features	71	4.2
H	Uncultivated lowlands, lacking freshwater features	70	4.3
I	Villages and associated cultivated lands (always by the sea)	38	7.8

Total number of 0.25 km^2 squares containing land = 676.

Total land area of the two islands = 160 km^2.

% includes lake shores, pols, rivers and multiple stream confluences.

order to provide a basis for stratification for vegetation sampling. Nine of the 16 Shetland land types were represented on the two Faröe islands, and these are summarised in Table 3.6.

In order to carry out the stratification, a 500 m × 500 m grid was superimposed on the 1:25,000 base maps of the islands. Each square was then assigned to one of the nine classes in Table 3.6 and three squares were then taken at random for each land class, with the constraint that one of the three should be from the smaller island of Mykines if the class was present there. The number of squares and the proportion of each surveyed – (3 / number of squares in each class) × 100% – are shown on the right of Table 3.6.

For each square selected in this manner, four sampling points were located at random within each quarter. This involved taking each quarter, gridding it and locating the point by random numbers. These points were then visited in the field and data were collected in a 200 m × 200 m area around each point using an approach similar to that recommended by Bunce and Shaw (1973):

(a) All vascular plants together with bryophytes and macro-lichens were listed.
(b) Estimation of the abundance of each species was made using the Domin scale (Table 3.4).
(c) A description of the soil profile at each site was completed with pH determination of samples from 100 mm depth.
(d) A record of the habitats within each plot was compiled.

The random point was then used as the centre of each 200 m × 200 m plot, and lines were pegged out to each of the four corners of the square. The diagonals were marked at certain intervals to give quadrats of 4 m × 4 m, 25 m × 25 m, 50 m × 50 m, 100 m × 100 m and 200 m × 200 m. The smallest quadrat was examined first and then by working outwards, complete coverage of the plot was ensured. In all, 104 such plots were sampled and 227 plant species recorded.

This study demonstrates a number of important principles relating to vegetation sampling. Firstly, a logical system of sampling was established, which approximated to a stratified random sample. This approach minimised the chances of omitting important vegetation types, while maintaining a high level of objectivity. Thus sampling was specifically designed to ensure that as much variation in the vegetation of the Faröes was included as possible. Secondly, the stratification process was based on a previous survey of similar environments in Shetland. Often, existing published information may be used to provide a basis for sampling and stratification. Thirdly, the expedition had only a relatively short period within which to work. The exact location of field plots was decided in advance of departure using the stratified random approach. Once in the field, this made data collection and planning of routes and camps much easier. Tomlinson states that because of this forward planning, they were able to complete 8–12 field plots in one day.

Sampling for the National Vegetation Classification (NVC) Survey of the Vegetation of Great Britain (Rodwell, 1991, 1992a,b, 1995a, 2000, 2006)

Perhaps surprisingly, until the year 2000, there was no complete description of the major plant communities of the British Isles. Sir Arthur Tansley (1911, 1939, 1949) and McVean and Ratcliffe (1962) in Scotland provided valuable early attempts at making an inventory of the vegetation types of Britain, but both were incomplete. In 1975, the National Vegetation Classification Project was established under Dr John Rodwell at Lancaster University, and for the following 25 years, work proceeded on collecting and analysing data for this project. The brief was to produce '. . . a classification with standardized descriptions of named and systematically arranged vegetation types . . . that was to be comprehensive in its coverage, taking in the whole of Great Britain but not Northern Ireland and including vegetation from all natural, semi-natural and major artificial habitats' (Rodwell, 1991: p. 4). The project was thus primarily phytosociological in nature.

The full methodology is described in Rodwell (2006). Data collection first involved making use of data from previous vegetation surveys, where data had been gathered that were judged to be of suitable quality. The collection of new data was based on training of field surveyors, who used subjective (and thus biased) sampling to locate their quadrats. Samples were taken from stands of vegetation that:

> . . .were chosen solely on the basis of their relative homogeneity in composition and structure. Thus crucial guidelines were to avoid obvious vegetation boundaries or unrepresentative floristic or physiognomic features. No prior judgments were necessary about the identity of the vegetation type, nor were stands ever selected because of the presence of species thought characteristic for one reason or another, nor by virtue of any observed uniformity of the environmental context. (Rodwell, 1991: p. 6)

The aim was thus to describe variability in species composition first and foremost.

Quadrat sizes depended upon the size of the dominant life-forms in the vegetation. Minimal area was not employed. Rather, quadrats of 2 m × 2 m were used for short, herbaceous vegetation and dwarf shrub heaths, 4 m × 4 m for taller and more open communities, sub-shrub heaths and low woodland field layers. For woodland field layers, dense scrub, species-poor or very tall herbaceous vegetation, 10 m × 10 m plots were used, rising to 50 m × 50 m for sparse scrub and woodland canopies. These are very

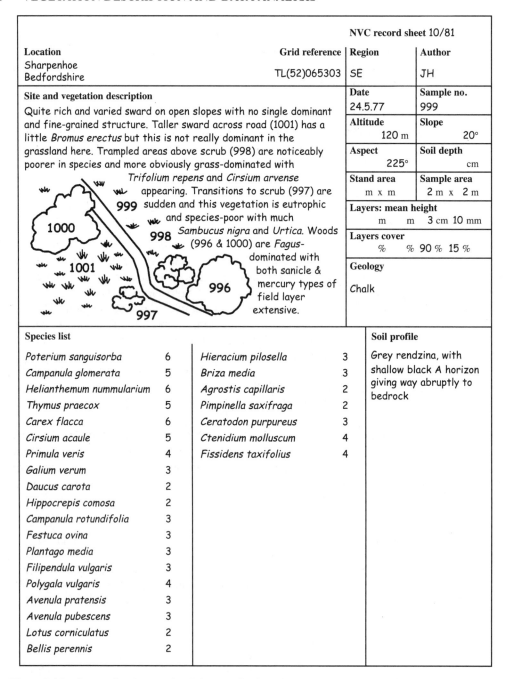

		NVC record sheet 10/81	
Location	**Grid reference**	**Region**	**Author**
Sharpenhoe Bedfordshire	TL(52)065303	SE	JH

Site and vegetation description

Quite rich and varied sward on open slopes with no single dominant and fine-grained structure. Taller sward across road (1001) has a little *Bromus erectus* but this is not really dominant in the grassland here. Trampled areas above scrub (998) are noticeably poorer in species and more obviously grass-dominated with *Trifolium repens* and *Cirsium arvense* appearing. Transitions to scrub (997) are sudden and this vegetation is eutrophic and species-poor with much *Sambucus nigra* and *Urtica*. Woods (996 & 1000) are *Fagus*-dominated with both sanicle & mercury types of field layer extensive.

Date 24.5.77	**Sample no.** 999
Altitude 120 m	**Slope** 20°
Aspect 225°	**Soil depth** cm
Stand area m x m	**Sample area** 2 m x 2 m
Layers: mean height m m 3 cm 10 mm	
Layers cover % % 90 % 15 %	
Geology Chalk	

Species list

				Soil profile
Poterium sanguisorba	6	Hieracium pilosella	3	Grey rendzina, with shallow black A horizon giving way abruptly to bedrock
Campanula glomerata	5	Briza media	3	
Helianthemum nummularium	6	Agrostis capillaris	2	
Thymus praecox	5	Pimpinella saxifraga	2	
Carex flacca	6	Ceratodon purpureus	3	
Cirsium acaule	5	Ctenidium molluscum	4	
Primula veris	4	Fissidens taxifolius	4	
Galium verum	3			
Daucus carota	2			
Hippocrepis comosa	2			
Campanula rotundifolia	3			
Festuca ovina	3			
Plantago media	3			
Filipendula vulgaris	3			
Polygala vulgaris	4			
Avenula pratensis	3			
Avenula pubescens	3			
Lotus corniculatus	2			
Bellis perennis	2			

Figure 3.21   A completed example of the standard National Vegetation Classification (NVC) record sheet. Redrawn from Rodwell (2006), with kind permission of the Joint Nature Conservation Committee.

similar to the guideline figures given in Table 3.3. Linear vegetation, such as stream and river margins, ditches, hedgerows or road verges, were sampled using 10 m strips with 30 m strips for hedgerow shrubs. Nomenclature for vascular plants originally followed *Flora Europaea* (Tutin *et al.*, 1964–93), but is now based on Stace (2010). Bryophyte identification was from Corley and Hill (1981), which has now been superseded by Purvis *et al.* (1992, 1994), and lichens were originally described from Dahl (1968), now replaced by Blockeel and Long (1998).

A 10-point Domin abundance scale, very similar to that in Table 3.4, was adopted as follows to record the above-ground living parts of all species:

Abundance score	Cover
10	91–100%
9	76–90%
8	51–75%
7	34–50%
6	26–33% cover
5	11–25% cover
4	4–10%
3	<4% with many individuals
2	<4% with several individuals
1	<4% with few individuals

In layered vegetation, such as heaths or woodland, species in the different layers were listed separately along with the height of each layer. The total cover was recorded, as well as the cover of any litter, bare rock or standing water. Selected environmental measurements were made, including altitude, slope, aspect, description of a small soil pit, and soil samples were collected for pH analysis, using a meter with temperature correction on a 1:5 soil water paste, as soon as possible after collection. Notes were made on aquatic features and conditions and on any biotic impacts, such as grazing, burning, moving or trampling. Further details to assist with describing biotic impacts in the British uplands are available in MacDonald *et al.* (1998). The main data were supported by notes, sketches and diagrams to illustrate any distinct features of the site and habitats. An example of a completed checksheet is shown in Figure 3.21.

Full details of the field description methods and methods for data analysis, together with further aspects of the presentation of results, are given in Rodwell (2006). Aspects of data analysis are included as a case study in Chapter 8.

Taken together, these case studies show the range of aims and objectives of vegetation description and demonstrate how methods almost always have to be adapted to suit particular scales and circumstances. Adaptability and flexibility are two key words when carrying out fieldwork in plant ecology and biogeography, but consistency in approach is also vital, once a particular sampling and vegetation description strategy has been decided upon.

Finally, once more, the importance of past literature, this time relating to vegetation description methods, is emphasised in this chapter. Many key papers and books on the subject date from 30–50 or even up to 100 years ago. Suggestion 2 of Belovsky *et al.* (2004) (Table 1.3) deserves to be re-emphasised in this context.

Chapter 4

The nature and properties of vegetation data

TABULATION AND CHECKING OF DATA

Virtually all methods for vegetation data analysis are applied to data based on floristics. Most physiognomic and structural data are presented in tabular or graphical form and are rarely suitable for any form of statistical manipulation.

The raw data matrix

When a series of vegetation samples or quadrats have been thrown and floristic data collected, the results are usually drawn up into a table known as the raw data matrix (Table 4.1). Here, samples or quadrats are displayed in columns and species in rows in the matrix, but it is equally valid to write the matrix in transposed form, with the quadrats in rows and species in columns (Table 4.2).

The ordering of species in a matrix depends upon the aims of the project and the type of analysis. When species are listed row-wise, the species names can be written horizontally at the start of each line. The Latin binomial should always be used and species are usually listed alphabetically, as in Table 4.1, although sometimes they may be shown in groupings such as trees, shrubs, grasses, herbs, lichens and bryophytes. However, most packages for computerised data analysis now expect the data to be presented in a computer spreadsheet (typically Excel® or Lotus 1-2-3®) with the samples or quadrats row-wise and the species names column-wise. In this case, a convention has developed that species Latin binomials are reduced to eight characters with two groups of four characters for each part of the binomial, and this eight character label is placed at the top of each species column, as in Table 4.2. Thus *Atriplex patula var hastata* in Table 4.1 becomes *Atripatu* in Table 4.2. Species can also be listed by constancy, which is defined as the number of samples or quadrats that each species occurs in. Species are then ranked downwards from high to low constancy and the commonness or rarity of species within the data-set is clearly displayed.

Within the matrix, species scores for either presence/absence or some form of abundance data may be entered. Presence/absence data are usually shown with a 1 to denote presence and 0 (zero)

Vegetation Description and Data Analysis: A Practical Approach, Second Edition. Martin Kent.
© 2012 John Wiley & Sons, Ltd. Published 2012 by John Wiley & Sons, Ltd.

Table 4.1   New Jersey salt marsh data (Fresco in Cottam *et al.*, 1978). Reproduced with permission from John Wiley & Sons, Inc.

Species	Quadrats

Species	1	2	3	4	5	6	7	8	9	10	11	12
1. *Atriplex patula var hastata*	1	10	2	1	1	2	5		1		5	2
2. *Distichlis spicata*		15	80	2	10	15	30	1	10	10	20	
3. *Iva frutescens*							5	1	2	1	20	10
4. *Juncus gerardii*			1			40	1					
5. *Phragmites communis*								1	10	20	5	30
6. *Salicornia europaea*	5	10	2	1	1		2			2		
7. *Salicornia virginica*				5	10							
8. *Scirpus olneyi*						5	20				1	
9. *Solidago sempervirens*									1	5	1	2
10. *Spartina alterniflora*	75	30	5	20	5	1		10	1	2		
11. *Spartina patens*								20	10	50	2	5
12. *Suaeda maritima*				20	10							

to indicate absence (Table 4.2), while with abundance data, the actual species scores within each quadrat are displayed, with zeros where a species is absent (Table 4.1).

Great care has to be taken in the preparation of a raw data table. The transfer of data from checksheets or a field notebook is often extremely error-prone. Mistakes are most common with those species that are found in only a few quadrats. Once completed, the table should always be double-checked against the original data.

An important characteristic of most vegetation data matrices is that they are sparse, meaning that there are a large number of zero entries indicating species absences. In most species data matrices, the number of zero entries far exceeds the number of presences, and is typically 70–85% of all cells. The sparse nature of most vegetation data matrices has important implications for many of the methods of data analysis described later. When quadrats or samples are compared, the ecological interpretation of joint absences is ambiguous. Some (dis)similarity or resemblance coefficients discussed below (e.g. χ^2 with 2×2 contingency tables and the Pearson product moment correlation coefficient) regard joint absences as indicative of a positive relationship, with the result that very dissimilar samples or quadrats can appear similar because of their common zeros. If two positively associated species have closely related abundance curves along an environmental gradient (Figure 4.1a), their distribution when plotted together in two dimensions is not linear but has a curved or ellipsoid shape (Figure 4.1b). Two negatively associated species (Figure 4.1c), when plotted together in the same manner, produce the plot in Figure 4.1d, where most values lie along the axes of the graph apart from the corner of the graph where the low scores for where the two species do occur together are found. This accumulation of points in the corner for negatively associated species has been termed 'the dust bunny effect' by McCune and Grace (2002), equivalent to the accumulation of dust and debris in the corner of a room that has not been cleaned.

The example data matrices

Three sets of vegetation data are used throughout the rest of the book to demonstrate methods of data analysis. All three sets have been chosen because they are small in size and can thus be used to demonstrate hand-worked examples of various techniques for data analysis. Nevertheless they contain interesting internal variation.

Table 4.2   The New Jersey salt marsh data as a spreadsheet file in transposed and in presence/absence form. Reproduced with permission from John Wiley & Sons, Inc.

New Jersey salt marsh data - Fresco in Cottam et al. (1978)

	Atripatu	Distspic	Iva frut	Juncgera	Phraaust	Salieuro	Salivirg	Scirolne	Solisemo	Sparalte	Sparpate	Suedmari
S1	1	0	0	0	0	1	0	0	0	1	0	0
S2	1	1	0	0	0	1	0	0	0	1	0	0
S3	1	1	0	1	1	1	0	0	0	1	0	0
S4	1	1	0	0	0	1	1	0	0	1	0	1
S5	1	1	0	0	0	1	1	0	0	1	0	1
S6	1	1	0	1	0	0	0	1	0	1	0	0
S7	1	1	1	1	0	1	0	1	0	0	1	0
S8	0	1	1	0	1	0	0	0	0	1	1	0
S9	1	1	1	0	1	0	0	0	1	1	1	0
S10	0	1	1	0	1	1	0	0	1	1	0	0
S11	1	1	1	0	1	0	0	1	1	0	1	1
S12	1	0	1	0	1	0	0	0	1	0	1	0

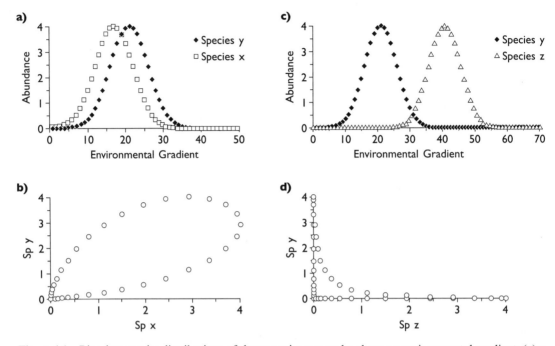

Figure 4.1   Bivariate species distributions of three species x, y and z along an environmental gradient: (a) x and y, with Gaussian response curves, are positively associated; (b) the plot of the bivariate distribution of the two positively associated species x and y, resulting in a curved (ellipsoid) shape; (c) y and z, with Gaussian response curves, are negatively associated; (d) the bivariate distribution of the two negatively associated species y and z, resulting in the 'dust bunny' effect. Adapted and redrawn from McCune and Grace (2002) with kind permission of MjM Software Design and Bruce McCune.

Data-set 1 – the New Jersey salt marsh data (Fresco, in Cottam et al., 1978) (Tables 4.1 and 4.2)

This small data-set has been chosen for two reasons. Firstly, it has been presented and used quite widely in the literature on vegetation analysis in order to demonstrate and test different analytical methods. Secondly, it is an American data-set, in contrast with the other two which are from Europe. The data are in the form of percentage cover values and are part of a much larger data-set.

Data-set 2 – garigue and maquis vegetation from the Garraf Massif, northeast Spain (Table 4.3)

The Garraf massif lies 20 km south of Barcelona in northeast Spain and comprises a large area of limestone overlain by degraded Mediterranean vegetation, known as maquis and garigue (Figure 4.2). The former climax forest was probably either Kermes Oak Forest (*Quercus coccifera*) and/or Aleppo pine (*Pinus halepensis*). The vegetation has been subject to clearance for viticulture in many places over the historical period, but the terraces and vineyards have subsequently been abandoned. Regular burning has taken place and the exposed *terra rossa* soil cover has largely been eroded and washed into the valley bottoms, where it is still cultivated. The vegetation represents a post-fire successional stage which is still subject to intermittent burning (Plate 4.1). Aspect is also a very significant control of floristic variation in the Mediterranean region. The matrix comprises a small subset of quadrats collected from a transect across an east–west orientated hill ridge, with

Table 4.3 Maquis and garigue data from Garraf, northeast Spain. Reproduced with permission from John Wiley & Sons, Inc.

Species	Quadrats														
	1	2	3	4	5	6	7	8	9	10	11	12	13	14	15
1. Bare rock/earth	3	3	6	4	3	4	6	6	6	3	6	3	6	5	6
2. *Brachypodium ramosum*	3			4											
3. *Ceratonia siliqua*			2				2		1						1
4. *Chamaerops humilis*	1			1										2	
5. *Cortaderia selloana*		1				2				2		4		2	
6. *Erica multiflora*	4			3	3					3					
7. *Euphrasia sp.*	2	1		2						1		2		2	
8. *Lavandula angustifolia*	1			1	1										
9. *Olea europaea*			3				1	4	2		4		3		3
10. *Phillyrea angustifolia*	2	3		3	2	3				3		1		4	
11. *Pinus halepensis*	3	2		3	3	2				2		3		3	
12. *Pistacea lentiscus*	3			2	3				4						
13. *Quercus coccifera*	5	4	1	6	4	5	2		1	4	1	4	1	5	1
14. *Rosmarinus officinalis*	3	1	2	2	3		2	3	3	1	2	2	1	1	2
15. *Salvia sp.*	1			3	3										
16. *Sedum sp.*		1				1			2			1		1	
17. *Smilax asper*			1				2	3			1		2	2	2
Aspect	N	N	S	N	N	N	S	S	S	N	S	N	S	N	S

Domin cover scale: 1 = present– 1 %; 2 = 2–5%; 3 = 6–10%; 4 = 11–20%; 5 = 21–50%; 6 = >50%

half the quadrats from the north-facing slope and half from the south-facing slope. A quadrat size of 5 m × 5 m was used to allow for the shrubby nature of the vegetation, and the abundance data were collected using a six-point Domin cover scale.

Data-set 3 – data on upland grassland and heathland at Gutter Tor, Dartmoor, South West England (Table 4.4)

These data comprise 25 quadrats collected along a transect up a tor on southwest Dartmoor, Devon, England (Plate 4.2). The vegetation is typical of the uplands of western and northern Britain, with grazed, short-cropped grassland, in some places invaded by bracken (*Pteridium aquilinum*), and patches of heath, represented by species of heather (*Calluna vulgaris*, *Erica cinerea*) and bilberry (*Vaccinium myrtillus*). At the bottom of the hill is a small valley bog, within which are found species such as the round-leaved sundew (*Drosera rotundifolia*) and bog asphodel (*Narthecium ossifragum*). The area is heavily grazed by sheep, cattle and horses and has been subject to burning at infrequent intervals. A quadrat size of 1 m × 1 m was used with subjective estimates of percentage cover.

MULTIVARIATE DATA AND MULTIVARIATE ANALYSIS

Floristic data recorded in the two-way (quadrat/sample by species/variable) matrix represent multivariate data. Each species/variable or quadrat/sample added to the data-set represents a potential source of variation, and each extra species/variable or quadrat/sample constitutes an additional dimension. The methods for analysing such data are thus techniques for multivariate analysis – multiple quadrats/samples and multiple species (Manly, 2005).

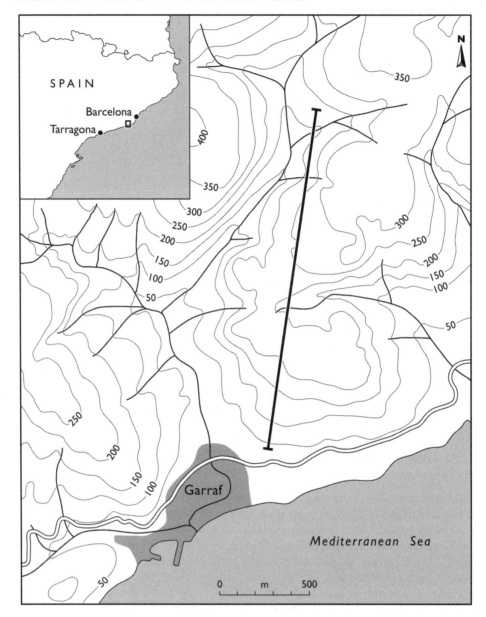

Figure 4.2   The location of the vegetation transect at Garraf, south of Barcelona, northeast Spain. Reproduced with permission from John Wiley & Sons, Inc.

In order to understand how multivariate methods of vegetation analysis work, it is necessary to explain the concept of species and sample space. Table 4.5 presents data for six species in three quadrats or samples. Figure 4.3a shows the species in the one-dimensional sample space of quadrat X, Figure 4.3b shows the species within the two-dimensional sample space of quadrats X and Y, while Figure 4.3c shows them within the three-dimensional space of quadrats X, Y and Z. The position of each species is thus determined by the coordinates of its scores in quadrats X, Y and

Table 4.4 Vegetation and environmental data from Gutter Tor, Dartmoor, southwest England. Reproduced with permission from John Wiley & Sons, Inc.

Quadrats (% cover)

Species	1	2	3	4	5	6	7	8	9	10	11	12	13	14	15	16	17	18	19	20	21	22	23	24	25
1. *Agrostis capillaris*			80						1	25			70	90				90		75				65	
2. *Agrostis curtisii*				80		70														15					
3. *Bryophytes*				15	1	2	15															5			
4. *Calluna vulgaris*				10	35						25					100					90				2
5. *Carex nigra*	15	10			5				5	25	10								5			5			10
6. *Cerastium glomeratum*				1														2							
7. *Cladonia portentosa*									10	10	5					15					20	10			
8. *Danthonia decumbens*													5	15						5					
9. *Drosera rotundifolia*	2																								2
10. *Erica cinerea*										5		2													
11. *Erica tetralix*											20	25					20			25	1	25			
12. *Festuca ovina*			10		40		40	40					30	20	30	5		20	10				50	50	
13. *Galium saxatile*			10		1	5	35	40	5	10			10	20	20	5	5	10	25				50	25	
14. *Juncus effusus*									5																20
15. *Juncus squarrosus*	5	15																		5					
16. *Luzula sylvatica*																				3					
17. *Molinia caerulea*	35	50		40		20				25	75	10					90					95			10
18. *Narthecium ossifragum*	20											10													5
19. *Plantago lanceolata*			20			5					5						10	5							
20. *Potentilla erecta*				3		5		10	2	10			10	1		10		5	2					2	
21. *Pteridium aquilinum*								90							60				100					15	
22. *Sphagnum* sp.	80	80							75			55											95		90
23. *Taraxacum officinale*																		1							
24. *Trichophorum cespitosum*		5										10												5	
25. *Trifolium repens*			3		50													3							
26. *Vaccinium myrtillus*										20				20							20	5			
27. *Viola riviniana*																5	5		5					5	5

(Continued)

Table 4.4 (*Continued*)

Species	Quadrats (% cover)																								
	1	2	3	4	5	6	7	8	9	10	11	12	13	14	15	16	17	18	19	20	21	22	23	24	25
Total cover	157	160	123	154	127	102	110	190	105	130	150	119	128	147	115	127	130	136	147	128	131	145	195	162	144
Number of species	6	5	5	7	5	5	5	5	8	8	7	8	7	6	4	6	5	8	6	6	4	6	3	6	8
Environmental data																									
Soil/peat depth (cm)	54	42	12	15	16	11	9	15	40	38	31	25	17	10	14	15	23	18	20	10	8	16	27	13	55
Slope angle°	3	1	20	6	18	10	9	20	4	3	4	4	15	12	25	9	10	15	17	5	7	2	12	13	1
pH	4.8	4.6	3.7	3.9	4.0	3.7	4.0	3.7	4.4	4.2	3.9	4.7	3.8	4.0	4.0	3.7	4.2	3.9	4.2	3.9	3.6	4.0	4.1	4.2	5.2
% soil moisture	95	110	34	75	55	52	41	31	95	76	67	105	35	43	15	50	72	36	21	43	50	72	33	21	112
Grazing (Number of faecal units)																									
Cattle	—	—	2	1	1	—	1	—	—	1	2	—	2	2	—	—	2	3	1	2	—	—	—	1	—
Sheep	—	—	4	1	4	3	2	3	—	1	1	—	6	5	—	—	—	6	2	4	3	—	1	1	—
Horses	—	—	1	2	—	—	—	—	—	1	—	—	—	1	—	1	—	1	—	2	—	1	1	2	—
Rabbits	—	—	—	1	—	—	—	1	—	—	1	—	—	1	1	1	—	1	—	1	—	1	—	—	—
Total	0	0	7	5	5	3	3	4	0	3	5	0	8	9	1	2	2	11	3	9	3	1	2	4	0

Table 4.5 A matrix of 3 quadrats/samples × 6 species to illustrate the geometric models of species in sample space, and samples in species space. Reproduced with permission from John Wiley & Sons, Inc.

Species	Quadrats/samples		
	X	Y	Z
a	2	7	2
b	6	0	9
c	8	1	5
d	1	9	6
e	5	3	5
f	9	4	4

Z. Any further quadrats added to these would increase the dimensions by one for each additional quadrat. However, although this is simple to describe mathematically, it cannot be shown visually and geometrically beyond three dimensions.

It is clearly possible to construct this geometrical diagram the other way around, with quadrats or samples defined in terms of species space. However, again this could only be done graphically for three species, although mathematically there is no limit – every further species included adds another dimension. Figure 4.3d shows the three quadrats X, Y and Z defined in terms of just the first three species, a, b and c – d, e and f, the fourth, fifth and sixth, cannot be used in the geometrical diagram but can be used mathematically.

Redundancy and noise

A two-way quadrats/samples by species matrix is often described as having a moderate to large amount of redundancy within it, in that many samples/quadrats are very similar to others and thus duplicate variation in the data. Equally, many species are like other species in their distribution across a series of samples or quadrats, and again the variation is effectively duplicated. A related concept is that most methods for analysing vegetation data are concerned with data reduction, whereby the original data, containing a lot of redundancy and duplication, are reduced down to the major sources of variation within them, eliminating the duplication and redundancy in the process.

Another property of vegetation data matrices is noise. Noise is best explained in terms of plant response to environment. If two samples are taken from the same local plant community, although they will be very similar to each other, they will rarely be identical. Thus a dominant species in one sample could be recorded as having 90% cover, while in another sample from the same community, only 70% cover. Similar variations will occur for all other species recorded in the two samples. The observed differences between the two samples can partly be explained by very local differences in environmental factors, the chance occurrence of individuals, local variations in biotic pressures, such as burning and grazing, or errors or variations in the recording of abundance. However, these minor sources of variation are of little interest compared with the much greater differences between the samples from this community type and those from other contrasting community types for which other data may have been collected. At this broader scale, the data from the original community might be described as 'noisy' because nearly identical samples from that local environment are nevertheless showing variation which is not of any great ecological significance at the broader scale. All data-sets thus contain two kinds of variability, one which is important and

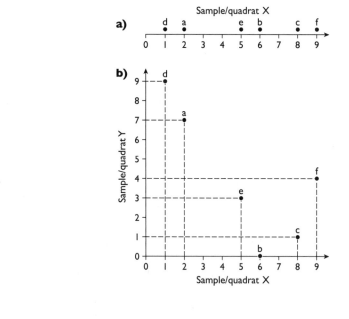

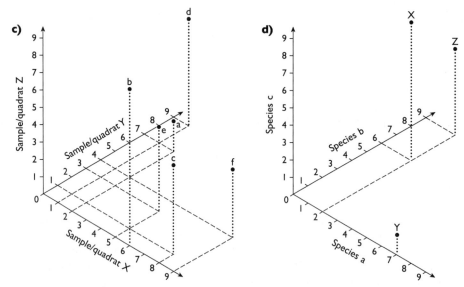

Figure 4.3   The geometric model of species in quadrat/sample space: (a) one dimension (quadrat/sample X); (b) two dimensions (quadrats/samples X and Y); three dimensions (quadrats/samples X, Y and Z); (d) samples X, Y and Z in three species space (species a, b and c). Reproduced with permission from John Wiley & Sons, Inc.

interesting and assists with explanation and understanding of plant–environment relationships, and other more detailed variation, which is of very little interest and can interfere with and blur the overall picture.

While this description of noise is somewhat simplified, it is an important concept in vegetation analysis. Further discussion of the subject can be found in Pignatti (1980) and Gauch (1982a,b). An important related issue is the possibility of deleting rare species (those with just one occurrence or only a very few occurrences) from a data-set. This is discussed in the next section under 'Transformations of data'.

TYPES OF DATA

The values held within raw data matrices vary in type, and this has important implications for the various statistical methods described in subsequent chapters. Of the range of data types, four main categories can be identified:

(a) *Nominal* – this type of measurement involves categorisation without numerical values or ranks. Presence/absence data coded as +/− or 1/0 is an example of this type. However, the data may also be in a series of divisions by name. Colours (green, brown, yellow) are a good example, or trees, shrubs, grasses, herbs and bryophytes. Such data cannot be manipulated according to the basic rules of arithmetic. One colour is not greater or less than another, for example – nor can a colour be divided. Such data are also sometimes called unordered multistate data.

(b) *Ordinal* – such data can be placed in rank order. Some of the simpler abundance scales for recording vegetation data are of this type, e.g. dominant, abundant, frequent, occasional and rare. All four basic arithmetical operations (addition, subtraction, multiplication and division) can be performed on ranked data. The Braun-Blanquet scale (Table 3.4) is another good example. Ordered multistate data is another name for this data type.

(c) *Interval* – interval data have a constant unit of measurement and thus differences between values can be compared. Temperature on the Celsius or Fahrenheit scale is the most widely quoted example. A difference of 3° means the same across the whole scale. However, the position of zero in an interval scale is arbitrary. A consequence of this is that a given difference can then be said to be twice great as another, but not that a certain object is twice as large as the other when the corresponding values are, for example, in the ratio 2:1, so that a temperature of 40°C is not twice a temperature of 20°C.

(d) *Ratio* – these are the most precisely defined type of data, which differ from the interval scale in one respect only: they have an absolute zero point. It is thus possible to count ratios. Again to illustrate the difference, if one quadrat is 2 m from a tree and another is 4 m, then the second is twice as far from the tree as the first. However, if the temperature one day is 10°C and 20°C the next, it is not twice as warm on the second day; those would be interval data.

For the purposes of this book, the differences between the interval and ratio scales are of little consequence, since most vegetation data are ratio data, but the differences between interval/ratio, ordinal and nominal types are very important. Also data collected on the interval/ratio scales usually take more time to collect than data on ordinal and nominal scales. This is because greater precision is required. However, in some situations, nominal data, such as presence/absence, may suffice and save a great deal of time and money.

TRANSFORMATIONS OF DATA

Many sets of both vegetation and environmental data may require transformation prior to data analysis. Some transformations of environmental data are needed to assist with statistical issues, such as ensuring normality of individual variable distributions, equality or homogeneity of variances or linearity. Standardisation of environmental variables is also often necessary to ensure comparability of measurement scales. For example, if one environmental variable, such as altitude, is measured in the range 1–10,000 m, while others such as soil pH are in the range from 3.00–10.00, certain techniques of analysis will require the differences in measurement units to be standardised to a common scale. The commonest, zero mean-unit variance standardisation, is described at the start of the section on principal component ordination in Chapter 6.

With species data, there are important ecological reasons for employing transformation. Certain of the (dis)similarity coefficients discussed below will perform better following transformation. The impact of significant variations in total abundances can be reduced or eliminated and the contributions of common and rare species can be equalised. The variation in species data-sets that is of real interest may be masked by either common species of high abundance or the occasional presence of rarities (van der Maarel, 1979). Even before considering transformations, Waite (2000), for example, suggested removing from a data-set either any common species that occur in >90% of samples, or/and rare species that occur in <10% of samples, although these thresholds are entirely arbitrary. There is further discussion of the advantages and disadvantages of deleting rare species in McCune and Grace (2002).

The first group of transformations are monotonic in nature (only increasing or decreasing over their range):

Power transformations: $b_{ij} = x_{ij}^p$ where p is the exponent of the transformation. $p^{0.5}$ is the square root, while p^2 is the square. The square root transformation compresses high values and spreads out lower values, although not by as much as logarithmic transformation. Higher roots are used more rarely. Roots above a power of three virtually change the data to presence/absence form.

Logarithmic transformation has the form $b_{ij} = \log(x_{ij})$, where logs are taken to any base, usually $\log_{10}$ or ln. High values are compressed and low values spread out. Where zeros are present, it is usual to employ $b_{ij} = \log(x_{ij} + 1)$, and 1 is added to all observations. However, various authors warn against log transformations (Wilson, 2007), and when the lowest non-zero observation differs from 1 by more than one order of magnitude, then the addition of 1 will distort the relationship between zeros and other values in the data-set. McCune and Grace (2002) suggest an alternative generalised transformation.

Arcsine-square root transformation $b_{ij} = 2/\pi * \arcsin(\sqrt{x_{ij}})$ spreads the ends of the range for proportional data (in proportions in the range 0–1) and compresses the centre. The arcsin square root is multiplied by $2/\pi$ to rescale the result back to 0–1.

The second group of transformations are better termed relativisations or standardisations (McCune and Grace, 2002). They stress that relativisations must only be used in the context of the research question and the properties of the species data. Species data can be relativised by column (species) total or by row (sample) total, or by column (species) maximum, which equalises the heights of species response curves along environmental gradients, or row (sample) maximum, which equalises the areas under the species response curves. If relativisation is applied to both species and samples, this is known as 'double relativisation'. An extensive discussion of most aspects of the transformation of species data is presented in McCune and Grace (2002).

MEASUREMENT OF ASSOCIATION AND SIMILARITY BETWEEN SPECIES AND SAMPLES

General

One of the simplest means of analysing floristic vegetation data is to look at the degree of association between species and the level of similarity between quadrats or samples. Such ideas are fundamental to most of the more complex multivariate methods covered later in this text.

Chi-square (χ^2) as a measure of association between species

In Chapter 1, the plant community was defined as an assemblage of plant species that show a definite affinity or association with each other. There is no doubt that certain species tend to grow together in certain locations, while others never coexist. These ideas can be represented by the concepts of positive and negative association. If two species are found to be positively associated, this means that they are found growing together more often than would be expected by chance or random events. Conversely, negative association means that one species is found growing without the other more often than would be expected by chance.

The degree of association between species in a set of quadrat samples can be quantified. One of the most widely used methods is that of chi-square (χ^2) using 2×2 contingency tables. Awareness of this method is important in understanding the various forms of correspondence analysis described in Chapter 6 on ordination.

The contingency table

In order to calculate χ^2 as a measure of association, the data on presence/absence of the two species from a series of quadrats need to be arranged into a 2×2 contingency table of the form shown below:

	Species X +	Species X −	
Species Y +	a	c	$a + c$
Species Y −	b	d	$b + d$
	$a + b$	$c + d$	N

The table is drawn up for two species X and Y whose presence/absence is measured in N quadrats. In each quadrat or sample, there are four possible situations in terms of presence and absence. Both species X and Y can be present ($+$X/$+$Y – cell a); X can be present but not Y ($+$X/$-$Y – cell b); Y can be present but not X ($-$X/$+$Y – cell c); and neither X nor Y may be present ($-$X/$-$Y – cell d). In the marginal totals, $a + b$ gives the number of occurrences of X, and $a + c$ the number of occurrences of Y; $c + d$ shows the number of quadrats not containing X and $b + d$ the number of quadrats not containing Y.

Calculation of χ^2 can be explained by an actual example. From a set of 180 quadrats thrown on acidic upland grassland in Britain, the following contingency table was constructed:

	Agrostis capillaris +	*Agrostis capillaris* −	
Festuca ovina +	112	15	127
Festuca ovina −	25	28	53
	137	43	180

χ^2 is calculated using the contingency table and the following formula:

$$\chi^2 = \frac{(|ad - bc| - 0.5N)^2 \times N}{(a + b)\,(c + d)\,(a + c)\,(b + d)} \qquad (4.1)$$

where a, b, c, d and N are as above. $|ad - bc|$ is the absolute difference of ad and bc regardless of sign. This formula includes Yates' correction for small samples, which is the $-0.5N$ on the top line of the formula. For large samples (over 500) this may be omitted. The χ^2 formula is based on the sum of the differences between the observed and expected frequencies of each cell in the contingency table. The method of calculating expected frequencies is explained below. In the above example, the formula gives:

$$\chi^2 = \frac{(|3136 - 375| - 90)^2 \times 180}{137 \times 43 \times 127 \times 53}$$

$$\chi^2 = \frac{1284163380}{39652321}$$

$$\chi^2 = 32.39$$

χ^2 significance tables available in many statistical textbooks (e.g. Hammond and McCullagh, 1978; Bishop, 1983; Lee and Lee, 1982; Wardlaw, 1985; Eddison, 2000; Wheeler $et\ al.$, 2004) show that with one degree of freedom, χ^2 has to exceed 3.84 to be significant at the 0.05 level and 6.64 to be significant at the 0.01 level. Clearly in this example ($\chi^2 = 32.39$), a very high level of significance is achieved, which means that the two species $Agrostis\ capillaris$ and $Festuca\ ovina$ are associated with each other more than would be expected by chance.

However, although a highly significant association between the two species has been found, the nature of the relationship, and in particular whether it is positive or negative, has not been determined. To do this, the expected frequency for the cell of the contingency table containing the joint occurrences of species X and Y ($+$X/$+$Y – cell a) is calculated and then compared with the value for the observed frequencies. The observed value in the above contingency table is 112. The expected value is calculated by taking the probability of $Agrostis\ capillaris$ occurring in the data-set ($a + b$) / $N = 137/180$, and the probability of $Festuca\ ovina$ occurring in the data set ($a + c$) / $N = 127/180$, and multiplying the two probabilities together to get the joint probability or expected frequency.

$$\text{Expected frequency of cell } a \text{ (joint occurrences)} = \frac{(a + b) \times (a + c)}{N} = \frac{137 \times 127}{180} = 96.7$$

Thus the observed frequency for joint occurrences of the two species is 112, while the expected frequency is 96.7.

If the observed value for joint occurrences exceeds the expected value, then the two species are positively associated. Clearly this is the case in the above example, and $Agrostis\ capillaris$ and $Festuca\ ovina$ tend to be either present together or absent together. Only infrequently does one species occur without the other. If, however, the observed value had been less than the expected value, then the association would have been negative and the species would rarely have been found together, or rather, when one species was found, the other was more likely than not to be absent. The larger the value of χ^2 between any two species, the stronger the association.

Exactly why two species should be positively associated is not always as obvious as might at first seem. Clearly the most common reason why two species regularly appear together is because they prefer the same environmental conditions. However, further investigation often reveals such explanations to be rather simplistic, and factors such as plant species strategies, competition and interaction also need to be taken into account (Chapter 2).

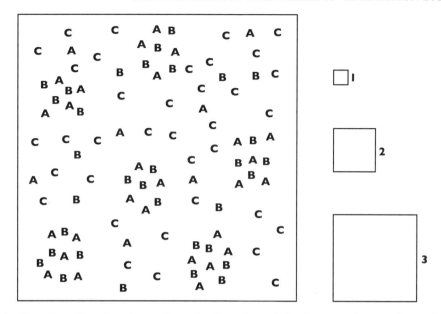

Figure 4.4   The effect of quadrat size on determination of association between plant species. Redrawn from Kershaw and Looney (1985), with kind permission of Edward Arnold/Hodder and Stoughton. Reproduced with permission from Arnold/Hodder and Stoughton.

Problems of using χ^2 for the analysis of association

Quadrat size is very important. In the above example, 1 m × 1 m quadrats were used. If quadrat size is too small and approaches the size of some individual plants, then negative associations will appear which mean nothing more than that two individuals of different species cannot occupy the same location (quadrat size 1, Figure 4.4). If the quadrat size is too large for the local pattern of the vegetation, then no information on the underlying associations would be obtained (quadrat size 3, Figure 4.4 – here every placing of the quadrat would include all three species). If the morphologies of some species for which associations are being calculated are very different, then spatial exclusion by the large species may cause positive associations among the smaller species. Thus quadrat size must be chosen in relation to the size of the dominant species, but different sizes may be needed for species of highly differing morphologies, and it may not be valid to calculate association between species that differ greatly in morphology. In Figure 4.4, quadrat size 2 is the optimal size and will show up the true associations between the species. This is because the area of quadrat 2 is approximately the same as that of the groups of species A and B, which are highly clustered.

Association between quadrats or samples

Just as it is possible to calculate χ^2 between species to examine association, so it can be calculated between quadrats or samples. To students, this often seems a strange concept at first, but the association in this case is best described as similarity and dissimilarity of floristic composition between samples. Remembering that association is only calculated with presence/absence data, the more species that two quadrats have in common, the more similar they are in terms of species composition, and a high χ^2 value coupled with positive association means that the quadrats must contain very similar species. Equally, a high χ^2 value can be obtained with negative association. This means that the quadrats have very few species in common and are very dissimilar or unlike. Also, the χ^2 test has been shown to be unreliable if the expected value in any one cell of the contingency table is less than 5. However, this is not a rigid rule, although it is worth checking for.

A value of χ^2 may be converted into a coefficient of similarity (S) between samples or quadrats (see below) with values lying between 0 and 1.0 by using the following formula:

$$S_X^2 = \frac{m + \chi^2}{2m} \tag{4.2}$$

where m is the total number of species found in the two samples.

However, compared with the coefficients presented below, the χ^2 coefficient performs poorly, because values depend not only on the number of species that samples have in common, but also on the number of species not found in either (cell d of the contingency table).

Measurement of similarity and dissimilarity (resemblance)

Many measures exist for the assessment of similarity between vegetation samples or quadrats. Some are qualitative and based on presence/absence data, while others are quantitative and will work on abundance data. Many will cater for both data types. Similarity coefficients measure the degree to which the species composition of quadrats or samples is alike. Dissimilarity coefficients assess the degree to which two quadrats or samples differ in composition. It follows that dissimilarity is the complement of similarity. Another term that is widely used to describe similarity is resemblance, and similarity coefficients are sometimes called resemblance functions.

Many similarity coefficients have been devised by ecologists and statisticians. Of these, the most useful are the Jaccard coefficient, which is generally applied to qualitative data, the Steinhaus (Sørensen/Czekanowski) coefficient, and the coefficient of Euclidean distance, both of which are suited to either quantitative or qualitative data.

The Jaccard coefficient

Jaccard (1901, 1912, 1928) developed a very simple mathematical expression, which although originally used to compare the general floras of larger areas, has subsequently been shown to be suitable for assessing the similarity of quadrat samples in terms of species composition. The formula is:

$$S_J = \frac{a}{a + b + c} \tag{4.3}$$

where a is the number of species common to both quadrats/samples, b is the number of species in quadrat/sample 1 only, and c is the number of species in quadrat/sample 2 only. Often the coefficient is multiplied by 100 to give a percentage similarity figure.

Dissimilarity (D_J) is then:

$$D_J = \frac{b + c}{a + b + c} \quad \text{or} \quad 1.0 - S_J \tag{4.4}$$

The figure can again be converted to a percentage if required.

The Steinhaus (Sørensen/Czekanowski) coefficient

Legendre and Legendre (1998) stated that this coefficient of similarity (S_S) was first proposed by a Polish mathematician, H. Steinhaus and also described by Motyka (1947), and it is thus increasingly referred to as the Steinhaus coefficient. However, a very similar coefficient was first proposed by Czekanowski (1909, 1913). It is defined using the same symbolism as:

$$S_S = \frac{2a}{2a+b+c} \qquad (4.5)$$

Dissimilarity is then:

$$D_S = \frac{b+c}{2a+b+c} \quad \text{or} \quad 1.0 - S_S \qquad (4.6)$$

Again, the figure can be calculated as a percentage if required. Both coefficients can be used on quantitative as well as qualitative data.

Sørensen first published his coefficient, which is also very similar to that of Steinhaus and Czekanowski, in 1948. Generally, the Steinhaus (Sørensen/Czekanowski) coefficient is preferred to the Jaccard, because it gives weight to the species that are common to both the quadrats or samples, rather than to those which are unique to either sample.

As an example, the Steinhaus coefficient is calculated for quadrats 11 and 12 of the New Jersey salt marsh data-set (Table 4.2). In this case, presence/absence data are being used.

Species	Quadrat 11	Quadrat 12
Atriplex patula	1	1
Distichlis spicata	1	0
Iva frutescens	1	1
Juncus gerardii	0	0
Phragmites communis	1	1
Salicornia europaea	0	0
Salicornia virginica	0	0
Scirpus olneyi	1	0
Solidago sempervirens	1	1
Spartina alterniflora	0	0
Spartina patens	1	1
Sueda maritima	0	0
Total occurrences	7 (b)	5 (c)
Number of joint occurrences	5 (a)	

$$S_S = \frac{2a}{2a+b+c} = \frac{2 \times 5}{10+7+5} = 0.45(45\%)$$

$$D_S = \frac{b+c}{2a+b+c} \quad \text{or} \quad 1.0 - S_S$$

$$D_S = \frac{7+5}{10+7+5} = 0.55\,(55\%)$$

The coefficient is then presented as:

$$S_S = \frac{2\sum_{i=1}^{m} \min(X_i, Y_i)}{\sum_{i=1}^{m} X_i + \sum_{i=1}^{m} Y_i} \qquad (4.7)$$

where X_i and Y_i are the abundances of species i, min (X_i, Y_i) is the sum of the lesser scores of species i where it occurs in both quadrats, and m is the number of species. The coefficient values range from 0 (complete dissimilarity) to 1 (total similarity).

Using the same quadrats (11 and 12) from the New Jersey data but with the quantitative values (Table 4.1), the Steinhaus (Sørensen/Czekanowski) coefficient is calculated below. The lower values of joint occurrences are underlined, and when summed give a total of 20.

Species	Quadrat 11	Quadrat 12
Atriplex patula	5	2
Distichlis spicata	20	0
Iva frutescens	20	10
Juncus gerardii	0	0
Phragmites communis	5	30
Salicornia europaea	0	0
Salicornia virginica	0	0
Scirpus olneyi	1	0
Solidago sempervirens	1	2
Spartina alterniflora	0	0
Spartina patens	2	5
Sueda maritima	0	0
Total cover	54	49
Sum of the lower scores of species common to both quadrats (underlined values)		20

Thus

$$S_S = \frac{2 \times 20}{54 + 49} = 0.39 (39\%)$$

A value for dissimilarity can be calculated by taking the similarity value away from 1.0 (or 100%). Thus dissimilarity = 0.61 (61%).

The coefficient is often also referred to as the 'Bray and Curtis' coefficient or the 'percentage (dis)similarity' coefficient because of its use in Bray and Curtis ordination (Chapter 6). In this case, it is often described as the '(2W/(A + B)) × 100%' similarity coefficient, where A is the sum of scores in the first quadrat being compared and B is the sum of scores in the second quadrat, with W as the sum of the lesser scores of species common to both quadrats. When the percentage similarity value is taken away from 100, the resulting value for percentage dissimilarity is taken as a distance value, and it is this distance value that is used to construct a Bray and Curtis (polar) ordination diagram (Chapter 6).

The Euclidean distance coefficient

This coefficient is based on the Euclidean properties of a right-angled triangle and the fact that the square on the hypotenuse is equal to the sum of the squares on the opposite two sides (Figure 4.5). If two species X and Y occur in two quadrats 1 and 2, the (dis)similarity or 'distance' between the two species in geometric space is defined as:

$$D_E = \sqrt{(X_1 - X_2)^2 + (Y_1 - Y_2)^2} \tag{4.8}$$

For more than two species, the generalised formula becomes:

$$D_{ij} = \sqrt{\sum_{k=1}^{m} (X_{ik} - X_{jk})^2} \tag{4.9}$$

where D_{ij} is the Euclidean distance between quadrats i and j, m is the number of species, X_{ik} is the value of the kth species in quadrat i, and X_{jk} is the value of the kth species in quadrat j.

The lower the value of the Euclidean distance between two quadrats, the more similar they are in terms of species composition. If quadrats 11 and 12 of the New Jersey data are again taken to illustrate the calculation, then:

Species	Quadrat 11	Quadrat 12	Difference	Difference^2
Atriplex patula	5	2	3	9
Distichlis spicata	20	0	20	400
Iva frutescens	20	10	10	100
Juncus gerardii	0	0	0	0
Phragmites communis	5	30	25	625
Salicornia europaea	0	0	0	0
Salicornia virginica	0	0	0	0
Scirpus olneyi	1	0	1	1
Solidago sempervirens	1	2	1	1
Spartina alterniflora	0	0	0	0
Spartina patens	2	5	3	9
Sueda maritima	0	0	0	0

$$\sum_{k=1}^{m} = 1145$$

$$D_E = \sqrt{1145} = 33.84$$

The lower limit of the coefficient is 0, representing complete similarity. However, there is no fixed upper limit for this coefficient. For this reason, it is known as a coefficient of dissimilarity.

(Dis)similarity matrices

If (dis)similarity (similarity or dissimilarity) is calculated between all pairs of quadrats/samples and/or between all species, then the matrix that results is known as a half-matrix of (dis)similarity coefficients. Such matrices are also referred to as resemblance matrices. As an example, Table 4.6

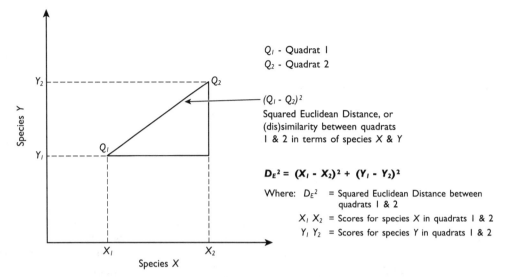

Figure 4.5 Euclidean distance between two quadrats/samples (1 and 2) in terms of two species (X and Y). Reproduced with permission from John Wiley & Sons, Inc.

contains a matrix of Bray and Curtis dissimilarity coefficients for the 12 quadrats in the New Jersey salt marsh data (Table 4.1). These types of matrices form the basis for some methods of ordination and classification discussed in Chapters 6 and 8.

MEASUREMENT OF CORRELATION AND SIMILARITY BETWEEN ENVIRONMENTAL AND BIOTIC VARIABLES

As with the example of data-set 3 from Gutter Tor on Dartmoor, southwest England (Table 4.4), in addition to recording data on species composition in quadrats, it is normal to collect paired data

Table 4.6 A half-matrix of Bray and Curtis dissimilarity coefficients calculated between the 12 quadrats of the New Jersey salt marsh data (Table 4.1). Reproduced with permission from John Wiley & Sons, Inc.

Quadrat number	Quadrat number											
	1	2	3	4	5	6	7	8	9	10	11	12
1	0.00											
2	0.51	0.00										
3	0.91	0.69	0.00									
4	0.66	0.58	0.87	0.00								
5	0.88	0.67	0.73	0.44	0.00							
6	0.97	0.72	0.75	0.93	0.76	0.00						
7	0.96	0.70	0.60	0.94	0.80	0.69	0.00					
8	0.81	0.75	0.89	0.69	0.80	0.95	0.77	0.00				
9	0.97	0.83	0.86	0.94	0.79	0.83	0.58	0.71	0.00			
10	0.93	0.73	0.79	0.89	0.66	0.79	0.79	0.84	0.60	0.00		
11	0.99	0.66	0.69	0.94	0.76	0.69	0.52	0.87	0.67	0.64	0.00	
12	0.99	0.97	0.97	0.98	0.98	0.96	0.82	0.81	0.69	0.48	0.61	0.00

on environmental and biotic variables at the same sampling points or locations. Data analysis will then usually seek to examine relationships and correlations between variation in the vegetation or species data, and variation in the environmental and biotic data. Where such paired data are gathered, there are thus three possible approaches to multivariate data analysis:

(a) analysis of the vegetation/species data alone;
(b) analysis of the environmental and biotic data alone;
(c) joint analysis of both the vegetation/species data and the environmental/biotic data.

In the case of environmental and biotic data, just as with the vegetation/species data, it is possible to look for similarities and correlations between the variables or between the samples. It is important to stress that this can be done by examining the environmental/biotic data alone, just as the vegetation/species data could be analysed on their own.

Correlation coefficients

Where data are measured on the interval and ratio scales, the relationships between the variables and/or samples are normally analysed using the Pearson product-moment correlation coefficient described in detail in Chapter 5. Where some data are in ranked or ordered form, then a method of rank correlation should be employed, of which Spearman's rank correlation and Kendall's tau coefficients are the most widely employed. Again, these are presented in detail in Chapter 5.

A common problem with ecological multivariate data-sets, however, is that where a set of environmental/biotic variables have been measured, different variables will have been measured on different data types, some ratio and interval, some ordinal (ranked or ordered multistate) and some nominal (unordered multistate). For obtaining a measurement of (dis)similarity between samples, this presents difficulties because none of the above measures of (dis)similarity or correlation coefficients will be appropriate.

Gower's general similarity coefficient

A way around this difficulty is provided by Gower's general similarity coefficient (Gower, 1971), which involves calculating a similarity value s_{jk} for each variable when two samples are being compared. The sum of the values gives the overall measure of similarity between samples, which is then scaled between 0 and 1.0. Each variable, regardless of data type, contributes equally to the coefficient value. The formula is:

$$G_{ijk} = \frac{\sum\limits_{i=1}^{n} S_{ijk}}{\sum\limits_{i=1}^{n} W_{ijk}} \qquad (4.10)$$

where the numerator is the sum of similarity values between sample j and sample k across all n variables (i), and the denominator is the sum of all weighting factors across all n variables (i). Thus the similarity values for each variable of each data type are calculated and then summed.

For binary variables, the calculation of the similarity value is the same as Jaccard's index of similarity (Equation 4.3). For multistate variables, if there are only two states, they are treated the same as binary variables. Where there are more states, each matching state is given a binary score (s_{ijk}) of 1.0, while mismatches are given a score of 0. The similarity scores are summed (Σs_{ijk}) and then divided by the total of the weighting factors (Σw_{ijk}), with each variable having been given a weighting factor of 1.0.

With quantitative variables (ratio and interval data) the following formula is used to calculate similarity scores:

$$S_{ijk} = 1 - D_{ijk} \qquad (4.11)$$

where

$$D_{ijk} = \sum_{i=1}^{n} \frac{1}{r_i} \left| x_{ij} - x_{ik} \right| = \sum_{i=1}^{n} d_{ik} \qquad (4.12)$$

For every variable, the absolute difference is calculated and this value is divided by a scaling factor r_i, which is equal to the range of each variable. Thus, if the range (r_i) of a particular variable i is from 4 to 40 and $x_{ij} = 10$ and $x_{ik} = 18$, then $r_i = 36$ and the dissimilarity value for variable i is $d_{ijk} = |10 - 18|/36 = 8/36 = 0.222$. This scaling makes the dissimilarity values for all variables lie between 0 and 1.0.

Each variable is given an equal weighting ($w_{ijk} = 1.0$) and the final index is calculated by summing the dissimilarity scores (Σd_{ijk}) and dividing that sum by the sum of the variable weightings (Σw_{ijk}), which is equivalent to the number of variables across all variable data types. This dissimilarity index is extremely useful for sets of variables with mixed data types, which is commonplace in environmental/biotic variable data-sets.

DIVERSITY AND SPECIES RICHNESS

The whole subject of species richness and diversity is fraught with problems and misunderstandings. There is confusion over the meaning of diversity (Hurlbert, 1971; Ghilarov, 1996), methods for measuring and assessing diversity, and over the ecological interpretation of different levels of diversity and in particular the relationship between diversity and stability in ecosystems. Schluter and Ricklefs (1993), Huston (1979, 2003), Gaston and Spicer (2004) and Magurran (2004) have provided valuable reviews of concepts of diversity, and stress that the word is hard to define. The word 'biodiversity' is now widely equated with and substituted for the word 'diversity' (Huston, 2003; Hamilton, 2005; Lepš, 2005). Species richness, meaning a count of the number of plant species in a quadrat, area or community, is also often equated with diversity. When ecologists talk of high diversity, they often mean a community containing a large number of different species. However, as Magurran (2004) states, most methods for measuring diversity actually consist of two components. The first is species richness, but the second is the relative abundance (evenness or unevenness) of species within the sample or community. Perfect evenness of five species in a quadrat would mean that if total cover was 100%, they would be distributed 20, 20, 20, 20, 20. Diversity is thus measured by recording the number of species and their relative abundances. The two components may then be examined separately or combined into some form of index.

Alpha, beta and gamma diversity

Whittaker (1965, 1972, 1975) made a distinction between three types of diversity – alpha, beta and gamma diversity. Alpha diversity refers to the number of species within the sample area or community, such as one type of woodland or grassland. Beta diversity is the difference in species diversity between samples or communities that correspond to 'pieces' in the landscape mosaic (Figure 2.2). Beta diversity is thus sometimes called habitat diversity because it represents differences in species composition between very different areas or environments and the rapidity of change of those habitats. It is also related to Whittaker's concept of plant communities as mosaics (Chapter 2). The smaller and more numerous the 'pieces' of the mosaic of different plant communities, the higher the beta diversity. Alpha diversity remains the number of species within each individual piece of the mosaic.

Gamma diversity represents the product of alpha and beta diversity within a landscape – the number and frequency of 'pieces' within the landscape mosaic (beta diversity) combined with the number of species within each 'piece' (alpha diversity). Whittaker *et al.* (2001) provided a valuable discussion on this subject, particularly in relation to the links of alpha, beta and gamma diversity with differing spatial scales (Chapter 2).

Diversity indices

A large number of indices of diversity have been devised, each of which seeks to express the diversity of a sample or quadrat by a single number. Some indices are also known by more than one name and are presented in different ways (e.g. Hill, 1973a; Southwood and Henderson, 2000). Of the various indices, the most frequently used is the simple totalling of species numbers to give species richness (Magurran, 2004). However, of the indices combining species richness with relative abundance, probably the most widely used are the Simpson index (D) and the Shannon index (H'). The other index occasionally used in plant ecology is McIntosh's diversity index (U). Anderson and Marcus (1993) discussed the implications of quadrat size for the collection of data on species richness and diversity, and these relate back to the discussion of quadrat size in Chapter 3.

The Simpson index

Simpson (1949) derived an index based on the probability that any two individuals taken at random from an infinitely large community will belong to the same species expressed in the following formula:

$$D = \sum p_i^2 \tag{4.13}$$

where p_i is the proportion of the number of individuals or the abundance of the ith species. The related index (D) for count data is calculated from the formula:

$$D = \sum \left(\frac{n_i \, [n_i - 1]}{N \, [N - 1]} \right) \tag{4.14}$$

where n is the number of individuals of the ith species and N is the total number of individuals. The index is usually presented as $1 - D$ (the complement) or $1/D$ (the reciprocal). As index values increase, diversity decreases. Magurran (2004) stated that the index emphasises the most abundant species in a sample or quadrat and is less sensitive to species richness than others. Simpson's index is probably the most widely used diversity index and is the most robust in that it is based on calculation of the variance of the species distribution of the sample. The index is most commonly expressed as the reciprocal ($1/D$), although Rosenzweig (1995) suggests that $-\ln(D)$ (Pielou, 1975) is preferable because the reciprocal form is sensitive to the size of the variances.

An evenness measure can also be obtained by dividing the reciprocal form of the index by the number of species in the sample (S) (Smith and Wilson, 1996; Krebs, 1999).

$$E_{1/D} = \frac{(1/D)}{S} \tag{4.15}$$

The result lies between 0 and 1 and is not influenced by species richness.

As an example of the use of the index, two sample quadrats are taken from the Gutter Tor, Dartmoor, data-set, where % cover abundance data have been collected (Table 4.4).

Quadrat 1 – 6 species present

Species	% cover	Proportion of total cover (p_i)	p_i^2
Carex nigra	15	0.0955	0.0091
Drosera rotundifolia	2	0.0127	0.0001
Juncus effusus	5	0.0318	0.0010
Molinia caerulea	35	0.2229	0.0496
Narthecium ossifragum	20	0.1274	0.0162
Sphagnum sp.	80	0.5096	0.2596
Total cover (%)	157		$\Sigma p_i^2 = 0.3359$

$D = 1 - 0.3359 = 0.6641$

Quadrat 25 – 8 species present

Species	% cover	Proportion of total cover (p_i)	p_i^2
Calluna vulgaris	2	0.0139	0.0002
Carex nigra	10	0.0694	0.0048
Drosera rotundifolia	2	0.0139	0.0002
Juncus effusus	20	0.1389	0.0193
Molinia caerulea	10	0.0694	0.0048
Narthecium ossifragum	5	0.0347	0.0012
Sphagnum sp.	90	0.6250	0.3906
Trichophorum cespitosum	5	0.0347	0.0012
Total cover (%)	144		$\Sigma p_i^2 = 0.4224$

$D = 1 - 0.4224 = 0.5776$

The evenness or equitability for the two quadrats can now be calculated:

For quadrat 1:

$$E_{1/D} = \frac{(1/0.6641)}{6} = 0.2510$$

For quadrat 25:

$$E_{1/D} = \frac{(1/0.5776)}{8} = 0.2164$$

Values of $E_{1/D}$ lie between 0 and 1 and the lower the index value, the more evenly distributed are the species.

The Shannon index

This is based on information theory and the concept that the diversity or information in a sample or community can be measured in a similar way to the information contained within a message or code. The index (H') is sometimes correctly called the Shannon-Wiener index but elsewhere is referred to wrongly as the Shannon-Weaver index (Spellerberg and Fedor, 2003). The index makes the assumption that individuals are randomly sampled from an 'infinitely large' population and also assumes that all the species from a community are included in the sample. This last assumption is not always easy to meet and presupposes that the ecologist knows exactly what is the species composition of the community – a very difficult question for most plant ecologists!

The Shannon diversity index is calculated from the formula:

$$\text{Diversity} H' = - \sum_{i=1}^{s} p_i \ln p_i \tag{4.16}$$

where s is the number of species, p_i is the proportion of individuals or the abundance of the ith species, and $\ln = \log \text{base}_n$.

Any base of logarithms may be taken, with $\log_2$ and $\log_{10}$ being the most popular choices. The choice of log base must be kept constant when comparing diversity between samples. Values of the index usually lie between 1.5 and 3.5, although in exceptional cases, the value can exceed 4.5. If a sample is used, then the true value of p_i is unknown but is estimated as n_i/N, the maximum likelihood estimator (Pielou, 1969).

It is also possible to calculate an equitability or evenness index (E) of the form:

$$E = \frac{H'}{H'_{\text{max}}} = - \frac{\sum_{i=1}^{s} p_i \ln p_i}{\ln s} \tag{4.17}$$

where s is the number of species, p_i is the proportion of individuals or the abundance of the ith species expressed as a proportion of total cover, and $\ln = \log \text{base}_n$.

As an example of the use of the index, two sample quadrats from the Gutter Tor, Dartmoor, data-set (Table 4.4) using $\ln$ are taken.

Quadrat 1 – 6 species present

Species	% cover	Proportion of total cover (p_i)	$\ln p_i$	$p_i \ln p_i$
Carex nigra	15	0.0955	−2.3482	−0.2243
Drosera rotundifolia	2	0.0127	−4.3631	−0.0556
Juncus effusus	5	0.0318	−3.4468	−0.1098
Molinia caerulea	35	0.2229	−1.5009	−0.3346
Narthecium ossifragum	20	0.1274	−2.0605	−0.2625
Sphagnum sp.	80	0.5096	−0.6742	−0.3436
Total cover (%)	157		$H' = 1.330$	

The index is best calculated by drawing up a table as above. The diversity of quadrat 1 (H') = 1.330. The formula for the Shannon index commences with a negative sign to cancel out the minus created when taking logarithms of the proportions.

Quadrat 25 – 8 species present

Species	% cover	Proportion of total cover (p_i)	$\ln p_i$	$p_i \ln p_i$
Calluna vulgaris	2	0.0139	−4.2767	−0.0594
Carex nigra	10	0.0694	−2.6672	−0.1852
Drosera rotundifolia	2	0.0139	−4.2767	−0.0594
Juncus effusus	20	0.1389	−1.9741	−0.2742
Molinia caerulea	10	0.0694	−2.6672	−0.1852
Narthecium ossifragum	5	0.0347	−3.3604	−0.1167
Sphagnum sp.	90	0.6250	−0.4700	−0.2938
Trichophorum cespitosum	5	0.0347	−3.3604	−0.1167
Total cover (%)	144		$H' = 1.291$	

The diversity of quadrat 25 (H') = 1.291. The equitability or evenness of the two quadrats can now be calculated using the formula above:

$$\text{Quadrat 1} \quad E = \frac{H'}{\ln s} = \frac{1.330}{\ln 6} = \frac{1.330}{1.792} = 0.742$$

$$\text{Quadrat 25} \quad E = \frac{H'}{\ln s} = \frac{1.291}{\ln 8} = \frac{1.291}{2.079} = 0.621$$

The higher the value of E, the more even the species are in their distribution within the quadrat. Thus quadrat 1 has a more even distribution than quadrat 25. On the basis of the index, quadrat 1 is more diverse than quadrat 25, even though quadrat 25 has 8 species as opposed to 6. This demonstrates the manner in which evenness as well as species richness are combined in the index.

Beilsel and Moreteau (1997) presented a simple formula for calculating the lower limit of the Shannon index, while Hutcheson (1970) and Magurran (2004) described a method for calculating the variance of H' and a method of calculating a t-value to test for significant differences between quadrats or samples.

McIntosh's diversity index (U)

McIntosh (1967a) presented an index (U) with the form:

$$U = \sqrt{\sum_{i=1}^{s} n_i^2} \tag{4.18}$$

where U is the McIntosh diversity index, s is the number of species in the quadrat/sample, and n is the number of individuals or the abundance of the ith species in the quadrat/sample

This index is related to the concept of Euclidean distance discussed earlier. The resulting form is dependent upon the total abundance of the quadrat sample (N), and it is possible to calculate an index of dominance (D) allowing for N (Pielou, 1975) as follows:

$$D = \frac{N - U}{N - \sqrt{N}} \tag{4.19}$$

and a measure of evenness (E) from:

$$E = \frac{N - U}{N - N/\sqrt{s}}$$

(4.20)

The calculation of the index for quadrat 1 of the Gutter Tor data is shown below:
Quadrat 1 – 6 species present

Species	n_i	n_i^2
Carex nigra	15	225
Drosera rotundifolia	2	4
Juncus effusus	5	25
Molinia caerulea	35	1225
Narthecium ossifragum	20	400
Sphagnum sp.	80	6400
Total cover (%)	157	

$$\sum_{i=1}^{S} n_i^2 = 8279$$

$$U = \sqrt{\sum_{i=1}^{S} n_i^2} = \sqrt{8279} = 90.99$$

For quadrat 25 – 8 species present

Species	n_i	n_i^2
Calluna vulgaris	2	4
Carex nigra	10	100
Drosera rotundifolia	2	4
Juncus effusus	20	400
Molinia caerulea	10	100
Narthecium ossifragum	5	25
Sphagnum sp.	90	8100
Trichophorum cespitosum	5	25
Total cover (%)	144	

$$\sum_{i=1}^{S} n_i^2 = 8758$$

$$U = \sqrt{\sum_{i=1}^{S} n_i^2} = \sqrt{8758} = 93.58$$

Account must be taken of total cover (sample size) and the dominance measure (D) and evenness index (E) are calculated as follows:

For Quadrat 1:

$$D = \frac{N - U}{N - \sqrt{N}} = \frac{157 - 90.99}{157 - \sqrt{157}} = \frac{66.01}{144.47} = 0.457$$

and a measure of evenness (E) from:

$$E = \frac{N - U}{N - N/\sqrt{s}} = \frac{157 - 90.99}{157 - 157\sqrt{6}} = \frac{66.01}{92.91} = 0.711$$

For Quadrat 25:

$$D = \frac{N - U}{N - \sqrt{N}} = \frac{144 - 93.58}{144 - \sqrt{144}} = \frac{50.42}{132.00} = 0.382$$

and a measure of evenness (E) from:

$$E = \frac{N - U}{N - N/\sqrt{s}} = \frac{144 - 93.58}{144 - 144/\sqrt{8}} = \frac{50.42}{93.09} = 0.542$$

The lower the dominance (D) value, the more uneven the species distribution is in the quadrat. The higher the evenness (E) value, the more even the distribution of species is in the quadrat. Thus quadrat 1 has a more even distribution than quadrat 25.

Evenness (equitability) indices

In addition to the evenness indices linked to the Simpson, Shannon and McIntosh diversity measures, various other evenness indices exist. These are well summarised in Smith and Wilson (1996) and they also introduce their own new measure (E_{var}). This has the form:

$$E_{var} = 1 - \left[\frac{2}{\pi \arctan \left\{ \sum_{i=1}^{S} \left(\ln n_i - \sum_{j=1}^{S} \ln n_j/S \right)^2 /S \right\}} \right] \tag{4.21}$$

where n_i is the number of individuals in species i, n_j is the number of individuals in species j, and S is the total number of species.

The index measures the variance of species abundances, dividing this variance by log abundance which gives proportional differences and makes the index independent of measurement units (Magurran, 2004).

New and updated diversity indices and measures of diversity and evenness continue to be produced (Heltshe and Forrester, 1983a,b, 1985; Molinari, 1989, 1996; Bulla, 1994; Champely and Chessel, 2002; Ricotta, 2004; Gorlick, 2006; Gregorius and Gillet, 2008). Various researchers provide guidelines on the selection of indices (Cousins, 1991; Harper and Hawkesworth, 1994; Smith and Wilson, 1996 [evenness only]; Guo and Rundel, 1997; Purvis and Hector, 2000; Gotellli and Colwell, 2001; Magurran, 2004).

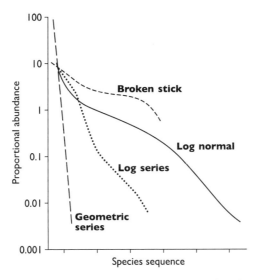

Figure 4.6   Dominance-diversity curves (rank abundance plots) showing the typical form of four species abundance models: geometric series; log series; log normal; and broken stick. In these graphs, the abundance of each species is plotted on a logarithmic scale from the most abundant to least abundant species. Redrawn from Magurran (2004) with kind permission of Wiley-Blackwell Publishing.

Dominance–diversity curves (rank abundance diagrams)

The application of single-figure diversity indices to characterise complex community structure can be criticised because so much of the original species information is lost. In consequence, various workers, notably Whittaker (1965, 1975), have plotted the graph of the proportional abundance of species in a sample or quadrat on a log scale against their rank from most to least abundant (Figure 4.6). The form of the resulting line or curve can be used to describe the evenness of species distribution and relative species dominance within a community. Dominance is the opposite of evenness. Various names such as 'geometric series', 'log series', 'log normal' and 'broken stick' derived from mathematical species abundance models are given to the resulting line (Figure 4.6). However, as noted by Wilson (1991a), the curves of the various models have rarely been fitted to empirical data, and he remedied this by providing methods for fitting the various models to dominance–diversity curves.

As an example of the use of this method, Hutchings (1983) plotted dominance–diversity curves for species within chalk grassland at a range of downland sites in southeast England (Figure 4.7). His aim was to study aspect and seasonality in relation to dominance and diversity within the plant communities. Figure 4.7 shows the dominance–diversity curves for south- and west-facing slopes. The steepest curves, indicating dominance and unevenness, were in July for north- and south-facing slopes and January for east- and west-facing slopes. His results are interesting, in that, in addition to showing how aspect affects dominance, dominance itself also changes seasonally.

Whittaker (1965) showed how these curves may be interpreted through the concept of the species niche. The term niche refers to the position of the species within the community, its position in vertical (above ground and below ground) space, horizontal space (internal mosaics and patterns within the plant community), community functional relationships (such as trophic structure and feeding relationships) and seasonal and daily variations in the species and their interactions with

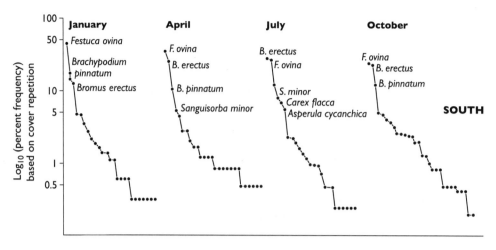

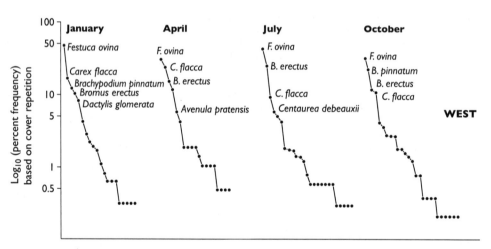

Figure 4.7 Dominance–diversity curves for chalk grassland sites of different aspects, showing how dominance and diversity vary over the year (Hutchings, 1983; redrawn with kind permission of Kluwer Academic Publishers).

other species. Dominance–diversity curves are thus a means of displaying how the resources of the local area where the plants are growing are partitioned amongst the species.

Functional diversity

Tilman (2001: p. 109) defined functional diversity as 'those components of diversity that influence how an ecosystem operates or functions'. The commonly observed relationship between ecosystem functioning and species richness can be ascribed to the larger number of functional groups found in richer assemblages (Diaz and Cabido, 1997, Tilman, 1999, 2001; Loreau *et al.*, 2002; Petchey and Gaston, 2002, 2006, 2007, 2009). In the context of vegetation, methods for measuring functional diversity are based on the ideas of plant functional types and grouping of species into types on the basis of selected traits or characteristics as described in Chapter 3. However, in this case, the traits are those that relate most closely to ecosystem function, for example leaf and canopy characteristics that link to photosynthesis and productivity. Like many aspects of the description of diversity, the topic is controversial and the exact definition of functional diversity is still a matter of debate (Mason *et al.*, 2005). Selection of which traits to use to define functional groups is a major issue (Petchey *et al.*, 2004), as is the number of species that need to have their traits measured (Pakeman and Quested, 2007). The methods of numerical classification described in Chapter 8 are used to define groups, but have required modification and adaption (Petchey and Gaston, 2007, 2009; Mouchet *et al.*, 2008). Once functional groups have been defined in terms of traits, vegetation samples are re-described in terms of the abundance of species within each of those functional type groups (Mouillot *et al.*, 2007; Laliberté and Legendre, 2010). Diversity and evenness measures can then be applied to either groups or on the trait information directly (Petchey and Gaston, 2006; Schmera *et al.*, 2009).

More attention is now being paid to functional diversity measures that are based on species traits, rather than on functional groups. Functional diversity has now been decomposed into three components: functional richness, functional evenness and functional divergence. Each of these components measures different aspects of the diversity of functional traits within a community (Mason *et al.*, 2005; Villéger *et al.* 2008). Mouchet *et al.* (2010) examined the utility of these measures against other possible diversity indices and they were shown to perform well in measuring individual aspects of functional diversity.

Perhaps because of the uncertainly over various aspects of its implementation, use of functional diversity in general vegetation description remains limited, although a good example is provided by Kuiters *et al.* (2009). Nevertheless, Lavorel *et al.* (2008) have provided a comprehensive evaluation of the whole approach and its use in the field, and research in this aspect of vegetation description clearly represents an important area for the future.

Interpretation of diversity indices

The Simpson, Shannon and McIntosh indices are based upon both species richness and evenness of species abundances. However, interpretation and comparison of indices from different samples and plant communities is often difficult. Despite arguments to the contrary, most interpretation of diversity indices is still based primarily on species richness rather than both richness and evenness (Spellerberg and Fedor, 2003). Separation of the evenness component of the index from the species richness component is not easy, particularly since in the case of the McIntosh index, for any given sample, the resulting index is a function of not only species richness (s) but also the overall abundance (N). For this reason, the Simpson and Shannon indices are often preferred because the species abundances are standardised to proportions (Keylock, 2005). Many other diversity indices exist, of which the most widely used are those of Berger-Parker and Brillouin. These and others are described with worked examples in Magurran (2004).

Debates over the ecological significance of diversity – diversity and ecosystem stability

Arguments over what the ecological significance of high as opposed to low diversity actually means still abound. Most of these arguments assume that diversity is equated solely with species richness and take no account of the relative species abundances and evenness. Most ecologists consider high species richness to be a desirable property of any community or ecosystem, and this criterion has dominated many attempts at ecological and conservation assessment (Usher, 1986; Spellerberg, 1992; Gove *et al.*, 1996; Huston, 2003; Gaston and Spicer, 2004). However, Magurran (1988) showed clearly how the Shannon diversity index was of limited value in assessing ten woodlands in Northern Ireland for their potential as nature reserves. The two woodlands that actually were nature reserves were bottom of the list in terms of diversity indices calculated on their ground floras. The reason for this was that the vegetation of these two woods was far more characteristic of the area in general habitat terms, and they were also two of the largest sites. Magurran made the point that diversity is only one of the various factors that may be used to select sites for conservation protection.

The debate on the relationship between diversity and stability continues (Goodman, 1975; Walker, 1989; Tilman, 1996; Tilman *et al.*, 1998; McCann, 2000). It has been argued that the more species-rich a community, the greater the 'portfolio' or 'insurance' effect. In other words, where there are many species, replacement of one species and its function in both community and ecosystem by another is more likely (Doak *et al.*, 1998; Lepš, 2005). There is no automatic relationship between high diversity or species richness and ecosystem/community stability. Many species-poor communities are extremely stable, for example some heather moorland in Britain, while many species-rich communities show considerable instability, for example, chalk grassland communities in southeast England. Many years ago, Pielou (1975) discussed this problem at length and showed that ecosystem and community stability is ultimately dependent on environmental stability. A stable environment will often encourage stability in the biota of a community or ecosystem, and over time, depending on factors such as position on a world scale and speciation (rate of production of new species through evolution), species diversity *may* increase. Even this argument needs to be treated with care. Tropical rainforest communities have the highest species diversities of all ecosystems and although this has often been attributed in part to the absence of glaciation in tropical regions during the Quaternary era, research has increasingly demonstrated that the tropics did undergo substantial climatic and environmental perturbation in that period (Hill and Hill, 2001). Prance (1977, 1978, 1996), working in the Amazonian Basin, showed that the old 'heartland' regions to which the forest retreated during the times of extreme climatic change in the Pleistocene are still the most diverse. However, although they were relatively the most 'stable' areas, they, and all other areas of tropical rain and moist forests such as in southeast Asia, still underwent considerable climatic and environmental change, which may also have promoted their diversity (Whitmore, 1998; Hill and Hill, 2001; Sodhi and Brook, 2006; Primack and Corlett, 2011).

Useful perspectives on complexity, diversity and stability are also presented by Kikkawa (1986), Rosenzweig (1995) and Lepš (2005), who stress that the difference between species richness and diversity in one component of an ecosystem, such as the plant community, is a very different concept from, for example, food web diversity, which defines the number of feeding links within and between all trophic levels in the same ecosystem. More recent discussion is to be found in Tilman and Pacala (1993), Huston (2003) and Lepš (2005).

Shmida and Wilson (1985) and Auerbach and Shmida (1987) presented one of the most interesting reviews of the factors affecting species diversity at different scales (Figure 4.8). Most discussion of diversity is at the local community scale, described by Shmida and Wilson as niche relations (Whittaker's alpha diversity). However, overall diversity is increased much more at the next scale of habitat diversity (Whittaker's beta diversity) and is further accentuated by mass effects, which represent the flow of propagules of individuals from areas of high diversity (core areas) to unfavourable areas, where they achieve viability. These marginal species from adjacent areas around one particular habitat or community type can greatly increase species diversity. This effect also explains the high species diversities of many transitional zones or ecotones between

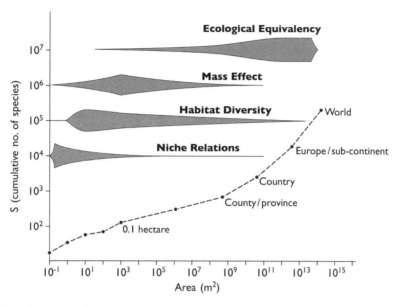

Figure 4.8 Biological mechanisms that determine species diversity (Shmida and Wilson, 1985; redrawn with kind permission of Wiley-Blackwell Publishing).

differing plant communities or habitat types (Kent *et al.*, 1997). Finally, at the large scale, diversity is increased by ecological equivalency. In one part of the world, a certain collection of plants will have evolved and comprise a particular community. Elsewhere in the world, in a very similar environment, an equivalent but totally different set of species has evolved to make up the communities in that region. Thus diversity is maintained simply by the spatial separation of the two locations and their very different species compositions. Problems now occur when species from one of these environments are transported by humans into the other similar environment. A good example is *Rhododendron*, of which many species have been brought to the British Isles for horticultural purposes but one species of which, in particular, *Rhododendron ponticum*, is now seen as a pest in natural communities such as woodland and moorland, having 'escaped' from suburban gardens. It was introduced from the Mediterranean in the mid-eighteenth century. Once established, *Rhododendron ponticum* has the ability to grow very rapidly and effectively sterilise the ground beneath itself, greatly reducing the diversity of the former community.

The development of the subject of macroecology (Gaston, 2003; Storch and Gaston, 2004; Blackburn and Gaston, 2004, 2006; Kent, 2005, 2007b) has also provided valuable insights into patterns and causes of variation in species diversity at regional and continental scales. Grubb (1987) demonstrated the importance of latitude in determining species richness, and this is an important theme that has been taken up by a number of macroecological researchers (Hawkins and Agrawal, 2005). Aspects of scale in relation to methodology are particularly important in such work, as has been reviewed by Beever *et al.* (2006).

Traditionally, most plant ecologists have collected and analysed data on alpha or species diversity. However, macroecologists have emphasised that diversity can also be described and analysed at the beta or habitat scale and at wide regional and continental scales (Koleff *et al.*, 2003). Whittaker (1960, 1972) proposed that beta diversity should be measured as the proportion by which the species richness of a region exceeds the average species richness of a single locality within that region. Another way of describing this is '...as the amount of turnover in species composition from one location to another' (Bacaro and Ricotta, 2007: p. 41). The result has been the derivation of a large number of indices of beta diversity based on the methodologies of both alpha diversity indices and coefficients of (dis)similarity. A total of 24 measures of beta diversity were reviewed by Koleff *et al.* (2003). New ones are still being developed, for example by Vellend (2001) and Bacaro and Ricotta (2007), who provide two of the comparatively few examples of their application to vegetation data, and emphasise the importance of accounting for spatial structure and spatial autocorrelation (see Chapter 5) in the data. Stohlgren *et al.* (1997) demonstrated the application of diversity assessment at the beta or landscape scale.

The relative contribution of commonness and rarity to patterns of species richness represents another interesting aspect in the attempts to explain variations in diversity. Using bird species, Lennon *et al.* (2004) found that across a range of spatial scales, a set of common species are likely to be better indicators of low species richness than a similar-sized set of rare species are to be of high species richness. Thus commoner species contribute more to species richness than rare ones because patterns of richness should be seen as being characterised by where common species are absent, rather than where rare species are present. Linked to this idea, Lennon *et al.* (2011), using studies on machair vegetation in Scotland, at a very detailed scale, have shown that common species often tend to underlie species richness–environmental correlations, and provide explanations for such patterns more than rare species. This result has important implications for many biodiversity surveys for conservation, where the focus has almost always been primarily on the documentation of rare species distributions (Austin, 1999b). However, Araújo *et al.* (2001), using higher plants and terrestrial invertebrates, have shown at the European continental scale that environmental diversity is frequently a poor predictor of species diversity.

One final area where the analysis of species richness and diversity data has become important is in monitoring the effects of climate change on species composition (Yoccoz *et al.*, 2001, 2003; Magurran, 2004). Buckland *et al.* (2005) have reviewed the potential use of diversity indices and associated problems in monitoring species population in this context. An example of a survey monitoring change in vegetation in Britain in response to climate and other human impact effects is presented in Bunce *et al.* (1999a,b).

The purpose of this chapter has been to introduce the basic nature of vegetation data and ideas of comparing samples in terms of floristic composition using similarity and dissimilarity indices. These concepts underlie many of the numerical methods for analysing vegetation data presented in Chapters 6 and 8. Ideas of diversity and species richness have also been introduced. Again, these are important in assisting with interpretation of the results obtained from those more complex methods described in later chapters.

Case studies

Simpson and Shannon diversity indices as measures of species richness and evenness in the early stages of succession of plant species on the volcanic island of Surtsey (Fridriksson, 1975, 1987, 1989)

The volcanic island of Surtsey appeared in the Atlantic, south of Iceland, following a volcanic eruption in 1965. Between 1965 and 1980, annual records of the numbers of individuals of each plant species were made. The changing pattern of diversity can be examined by calculating the Shannon and Simpson diversity indices for each year. Table 4.7 shows the raw data and the calculated Simpson and Shannon diversity indices

Table 4.7 Calculation of diversity indices for colonising vascular plants on the island of Surtsey between 1965 and 1980, together with additional data for 1988 (extracted from Fridriksson, 1975, 1989).

	1965	1966	1967	1968	1969	1970	1971	1972	1973	1974	1975	1976	1977	1978	1979	1980	1988
Cakile arctica	23	5	22	0	2	0	0	1	33	3	5	0	1	0	1	1	0
Elymus arenarius	0	2	4	6	5	4	3	0	66	26	12	10	8	14	5	5	1000
Honckenya peploides	0	0	24	103	52	63	52	71	548	857	428	500	632	3080	24000	50000	∞
Mertensia maritima	0	0	1	4	0	0	0	15	25	44	11	6	3	9	8	7	400
Cochlearia officinalis	0	0	0	0	4	30	21	98	586	372	863	501	286	160	91	75	25
Stellaria media	0	0	0	0	0	0	2	2	1	0	0	0	0	0	0	0	2
Cystopteris frogilis	0	0	0	0	0	0	3	4	3	3	2	2	2	9	5	5	1
Angelica archangelica	0	0	0	0	0	0	0	2	2	0	0	0	0	0	0	0	0
Carex maritima	0	0	0	0	0	0	0	1	1	1	0	2	1	5	2	1	2
Puccinellia retroflexa	0	0	0	0	0	0	0	2	1	9	8	8	2	6	40	7	30
Triplospermum maritimum	0	0	0	0	0	0	0	1	5	2	2	2	1	4	1	1	1
Festuca rubra	0	0	0	0	0	0	0	0	1	1	2	1	1	5	3	3	2
Cerastium fontanum	0	0	0	0	0	0	0	0	0	0	106	99	19	6	97	150	30
Equisetum arvense	0	0	0	0	0	0	0	0	0	0	2	0	0	0	0	0	0
Silene vulgaris	0	0	0	0	0	0	0	0	0	0	0	0	0	0	0	0	0
Juncus sp.	0	0	0	0	0	0	0	0	0	0	1	0	0	0	0	0	0
Atriplex patula	0	0	0	0	0	0	0	0	0	0	1	0	1	0	0	0	0
Rumex acetosello	0	0	0	0	0	0	0	0	0	0	0	0	0	124	31	40	500
Cardaminopsis petraea	0	0	0	0	0	0	0	0	0	0	0	0	0	5	6	8	25
Poa pratensis	0	0	0	0	0	0	0	0	0	0	1	0	0	0	0	0	25
Sagina selaginoides	0	0	0	0	0	0	0	0	0	0	1	0	0	0	0	0	1000
Armeria maritima	0	0	0	0	0	0	0	0	0	0	0	0	0	0	0	0	1
Poa annua	0	0	0	0	0	0	0	0	0	0	0	0	0	0	0	0	15
Agrostis stolonifera	0	0	0	0	0	0	0	0	0	0	0	0	0	0	0	0	1
Unidentified plants	0	0	0	1	0	4	2	1	1	2	0	1	0	0	0	0	0
Total number of plants	23	5	51	114	63	105	83	198	1273	1320	1448	1132	962	3427	24290	50303	∞
Species richness	1	2	4	4	4	5	6	11	13	11	16	11	12	12	13	13	18
Shannon diversity [H']	0	0.5	0.994	0.406	0.644	1.038	1.06	1.236	1.12	0.92	1.054	1.085	0.855	0.471	0.085	0.046	
Evenness (E)	0	0.722	0.717	0.293	0.465	0.645	0.592	0.516	0.436	0.384	0.38	0.453	0.344	0.19	0.033	0.018	
Simpson diversity (D)	0	0.32	0.536	0.18	0.307	0.554	0.54	0.62	0.599	0.498	0.552	0.601	0.48	0.189	0.024	0.012	

over that period. In the early period, although it is clear that species richness increased, the indices are variable. They nevertheless show a definite increase, particularly in the middle of the study period, declining once again towards the end. This is presumably closely related to the increase of species richness, but moderated towards the end by changes in evenness and dominance. The evenness figures (E) are even more variable, with a trend towards greater unevenness as time progresses (the lower the value of E, the more uneven the species distribution). This is partly a response to the greater number of species in later years, but also illustrates the problems of interpretation discussed earlier in this chapter. The value of the extra information obtained from the values for the diversity index and the evenness statistic over and above the simple statistics on species richness could be said to be questionable.

By 1980 and 1988, the number of species and individuals had changed dramatically. Further new species had arrived, while some early colonisers had gone. The most spectacular change was the explosive growth of *Honckenya peploides* from just 548 plants in 1973 to over 50,000 in 1980. By 1988, even an estimate of the number of plants was impossible. Since the diversity indices are based on counts of the number of individuals, after 1980 the calculation of diversity indices became impossible. Another excellent example of this type of research is presented in Whittaker *et al.* (1989), who examined early successional patterns on the volcanic islands of Krakatau in Indonesia.

Calculations of the Shannon and Simpson diversity index values for 1973 are as follows:

Species	Number of individuals in 1973	Proportion (p_i)	$\ln p_i$	$p_i \ln p_i$	p_i^2
Cakile arctica	33	0.0259	−3.6526	−0.0947	0.000672
Elymus arenarius	66	0.0518	−2.9595	−0.1534	0.002688
Honckenya peploides	548	0.4305	−0.8428	−0.3628	0.185312
Mertensia maritima	25	0.0196	−3.9303	−0.0772	0.000386
Cochlearia officinalis	586	0.4603	−0.7758	−0.3571	0.211904
Stellaria media	1	0.0008	−7.1491	−0.0056	6.17E-07
Cystopteris fragilis	3	0.0023	−6.0505	−0.0143	5.55E-06
Angelia archangelica	2	0.0016	−6.4560	−0.0101	2.47E-06
Carex maritima	1	0.0008	−7.1491	−0.0056	6.17E-07
Puccinellia retroflexa	1	0.0008	−7.1491	−0.0056	6.17E-07
Triplospermum maritimum	5	0.0039	−5.5397	−0.0218	1.54E-05
Festuca rubra	1	0.0008	−7.1491	−0.0056	6.17E-07
Unidentified plants	1	0.0008	−7.1491	−0.0056	6.17E-07
Total	1273	1.0000		−1.1195	0.4010

Species richness (s) = 13
Shannon diversity (H') = 1.1195
Shannon (E) = (H'/ln s) = 1.1195/2.5649 = 0.4365
Simpson diversity (1.0 − D) = 1.0 − 0.4010 = 0.5990

Dominance–diversity curves and the study of urban grassland (Wathern, 1976)

An increasingly important part of landscape architecture and design involves the sowing of grassland using standard seed mixes. Building projects, such as new housing schemes, industrial estates and road and motorway construction, often require large areas to be reinstated and resown following the total removal of the previous vegetation cover and ecosystem. In Britain, usually a standard seed mixture, dominated by grasses such as *Lolium perenne* (perennial ryegrass), *Agrostis capillaris* (common bent), *Festuca* species (fescues) and *Poa* species (meadow grasses), is sown, perhaps accompanied by a legume, such as *Trifolium* species (clovers) to encourage nitrogen fixation in the soil. Such swards are inexpensive to produce and large areas can be sown rapidly. These reinstated areas offer considerable potential for wildlife conservation. One of the most obvious possibilities is the opportunity to increase the species diversity within these simplified swards. This will happen naturally over time, but success depends upon many factors such as management practice, fertiliser treatment and the supply of natural propagules of potential new colonisers.

Wathern (1976) studied 69 grasslands from 17 sites in the city of Sheffield, Yorkshire, ranging in age from 1 to approximately 200 years. At each site, a number of random 1 m × 1 m quadrats were placed to sample the range of grasslands present. Within each quadrat, rooting frequency was recorded using 10 × 10 subdivisions. As part of a larger study of the structure of these communities, Wathern used dominance–diversity curves to demonstrate differences between the various grassland types. Three types of grassland were identified in the survey and dominance–diversity curves were drawn from samples of each (Figure 4.9). They show very clearly the simplified nature of the recently created swards (Figure 4.9a), containing between 5–8 species, compared with the 12–16 species in the old neutral grassland (Figure 4.9b). The curves for the recent swards are very much steeper, indicating the one or two dominant sown species, together with only a few others. In contrast, those for the old neutral grassland show a much flatter curve, indicating a more even distribution of species as well as greater species richness. Interestingly, however, some of the old acidic grasslands that were also included in the study had curves which were much closer to those of the recent swards (Figure 4.9c).

These graphs demonstrate two important points. Firstly, a number of factors control grassland composition and diversity, but nutrient status is one of the most critical of all. Secondly, although the most obvious characteristic of sown swards is their relatively low species diversity and unevenness, as shown by the steep dominance–diversity curves, some natural communities exhibit the same properties to an even more marked degree. Thus in terms of management and conservation goals, diversification of all swards is not necessarily always desirable. Careful consideration has to be given to the local environment, particularly nutrient status, and to the management strategies that will or will not be employed once the grassland is established.

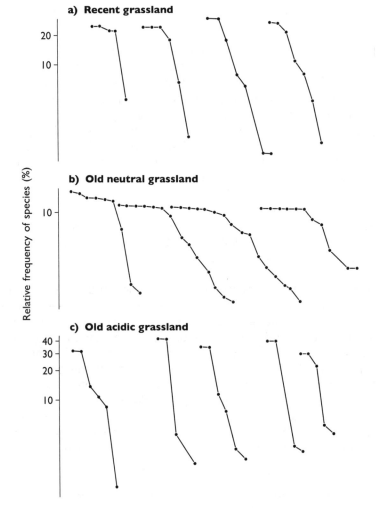

Figure 4.9 Dominance–diversity curves for town/urban grasslands in Sheffield, Yorkshire (Wathern, 1976). Reproduced with permission from John Wiley & Sons, Inc.

Chapter 5
Basic statistical methods for understanding multivariate analysis

INTRODUCTION

The aim of this chapter is to introduce and stress the importance of a range of basic statistical techniques that underpin the multivariate methods that are presented later to explore and analyse both vegetation and environmental/biotic data. Ultimately, the application of all statistical methods is part of problem-solving, and a valuable introduction to ideas of problem-solving is presented in Chatfield (1995). Rather than repeat information that is now presented in many introductory statistical textbooks in ecology and biology, key concepts and methods will be introduced, but for detailed understanding the reader will need to consult highlighted texts. Beginning at the more elementary level, the texts by van Emden (2008) and Dytham (2010) provide an excellent starting point, with more sophisticated and advanced treatment in Waite (2000), Quinn and Keough (2002) and Gotelli and Ellison (2004).

CLASSICAL (CONFIRMATORY), EXPLORATORY, AND BAYESIAN DATA ANALYSIS

Most statisticians now believe that it is helpful to make a distinction between exploratory data analysis (EDA), classical (confirmatory) (CDA) and Bayesian data analysis (BDA) (Chatfield, 1995; Quinn and Keough, 2002; Gotelli and Ellison, 2004; McCarthy, 2007; Kéry, 2010). The three approaches start in a similar manner but differ in the ordering of the steps necessary to reach a conclusion:

For exploratory data analysis, the steps are:

Problem ▸ *Data* ▸ *Analysis* ▸ *Model* ▸ *Conclusions*

For classical (confirmatory) analysis, the steps are:

Problem ▸ *Data* ▸ *Model* ▸ *Analysis* ▸ *Conclusions*

Vegetation Description and Data Analysis: A Practical Approach, Second Edition. Martin Kent.
© 2012 John Wiley & Sons, Ltd. Published 2012 by John Wiley & Sons, Ltd.

For Bayesian analysis, the steps are:

Problem ▶ *Data* ▶ *Model* ▶ *Prior Distribution* ▶ *Analysis* ▶ *Conclusions*

EXPLORATORY DATA ANALYSIS (EDA)

Exploratory data analysis (EDA) is concerned with a completely new set of data about which there is little knowledge and which has not been subject to previous analysis. A classical or confirmatory analysis, however, is used to check for the presence or absence of phenomena observed in a previous analysis or to test hypotheses derived from existing results or established theories. In most reported scientific literature in biology and ecology, 'significant' results are derived from one-off exploratory data-sets, rather than from data-sets specifically collected for confirmatory analysis. Furthermore, it is also commonplace for a researcher to carry out exploratory data analysis followed by hypothesis generation and confirmatory analysis on the same data-set, having noted some very interesting feature of it in the exploratory analysis. As Nelder (1986) pointed out, there is too much emphasis on analysing single data-sets in isolation. Thus, exploratory data analysis is now seen as being of importance to the application of statistics in many areas of science.

Although much statistical analysis is concerned with hypothesis generation and testing, some reservations about the dominance of this approach have been expressed over the past 40 years (Tukey, 1969, 1977; Hoaglin, *et al.*, 1983; Sibley, 1987; Erickson and Nosanchuk, 1992; Marsh and Elliott, 2008). While rigorous statistical testing and confirmation is important, these authors make the point that a great deal of data analysis is exploratory in approach and is primarily concerned with the search for pattern and order in data. Sibley (1987) argued that the process of data exploration should be seen as circular, rather than as a series of steps. Also, it is a process of 'successive deepening'. Examination of a set of data by one form of preliminary analysis will lead to further refinement of ideas and hypotheses and will throw up new ideas and show patterns that had not previously been apparent. When these patterns are explored further, still more ideas are generated. Analogies have been made with peeling the layers off an onion, with each successive layer increasing understanding about the patterns and properties of the data.

Given this basic idea of EDA, a whole set of 'robust' techniques have been evolved which are designed to explore rather than rigorously test data. An important aspect of these techniques is that, in common with many non-parametric methods, they are much less rigorous in their assumptions in terms of both sampling and statistical properties. This is because of the removal of the need for confirmatory analysis based on hypothesis generation and testing and the probabilistic approach.

Exploratory data analysis as description and induction

As explained in Chapter 1, a large amount of work in vegetation science is inductive and descriptive in approach and is concerned with the examination of variation within a set of data and the search for patterns and trends (Figure 1.6a). Thus when collecting and analysing floristic data for the purpose of defining plant communities, there will often be no clear idea at the outset of exactly what the result will be. Instead, the aim is simply to show how vegetation varies from place to place or within a certain area or region, to make an initial assessment of primary environmental/biotic gradients and to present these data in summary form. All the multivariate methods of ordination and classification presented in Chapters 6 to 8 are often used in this inductive manner, and in many situations can also be seen as forms of exploratory data analysis. Once variation has been summarised and pattern displayed, then hypothesis generation can occur.

Quinn and Keough (2002) stressed that much EDA involves graphical analysis of data as part of an initial screening process. These include:

- Histograms
- Dotplots (Cleveland dotplots)
- Box and whisker plots
- Stem and leaf plots
- Scattergrams
- Data transformation
- Rank correlation
- Robust regression

Sources of further detail on the above are Kent and Coker (1992), Cleveland (1993), Waite (2000), Quinn and Keough (2002), Gotelli and Ellison (2004), and Dytham (2010).

The importance of performing exploratory data analysis in ecology has been emphasised by Zuur *et al.* (2010), and this paper should be essential reading for all vegetation scientists.

CLASSICAL (CONFIRMATORY) DATA ANALYSIS (CDA)

Hypothesis testing and inferential statistics

The kinds of questions that emerge from vegetation analysis and which lead to hypothesis generation often take the form of 'Why?' Why does the vegetation vary in the way that it does? Can a reasonable explanation be offered for any of the observed variations and differences? These then lead to a need for further understanding of the biological and ecological processes behind the observed variations. Once hypotheses have been generated, then the deductive approach to scientific enquiry is being applied and different approaches to data analysis have to be employed (Figure 1.6b). This will involve the use of inferential statistical analysis, where data are collected and analysed with the aim of proving or disproving a hypothesis. Inferential statistics are concerned with mathematical probabilities and in the context of scientific investigation involve a search for principles that have a degree of generality. Using inferential statistics enables the vegetation scientist to apply results taken from a small sample of reality to a much larger environment or population. Making generalisations from a sample to a population is therefore known as statistical inference and is a vital part of scientific method.

Hypotheses

A hypothesis can be defined as a preliminary explanation of observed facts or phenomena, which is usually then tested for its validity. Most hypotheses are either concerned with observed differences in plant species, vegetation or environmental factors at different locations in space or points in time, or else with relationships between phenomena – plant species, vegetation and environment, or between species or environmental factors themselves. In practice, hypotheses are stated with varying degrees of precision. Many statistical texts distinguish three forms of hypothesis:

(a) the research hypothesis or general ecological statement relating to plant species, vegetation and/or environment;
(b) the null hypothesis of no difference or no relationship (H_0);
(c) the alternative hypothesis stating the nature of the difference or the relationship in some detail (H_1).

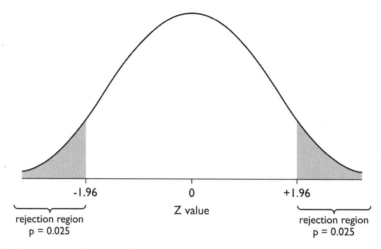

Figure 5.1   Rejection areas on the normal curve for the 0.05 significance level. Redrawn from Wheeler *et al.* (2004), with kind permission of David Fulton Publishers.

As a simple example, if, having made a preliminary study of the vegetation of an area in terms of both vegetation cover and its environmental controls, an idea emerges that the distribution of one species (X) is closely related to soil drainage quality, a general research hypothesis may then be formulated that is 'there is a relationship between the distribution of species X and the quantity of moisture in the underlying soil'. For all types of statistical analysis, the strategy is broadly similar. Next the hypothesis can then be stated in the negative, known as the null hypothesis (H_0). In the example, this would be that there is no relationship between species X and soil moisture. Then the alternative hypothesis is stated, usually in more detail than the original general research hypothesis, because it must be defined in relation to the particular statistical test being carried out.

The normal distribution and levels of significance – The sampling distribution of test statistics

A sampling distribution is a theoretical distribution, defined as the distribution that would result by randomly taking all possible samples from a specified population. With inferential statistical testing, if it was possible to repeat any test using different repeat samples a large number of times, each time with a test statistic being calculated, the distribution of all those test statistics could be plotted. The result would be a normal, Gaussian or unimodal distribution as shown in Figure 5.1. Obviously it is not possible or practicable to repeat a test using a large number of different repeat samples in a piece of real world research, so mathematicians have devised theorems that make use of known distributions, such as the normal distribution. Some distributions may be closer to other theoretical distributions, such as the Poisson distribution. These mathematically calculated distributions provide either a close approximation to the presumed distribution or, in some cases, an exact probability distribution.

The key point is that from a known probability distribution of a test statistic, it is possible to calculate the probability with which any value of the test statistic will occur. This characteristic underlies the whole basis of inferential statistics. Further details of the normal and Poisson distributions are given in most elementary textbooks (Waite, 2000; Quinn and Keough, 2002; Gotelli and Ellison, 2004; Dytham, 2010).

Directional and non-directional hypotheses

A further important aspect of hypothesis statements is whether the hypothesis is stated in a directional or non-directional manner. In the case of the soil moisture example above, a directional hypothesis states the anticipated direction of the relationship – perhaps that the abundance of species X is highest where soil moisture values are lowest. A non-directional hypothesis would not go as far, saying simply that some form of relationship exists between the two variables, the abundance of species X and soil moisture. Whether the alternative hypothesis is stated in directional or non-directional terms is very important in the interpretation of probability values at the end of the statistical analysis. Where the direction of the test has been stated, a one-tailed significance test is applied, which assesses the probability of the result in terms of just one end of the normal distribution shown in Figure 5.1 (probability $p = 0.025$). Where the hypothesis cannot be stated with such precision, then the relationship is non-directional, a two-tailed test is used and the probability of the result occurring in either of the tails of the distribution of Figure 5.1 is assessed (probability $p = 0.05$).

As explained above, each statistical method or test has its own statistic and associated probability distribution. From the latter it is possible to assess the probability of the test statistic in relation to a random variable: that is, what is the chance of that test figure occurring randomly or by chance? When the test statistic is shown to be unlikely to have occurred by chance, the null hypothesis can be rejected.

Usually, it can be assumed that the probability of the test statistic is also the probability of H_0 being correct. Probability values are usually expressed in the range 0–1 or 0–100%. Low probabilities of the test statistic, such as 0.01 or 1%, would imply that H_0 can be rejected, but a higher figure of 0.5 or 50% would mean that it should be accepted. In rejecting the null hypothesis H_0, it follows that the alternative hypothesis H_1 must be accepted. Before the calculation of any test statistic, the rejection levels, beyond which the probability of the null hypothesis being correct is unacceptable, must be decided. This has been a matter of tradition, with values of 0.01 (1%) and 0.05 (5%) typically being used. Figure 5.1 shows the rejection regions of the normal distribution for 0.05 (5%). These are also known as significance levels. Increasingly, these 'standard' significance levels are being questioned, however, and they are thought of as being too conservative (Quinn and Keough, 2002; Gotelli and Ellison, 2004; Zuur *et al.*, 2010). As suggested in Chapter 1, the advocates of Bayesian statistical analysis are particularly critical of the unquestioning acceptance of the 'standard' significance levels (Stephens *et al.*, 2005, 2007; Lukacs *et al.*, 2007; McCarthy, 2007).

Once a test statistic has been calculated, for most tests, if the value of the test statistic is greater than the critical value for a given confidence level, then there is only a small probability that the null hypothesis will be wrongly rejected and the rejection of the null hypothesis is justified. However, if the test statistic is less than the critical value for a given rejection level, then the probability that the null hypothesis will be wrongly rejected is high, the null hypothesis cannot be rejected and must be accepted.

Type I and Type II errors

Depending on the nature of the hypothesis being tested and the statistical test involved, it is possible to make one or other of two types of error in statistical testing:

A *Type I* error occurs when a null hypothesis that is true is rejected. This type of error can been seen as a 'false positive' and thus potentially serious. One situation where the risk of

Type I errors is increased is where spatial or temporal autocorrelation (see below) is present in the data (Lennon, 2000; Legendre *et al.*, 2002).

A *Type II* error occurs when there is a failure to reject the null hypothesis when it is untrue. This type of error is probably less serious than the Type I error but nevertheless must be avoided if possible. In the methods of multivariate analysis, particularly ordination, that are central to vegetation data analysis, a possible cause of Type II errors is the presence of collinearity (inter-correlation) between independent variables in multiple regression (see case study) or canonical correspondence analysis ordination (Chapter 6).

Degrees of freedom

Associated with all inferential statistical tests is the quantity known as degrees of freedom, which is defined as the number of observations in a particular statistical analysis that can take on any value but within the limitations imposed by any calculations on those values. As an example, suppose that the density of a plant species (X) has been measured in five different localities: a–e. The mean density is 36; then:

$$\text{If } X = 36; \text{ and } N = 5; \text{ then } \Sigma X = 36.0 \times 5 = 180$$

Given any four density values, such as for areas a–d: 75, 25, 50 and 20, then;

$$(a + b + c + d + e) = 180$$

$$(75 + 25 + 50 + +20 + e) = 180$$

$$e = 180 - (75 + 25 + 50 + 20) = 10$$

Thus, when $(N - 1)$ numbers are specified, the Nth is determined and the degrees of freedom in this case are the number of observations minus 1, $(N - 1)$. The basic rule in applying degrees of freedom is that one degree of freedom is lost for every fixed value. The fixed or known value in this case is the mean, and knowledge of the mean always enables the value of the last observation to be determined. Hence one degree of freedom was lost.

Sampling and inferential statistics

All inferential statistical tests make assumptions about the nature of the data being analysed and the manner in which they have been collected. Many of these assumptions relate to the idea of random sampling, particularly if parametric methods are being used (see below). Very rarely are statistical analyses based on the total population of individuals. Instead, they are based on the premise that a subset of the population can be measured and that this subset may exhibit the same properties as the total population from which it was drawn. With random sampling, each sample measurement should be independent of any other and on every sampling occasion, every individual should have an equal chance of being selected. If these principles are not followed, the sample becomes biased. In order to prevent such bias, the nature of the overall population must be known, but often problems arise in assessing how representative any sample is of its population. In general, the larger the sample, the more likely it is to be representative of the population from which it is drawn. Unfortunately, comparatively few vegetation data are collected at random (see Chapter 3), partly because of the practical difficulties involved in locating a truly random sample in the field and because of the time involved.

Spatial and temporal autocorrelation

This subject was introduced in Chapter 3. A further difficulty that affects the independence of samples is that of spatial autocorrelation. This is a difficult and complex subject which has attracted the attention of geographers and spatial analysts for many years. Although the basic concepts are fairly readily understood (Cliff and Ord, 1973, 1981; Legendre and Fortin, 1989; Koenig, 1999; Fortin and Dale, 2005), solutions to the problem are much more difficult. The problem is present in all spatial sampling (all vegetation sampling is, by definition, spatial) and is due to the inevitable relationships between points in space and, in particular, their proximity to each other. As an example, if four quadrats A–D are thrown along a transect at 50 m intervals, it is impossible for those samples to be completely independent of each other. The observed species composition of quadrat B will be affected by the composition of quadrat A simply because of its proximity in space to quadrat A. Similar relationships and varying degrees of influence will exist between the other quadrats because of their proximity to each other. Statisticians and geographers have begun to look at special forms of statistical analysis to examine this problem, and new methods which are beginning to be more widely applied are now available (Lennon, 2000; Legendre *et al.*, 2002; Fortin and Dale, 2005; Kent *et al.*, 2006). The same problem exists if samples are collected from one location but at successive points in time. This is known as temporal autocorrelation.

Vegetation scientists have often preferred to ignore these problems, since a great deal of their work is inductive and descriptive rather than deductive and does not involve the use of inferential statistics. Nevertheless, this does represent an important area for future research, and if plant ecologists are to be encouraged to apply more inferential statistical analysis and hypothesis testing, they will have to take greater regard of the problem (Fortin *et al.*, 1989; Legendre and Fortin, 1989; Gustafson, 1998; Sokal *et al.*, 1998a,b; Koenig, 1999; Fortin, 1999; Lennon, 2000; Mistral *et al.*, 2000; Fortin and Dale, 2005; Jetz *et al.*, 2005).

Parametric and non-parametric statistics

Parametric tests make certain assumptions about the background populations from which samples are drawn. Some of these assumptions may be relaxed by using non-parametric as opposed to parametric tests. The most important of these is that the background population is approximately normally distributed (Figure 5.1) and the smaller the sample size being tested, the more important it is that the background population approximates to normality. For many variables, it may not be reasonable to assume that the background population is normally distributed, and in many circumstances it is not possible to test this assumption by collecting a very large sample. Another assumption is that the variances of the variables involved are more or less equal. These problems can be overcome by using distribution-free (non-parametric) tests, which make no such assumptions about the distribution of the background population (Johnson, 1995; Smith, 1995). A number of the tests outlined or described in the remainder of this chapter are distribution-free and non-parametric.

Siegel (1956) was a psychologist and statistician who was one of those originally responsible for the widespread dissemination of non-parametric methods. In addition to the removal of the assumption of normality in the population distribution, he also listed other advantages of non-parametric techniques. They can deal with very small samples; can be used on a variety of measurement scales; suitable non-parametric tests exist for analysing samples drawn from several different populations; and finally, the methods are usually easier to understand and to apply than parametric ones. A further important concept is the power-efficiency of a non-parametric test. In statistical terms, the power of a test is related to its ability to state correctly whether a hypothesis is true or false. Parametric tests are usually more powerful than non-parametric, and the power of a test is influenced by the size of a sample. A non-parametric test of low power-efficiency requires a larger sample to achieve the same level of power as a parametric test with a relatively high power-efficiency ratio. As an example, the parametric method for testing differences between the means of two independent sets of observations or samples on one variable is the t-test. The non-parametric equivalent is the Mann-Whitney U test which is around 95% power-efficient, when compared with the t-test.

Descriptive statistics

Firstly, students and researchers should have a good understanding of measures of central tendency: the mean, mode and median. Secondly, a knowledge of measures of dispersion – the interquartile range, the variance, the standard deviation and the standard error of the mean, together with the coefficient of variation – is essential, in addition to the descriptive power of histograms and frequency tables. Good introductions to these topics are given in Townend (2002), Wheeler *et al*. (2004), van Emden (2008) and Dytham (2010), with more advanced discussion in Waite (2000), Quinn and Keough (2002) and Gotelli and Ellison (2004).

As discussed above, basic methods of exploratory data analysis are also important in data description, notably stem and leaf plots, dot plots and box plots (Tukey, 1977; Marsh and Elliott, 2008). Identification and examination of extreme data points or outliers is also very important (Zuur *et al.*, 2010).

Comparison of samples

Parametric and non-parametric tests for two samples

In plant ecology, it is often necessary to test whether two samples of the same phenomenon are derived from the same parent population. As an example, hypotheses may have been formulated that the abundance of one species of plant may differ significantly between two rock types or slope aspects, or the numbers of grazing animals may vary between two different plant community types. Data will then be collected on the plant species in the two environments or on the different slopes, or on the numbers of grazing animals on the two different vegetation types. Where the samples are independent of each other, then the t-test may be applied as a parametric test and the Mann-Whitney U test as a non-parametric test. Where the samples are not independent but paired, then the paired t-test is applied for data which are up to parametric standards and the Wilcoxon signed rank test for paired samples in the non-parametric case. The application of all these tests is fully described in Waite (2000), Quinn and Keough (2002), Gotelli and Ellison (2004), Wheeler *et al*. (2004), van Emden (2008) and Dytham (2010).

Analysis of variance

Where more than two sets of samples are being compared, then the simplest form of analysis of variance (ANOVA) – one-way analysis of variance – is used. However, beyond this, the whole field of analysis of variance is extensive and lies beyond the scope of this book. Use of ANOVA is closely linked to ideas of experimental design and the manner in which data and samples are collected. Once again, introductions to the simpler aspects of ANOVA are detailed in van Emden (2008) and Dytham (2010), with more thorough discussion in relation to experimental design in Waite

(2000), Quinn and Keough (2002) and Gotelli and Ellison (2004). A particularly exhaustive text on the whole subject is by Underwood (1997), which ranges widely over all aspects of sampling, experimental design and ANOVA in ecology. However, it is definitely not for beginners and is best suited to advanced researchers. A more straightforward general guide to the use of ANOVA is provided in Roberts and Russo (1999).

Correlation and regression analysis

An appreciation of methods for correlation and regression is very important both for classical (confirmatory) hypothesis testing and exploratory data analysis of floristic and environmental/biotic data, but is also crucial as a basis for understanding the more complex methods of ordination and classification described in Chapters 6 to 8. For this reason, it is covered in depth here.

Correlation analysis is a set of methods used to determine the strength of relationship between variables. The result of a correlation analysis is a statistic lying between -1.0 and $+1.0$ which describes the degree of relationship between the two variables. Regression analysis takes this a stage further by measuring and describing the form of the relationship between two variables and allowing prediction of values of one variable in terms of variation in the other. Correlation and regression are closely related techniques.

Both parametric and non-parametric methods exist for correlation and regression. For correlation, the major parametric method is Pearson's product-moment correlation coefficient, and for regression, the equivalent is the least-squares technique. However, non-parametric alternatives, which are often better suited to the quality of data generated by vegetation scientists, also exist. For correlation there are several possibilities, but the most widely used are Kendall's rank order correlation, known as Kendall's tau, and Spearman's rank correlation coefficient. For regression a simple alternative for describing the form of the regression relationship is the method of semi-averages, although it does not enable any significance testing to be carried out. A more useful non-parametric regression method still, which is derived from techniques of exploratory data analysis (Tukey, 1977; Marsh and Elliott, 2008), is the fitting of resistant lines.

Once again, the decision of whether to use parametric or non-parametric tests depends upon the quality of the data. In correlation, the Pearson product-moment correlation coefficient (r) is the most powerful, while Spearman's rank correlation coefficient (r_s) is only 91% as power efficient as Pearson's r. This means that if Pearson's r is significant in a sample of 100 cases taken from two normally distributed variables, it will require a sample of 110 cases of the same data to achieve the same level of significance from Spearman's rank order coefficient r_s.

Scattergrams

At the start of any correlation and regression analysis, it is always worthwhile plotting a graph or scattergram of the relationship between the two variables. Values of correlations vary from -1.0 through 0.0 to $+1.0$. The values $+1.0$ and -1.0 represent a perfect relationship between the two variables. Thus Figure 5.2a shows a perfect positive relationship ($r = +1.0$) between two variables X and Y, which means that an increase in the amount of X is directly matched by an increase in the amount of Y. Figure 5.2b shows a perfect negative relationship ($r = -1.0$), with an increase in the amount of X being matched exactly by a decrease in the amount of Y. It is also important to note that in either case, the relationship is linear. Figure 5.2c shows an example where there is no correlation. The closer a value is to $+1.0$ or -1.0, the stronger is the relationship and the correlation.

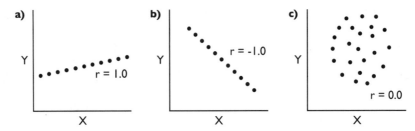

Figure 5.2 Scattergrams of (a) perfect positive correlation; (b) perfect negative correlation; and (c) no correlation between variables X and Y.

However, the fact that two variables have a strong relationship does not necessarily mean that one variable is causing the variation in the other. A causal relationship cannot be deduced from a correlation coefficient alone. Cause and effect can only be determined through other evidence and the judgement of the researcher.

Table 5.1 Calculation of the product-moment correlation coefficient for data on plant species richness and site age from 26 vacant urban lots in Chicago (Crowe, 1979).

1 Lot	2 Species numbers (Y_i)	3 Lot age (months) (X_i)	4 Y_i^2	5 X_i^2	6 $X_i Y_i$
1	11	3	121	9	33
2	9	3	81	9	27
3	16	7	256	49	112
4	27	15	729	225	405
5	21	18	441	324	378
6	32	20	1024	400	640
7	22	20	484	400	440
8	16	20	256	400	320
9	15	20	225	400	300
10	26	22	676	484	572
11	25	25	625	625	625
12	15	30	225	900	450
13	30	30	900	900	900
14	45	40	2025	1600	1800
15	22	50	484	2500	1100
16	56	65	3136	4225	3640
17	47	70	2209	4900	3290
18	20	70	400	4900	1400
19	45	80	2025	6400	3600
20	44	90	1936	8100	3960
21	54	100	2916	10000	5400
22	47	100	2209	10000	4700
23	37	100	1369	10000	3700
24	69	113	4761	12769	7797
25	46	120	2116	14400	5520
26	47	150	2209	22500	7050
	$\sum_{i=1}^{n} Y_i\ 844$	$\sum_{i=1}^{n} X_i\ 1381$	$\sum_{i=1}^{n} Y_i^2\ 33838$	$\sum_{i=1}^{n} X_i^2\ 117419$	$\sum_{i=1}^{n} X_i Y_i\ 58159$

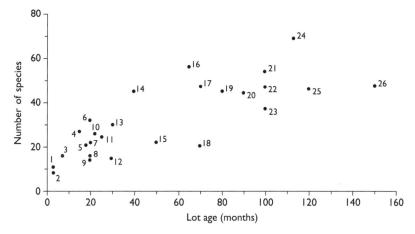

Figure 5.3 Scattergram of the two variables, species richness (Y) and abandoned urban lot age (X) in Chicago, USA. Redrawn from Crowe (1979) with kind permission of Wiley-Blackwell Publishers.

In order to demonstrate the use and calculation of correlation and regression analysis, an example of plant species richness in relation to site age is taken for 26 vacant urban lots in Chicago, USA (Crowe, 1979). Crowe recorded the number of plant species growing on abandoned urban plots ranging in age following abandonment from 3–150 months. The data are shown in columns 2 and 3 of Table 5.1. In Figure 5.3, these data are plotted as a scattergram. Inspection of this graph shows evidence of a possible positive linear relationship between the two variables.

The Pearson product-moment correlation coefficient

As with all parametric tests, the product-moment correlation coefficient is based on certain assumptions about the data to which it can be applied. Firstly, data must be continuous and measured on either interval or ratio scales. Secondly, the background populations of each set of data should fit the normal distribution (Figure 5.1). The computational formula for the product-moment correlation coefficient is as follows:

$$r = \frac{\sum\limits_{i=1}^{n} X_i Y_i - \left(\sum\limits_{i=1}^{n} X_i\right)\left(\sum\limits_{i=1}^{n} Y_i\right) \bigg/ n}{\sqrt{\left[\sum\limits_{i=1}^{n} X_i^2 - \left(\sum\limits_{i=1}^{n} X_i\right)^2 \bigg/ n\right]\left[\sum\limits_{i=1}^{n} Y_i^2 - \left(\sum\limits_{i=1}^{n} Y_i\right)^2 \bigg/ n\right]}} \tag{5.1}$$

where

　　　r = product–moment correlation coefficient
　　　n = the number of pairs of observations
$\sum\limits_{i=1}^{n} X_i$ = sum of observations on X

$\sum\limits_{i=1}^{n} Y_i$ = sum of observations on Y

Table 5.2 Calculation of Spearman's rank correlation coefficient between plant species richness and site age from 26 vacant urban lots in Chicago (Crowe, 1979).

1 Lot	2 Species numbers	3 Rank	4 Lot age (months)	5 Rank	6 Difference (d)	7 Difference^2 (d^2)
1	11	2.0	3	1.5	0.5	0.25
2	9	1.0	3	1.5	0.5	0.25
3	16	5.5	7	3.0	2.5	6.25
4	27	13.0	15	4.0	9.0	81.00
5	21	8.0	18	5.0	3.0	9.00
6	32	15.0	20	7.5	7.5	56.25
7	22	9.5	20	7.5	2.0	4.00
8	16	5.5	20	7.5	2.0	4.00
9	15	3.5	20	7.5	4.0	16.00
10	26	12.0	22	10.0	2.0	4.00
11	25	11.0	25	11.0	0.0	0.00
12	15	3.5	30	12.5	9.0	81.00
13	30	14.0	30	12.5	1.5	2.25
14	45	18.5	40	14.0	4.5	20.25
15	22	9.5	50	15.0	5.5	30.25
16	56	25.0	65	16.0	9.0	81.00
17	47	22.0	70	17.5	4.5	20.25
18	20	7.0	70	17.5	10.5	110.25
19	45	18.5	80	19.0	0.5	0.25
20	44	17.0	90	20.0	3.0	9.00
21	54	24.0	100	22.0	2.0	4.00
22	47	22.0	100	22.0	0.0	0.00
23	37	16.0	100	22.0	6.0	36.00
24	69	26.0	113	24.0	2.0	4.00
25	46	20.0	120	25.0	5.0	25.00
26	47	22.0	150	26.0	4.0	16.00
					$\sum_{i=1}^{n} d^2$	620.50

In the Chicago example, the general research hypothesis is that there is a relationship between species numbers and age of urban lot. The null hypothesis (H_0) is that there is no such relationship. The alternative directional hypothesis (H_1), derived from examination of Figure 5.3, is that there is a positive correlation between the two variables, with plant species richness increasing as lot age increases.

Using the above formula enables the calculation to be broken down into a number of simple steps, as shown in Table 5.1. Using the subtotals from the table, the following equation results:

$$r = \frac{58159 - (1381)(844)/26}{\sqrt{[117419 - (1381)^2/26]}\sqrt{[33838 - (844)^2/26]}}$$

$$r = \frac{58159 - 44829.38}{209.92 \quad 80.25} = \frac{13329.62}{16846.08} = 0.791$$

Thus the existence of a positive relationship between species richness and age is demonstrated with a correlation coefficient of 0.791. However, this must then be tested for significance.

Testing the significance of r

When sampling from bivariate populations, there is always the possibility that an entirely spurious correlation coefficient may be derived, particularly where small samples have been collected. The significance test is designed to calculate the probability that for the given sample size, the correlation coefficient could have been derived by chance. The test is based on the use of t-tables printed in many statistical textbooks. When using most present-day computer packages, the significance test and associated p value is usually calculated automatically. Since a directional alternative hypothesis (H_1) has been stated, the significance test is one-tailed.

The value of t is found by the formula:

$$t = r\sqrt{\frac{n-2}{1.0-r^2}} \tag{5.2}$$

thus

$$t = 0.791\sqrt{\frac{26-2}{1.0-0.626}} = 0.791\sqrt{\frac{24}{0.374}} = 6.336$$

Degrees of freedom for correlation coefficients are the number of pairs of observations less two ($n-2$). Reference to tables of Student's t shows that with 24 degrees of freedom, t must exceed 2.80 to be significant in a two-tailed test at the $p = 0.01$ level, and would have to exceed 2.49 in a one-tailed test. The calculated value was 6.336. The hypothesised positive relationship between the two variables is thus highly significant.

However, although this ecologically interesting relationship between plant species richness and age of site in Chicago has been found, it is important to be very careful in interpretation of the result and in assigning causality. Undoubtedly, the length of time during which colonisation has been possible is an important causal variable, but there must be others. A measure of how much of the variation in the species richness data is explained by the variation in lot age is obtained by squaring the correlation coefficient to give r^2 – in this case $r^2 = 0.791^2 = 0.63$. This means that 63% of the variation in species richness in the 26 urban lots is accounted for by variation in lot age. This value is also known as the coefficient of determination. However, 37% of the variation still remains unexplained and may be attributable to other possible factors, for example, lot size, isolation, substrate/geology or degree of human interference. The implications of this are discussed further in the case study at the end of the chapter.

Spearman's rank correlation coefficient

This coefficient, along with Kendall's tau (see below), is the most widely used non-parametric coefficient. The data from Chicago are again used to illustrate the calculation of Spearman's rank correlation coefficient. The following formula is used:

$$r_s = 1.0 - \frac{6\Sigma d^2}{n^3 - n} \tag{5.3}$$

where

    d = difference between paired ranks (see text)
    n = number of pairs of observations

For the Chicago example, the hypotheses are laid out in exactly the same way as for the product-moment coefficient and the calculation is shown in Table 5.2. The two variables are ranked separately and consistently (that is, from high to low or low to high) and each observation is shown by the ranked values in Table 5.2. Where two or more values have the same value, they are said to be 'tied'. In this case, the rank scores that would have been given to the values are taken and averaged and that value is then given to all those observations with the same value in the original data. As an example, in the species richness data, there are three lots with the score of 47 species – numbers 17, 22 and 26 (Table 5.2). These would be ranked 21, 22 and 23. When these ranks are summed $(21 + 22 + 23 = 66)$ and averaged $(66/3 = 22)$, the average rank of 22 is given to all three scores of 47 species.

Once the ranks have been assigned, the difference between each pair of ranks is taken (d) and this value is then squared (d^2). The sum of the d^2 values is then calculated to give Σd^2. Using the figure for Σd^2 from the table, the following calculations are made:

$$r_s = 1.0 - \frac{6 \times 620.5}{26^3 - 26}$$

$$r_s = 1.0 - \frac{3723}{17550} = 0.787$$

The value of r_s (0.787) is less than that of the product-moment correlation coefficient (0.791) but only marginally so.

Significance testing

As before, the r_s value must be tested for significance using the t statistic and the tables available in most statistical textbooks, although once again, when using most present-day computer packages, the significance test and associated p value are usually calculated automatically. Exactly the same formula as for the product-moment coefficient is applied:

$$t = r_s \sqrt{\frac{n - 2}{1.0 - r_s^2}}$$

thus

$$t = 0.787 \sqrt{\frac{26 - 2}{1.0 - 0.619}}$$

$$t = 0.787 \sqrt{\frac{24}{0.381}} = 6.249$$

The value of t has to exceed 2.49 in a one-tailed test at the $p = 0.01$ level. This is achieved with the t value of 6.249, and once again the positive correlation is highly significant.

Kendall's rank correlation coefficient (Kendall's tau)

While Spearman's rank is probably the most widely used rank correlation coefficient, Kendall's tau has also been applied extensively in ecology. Whereas Spearman's rank correlation coefficient is closely related to the product-moment coefficient, Kendall's tau is calculated in an entirely different manner. Initially, the observations are ranked in the same way as for Spearman's r_s. Then tau is determined by comparing both sets of ranks and looking for concordances and discordances. All pairs of ranked observations are compared and the number of concordant and discordant pairs are counted. If there are n observations, then there are $[n(n - 1)/2]$ comparisons to be made. Kendall's tau then uses the formula:

$$\tau = \frac{N_c - N_D}{n(n-1)/2} \tag{5.4}$$

where N_C is the number of concordant pairs and N_D is the number of discordant pairs.

Further details and information on significance tests are given in Burt and Barber (1996). The calculation of Kendall's tau rank correlation on the Chicago data with species richness against lot age produces 247 concordant pairs, 59 discordant pairs and 19 tied pairs, resulting in a value of tau of 0.596, with a p value of 0.01.

Interpretation of correlations

Great care should be taken in the interpretation of correlation coefficients. A significant result in a correlation analysis does not necessarily mean that there is a causal relationship between the variables. While this may appear to be the case for example in the example of urban lot age and species richness in Chicago, other factors may also be important. Many correlations are examples of size relationships. A correlation between plant size or height and productivity which gave a significant positive correlation would be a good demonstration of this. Still others are best described as mutual interaction between the variables. Rather than one causing variation in the other, it is more realistic to talk of these variables 'varying together', rather than a one-way causal relationship.

Finally, there is the problem of closed number systems. This relates to the use of percentage data or proportions in correlation. As an example, take a situation where the extent of certain habitat types is being measured in four different regions of a country. The regions are A, B, C and D. The habitat types are woodland grassland and scrub. In Table 5.3a, the areas of the habitats are shown as hypothetical values in units of 100 km^2. Here the number system is open and if the areas of each habitat type across the four regions are correlated with each other, the correlations in Table 5.3a

Table 5.3   Closed and open systems in correlation analysis (based on Silk, 1979: p. 209).

(a) open number system – habitat types in units of 100 km^2

Region	Woodland (1)	Habitat type Scrubland (2)	Grassland (3)
A	5	2	1
B	7	4	2
C	8	6	4
D	10	8	5

$$r_{1,2} = 0.99;\ r_{1,3} = 0.96;\ r_{2,3} = 0.89$$

(b) closed number system – habitat types expressed as % (region = 100%)

Region	Woodland (1)	Habitat type Scrubland (2)	Grassland (3)
A	63	25	12
B	54	31	15
C	44	33	23
D	43	35	22

$$r_{1,2} = -0.97;\ r_{1,3} = -0.98;\ r_{2,3} = 0.89$$

are found. However, if the areas of habitat are now expressed as percentages, taking the total area of each region as 100%, the values in Table 5.3b result. These values occur within a closed number system and the corresponding correlation coefficients can be seen to be very different. It has been demonstrated that where there are three variables involving values that sum to a fixed total (in this case percentages summing to 100%), two of the correlations will always be negative and one positive, regardless of what the correlations are between the set of open numbers from which the closed numbers were obtained. Thus great caution should be taken when applying correlation to numbers based on percentages in different categories which sum to 100%, or with proportions summing to 1.0. Further discussion of closed number sets is presented in Silk (1979).

Regression analysis

The purpose of simple (two-variable or bivariate) regression analysis is to determine the relationship between two variables by fitting a mathematical function to the set of data. All the methods described below are for linear regression, which aims to fit a straight line to the graph or scattergram showing the relationship between two variables, as in Figure 5.3. As with correlation, there are both parametric and non-parametric methods for regression. Least-squares regression is the major parametric technique for linear regression. Several alternative non-parametric and essentially descriptive methods also exist, for example semi-averages and various types of techniques for deriving resistant lines. When developing a research hypothesis involving the use of regression, it is important to decide on dependent and independent variables, which introduces the concept of causality and the idea that variation in one variable (independent) causes variation in the other (dependent). In the example of the Chicago data, used to explain correlation (Table 5.1), the dependent variable is species richness and the independent variable is urban lot age. Lot age is also known as the causal variable, since it is age that may cause variation in species richness. It is most unlikely that lot age could be dependent on species richness. Normally, the dependent variable is plotted on the Y axis of a scattergram and the independent variable on the X axis

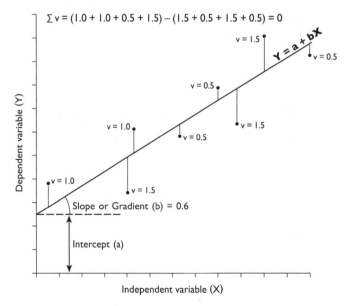

Figure 5.4 Properties of the linear regression line. Redrawn from Wheeler *et al.* (2004), with kind permission of David Fulton Publishers.

(Figure 5.4). Unfortunately, in some situations, it is less easy to determine which variable is the dependent. Many variables, for example, are size-related. Measurement of tree properties, such as height and trunk diameter breast high (dbh usually at 1.3 m) will usually give significant regression and correlation results, but causality cannot be assigned. Both sets of measurements are simply related to tree size. Nevertheless, it is still possible to use regression to describe the relationship.

Least-squares regression

Least-squares regression is a parametric method and thus makes a number of assumptions about the data, notably that they are measured on the interval or ratio scale and that the background population of each variable is normally distributed. The method is best described in two sections: firstly, the fitting of the regression line and testing for significance; and secondly, the use of the method for prediction and the analysis of residuals.

The linear regression model

Any straight-line graph drawn on X and Y axes can be represented by an equation of the form:

$$Y = a + bX \tag{5.5}$$

where a is the intercept term and represents the point on the graph where the straight line intersects the Y axis (Figure 5.4), and b is the slope term or gradient of the line. The gradient determines the rate at which the line rises or falls as X increases. In Figure 5.4, the line has a slope of 0.6. This means that for every unit (1.0) increase of X, there is an equivalent increase of 0.6 in the value of Y.

The regression line is fitted through a scattergram of points in such a way that the positive and negative deviations of individual points, measured in terms of the Y axis, from the line must sum to zero ($\Sigma v = 0$), and the sum of the squared deviations of the individual points from the line must be smaller than for any other line ($\Sigma v^2 = $ minimum). In Figure 5.4, both of these conditions are met for the eight points in the scattergram. It is important to realise that the line is fitted in terms of predicted values of Y for given values of X.

The differences between actual points, in terms of Y, and the points on the calculated line in terms of Y ($\hat{Y}$), are known as the residuals, and the least squares method involves finding the line such that the sum of the squares of the residuals (i.e. of differences between the actual and predicted line values) is minimised (Figure 5.4). Computation of the line requires values of a (intercept) and b (slope term).

b is determined from the formula:

$$b = \frac{\sum_{i=1}^{n} X_i Y_i - \left[\left(\sum_{i=1}^{n} X_i \right) \left(\sum_{i=1}^{n} Y_i \right) \right] \Big/ n}{\sum_{i=1}^{n} X_i^2 - \left(\sum_{i=1}^{n} X_i \right)^2 \Big/ n} \tag{5.6}$$

where

$\quad\quad$ n = number of pairs of observations

$\sum_{i=1}^{n} X_i$ = sum of observations on X

$\sum_{i=1}^{n} Y_i$ = sum of observations on Y

Table 5.4   Calculation of least-squares regression for data on plant species richness and site age from 26 vacant urban lots in Chicago (Crowe, 1979).

1 Lot	2 Species numbers (Y_i)	3 Lot age (months) (X_i)	4 Y_i^2	5 X_i^2	6 $X_i Y_i$	7 $\hat{Y}_i$	8 $Y_i - \hat{Y}_i$	9 Standardised residuals $\frac{Y_i - \hat{Y}_i}{\hat{\sigma}_e}$
1	11	3	121	9	33	19.72	−8.72	−0.66
2	9	3	81	9	27	19.12	−10.12	−0.87
3	16	7	256	49	112	21.23	5.23	−0.26
4	27	15	729	225	405	24.56	2.44	0.63
5	21	18	441	324	378	22.75	−1.75	−0.08
6	32	20	1024	400	640	26.07	5.93	0.99
7	22	20	484	400	440	23.05	−1.05	−0.05
8	16	20	256	400	320	21.23	−5.23	−0.66
9	15	20	225	400	300	20.93	−5.93	−0.76
10	26	22	676	484	572	24.26	1.74	0.30
11	25	25	625	625	625	23.96	1.04	0.11
12	15	30	225	900	450	20.93	−5.93	−1.07
13	30	30	900	900	900	25.47	4.53	0.46
14	45	40	2025	1600	1800	30.01	14.99	1.68
15	22	50	484	2500	1100	23.05	−1.05	−0.96
16	56	65	3136	4225	3640	33.33	22.67	2.03
17	47	70	2209	4900	3290	30.61	16.39	0.96
18	20	70	400	4900	1400	22.44	−2.44	−1.79
19	45	80	2025	6400	3600	30.01	14.99	0.45
20	44	90	1936	8100	3960	29.70	14.30	0.04
21	54	100	2916	10000	5400	32.73	21.27	0.77
22	47	100	2209	10000	4700	30.61	16.39	0.04
23	37	100	1369	10000	3700	27.59	9.41	−1.01
24	69	113	4761	12769	7797	37.27	31.73	1.96
25	46	120	2116	14400	5520	30.31	15.69	−0.72
26	47	150	2209	22500	7050	30.61	16.39	−1.70
	$\sum_{i=1}^{n} Y_i$ 844	$\sum_{i=1}^{n} X_i$ 1381	$\sum_{i=1}^{n} Y_i^2$ 33838	$\sum_{i=1}^{n} X_i^2$ 117419	$\sum_{i=1}^{n} X_i Y_i$ 58159			

and a from:

$$a = \bar{Y} - b\bar{X} \qquad (5.7)$$

where

$\bar{X}$ = mean of X
$\bar{Y}$ = mean of Y

Using these formulae and the Chicago data from Table 5.4 below, the least-squares line can be calculated as follows:

$$b = \frac{58159 - [(1381)\,(844)]\,/26}{117419 - (1381)^2\,/26}$$

$$b = \frac{58159 - 44829}{117419 - 73352} = \frac{13330}{44067}$$

$$b = 0.302$$

$$a = 32.46 - (0.302 \times 53.12) = 16.42$$

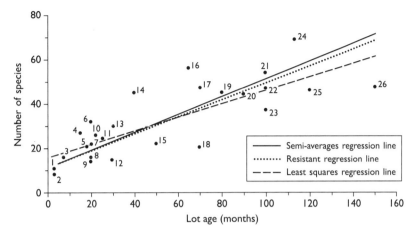

Figure 5.5 Least-squares, semi-averages and resistant regression lines plotted on the scattergram of the Chicago urban lot data. Redrawn from Crowe (1979) with kind permission of Wiley-Blackwell Publishers.

Thus the regression equation is:

$$Y = 16.42 + 0.302X$$

The resulting line has been plotted on the scattergram of Figure 5.3 in Figure 5.5.

There are two limitations to the use of the regression equation:

(a) The line is a line of best fit only within the range of the values of X in the analysis. The line should therefore never extend beyond the lowest and highest X values on the graph, since the relationship defined by the regression may not hold good outside the range of X values.
(b) The regression line giving estimated values of Y (species richness) on X (urban lot age) is not reversible. The line cannot be used to predict or estimate values of X for a given values of Y. This is because, unless the line just happens to be at 45°, the sum of squares of the Y (vertical) variation is different to the sum of squares of the X (horizontal) variation (Figure 5.6). Thus if it was necessary to estimate values of X from Y, then the regression would have to be recalculated and a different line would result.

Explained and unexplained variation in the Y values

The total variation in Y (Figure 5.7a) is:

$$\sum_{i=1}^{n} \left(Y_i - \bar{Y}_i\right)^2$$

or the sum of the squared deviations from the mean of Y. This total variation comprises two parts:

(a) the explained variation (Figure 5.7b) which is

$$\sum_{i=1}^{n} \left(\widehat{Y}_i - \bar{Y}\right)^2$$

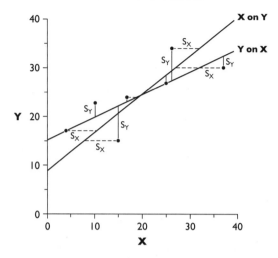

Figure 5.6   Vertical and horizontal deviations from a regression line and regressions of X on Y and Y on X (Hammond and McCullagh, 1978; redrawn with kind permission of Oxford University Press).

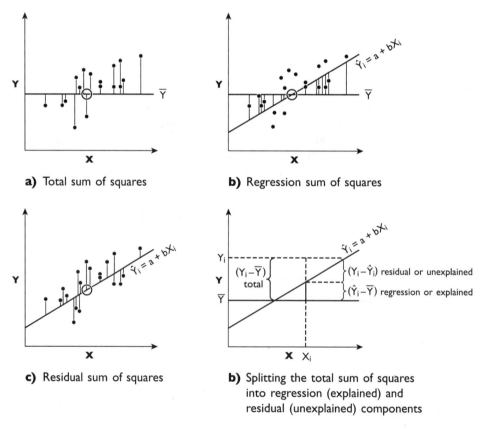

a) Total sum of squares

b) Regression sum of squares

c) Residual sum of squares

b) Splitting the total sum of squares into regression (explained) and residual (unexplained) components

Figure 5.7   Components of sums of squares and explained (regression) and unexplained (residual) variation in least-squares linear regression (Silk, 1979; redrawn with kind permission of Chapman and Hall).

representing the deviation of the regression line from $\bar{Y}$. This indicates the extent to which the prediction or estimation of the line has been improved by using $\hat{Y}_i$ instead of $\bar{Y}$, which would have been the best estimate of Y values if there was no knowledge of X and the regression equation.

(b) the unexplained variation (Figure 5.7c) which is

$$\sum_{i=1}^{n} \left(Y_i - \hat{Y}_i \right)^2$$

representing the deviation of the original Y values (Y_i) from the estimated or calculated line ($\hat{Y}_i$). Table 5.4, column 7, shows the estimated values of Y ($\hat{Y}_i$) for the original values of X. These are the values of $\hat{Y}$ calculated along the regression line for each value of X. The difference between Y_i and $\hat{Y}_i$ (column 8) is the residual or unexplained variation in each case. The sum of squares of these values is the total unexplained variation or the residual variation (Figure 5.7c).

For a single observation (X_i/Y_i), the contributions to total variation, explained (regression) and unexplained (residual) variation, are shown in Figure 5.7d.

The coefficient of determination

The explained and unexplained variation is used to calculate the coefficient of determination, which is given by the formula:

$$\frac{\text{Explained variation}}{\text{Total variation}} = \frac{\sum\limits_{i=1}^{n} \left(\hat{Y}_i - \bar{Y}_i \right)^2}{\sum\limits_{i=1}^{n} \left(Y_i - \bar{Y} \right)^2} \tag{5.8}$$

In the case of the Chicago data:

$$\text{Coefficient of determination} = \frac{4032.0}{6440.5} = 0.63 \quad (63 \text{ per cent})$$

Significance testing of least-squares regression lines

The $\hat{Y}$ values obtained from the regression line are only estimates based on samples drawn from a much larger and usually unknown population. As such, they are subject to sampling variations, and it is important to identify the reliability of sample estimates for the regression coefficients a and b and the estimated values of $\hat{Y}$. This can be done by using analysis of variance (ANOVA) on the explained and unexplained components of the variation described above.

An F (variance ratio) test is used to examine the ratio between the explained (regression) variance and the unexplained (residual) variance. The variance has to be related to degrees of freedom, as in Table 5.5.

A null hypothesis can be erected which is that there is no explanation of the variability of Y (the dependent variable) in terms of X (the independent or predictor variable). It follows that greater F-ratios are provided by higher proportions of explained variance. The associated degrees of freedom and predetermined significance level are used to determine critical F values from published significance tables. When using most present-day computer packages, the significance

Table 5.5 Regression equation analysis of variance (ANOVA) table.

Source of variation	Sums of squares	General description	Degrees of freedom
Explained (Regression)	$\sum_{i=1}^{n}\left(\hat{Y}_i - \bar{Y}\right)^2$	Sum of squared deviations of predicted (estimated) values from sample mean	k
Unexplained (Residual)	$\sum_{i=1}^{n}\left(Y_i - \hat{Y}_i\right)^2$	Sum of squared differences between observed and predicted (estimated) values	$n - k - 1$
Total	$\sum_{i=1}^{n}\left(Y_i - \bar{Y}\right)^2$	Sum of squared deviations of observations from sample mean	$n - 1$

where: n = number of observations
  $\bar{Y}$ = mean of observed Y values
  Y_i = individual Y values
  $\hat{Y}_i$ = estimated (predicted Y values)
  k = number of predictors (always 1 in simple regression)

test and associated p value is usually calculated automatically. Table 5.6 shows the figures for the Chicago case. Here the F statistic is calculated as:

$$\text{Explained variance (Regression)} = \frac{\sum_{i=1}^{n}\left(\hat{Y}_i - \bar{Y}\right)^2}{k} = \frac{4032.0}{1} = 4032.0 \qquad (5.9)$$

where

  k = the number of predictors (always 1 in simple regression)

$$\text{Unexplained variance (Residual)} = \frac{\sum_{i=1}^{n}\left(Y_i - \hat{Y}_i\right)^2}{n - k - 1} = \frac{2408.4}{24} = 100.4 \qquad (5.10)$$

$$F = \frac{4032.0}{100.4} = 40.18$$

Examination of statistical tables for critical values of F at the 0.01 level shows that with 1 and 24 degrees of freedom (v_1 is for the greater variance estimate; v_2 is for the lesser variance estimate), the F value must exceed 7.82. Thus the regression is highly significant.

Confidence limits of least-squares regression lines

Regression lines are used for prediction of $\hat{Y}$ values for given values of X. However, the $\hat{Y}$ values are only estimates based on a sample of a much larger population. It is important therefore to

Table 5.6 Regression analysis of variance from the Chicago study.

Source of variation	Sums of squares	Degrees of freedom	Mean square	F
Explained (Regression)	4032.0	1	4032.0	40.18
Unexplained (Residual)	2408.4	24	100.4	
Total	6440.4	25		

calculate the standard deviation of the estimated values $(\hat{Y})$ from the observed values (Y). This is the same as the standard deviation of the residuals. The formula is:

$$S_{YX} = \sqrt{\frac{\sum\limits_{i=1}^{n} \left(Y_i - \hat{Y}_i\right)^2}{n}} \tag{5.11}$$

where

S_{YX} = standard deviation of error of the residuals

An alternative formula if the product moment correlation coefficient has been calculated is:

$$S_{YX} = \sigma Y \sqrt{1 - r^2} \tag{5.12}$$

where

σY = the standard deviation of y
  r = the product–moment correlation coefficient between X and Y

In the Chicago example:

$$S_{YX} = \sqrt{\frac{2408.4}{26}} = 9.62$$

Where there are a small number of observations (<30), as in the Chicago example ($n = 26$), an alternative formula is applied to allow for understanding of variances. This 'best estimate' is found from:

$$\hat{\sigma}_e = S_{YX} \sqrt{\frac{n}{n - k - 1}} \tag{5.13}$$

where

S_{YX} = the standard deviation ot the residuals
  n = the number of observations
  k = the number of predictors (always 1 in simple regression)

In the Chicago case:

$$\hat{\sigma}_e = 9.62 \times \sqrt{26/24} = 10.01$$

This value is a measure of the spread of the observed points about the regression line. It is then possible, if desired, to plot confidence limits to the regression line. However, these will be curved, owing to sources of error in the estimation of both the regression coefficients a and b. The methods for this, and for allowing for sampling errors in a and b, are described well in Waite (2000), Wheeler *et al.* (2004), Quinn and Keough (2002) and Gotelli and Ellison (2004).

Problems in the application of least-squares regression

A number of assumptions underlie the application of the linear regression model. These are only briefly mentioned here, but are explained in much greater detail in Waite (2000), Wheeler *et al.* (2004), Quinn and Keough (2002) and Gotelli and Ellison (2004).

(a) The data for each variable should approximate to a normal distribution.

(b) The method assumes that the data are best fitted to a linear model – in the case of the Chicago data, transformation of the data or curve-fitting could improve the degree of fit of the relationship, although a better linear fit may be obtained by log-transforming the data (see case study).

(c) Autocorrelation – this can be either spatial or temporal in nature. A further assumption of least-squares regression analysis is that each observation is independent of all others. The idea of spatial and temporal autocorrelation was introduced earlier in this chapter and in Chapter 3 on sampling. The spatial positioning of sample points with respect to each other almost inevitably results in spatially autocorrelated data. Similar problems occur when data are collected from successive points in time.

(d) Lack of measurement error – it is assumed in least-squares regression that both X and Y are measured without error. If this is not the case, then the coefficients of the regression equation may be biased. It is, however, extremely difficult to determine the magnitude of measurement error in most instances.

(e) Homoscedasticity – which means 'equally scattered' and refers to the standard error of the residuals in a regression analysis. Ideally, this standard error, describing the scatter of observations around the regression, should remain the same along the whole length of the line, as in Figure 5.8. If there is considerable variation in the values of

$$\sum_{i=1}^{n} \left(Y_i - \hat{Y}_i \right)^2 \Big/ n$$

for each value of X, then the coefficients of the regression equation may be severely biased.

(f) The means of the conditional distributions should be zero – for every value of X, the mean of $(Y_i - \hat{Y}_i)$ should be zero. If not, the coefficients of the coefficients of the regression may be biased estimates.

Prediction

Once the least-squares regression line has been calculated and confidence limits have been set, the equation and the line can be used for prediction. It is important to remember that predictions can

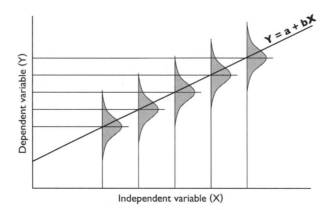

Figure 5.8 Representation of the principle of homoscedasticity where observations have constant variance along the regression line. Redrawn from Wheeler *et al.* (2004), with kind permission of David Fulton Publishers.

only be made for values of Y in terms of X from the regression line of Y on X. Also no predictions should be made beyond the limits of X.

The analysis of residuals

Analysis of residuals, the differences between the observed values of Y and those predicted by the regression equation ($\hat{Y}$), is a very important further step in regression analysis which is nevertheless often neglected. Study of the residuals can assist with understanding of the unexplained variance between X and Y, and enable information to be gained on possible new and additional variables causing variation in Y. This in turn may help with both refinement and modification of existing hypotheses and the generation of new ones.

The study of residuals involves standardisation of the residual values because otherwise they would be expressed in units of the original Y variable, which would depend on the particular measurements for the Y variable. Given that one of the assumptions of the regression model is that the residuals are normally distributed about the regression line (homoscedasticity) with uniform variance along the line (Figure 5.8), the process of standardising the residuals is achieved by making a best estimate of the standard error of the residuals ($\hat{\sigma}_e$), which was described in Equation (5.11). An individual residual value can then be standardised by the formula:

$$\text{Standardised residual} = \frac{Y_i - \hat{Y}_i}{\hat{\sigma}_e} \qquad (5.14)$$

The standardised residuals for the Chicago data are shown in Table 5.4, column 9. Examination of these values shows that urban lot number 16, in particular, is worth close examination with a standardised residual of 2.03. Other sites with large standardised residuals are numbers 14, 18, 24 and 26, all with values between 1.5–2.0 times the standard error. The position of these points on the scattergram in Figure 5.5 also shows their deviant nature. All four sites would almost certainly merit more detailed ecological survey to ascertain other possible factors affecting species richness. Further analysis of residuals is possible using residual plots and analysis for autocorrelation (Wheeler *et al.*, 2004).

As a parametric method, the use of least-squares regression involves making a number of assumptions and has many limitations which are often conveniently ignored by many researchers. Despite having used the Chicago data-set as an example, some questions could be asked as to its limitations for least-squares regression analysis. For example, the data do not appear to show strong homoscedasticity. As a response to these difficulties, various other non-parametric methods have been devised, which although not widely used, are frequently better suited to the quality of data generated by vegetation scientists. A good review of the use of regression in biology is presented in Fry (1993).

Non-parametric regression

Semi-averages regression

The simplest means of finding a best-fit line through a scattergram of points is by using semi-averages regression. Taking the Chicago data as an example, the procedure is as follows:

(a) Calculate the overall means (averages) for both variables:

Overall mean of X (urban lot age) = 53.12
Overall mean of Y (species numbers) = 32.46

(b) Calculate the first semi-average:

X coordinate = mean of all values below the overall mean of X = 21.53 (15 values)
Y coordinate = mean of all values below the overall mean of Y = 20.47 (15 values)

(c) Calculate the second semi-average:

X coordinate = mean of all values above the overall mean of X = 96.18 (11 values)
Y coordinate = mean of all values above the overall mean of Y = 48.81 (11 values)

These three sets of coordinates (x : y – 21.53 : 20.47; 53.12 : 32.46; 96.18 : 48.81) for the lower semi-average, the overall mean and the upper semi-average are then plotted on the scattergram together with the original data points (Figure 5.5). The three points should very nearly form a straight line and the 'best-fit' line can then be drawn as close as possible to the three points, although the line need not actually join them.

Resistant lines and robust regression

The numerous assumptions of the least-squares regression model and the evolution of techniques for exploratory data analysis have led to the development of methods for resistant or robust lines. Such methods are based on ranking of the regression data pairs on the basis of the X variable, partitioning the data into two or three groups, calculating the medians of those groups and using those values to derive a regression line. Various authors have suggested this approach (Bartlett, 1949; Brown and Mood, 1951; Quenouille, 1959). The technique described below and variations are presented in Daniel (1978), Sibley (1987) and Marsh and Elliott (2008), with further comment on resistant lines in Besag (1981).

An important aspect of this approach is, however, that it is exploratory in nature and is concerned as much with studying variation in residuals as the fitting of the regression line. With least-squares regression, correlated residuals violate the assumption of the model that the error term is a random component. Also, any exceptions or 'rogue' data points can influence the positioning of the line such that it becomes a somewhat meaningless summary of the overall relationship between the variables. In the Chicago example, analysis by least-squares regression is considerably influenced by extreme data values such as urban lots 16 and 26. The problem of these 'rogue' points can be dealt with in several ways. With least-squares regression, it is possible to calculate the regression line with all the data and then to recalculate the equation with the 'rogue' points omitted. Any differences between the two lines is then a measure of the influence of the 'rogue' points on the overall relationship between X and Y. Another approach is to weight observations with large residuals to reduce their influence on the line. However, neither of these alternatives is entirely satisfactory, particularly given the assumptions of the least-squares regression model. Also, in both cases, the definition of a 'rogue' point is somewhat subjective. The use of resistant lines provides a much better alternative and will generally give a fit that is far less influenced by extreme values.

The technique is based on dividing the X data into three roughly equal groups and calculating the medians of X and Y for each group. Either the two outer points or all three points are then used to derive a 'best-fit' line. It is important to realise that the aim is still to fit a linear relationship and the equation for a straight line ($Y = a + bX$) still applies. The procedure is as follows:

(a) Rank the X values in the order in which they occur from lowest to highest along the X axis.

(b) Split the rank order of X into three groups, left (L), middle (M) and right (R), as nearly equally as possible. If even groups are not possible, then use the following to achieve a balanced distribution:

Size of group (k = group size)

Group	n = 3k	n = 3k + 1	n = 3k + 2
Left	k	k	k + 1
Middle	k	k + 1	k
Right	k	k	k + 1

Where ranks are tied, however, all X tied values are allocated to the same group. This can be a serious problem if there are a large number of ties and the method becomes inaccurate. Exactly how many ties are needed to invalidate an analysis is uncertain, however.

(c) Calculate the summary points which are the coordinates of the medians of X and Y in each group. Thus for left, middle and right groups these are $(X_L:Y_L)$, $(X_M:Y_M)$ and $(X_R:Y_R)$.

(d) The slope of the line is then given by:

$$b = \frac{Y_R - Y_L}{X_R - X_L} \tag{5.15}$$

and the intercept by:

$$a = 1/3\{(Y_L - bX_L) + (Y_M - bX_M) + (Y_R - bX_R)\} \tag{5.16}$$

(e) On the assumption that a linear fit is appropriate, the line can be plotted. For the Chicago data this gives the following calculation based on Table 5.7. With 26 observations there are three groups of X values with 9 observations in the first group, 8 in the second and 9 in the third.

Slope (b) is calculated as follows:

$$b = \frac{Y_R - Y_L}{X_R - X_L} = \frac{46.0 - 16.0}{100.0 - 18.0} = \frac{30.0}{82.0} = 0.366$$

Table 5.7 Derivation of three groupings from the Chicago data for calculations of medians in resistant regression (n = 26).

Group L 9 observations			Group M 8 observations			Group R 9 observations		
X	Y	Y(ordered)	X	Y	Y(ordered)	X	Y	Y(ordered)
3	11	9	22	26	15	70	20	20
3	9	11	25	25	22	80	45	37
7	16	15	30	15	25	90	44	44
15	27	16	30	30	26	100	54	45
18	21	16	40	45	30	100	47	46
20	32	21	50	22	45	100	37	47
20	22	22	65	56	47	113	69	47
20	16	27	70	47	56	120	46	54
20	15	32				150	47	69
Medians								
$X_L = 18.0$	$Y_L = 16.0$		$X_M = 35.0$	$Y_M = 28.0$		$X_R = 100.0$	$Y_R = 46.0$	

The intercept is calculated from:

$$a = 1/3\{(Y_L - bX_L) + (Y_M - bX_M) + (Y_R - bX_R)\}$$
$$a = 1/3\{9.41 + 15.19 + 9.40\} = 11.33$$

The equation for the resistant regression line is:

$$\hat{Y} = 11.33 + 0.366X$$

This line has been plotted on Figure 5.5. The last, very important step in the analysis is to calculate and examine the residuals, which can be calculated from the predicted values:

$$r_i = Y_i - (a + bX_i) \tag{5.17}$$

Analysis of residuals

As with the least-squares method, analysis of the residuals from resistant regression lines can be extremely useful for further hypothesis generation and modification. The values of the residuals from the Chicago data can be inspected for pattern. The same points, 14, 16, 18, 24 and 26, are shown to have the highest deviations from the computed regression line.

BAYESIAN DATA ANALYSIS (BDA)

The Bayesian approach to hypothesis testing was introduced in Chapter 1. In the conclusion to that introduction, the point was made that thus far, very little Bayesian analysis has been employed in vegetation science. This has been partly because a great deal of vegetation description work is inductive and descriptive. Nevertheless, the approach will almost certainly have an increasingly important role in deductive research in the future. The complexity of Bayesian analysis precludes detailed discussion here, but examples of the whole approach applied in the context of correlation and regression are presented in McCarthy (2007), Kéry (2010) and Link and Barker (2010). The computational challenges are very substantial, and user-friendly software is only gradually becoming available. The key package available to ecologists is WinBUGS (Windows application of Bayesian inference Using Gibbs Sampling). This project has developed flexible software for the Bayesian analysis of complex statistical models using Markov Chain Monte Carlo (MCMC) methods (http://www.mrc-bsu.cam.ac.uk/bugs/welcome.shtml). However, a thorough understanding of Bayesian analysis is necessary, and the package must be used with care. Kéry (2010) provides an excellent starting point for Bayesian analysis with WinBUGS.

The biggest problems with Bayesian analysis are the decisions over the 'prior probabilities' that need to be allocated based on existing knowledge and previous research (Dennis, 1996; Gotelli and Ellison, 2004; McCarthy and Masters, 2005; McCarthy, 2007; de Valpine, 2009; Lele and Dennis, 2009; Kuhnert *et al.*, 2010). Within many research hypotheses in ecology and vegetation science, it is hard to set these. Nevertheless, there is no doubt that with further advances, the Bayesian approach will increasingly find a place in vegetation science.

Case study

Multiple correlation and regression analysis and urban plant ecology in Chicago (Crowe, 1979)

The study of the vegetation of towns and cities is now a very important area of plant ecology. Apart from the pioneering work of Shenstone (1912) and Salisbury (1943) in London, until comparatively recently, most ecologists ignored the great variety of plants that have adapted to the highly modified urban ecosystems of the world. However, since the mid-1970s, the vegetation of urban environments has been studied in more detail (Davis, 1976; Nature Conservancy Council, 1979; Haigh, 1980; Whitney and Adams, 1980; Kunick, 1982; Whitney, 1985; Emery, 1986; Gilbert, 1989; Sukopp *et al.*, 1990; Kent *et al.*, 1999; Pyšek *et al.*, 2004; Thompson *et al.*, 2004; Loram *et al.*, 2008).

Many aspects of urban plant ecology merit study, and philosophically there are important reasons for studying wildlife in the city (Harrison *et al.*, 1987). The recognition and definition of plant communities is an interesting topic, as well as surveys into the origins of different urban species. Some are found in surrounding countryside, others occur only in city environments, while still others are garden escapes. Further work can include diversity, succession and productivity, similar to the case study of Wathern (1976) in the previous chapter. More specialist topics are also possible, such as the study of the flora of walls or pavements (Darlington, 1981).

The study by Crowe (1979) on urban lots in Chicago, which was used to introduce techniques of correlation and regression, is a good example of urban plant ecology. Crowe was interested in ideas of island biogeography (MacArthur and Wilson, 1967; Whittaker and Fernández-Palacios, 2007), and in particular the idea that species richness is related to island area. Crowe makes the analogy between true oceanic islands and the abandoned urban lots of big cities, arguing that urban lots represent islands for potential colonisation within the 'ocean' or 'sea' of concrete or tarmac in the urban environment. Concepts of island biogeography and the problems of making these extrapolations of the theory to terrestrial environments are discussed in Kent (1987b) and Schrader-Frechette and McCoy (1993). Crowe was particularly interested in the higher plant species richness within abandoned urban lots in Chicago and the factors that determine species numbers. A total of 26 abandoned urban lots were examined, and in addition to collecting data on species richness, various variables relating to the area and distance effects predicted by island biogeographical theory were measured. The simple correlation and regression analyses already described had demonstrated that lot age (time since last major disturbance) was very important and explained 63% of the variation in species richness. However, other island biogeographical variables could be significant as well, and help to explain some of the remaining 37% of the variance. Thus measurements were taken of lot area, distance to the nearest older lots, the number of other lots within 1 km, distance to the oldest lot, distance to the nearest other lot and distance to the largest lot. Clearly all of these distance measures would affect potential colonisation, while area is related to extinction and the maximum number of species that a site can hold at one time.

One of the first interesting features of Crowe's paper was that having found a correlation of 0.79 ($r^2 = 0.63$) for species richness and age, this was increased to 0.85 ($r^2 = 0.72$) by transforming the data and taking the logarithms of both the data on species richness and age. Crowe also calculated correlation coefficients between the other variables to give a correlation matrix. The correlation of species richness and lot area was also highly significant with a coefficient of 0.48 ($r^2 = 0.23$) for untransformed data, significant at the 0.05 level, but increasing to 0.67 ($r^2 = 0.45$), highly significant at the 0.01 level, when logarithmic transformations were applied, although the strength of the relationship was much less strong than for lot age. A well-established part of island biogeographical theory predicts a log–log relationship between species and area, and Crowe was thus able to demonstrate this for the Chicago data. The fact that it applied for age as well, and age was even more highly correlated with species richness, was also a very valuable result.

The obvious question to ask next was 'What are the combined effects of age and area?' Clearly, the joint explanation is not the sum of the two coefficients of determination (r^2 age = 0.63; r^2 area = 0.23; sum= 0.86) because there is correlation between age and area ($r = 0.288$) and thus they overlap in some of their explanation of the variance.

In order to examine the joint effects of more than one variable, multiple correlation and regression need to be applied. There is a very large literature on these methods, and they have to be used carefully (Waite, 2000; Quinn and Keough, 2002; Gotelli and Ellison, 2004). In multiple correlation, one variable, in this case species richness, is correlated with two or more others, and the degree of correlation with the other variables in combination is determined. In the case of the Chicago data, if the scores for species richness are correlated against both age and area with no transformation, the correlation coefficient rises to 0.83 and the coefficient of determination (r^2) to 0.69. With log transformation, again this rises to 0.92 ($r^2 = 0.84$).

Multiple regression analysis similarly regresses the dependent variable (Y) on more than one independent (X) variable (X_1; X_2; X_3; X_n). With the Chicago data, using both lot age (X_1) and lot area (X_2) as independent variables and species richness (Y) as the dependent, the regression equation becomes:

$$Y = 14.5 + 0.273\,X_1 + 0.00281\,X_2$$

The figure of 14.5 is the intercept or constant, 0.273 is the partial regression coefficient for lot age (X_1), and 0.00281 is the partial regression coefficient for lot area (X_2). The regression coefficients then have to be tested for significance. This is done using t-tests and is described in Johnston (1978), Waite (2000), Quinn and Keough (2002), Gotelli and Ellison (2004) and Wheeler et al. (2004). Results are presented in the following table:

Predictor	Coefficient	Standard deviation	t-ratio	p
Constant	14.46	3.091	4.68	0.001
Age (X_1)	0.273	0.046	5.91	0.001
Area (X_2)	0.0028	0.0013	2.24	0.030

The p values indicate that all three coefficients are significant at the 0.05 level. Also a table is drawn up in a similar manner to Tables 5.5 and 5.6 for testing the significance of r^2 (0.69) (Table 5.8). The F-ratio is looked up with 2 and 23 degrees of freedom in appropriate statistical tables at the 0.01 level: the value has to exceed 5.66 to be significant. Once again, when using most present-day computer packages, the significance test and associated p values are usually calculated automatically. Clearly, the calculated value of 25.95 is highly significant and the hypothesised relationship between species richness

Table 5.8   Multiple regression analysis of variance tables from the Chicago study.

Untransformed data

Source of variation	Sums of squares	Degrees of freedom	Mean square	F
Explained (Regression)	4462.0	2	2231.3	25.95
Unexplained (Residual)	1177.9	23	86.0	
Total	6440.4	25		

Transformed data

Source of variation	Sums of squares	Degrees of freedom	Mean square	F
Explained (Regression)	6.2343	2	3.1171	60.31
Unexplained (Residual)	1.1888	23	0.0517	
Total	7.4231	25		

and the combined effects of age and area is accepted. This is a very interesting result in terms of island biogeographical theory and the possible analogies that have been made between urban lots and oceanic islands. It also shows how both time (age) and space (lot area) are important in successional processes and in determining species richness in these superficially uninteresting urban areas.

Again, the multiple regression can be calculated with the three variables transformed using logarithms. In this case, the regression equation becomes:

$$Y = 0.759 + 0.350 \, X_1 + 0.203 \, X_2$$

and the table for testing regression coefficients is:

Predictor	Coefficient	Standard deviation	t-ratio	p
Constant	0.7595	0.2978	2.55	0.020
Age (X_1)	0.3504	0.0464	7.56	0.001
Area (X_2)	0.2029	0.4834	4.20	0.001

The table for the F-ratio test is given in Table 5.8 and the multiple regression is even more highly significant (F = 60.3 with 2 and 23 degrees of freedom). Transformation also assists with normalising the variables and thus brings the data closer to the assumptions of the multiple regression model.

Multiple correlation and regression are complex methods that need to be handled with care. More detailed explanations are given in Jongman *et al.* (1995), Waite (2000), Quinn and Keough (2002) and Gotelli and Ellison (2004). There are many assumptions in the use of the multiple regression model. All of those discussed previously for simple (two variable/bivariate) regression apply, plus the assumption of absence of multicollinearity between the independent variables. This means that there should not be correlation between the independent variables, or at least those correlations should be very low. Thus a matrix of correlations between the independent variables should always be calculated and examined or the Variance Inflation Factors (VIFs) should be calculated for each independent variable in the particular multiple regression model being examined (Zuur *et al.*, 2010). This is defined as:

$$VIF = \sqrt{\{1/(1 - r^2)\}}$$

where r^2 is from the multiple linear regression model. Independent variables with Variance Inflation Factors of >3.0 should be looked at carefully, and any with very high scores should be removed from the model. As demonstrated here, multiple regression should therefore be seen as a modelling technique, whereby varying combinations of independent variables are assessed in relation to the dependent in both transformed and untransformed forms.

Another feature of many computer programs for multiple regression is the possibility of using a 'stepwise' procedure for the inclusion of independent variables, which selects that combination of independent environmental/biotic variables that gives the highest overall explained variance in the original, based on their relative combined explanatory power. Stepwise procedures should be used with caution, however, and should not be an excuse for failing to think carefully about the nature, causality and inter-relationships of the various explanatory variables (MacNally, 2002; Blanchet et al., 2008).

Multiple correlation and regression analyses always need to be carried out by computer and it is best to use one of the well-known packages, such as MINITAB or SPSS (Statistical Package for the Social Sciences) (see Chapter 9).

Regression and multiple regression are important methods for plant ecologists and are related to some of the more recent methods of ordination discussed in Chapter 6, particularly canonical correspondence analysis (CCA) and redundancy analysis (RDA). Austin (1971) provided a very useful review and example of its application in plant ecology. He used multiple regression to study the causes of variation in abundance of Eucalyptus rossii in relation to a number of environmental factors in the Southern Tablelands of New South Wales, Australia.

CONCLUSION

This is a very important chapter of this book. The description and analysis of vegetation and environmental/biotic data has tended to concentrate on the methods of ordination and classification described in the following chapters and to ignore the more simple but equally important methods of statistical analysis in relation to hypotheses and data exploration. The need to ask more specific questions about vegetation and its environment, and to generate and test hypotheses about individual species and their relationships to each other and to their environment, has already been stressed (Kent and Ballard, 1988; Keddy, 2001; von Wehrden et al., 2009; Zuur et al., 2010). The limited range of methods for exploratory data analysis which have been introduced here will be of value to many vegetation scientists and their students in addition to the more traditional confirmatory approach of hypothesis generation and testing. The techniques presented here are by no means the only ones appropriate to the simple analysis of vegetation data, and reference to the many textbooks mentioned will show how all these ideas can be taken further, particularly in relation to the future potential application of Bayesian approaches to statistical data analysis. Various statisticians have reviewed the problems of applying methods of statistical analysis, and the advanced student should read the articles by Nelder (1986), Cormack (1988), McPherson (1989) and Zuur et al. (2010), and the books by Chatfield (1995), Waite (2000) and Quinn and Keough (2002), which deal with the role of statistical analysis in general and in the biological and environmental sciences in particular.

Plate 1.1 Evergreen broadleaved forest at Tiantong Forest Reserve near Ningbo, Eastern China (photo: Martin Kent).

Plate 1.2 The Storbreen glacier foreland in southern Norway (photo: Jane Robbins).

Plate 1.3 *Lobelia urens* (L) – the heath lobelia in flower during July at one of the six sites in southern England (photo: Martin Kent).

Plate 1.4 Grazing wildebeest in the Serengeti National Park Tanzania (photo: Martin Kent/Chris Breen).

Plate 3.1 Use of 5 m × 5 m quadrats marked out with tapes to describe vegetation in *Calluna vulgaris* (heather or ling) moorland in northern England (photo: Martin Kent).

Plate 3.2 A 1 m × 0.5 m subdivided quadrat. The quadrat is turned over on itself to give a 1 m × 1 m area (photo: Martin Kent).

Plate 3.3 Recording percentage cover within a 0.5 m × 0.5 m quadrat using a quadrat made from brightly coloured cord and knitting needles (photo: Martin Kent).

Plate 3.4 The line-intercept method being used to describe semi-arid Mediterranean vegetation in Murcia, southeast Spain (photo: Martin Kent).

Plate 4.1 Garigue vegetation near Garraf, south of Barcelona, northeast Spain (photo: Martin Kent).

Plate 4.2 Vegetation at Gutter Tor, Dartmoor, southwest England (photo: Martin Kent).

Plate 8.1   The five volumes of the British National Vegetation Classification (NVC) (Rodwell, 1991; 1992a,b; 1995a; 2000). With kind permission of Cambridge University Press.

Chapter 6
Ordination methods

INTRODUCTION

The word 'ordination' means 'to set in order' and was first used by Goodall (1954), who showed that it is derived from the German word *Ordnung* applied to this approach by Ramenskii (1930). Here, ordering means the arrangement of vegetation samples or quadrats in relation to each other in terms of their similarity of species composition and/or their associated environmental controls. Ordination methods are also part of gradient analysis. In gradient analysis, variation in species composition is related to variation in associated environmental/biotic factors that can usually be represented by environmental/biotic gradients (Chapter 2). The emphasis in ordination methods is thus on individual samples or species and their degrees of similarity to each other, and on determining how the order of the individuals is correlated with underlying environmental/biotic controls. Usually the environmental/biotic data are collected at the same location as the species data. Such data are said to be 'paired'.

Gradient analysis and ordination techniques are a group of methods for data reduction and explanation leading to hypothesis generation (Figure 6.1). The methods are primarily descriptive and enable researchers to formulate ideas about plant community structure, as well as possible causal relationships between variation in vegetation and its environment.

Within plant ecology, gradient analysis and ordination methods can help with one or more of the following areas of research:

(a) Summarising plant community data and providing an indication of the true nature of variation within the vegetation of the area under study.
(b) Enabling distributions of individual species within communities to be examined and compared.
(c) Providing summaries of variation within sets of vegetation samples which can then be correlated with environmental/biotic controls to define environmental/biotic gradients.

DIRECT AND INDIRECT ORDINATION OR GRADIENT ANALYSIS

Direct gradient analysis is used to display the variation of vegetation in relation to environmental/biotic factors by using environmental/biotic data to order the vegetation samples. As the name suggests, in the early direct methods discussed below, the paired environmental/biotic data are

Vegetation Description and Data Analysis: A Practical Approach, Second Edition. Martin Kent.
© 2012 John Wiley & Sons, Ltd. Published 2012 by John Wiley & Sons, Ltd.

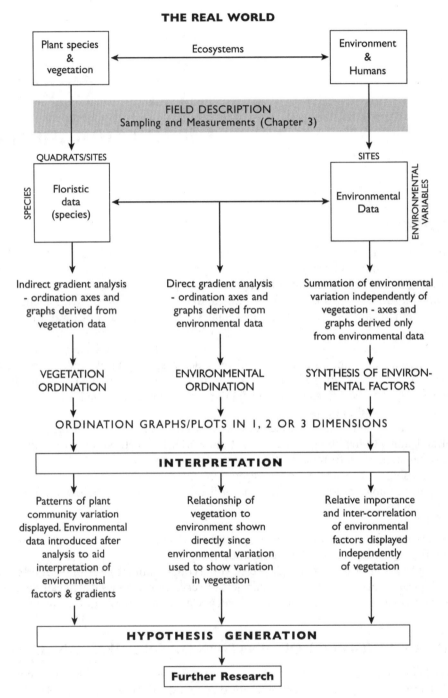

Figure 6.1   Approaches to ordination in plant ecology. (a) Indirect ordination, where vegetation data are analysed independently of environmental/biotic data and environmental/biotic data are only introduced after the ordination diagram has been produced (vegetation ordination). (b) Direct ordination, where vegetation samples are ordinated by using paired environmental/biotic data to construct the ordination diagram (environmental ordination). (c) Summarisation of environmental variation, where the variability of the paired environmental/biotic data is analysed independently of the floristic vegetation data (synthesis of environmental/biotic factors). Reproduced with permission from John Wiley & Sons, Inc.

used directly to organise the information on vegetation (central pathway, Figure 6.1). The methods necessarily assume that the underlying environmental/biotic gradients are known, and can be seen as quite distinct from the methods of indirect gradient analysis or ordination described below.

The term 'indirect' ordination methods or gradient analysis is applied to techniques that operate on a set of vegetation data by first examining the variation within it (left pathway, Figure 6.1). This is done independently of the paired environmental/biotic data. A second stage of analysis is then performed once the major sources of variation in the vegetation data have been described and summarised. Only then are the environmental/biotic data compared and correlated with the summarised vegetation data in order to detect possible environmental and biotic gradients. The environmental/biotic interpretation is thus indirect. These methods can be used in situations where the underlying environmental/biotic gradients are unknown or are unclear, although they are equally applicable where the environmental/biotic gradients are known (Whittaker, 1967, 1978a).

Historically, direct ordination techniques represent some of the oldest methods and were particularly prominent in America, where they were largely developed and where there was, and still is, a much stronger tradition of seeing vegetation as a continuum, rather than as a collection of distinct communities. Elsewhere in the world, particularly in Europe, indirect methods have assumed much greater importance until recently.

Some confusion has arisen over the direct/indirect gradient analysis terminology. Gauch (1982b), for example, reserved the word ordination just for indirect gradient analysis. However, most ecologists would probably agree with the definitions given here. Austin (1968, 2005) has assisted with this in calling methods for direct gradient analysis 'environmental ordination' and those for indirect gradient analysis 'vegetation ordination'.

A third approach to ordination involves the separate analysis of the environmental/biotic data collected in a set of quadrats. This can be described as summarisation of environmental/biotic variation (right pathway, Figure 6.1). Once the patterns of environmental/biotic variation have been analysed, species data may be introduced to examine relationships between species distributions and environmental/biotic factors or gradients.

EARLY DIRECT ORDINATION OR GRADIENT ANALYSIS METHODS

The simplest form of direct gradient analysis is a graph of species response to one environmental/biotic factor. A good example is provided in Tivy (1993), where species abundances along a transect up a hillslope have been plotted, with distance up the hillslope corresponding to increase in altitude (Figure 6.2). This graph demonstrates all the basic principles of gradient analysis and ordination – the nature of community variation up the slope is summarised, individual species distributions are displayed and relationships between the community composition and the environmental factor of altitude are shown. As such it is an example of a coenocline. However, it does represent a very basic form of gradient analysis. The gradient itself – altitude – is an example of a complex gradient, within which many other more specific environmental factors such as temperature, exposure, slope, drainage quality and even biotic factors are operating.

Mosaic diagrams of two factors

Burnett (1964) provided an example of a direct ordination based on two factors. He was studying variation in the upland pastures of Scotland and believed that he knew the principal environmental gradients determining the species composition of the pastures in advance, and that those factors

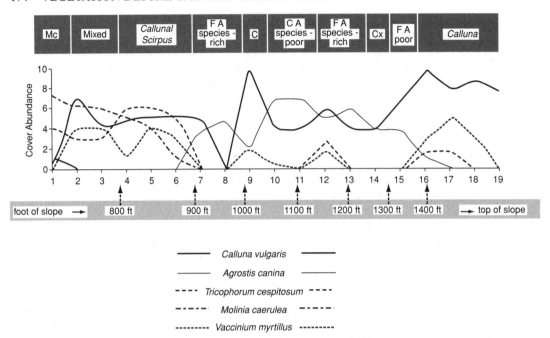

Figure 6.2   A simple one-dimensional direct ordination of selected species recorded in 1 m² quadrats along a continuous transect up a hillslope near Creetown, Kirkcudbrightshire, Scotland. F, *Festuca ovina* (sheep's fescue); A, *Agrostis capillaris* (common bent); C, *Calluna vulgaris* (heather or ling); Mc, *Molinia caerulea* (purple moor grass). After Mitchell (1977), redrawn from Tivy (1993) with kind permission of Longman Publishers Ltd.

were drainage quality and the pH (acidity/alkalinity) of the soil. He collected a large number of vegetation samples, together with data on soil moisture content and pH at each site. Burnett was then able to construct the ordination diagram of Figure 6.3 by placing each quadrat on the graph defined by soil moisture content and pH. On the graph, the points representing quadrats formed a number of groups, and by examining the species composition of the quadrats in each group he was able to define the particular types of grassland that characterise the uplands of Scotland and examine their tolerance ranges and degrees of overlap. This approach is similar to that of Whittaker (1960) in describing the vegetation of the Siskiyou Mountains in southwestern Oregon (Figure 6.4). Whittaker carried out a number of other similar studies, for example in the Great Smoky Mountains (1948, 1956). The success of the approach depends entirely on the choice of environmental axes. In Whittaker's studies (Whittaker, 1960), with geology, altitude and local terrain type, the gradients were pronounced and obvious. However, in many other situations, the environmental/biotic gradients will be less clear and the application of this method will be more difficult.

**The continuum index approach and weighted averages ordination
(Curtis and McIntosh, 1950, 1951)**

Some of the earliest developments in ordination were made by Curtis and McIntosh (1950, 1951). They studied the distribution of forest trees in southern Wisconsin, USA. For each stand or forest

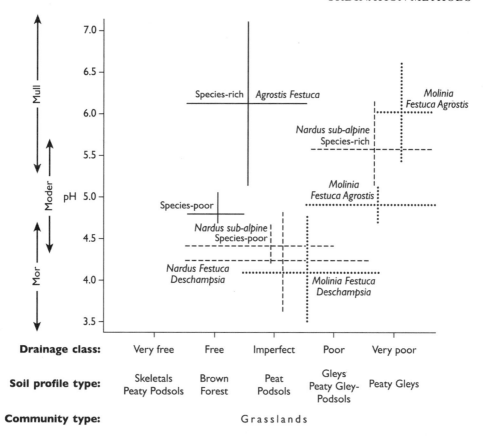

Figure 6.3   A two-dimensional direct ordination of grass-moorland community types in Scotland. Dominant species: 1. *Agrostis capillaris* (common bent)/*Festuca ovina* (sheep's fescue); 2. *Nardus stricta* (mat grass); 3. *Molinia caerulea* (purple moor grass). Adapted and redrawn from Burnett (1964) and Tivy (1993); reproduced with kind permission of Longman Publishers Ltd.

sample, an 'importance value' was calculated with three components, as follows:

$$1.\ \text{Relative density} = \frac{\text{Number of individuals of species} \times 100}{\text{Total number of individuals}}$$

$$2.\ \text{Relative dominance} = \frac{\text{Dominance}^* \text{ of a species} \times 100}{\text{Dominance of all species}}$$

$$3.\ \text{Relative frequency} = \frac{\text{Frequency of a species} \times 100}{\text{Frequency of all species}}$$

Importance value for each species = Relative density + Relative dominance + Relative frequency

(*Dominance is defined as the mean basal area per tree, multiplied by the number of trees of the species.)

Using the importance values for each species, samples/stands were arranged in a sequence, so that, as far as possible, the amount of each species showed a single peak or increased or decreased

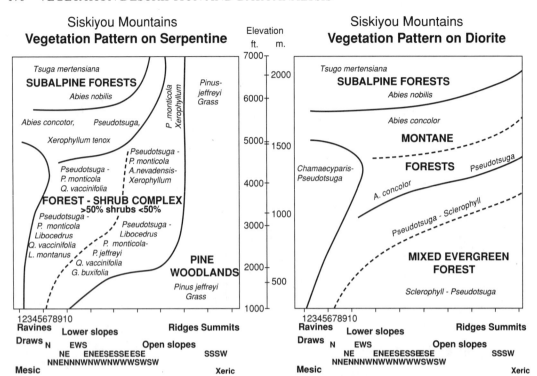

Figure 6.4 Mosaic diagrams of the plant communities of the Siskiyou Mountains, southwestern Oregon, USA. Different diagrams were produced for the two major rock types (serpentine and diorite), with separation of the plant communities on the basis of altitude and topographic position (Whittaker, 1960; redrawn with kind permission of the Ecological Society of America).

monotonically along the series. Then each tree species was allocated a weight according to the position of the peak, and these weightings were then used to provide a more accurate position of the stands or samples on the axis corresponding to the initial ordering. The weightings were calculated by initially grouping the stands according to the four 'leading dominant' species (the most abundant species) and arranging those groups in order (Table 6.1a). The average importance values of the other species (Table 6.1b) were then taken to calculate weightings or climax adaptation numbers for all species as in Table 6.2, where 1 = pioneer species and 10 = climax species.

Then a weighted average was calculated for each stand using the abundance of each tree species in the sample/stand and the species weight. Thus if A_{ij} is the abundance of the tree species in stand j and W_i is the species weight or climax adaptation number (Table 6.2), then an ordination score for each sample (S_j) on a single axis can be computed using the formula:

$$S_j = \frac{\sum_{i=1}^{n} A_{ij} W_i}{\sum_{i=1}^{n} A_{ij}} \tag{6.1}$$

where

$\sum_{i=1}^{n}$ = summation across all species in the stand or sample

n = number of species

Table 6.1 Derivation of the continuum indices for upland hardwood forests in Wisconsin (Curtis and McIntosh, 1951; reproduced with kind permission of the Ecological Society of America).

(a) Average importance values (IV) and constancy % of the four trees selected as the leading dominants in the four groups of stands recognised as the first stage of the analysis (for species with the highest importance potential only—80 stands)

Species	Leading dominant in stand			
	Q. velutina	Q. alba	Q. rubra	Acer saccharum
Q. velutina				
Average IV	165.1	39.6	13.6	0.0
Constancy (%)	100.0	72.3	38.3	0.0
Q. alba				
Average IV	69.9	126.8	52.7	13.7
Constancy (%)	100.0	100.0	97.1	66.7
Q. rubra				
Average IV	3.6	39.2	152.3	37.2
Constancy (%)	25.0	94.5	100.0	76.3
Acer saccharum				
Average IV	0.0	0.8	11.7	127.0
Constancy (%)	0.0	5.6	29.4	100.0

(b) Average importance value (IV) and constancy (%) of associated trees in the four groups of stands with the above four species as leading dominants

Species		Leading dominant in stand			
		Q. velutina	Q. alba	Q. rubra	A. saccharum
Q. macrocarpa	IV	15.6	3.5	4.2	0.1
	Constancy (%)	50.0	38.9	20.6	4.8
Prunus serotina	IV	21.4	21.8	5.9	1.4
	Constancy (%)	87.5	89.0	64.8	19.0
Carya ovata	IV	0.3	8.8	5.2	5.9
	Constancy (%)	12.5	61.2	38.3	33.3
Juglans nigra	IV	1.5	1.2	2.2	1.9
	Constancy (%)	12.5	11.1	20.6	23.8
Acer rubrum	IV	3.9	2.3	2.4	1.0
	Constancy (%)	12.5	33.3	23.5	4.8
Juglans cinerea	IV	0.0	2.7	1.7	4.8
	Constancy (%)	0.0	11.1	20.6	47.6
Fraxinus americana	IV	0.0	1.9	5.1	7.6
	Constancy (%)	0.0	11.1	20.6	42.8
Ulmus rubra	IV	4.6	7.7	8.3	32.5
	Constancy (%)	25.0	27.8	53.3	85.7
Tilia americana	IV	0.3	5.9	19.0	33.0
	Constancy (%)	12.5	16.7	73.5	100.0
Carya cordiformis	IV	2.5	5.8	4.1	8.2
	Constancy (%)	12.5	33.3	41.2	66.7
Ostrya virginiana	IV	0.0	2.4	5.5	16.2
	Constancy (%)	0.0	22.2	41.2	95.3

Table 6.2   Climax adaptation numbers for the tree species found in stands of upland hardwood forests in Wisconsin (Curtis and McIntosh, 1951; reproduced with kind permission of the Ecological Society of America).

	Climax adaptation number
Quercus macrocarpa	1.0
Populus tremuloides	1.0
Acer negundo	1.0
Populus grandidentata	1.5
Quercus velutina	2.0
Carya ovata	3.5
Prunus serotina	3.5
Quercus alba	4.0
Juglans nigra	5.0
Quercus rubra	6.0
Juglans cinerea	7.0
Ulmus thomasi	7.0
Acer rubrum	7.0
Fraxinus americana	7.5
Gymnocladus dioica	7.5
Tilia americana	8.0
Ulmus rubra	8.0
Carpinus caroliniana	8.0
Celtis occidentalis	8.0
Carya cordiformis	8.5
Ostrya virginiana	9.0
Acer saccharum	10.0

*The climax adaptation number of these species is tentative owing to their low frequency of occurrence in the survey.

The result was a single-axis ordination, which positioned each stand or sample along the gradient. This could then be used to show individual species distributions or values of environmental factors.

There were two problems with the Curtis and McIntosh technique. Firstly, the method only derived one axis of variation and thus only worked well where there was one primary environmental gradient, as was the case in the woodlands of southern Wisconsin, where a marked successional series was evident. Secondly, the precise weight given to each tree species (the climax adaptation number) was based on a subjective assessment of its position in a successional series, even though that was based on information provided by the importance values. The climax adaptation numbers would thus vary from one worker to another and depend upon the familiarity and experience that the worker had of the vegetation and its environmental controls. Further refinement of the method to overcome some of these problems was achieved by Brown and Curtis (1952) and Goff and Cottam (1967).

INDIRECT ORDINATION OR GRADIENT ANALYSIS

Indirect ordinations are based on the analysis of floristic data independently of any preconceived notions of controlling environmental/biotic factors or successional sequences. The concern is with the internal variability of the data and the assumption that examination of variability in floristics will inevitably reflect variation in environment and controlling biotic factors, and the relationships between vegetation and its environment can be explored after the floristic variation has been analysed and displayed. Thus environmental/biotic data are not used at any stage of analysis and are only introduced at the interpretation stage.

Methods of indirect ordination can be seen most simply as a means of summarising the information in a species data matrix, such as Tables 4.1 to 4.4, and in scatter diagrams (Figures 6.6 and 6.8). As such, they are known as methods of matrix approximation. However, Prentice (1977), ter Braak and Prentice (1988) and ter Braak (1995) presented another more sophisticated view that there is an underlying or latent structure in the data matrix, and the occurrences of all species in the matrix are determined by a few environmental/biotic variables (the latent variables). The purpose of ordination is then to recover that latent structure in the species data and, in the process, show how species are responding to prevailing environmental/biotic factors. In this case, ordination becomes more like regression, except that in ordination, the explanatory variables are not known environmental/biotic variables but theoretical variables (latent variables), which are constructed so as to explain the species data best. Thus, as with regression, each species represents a response variable, but in ordination the response variables are analysed simultaneously.

Constrained (direct ordinations)

This idea of ordination as a form of regression has been very important in some of the more recent developments in ordination methods that are known as 'constrained' ordinations, where the summary of variation in the species data is constrained by multiple regression with environmental/biotic variables (Chapter 5), as is explained later in this chapter. These 'constrained' methods' are now usually seen as forms of direct ordination because the environmental/biotic data are incorporated directly into the ordination analysis itself (pp. 237–248).

Linear and unimodal species response models

A problem with the idea of regression, discussed in the previous two sections, is that most people think of linear regression and linear response models (Chapter 5), but species response to an environmental factor has usually been conceptualised as the Gaussian, bell-shaped unimodal response model introduced in Chapter 2 (Figure 2.4). Ter Braak (1995) discussed the implications of these two very different types of species response models. Firstly, a model assumes a rectilinear (straight line) relationship between two species A and B in response to an environmental/biotic factor – X. If the abundances of A and B are recorded in a series of quadrats, these can be plotted against X (Figure 6.5a). If the relationship between the two species is rectilinear (i.e. a straight line), then the graph showing the relationship between the two species will be as in Figure 6.5b. However, if the other species response model is taken, corresponding to the Gaussian bell-shaped unimodal curve (Figure 6.5c), then the joint plot of the two species A and B produces the contorted curve of Figure 6.5d. With indirect gradient analysis, the aim is to make an inference about the relationships with the latent environmental variable X (Figure 6.5a,c), using only the species data (species A and B) (Figure 6.5b,d). The form of the relationship (linear or unimodal) clearly has a very substantial effect on the resulting graphs and hence the manner in which the relationships between the species in response to the environmental/biotic factor can be analysed.

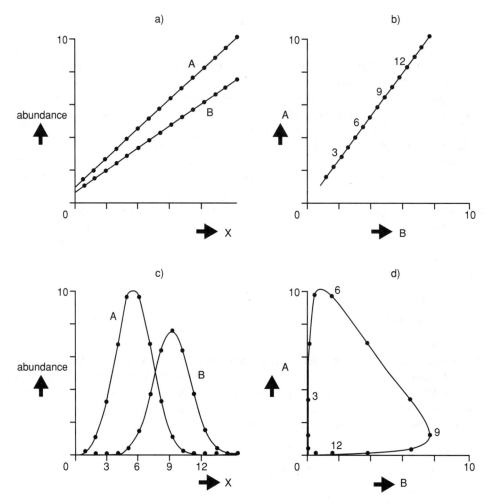

Figure 6.5   Two models of species response to an environmental/biotic factor X and the resulting graphs for the joint distributions of species A and B. (a) A rectilinear response model of two species A and B in relation to X; (b) the resulting line when the two species are plotted together; (c) a unimodal bell-shaped curve response model of two species A and B in relation to X; (d) the resulting contorted curve when the two species are plotted together (ter Braak, 1995; redrawn with kind permission of Cambridge University Press).

ORDINATION DIAGRAMS – THE END PRODUCT OF ORDINATIONS

All ordinations examine the similarity or dissimilarity of floristic composition of vegetation samples. This similarity or dissimilarity is expressed in graph form with plots of points in one, two or three dimensions, where each point represents a vegetation sample or quadrat (Figure 6.6). The distances between the points on the graph are taken as a measure of their degree of similarity or difference. Points that are close together will represent quadrats which are similar in species composition; the further apart any two points are, the more dissimilar or different the quadrats will be. Most published ordination diagrams are two-dimensional, simply because such graphs are easiest to read on a flat page and a two-dimensional graph will show more information than a

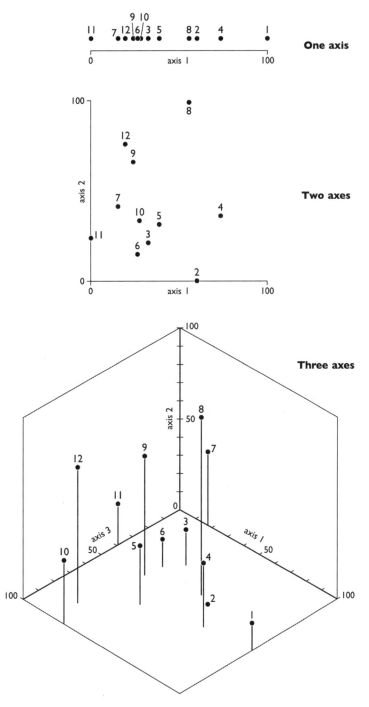

Figure 6.6 Ordination graphs in one, two and three dimensions from a Bray and Curtis (polar) ordination of the New Jersey salt marsh data (Table 4.1). Reproduced with permission from John Wiley & Sons, Inc.

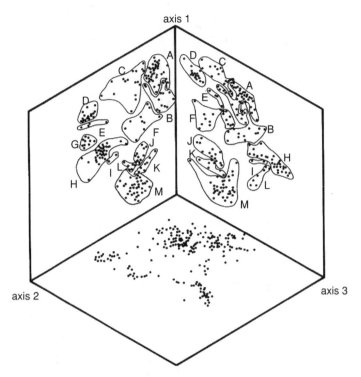

Figure 6.7 A three-dimensional species ordination plot of savanna vegetation from northern Nigeria. Graphs are drawn in pairs: left axis 1 v. axis 2; right axis 1 v. axis 3; bottom axis 2 v. axis 3; letters denote species groupings (Kershaw, 1968; reproduced by kind permission of the British Ecological Society and Wiley-Blackwell).

one-dimensional one. Three-dimensional plots are produced and can be drawn in perspective on a flat page (Figure 6.7) or as a perspective drawing of a three-dimensional cube (Figure 6.8).

With the exception of the method of non-metric multidimensional scaling (NMS), the axes of the graphs produced by an ordination method come out in descending order of importance, with the first axis summarising more variation than the second, the second more than the third, and so on. It is clearly also possible to plot ordination axes in pairs as two-dimensional graphs, axis 1 v. axis 2; axis 1 v. axis 3 and axis 2 v. axis 3, as in Figure 6.7. In most analyses, only the first three or four axes are of significance, although in a quadrat ordination it is normally possible to carry on extracting axes mathematically up to the number of species in the analysis. However, the summarising power of higher axes is minimal, and they make a trivial contribution to the analysis of overall variation in the vegetation data.

Figure 6.9 is a two-dimensional ordination plot for the vegetation data from Gutter Tor, Dartmoor (Table 4.4). This is known as a quadrat, site or sample ordination. Four groups of quadrats can be recognised: Group 1 – quadrats 18, 3, 20, 13, 14 and perhaps 24; Group 2 – 15, 8, 23 and 19; Group 3 – 21, 16, 10, 6, 4, 11, 17 and 22; and Group 4 – 1, 2, 12, 9 and 25. In groups 1, 2 and 4, quadrats are closer to each other than to quadrats in any other group. Thus within each group, quadrats must be relatively similar floristically. Group 3 is more spread out but nevertheless appears distinct, while quadrats 5 and 7 do not belong to any of the four groups and are transitional between groups 1, 2 and 3. This diagram shows the nature of vegetational variation

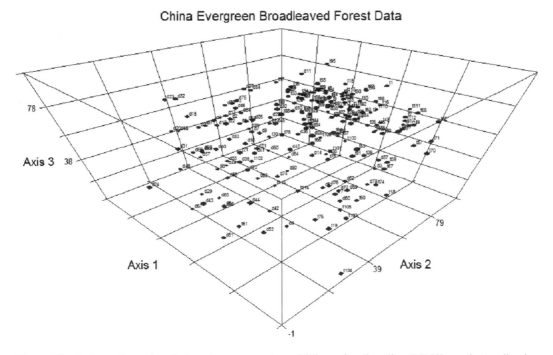

China Evergreen Broadleaved Forest Data

Figure 6.8 A three-dimensional plot of a non-metric multidimensional scaling (NMS) quadrat ordination of data from 199 broadleaved evergreen forest plots at Tiantong, southeast China (case study in Chapter 1). Analysis and plot produced using the *PC-ORD* computer software (McCune and Mefford, 1999, 2010). Original work of Martin Kent.

in the data. There are four reasonably distinct community types but some quadrats are located between these types. Also it demonstrates that group 3 is the most diffuse group. It could be argued that quadrats 16, 21 and perhaps 10 could be separated from 6, 4, 11, 17 and 22. However, this is an entirely subjective decision. This introduces the most important principle of interpretation, which is that, although the method is objective in the sense of repeatability, the interpretation still involves a considerable degree of subjectivity, based on the expertise and ecological knowledge of the researcher.

Species ordinations

Ordination can also be carried out for species, with the vegetation data matrix transposed, producing a one-, two- or three-dimensional graph where each point represents a species and the distances between the points are an expression of how similar the species are in their distribution across the quadrats (Figure 6.10). Such an ordination is called a species ordination. The two-dimensional species ordination plot for the Gutter Tor data is presented in Figure 6.11.

Interpretation of ordination diagrams

Regardless of which technique is used, the process of interpretation of ordination diagrams is similar. Also, the procedure is the same, whether the diagram is drawn in one, two or three dimensions.

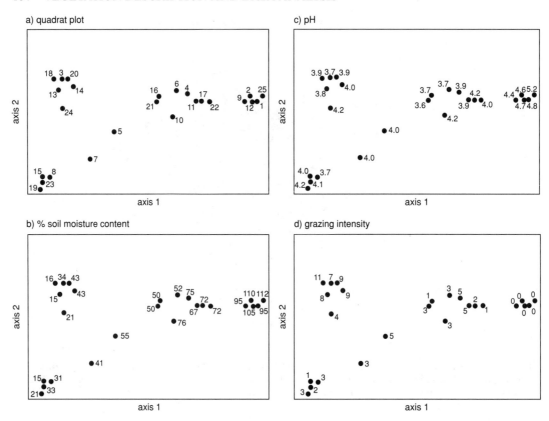

Figure 6.9   (a) A two-dimensional quadrat ordination plot derived from detrended correspondence analysis of the Gutter Tor, Dartmoor data (Table 4.4); (b) percentage soil moisture (gravimetric method) plotted on the quadrat ordination diagram; (c) soil pH plotted on the quadrat ordination diagram; (d) grazing intensity (0 min – 11 max) plotted on the quadrat ordination diagram. Reproduced with permission from John Wiley & Sons, Inc.

The species ordination diagram

Most interpretation is carried out in relation to the quadrat ordination diagram, but the species ordination diagram is always worth examination as well. As an example, the species ordination from detrended correspondence analysis (DCA) of the Gutter Tor data (Table 4.4) is presented in Figure 6.11. Each point on the graph corresponds to a species and the distances between points on the graph are an approximation to their degree of similarity in terms of their distribution within the quadrats. Thus two species occurring with exactly the same abundances in the same quadrats would occupy the same point on the graph. As species distributions diverge, so the distances between points on the species ordination diagram increase. For purposes of interpretation, it is useful to look for groupings of species. Thus in Figure 6.11, four general groupings of species emerge. Firstly, a group to the top left centre of the diagram contains species typical of improved upland pastures, such as *Agrostis capillaris*, *Trifolium repens* and *Cerastium glomeratum*; secondly, a bracken-invaded pasture group in the lower half contains *Pteridium aquilinum*, *Galium saxatile*, *Viola riviniana* and *Festuca ovina;* and thirdly, a heathland group on the right-hand side of the

Quadrat Ordination

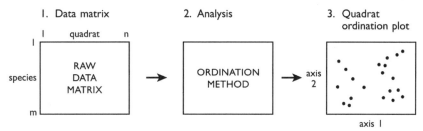

On the ordination plot, each point represents a **quadrat** and the greater the distance between any two points, the greater the difference in floristic composition of the quadrats which they represent.

Species Ordination

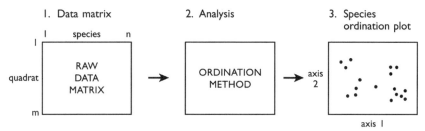

On the ordination plot, each point represents a **species** and the greater the distance between any two points, the greater the difference in the distribution of the species across all the quadrats.

Figure 6.10 The concept of species ordination. Reproduced with permission from John Wiley & Sons, Inc.

diagram can be split into dry heath – *Calluna vulgaris*, *Agrostis curtisii* and *Erica cinerea* – and wet heath – *Molinia caerulea*, *Erica tetralix*, *Juncus effusus*; grading into fourthly, valley bog, including *Narthecium ossifragum*, *Drosera rotundifolia* and *Sphagnum* species.

The quadrat ordination diagram

One of the principal aims of ordination is to define the underlying environmental gradients within a set of vegetation/environmental data. In indirect ordination, the analysis is carried out using only the vegetation/floristic data. Environmental relationships are determined once the ordination is complete and the graphs have been constructed. Most interpretation is carried out on two-dimensional graphs by superimposing environmental/biotic data collected at the same time as the floristic data in each quadrat onto the quadrat ordination diagram (paired data).

As an example, the two-dimensional quadrat ordination resulting from detrended correspondence analysis of the Gutter Tor data is presented in Figure 6.9a. From the data in Table 4.4, environmental plots have been produced for percentage soil moisture (Figure 6.9b); soil pH (Figure 6.9c) and grazing intensity (Figure 6.9d). Either raw data can be plotted, as in Figure 6.9b–d, or the data can be recoded as quartiles or quintiles as in Figure 6.12. This makes interpretation of trends very much easier.

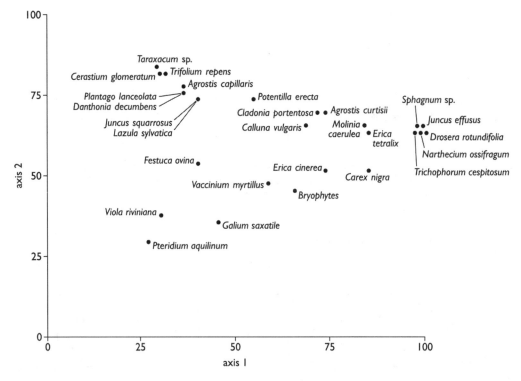

Figure 6.11   A two-dimensional species ordination plot derived from detrended correspondence analysis of the Gutter Tor, Dartmoor data (Table 4.4). Reproduced with permission from John Wiley & Sons, Inc.

By visual inspection, it is clear that % soil moisture values are generally lower on the left of the quadrat ordination diagram and higher on the right. A soil moisture gradient along the first axis is indicated. This trend is confirmed by the quintile plot (Figure 6.12a). Inspection of the plot of soil pH values, however, shows a more complicated picture (Figure 6.9c), although the quintile plot (Figure 6.12b) indicates that the highest values are all found in the group to the right of the diagram. This group would thus merit further investigation. Grazing pressure (Figures 6.9d and 6.12c) also shows a pattern, with high values in the top left group and lower values in the right-hand group. This is another pattern that almost certainly deserves further study.

Inspection of these graphs enables trends in the data to be recognised and hypotheses to be generated concerning vegetation and plant community variation in relation to the environment and other controlling biotic factors such as grazing. If there are clear gradients across the graph, then environmental/biotic gradients may be assigned to axes. However, such patterns are rarely clear-cut. Alternatively, patterns of high or low distribution may be observed in different parts of the diagram and interpreted accordingly. If trends are detectable, particularly if parallel to the axes, then rank correlation coefficients may help to display and explore any possible relationships.

In Table 6.3, Kendall's tau and Spearman's rank correlation coefficients have been calculated between the first two axes of the detrended correspondence analysis (DCA) of the Gutter Tor data and the environmental data collected for each of the 25 quadrats on percentage soil moisture, soil pH and grazing intensity. The first point to note is that the ordination axes are uncorrelated with each other, indicating that they may reflect different sources of variation within the vegetation data.

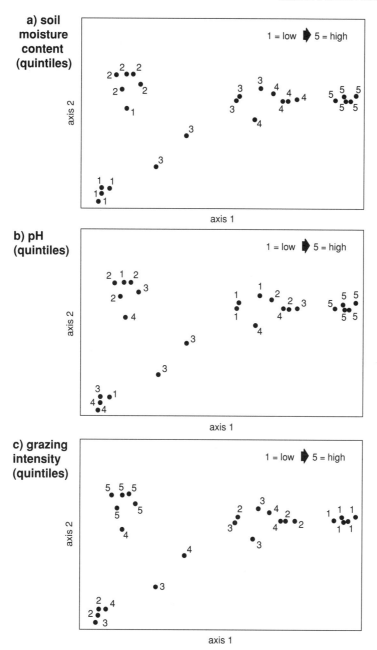

Figure 6.12 The quadrats ordination plots from Gutter Tor, Dartmoor, with overlaid values for environmental/biotic data for each quadrat plotted as quintiles: (a) percentage soil moisture; (b) soil pH; and (c) grazing intensity in order to enable patterns to be seen more clearly. Reproduced with permission from John Wiley & Sons, Inc.

Table 6.3 Matrices of Kendall's tau rank (upper triangle) and Spearman's rank (lower triangle) correlation coefficients between the quadrat scores on the first two axes of the detrended correspondence analysis of the Gutter Tor data (Table 4.4) and the environmental data for percentage soil moisture, soil pH and grazing estimates for those quadrats. A correlation above/below ±0.61 is significant at $p = 0.001$; ±0.45 at $p = 0.01$ and ±0.33 at $p = 0.05$. Original work of author.

		Ordination axes		Percentage soil moisture	Soil pH	Grazing intensity
		1	2			
Ordination axes	1	–	0.167	0.817	0.364	−0.452
	2	0.217	–	−0.145	0.168	0.242
Percentage soil moisture		0.953	0.234	–	0.367	−0.368
Soil pH		0.464	−0.310	0.480	–	−0.457
Grazing intensity		−0.582	0.357	−0.517	−0.613	–

Secondly, for both coefficients, the first axis is very highly correlated with percentage soil moisture, confirming the visual relationship described above. However, both of the other environmental factors are also significantly correlated, although pH only at the 0.05 level. The second axis is only significantly correlated with grazing intensity, which is already more highly correlated with the first axis. This demonstrates the point that application of correlation methods may be useful, but they will only find significant linear relationships parallel to each axis. A diagonal trend in an environmental factor across an ordination plot will thus result in significant correlations on both axes, as is the case with grazing intensity.

An important concluding point is that such interpretations are only meant to be a beginning. From the descriptive trends identified here, hypotheses concerning variation in the vegetation and between the vegetation and environmental factors should emerge, resulting in ideas for more detailed and specific research work involving the collection of new data. As an example, while the importance of soil moisture as a primary gradient on Gutter Tor has been clearly demonstrated, questions emerge as to the nature and intensity of grazing pressure on the different communities and how this may be reflected in the soils and their pH. A separate study of grazing habits could be valuable, as well as a study of nutrient turnover by animals as expressed in their dunging patterns.

MORE RECENT DEVELOPMENTS IN METHODS OF INDIRECT AND DIRECT ORDINATION

A number of increasingly sophisticated techniques for both indirect and direct ordination have been devised and applied over the past 50 years. Although the methods differ in their approach to analysis, it is important to realise that they all result in ordination diagrams of the type described above, with graphs in one, two or three dimensions.

Table 6.4 summarises the major methods of indirect ordination but also includes canonical correspondence analysis and redundancy analysis, which are best seen as methods of 'constrained' direct ordination. An important feature of this table is the process of evolution in methodology. The earliest method was that of Bray and Curtis (polar) ordination (PO) which was originally devised in 1957. This was then followed by principal component analysis (PCA) (Orlóci, 1966; Austin and Orlóci, 1966; Gittins, 1969), non-metric multidimensional scaling (NMS) (Fasham, 1977; Prentice, 1977); correspondence analysis and reciprocal averaging (CA/RA) (Benzécri, 1969, 1973; Hill, 1973b, 1974) and then detrended correspondence analysis (DCA) (Hill, 1979a; Hill

Table 6.4   A summary of methods of ordination. Original work of author.

Method	Author	Date	Comment	Application
Bray and Curtis polar ordination (PO)	Bray and Curtis	1957	Originally calculated and drawn by hand using compass construction	Widely used between 1960 and 1970. Superseded by later developments
Principal component analysis (PCA)	Gittins, Orlóci	1966	Relatively complex, requiring computer analysis	Only used on short linear response gradients due to distortion ('horseshoe' effect)
Correspondence analysis/reciprocal averaging (CA/RA)	Benzécri, Hill	1969, 1973	Simple calculation for one axis	Now largely replaced by DCA, NMS and CCA
Non-metric multidimensional scaling (NMS)	Fasham, Prentice	1977	A radically different approach – computationally intensive for large data-sets	Increasingly widely used
Detrended correspondence analysis (DCA)	Hill and Gauch	1979	'Improved' version of CA/RA with 'artificial' removal of 'arch'	Widely used but criticised
Canonical correspondence analysis (CCA)	ter Braak	1986, 1988	Really a form of direct ordination, since it is a revised version of correspondence analysis, constrained by multiple regression with environmental/ biotic factors	Widely used but criticised
Redundancy analysis (RDA)	Rao, ter Braak	1964, 1988	Similar to CCA but uses PCA instead of CA/RA constrained by multiple regression with environmental/ biotic factors	Only suitable for data on short linear response gradients

and Gauch, 1980). Canonical correspondence analysis (CCA) (ter Braak, 1985, 1986a, 1987, 1988a,b,c, 1994) and the related method of redundancy analysis (RDA) (Rao, 1964; ter Braak, 1988a,b) represent two of the more recent and widely applied ordination methods in plant ecology. Canonical correspondence analysis and redundancy analysis are, however, usually now seen as direct methods (Økland, 1996), as will be explained later.

SPECIES DATA TYPES IN ORDINATION ANALYSES

An interesting issue was raised by Podani (2005) in relation to the types of species data input to both ordination and also classification methods (Chapter 8). All of the ordination methods listed in Table 6.4 are usually applied to both presence/absence and abundance data, but abundance data can exist in a range of types, some of which can be described as ordinal (ranked multistate) data (e.g. Domin and Braun-Blanquet scales – Table 3.4). Podani argued that there may be some implications for analysis by some of the methods of ordination and classification on these types of data. Most researchers have chosen to ignore this issue, although Ricotta and Avena (2006) and Van der Maarel (2007) provide some useful further discussion on this question.

SPECIES DATA ARE OFTEN EASIER TO COLLECT THAN ENVIRONMENTAL/BIOTIC DATA

Species data are generally very much easier to collect than environmental/biotic data. Within any specific research problem, there are many environmental/biotic variables that could be measured and it is often not easy to predict in advance which ones are going to be of importance. Also, many environmental/biotic variables do not lend themselves easily to measurement, for example biotic pressures such as grazing, burning and trampling. Many environmental/biotic measures are indirect in terms of their effects on plants. A good example of this is soil pH, which is easy to measure with relative accuracy and is widely used as a general index of soil nutrient status. However, in terms of detailed variation in species composition, it may be a rather crude variable, with more detailed measurements of soil chemistry being of greater importance. Finally, environmental/biotic measurement is often expensive in terms of time, resources and money. For all these reasons, data on species composition is often much easier and more rapid to collect and to analyse. Where species data alone have been collected, then only indirect methods may be applied.

DETAILED DESCRIPTIONS AND EXPLANATIONS OF ORDINATION METHODS

Bray and Curtis (polar) ordination (PO)

Bray and Curtis or polar ordination (PO) was the first widely used method of indirect ordination devised in the 1950s before the advent of modern computers (Bray and Curtis, 1957). Although it has now been superseded, research indicates that it may still be a relatively efficient method of ordination (Beals, 1973, 1984; McCune and Beals, 1993; McCune and Grace, 2002; Huerta-Martínez et al., 2004). More importantly, however, of all methods, it is the one that is easiest to understand and to teach, since it is based on simple calculation and drawing compass construction. For this reason it is always a valuable exercise for newcomers to ordination to perform a hand-worked Bray and Curtis ordination on a small data-set (10–20 quadrats/samples).

The method

The method will be described using the New Jersey salt marsh data (Table 4.1).

(1) The raw data matrix, containing either presence/absence or abundance data, is tabulated (Tables 4.1 and 4.2).
(2) For a quadrat (sample) ordination, a dissimilarity matrix is computed between all quadrats in all combinations. Several coefficients are available, but Bray and Curtis used the Steinhaus (Sørensen/Czekanowski) coefficient in their original method. As described in Chapter 4, this is a similarity coefficient (S_C) that calculates values between 0 and 1. If two quadrats are identical, then the coefficient = 1.0. If two quadrats have nothing in common, the score is

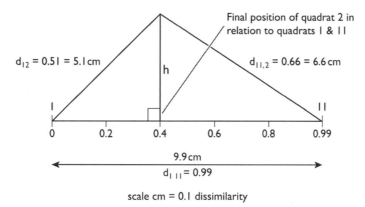

Final position of quadrat 2 in
relation to quadrats 1 & 11

$d_{12} = 0.51 = 5.1$ cm $d_{11,2} = 0.66 = 6.6$ cm

h

1   11

0   0.2   0.4   0.6   0.8   0.99

9.9 cm

$d_{1\ 11} = 0.99$

scale cm = 0.1 dissimilarity

Figure 6.13   Location of end points (quadrats 1 and 11) and quadrat 2 on the first axis of the Bray and Curtis (polar) quadrat ordination of the New Jersey salt marsh data. Reproduced with permission from John Wiley & Sons, Inc.

zero. These similarity coefficients are converted into dissimilarity coefficients (D_C) by taking the values away from 1.0, so that identical quadrats then have a score of 0.0 and quadrats with no species in common have a score of 1.0 (maximum dissimilarity). The dissimilarity measures are then used as distances in the construction of the ordination diagram.

Thus
$$D_C = 1.0 - S_C \tag{6.2}$$

For the New Jersey salt marsh data, the matrix of dissimilarity coefficients for quadrats is presented in Table 4.6.

(3) The ordination diagram is then constructed geometrically on the basis of the dissimilarity matrix, using a drawing compass and ruler. The two quadrats with the highest dissimilarity value are taken. If two quadrats have no species in common, then the highest possible value will be 1.0. In the case of the New Jersey data, no two quadrats are totally different, but quadrats 1 and 11 and 1 and 12 both have dissimilarity scores of 0.99. Either of these pairs of quadrats can be taken as reference quadrats to be placed at either end of the first axis. Taking quadrats 1 and 11, the first axis itself is scaled to the dissimilarity value between 1 and 11 (0.99). Thus on a piece of graph paper, if a scale of 1 cm = 0.1 dissimilarity is taken, a line representing the first axis 9.9 cm long would be drawn with quadrat 1 at one end and quadrat 11 at the other (Figure 6.13). It makes no difference which quadrat is placed at which end of the axis. The same plot will be produced, except that, depending on which quadrat is at which end, one plot will be a mirror image of the other. The distances between the points representing quadrats will be the same.

(4) The remaining ten quadrats are then positioned on the first axis using compass construction. Quadrat 2 is located on the first axis with respect to quadrats 1 and 11 by first taking the dissimilarity values between quadrats 1 and 2 (0.51 – Table 4.6) and between quadrats 2 and 11 (0.66 – Table 4.6). With the drawing compass point located at the end of the axis where quadrat 1 is located, an arc is drawn with its radius proportional to the dissimilarity between quadrats 1 and 2 (0.51 = 5.1 cm – Figure 6.13). Similarly, a second arc is drawn with the compass point positioned at the other end of the axis, where quadrat 11 is located, with a radius proportional to the dissimilarity between quadrats 2 and 11 (0.66 = 6.6 cm). At

the intersection of the two arcs, a perpendicular is dropped to the axis or baseline, and the point at which it intersects the line represents the final position of quadrat 2 on the first axis (Figure 6.13).

(5) The other quadrats are then positioned with respect to quadrats 1 and 11 in exactly the same manner, resulting in the single axis ordination shown in Figure 6.14a.

(6) The second axis is now constructed. Two reference quadrats for the second axis are chosen. The general rule for those quadrats that qualify is that they must be close to the centre of the first axis but still have a high dissimilarity value between them. In the New Jersey case, quadrats 2 and 8 are near to the centre of the first axis and yet have a dissimilarity value of 0.75. Thus they are suitable as reference quadrats for the second axis.

(7) To construct the second axis, a line is drawn proportional to the dissimilarity between quadrats 2 and 8 (0.75 = 7.5 cm) and the other ten quadrats are positioned by compass construction in the same manner as for the first axis (Figure 6.14b).

(8) The last stage involves the plotting of the points representing quadrats with respect to their coordinates on both axes. These are usually placed at right angles to each other (Figure 6.14c). Orlóci (1974, 1978) pointed out that the true angle between the axes is usually oblique and presented an analytical method that calculated the true angle between the axes and which corrects the rotated coordinates on the second axis back to rectangular coordinates for plotting (Kent, 1977). The differences attributable to the correction are usually marginal, however.

(9) Construction of a third axis to give a three-dimensional ordination is possible. The procedure for the selection of reference quadrats is the same as on the second axis, except that selection is based on combined assessment of quadrats in the middle of both the first and the second axes, which still have relatively high dissimilarity values between them. Quadrats 7 and 10 would be suitable, being relatively close together and near to the centre of the second axis but still having a dissimilarity value of 0.79. The resulting three-dimensional ordination diagram is shown in Figure 6.6c.

A problem in using Bray and Curtis ordination is that occasionally the positioning of a quadrat by the intersection of the two arcs may be impossible and the arcs do not intersect. The reason for this is the failure of the Steinhaus (Sørensen/Czekanowski/Bray and Curtis) dissimilarity coefficient to fulfil all the properties of a metric coefficient of similarity. The true distances between the quadrats are distorted when they are projected into a space of fewer dimensions. To position points accurately in relation to each other requires two dimensions of space, and when these distances are projected into one dimension on a particular axis, the distances are foreshortened. However, relatively, the distances are still preserved and this is why the ordination still works. If axes do not intersect, it is legitimate to select a different pair of reference quadrats and to reconstruct the ordination using those. The choosing of different pairs of reference quadrats only has the effect of 'viewing' the scatter of points and their relative dissimilarities from a different 'angle'. However, this also introduces the idea of 'distortion' of the true relationships between points on an ordination diagram due to the computational process of the ordination itself.

A further solution is to use a different similarity coefficient such as those recommended by Orlóci (1974, 1978), Legendre and Legendre (1998) or Borcard et al. (2011).

For computerised analysis, the compass construction can be replaced by the following formulae (Causton, 1988).

In Figure 6.15, for triangle $Q_1Q_iP_i$, Pythagoras' theorem gives:

$$h_i^2 = d_{1i}^2 - X_{1i}^2 \qquad (6.3)$$

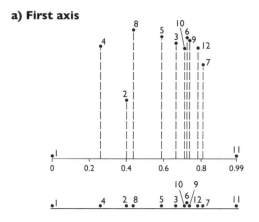

a) First axis

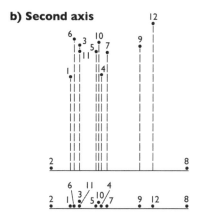

b) Second axis

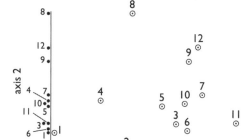

c) Joint plot of axes 1 & 2

Figure 6.14 (a) Positioning of all quadrats on the first axis of the Bray and Curtis (polar) ordination of the New Jersey salt marsh data using compass construction. (b) Positioning of all quadrats on the second axis of the Bray and Curtis (polar) ordination of the New Jersey salt marsh data using compass construction. (c) The final ordination graph with the two axes drawn at right angles (orthogonal) to each other. Note that this is the same as in Figure 6.6, but the endpoints have been inverted on the first axis. This makes no difference to the result and is only a 'mirror image'. The relative distances between the points are preserved. Reproduced with permission from John Wiley & Sons, Inc.

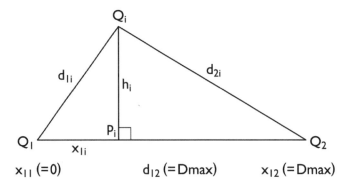

Figure 6.15   General principles for calculating the position of a given quadrat with respect to two reference quadrats (Q_1 and Q_2) which define the axis (equations 6.3–6.6). Redrawn from Causton (1988), with kind permission of Chapman & Hall/Unwin Hyman Publishers.

For triangle $Q_2Q_iP_i$, Pythagoras gives:

$$h_i^2 = d_{2i}^2 - (d_{12} - x_{1i})^2 \tag{6.4}$$

Multiplying out the bracket in Equation (6.4) and elimination of h_i^2 between the two equations gives:

$$d_{1i}^2 - x_{1i}^2 = d_{2i}^2 - d_{12}^2 + 2d_{12}x_{1i} - x_{1i}^2 \tag{6.5}$$

Re-arrangement following cancellation of x_{1i}^2 terms gives:

$$X_{1i} = \frac{d_{1i}^2 - d_{2i}^2 + d_{12}^2}{2d_{12}} \tag{6.6}$$

The advantage of this is that the calculation of quadrat position on the axis (x_{1i}) is defined entirely from the dissimilarity values and thus can be computed from the original dissimilarity matrix.

In conclusion, Bray and Curtis (polar) ordination is still seen as an effective ordination method, which compares favourably with some subsequent methods such as principal component analysis and correspondence analysis/reciprocal averaging (Gauch and Scruggs, 1979). However, polar ordination is not without its problems, the principal of which is the range of choice of pairs for endpoints of each axis. In small data-sets this does not matter but, as amounts of data become larger, the chances of obtaining consistent ordinations from different endpoint pairs are greatly reduced. The main advantage of polar ordination is its value in teaching, since if carried out manually using compass construction, non-mathematically orientated students learn a great deal about multidimensionality and the ordination process.

Principal component analysis (PCA)

The method of principal component analysis (PCA) was first described by Karl Pearson (1901). However, a fully practical computational method was only devised much later by Hotelling (1933). Even then, the process of calculation was very daunting other than for very small data-sets because it had to be carried out by hand. Thus only when computers became widely available did the technique come into general use in plant ecology. Although Goodall (1954) first applied

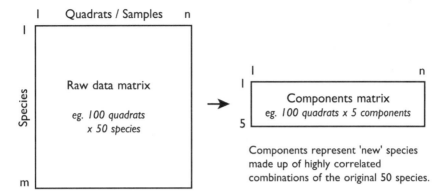

Figure 6.16   Reduction of many species or environmental/biotic variables into a few components.

the method under the erroneous title of factor analysis, PCA only became popular after the publication of a paper by Orlóci in 1966.

The use of PCA as an ordination method on species data is now a contentious issue. Nevertheless, the method is important because it has been widely used on species data in the past and there are many examples of its application in the literature. Much more important, however, is the use of PCA to synthesise environmental/biotic data and to produce an ordination of quadrats based on environmental/biotic variables alone (Figure 6.1c). Here it is still one of the most effective methods of analysis.

PCA was the first ordination technique in which the ordination axes were computed from the data matrix alone and the researcher did not have to supply weights, endpoints or other subjective information during the computation process. For most non-mathematical students, PCA is best explained geometrically, rather than by matrix algebra. The basic idea of PCA is that if data to form a matrix of n quadrats and m species or m environmental/biotic variables are collected, as in Figure 6.16, there will be a large amount of duplication or correlation in the variability of the species or environmental/biotic variables across the quadrats. Thus an original data matrix of, for example, 100 quadrats and 50 species/variables can be reduced to 100 quadrats and 5 or even fewer components. These components can be regarded as new 'super-species' or 'super-variables' made up of highly correlated combinations of the original 50 species or environmental/biotic variables (Figure 6.16).

One of the most important features of the components is that, whereas all the former species/variables were almost certainly highly intercorrelated amongst each other, the new components are completely uncorrelated and are said to be orthogonal. The problems of intercorrelation and duplication within the original species/variables in the data-set are removed by the analysis.

Another important concept is that, just as in the original species/variable matrix, where each quadrat/sample had a score for each species or environmental/biotic variable, so in the new components matrix, each quadrat/sample has a score for each component, which taken together are known as the component scores.

The core of any PCA are eigenvectors and eigenvalues.

Eigenvectors

Eigenvectors are sets of scores, each of which represents the weighting of each of the original species or variables on each component. The eigenvector scores are scaled like correlation coefficients

(Chapter 5) and range from +1.0 through zero to −1.0. For each component, every species or variable has a corresponding set of eigenvector scores, and the nearer the score is to +1.0 or −1.0, that is, the furthest away from zero, the more important that species or variable is in terms of weighting that component.

As an example, if a set of data collected in *n* quadrats contained six species or environmental variables A–F, following PCA, the eigenvector scores for the first two components could be:

| | Eigenvectors | |
Species/variable	Component I	Component II
A	0.91	0.21
B	0.79	0.19
C	0.75	0.29
D	0.01	−0.90
E	−0.12	−0.86
F	−0.01	−0.93

Note that conventionally component numbers are denoted by Roman numerals.

On component I, the highest weightings (eigenvector scores) are for species/variables A, B and C, and on component II the highest weightings are for species D, E and F. It is on the basis of these eigenvector scores that the interpretation of a component is made. In this case, component I would be a 'new' species/variable with properties of the original species/variables A, B and C, and component II would be interpreted as a combination of the original species/variables D, E and F. If the similarity or correlation matrix between the species/variables was calculated, it would be clear that species/variables A, B and C were very similar in their distribution amongst the quadrats, as were D, E and F. Note that in some analyses, eigenvectors and their scores are known as component or factor loadings.

Eigenvalues

Eigenvalues (as opposed to eigenvectors) are values that represent the relative contribution of each component to explanation of the total variation in the data. There is one eigenvalue for each component, and the size of the eigenvalue for a component is a direct indication of the importance of that component in explaining the total variation within the data-set.

A geometrical explanation of principal components analysis

The following geometrical explanation is adapted from Gould and White (1974, 1986):

The raw data matrix, centring and standardisation   The starting point of PCA is a normal data matrix with quadrats and species as below. The data could be either the cover of each of seven species measured on a 10-point Domin scale for ten quadrats in a meadow, or for seven environmental variables that have been scaled between 0 and 10. This matrix has been deliberately contrived to assist with explanation of the method:

Species or variables		Quadrats									
		1	2	3	4	5	6	7	8	9	10
	A	1	5	7	9	10	8	6	4	3	2
	B	8	10	7	9	6	5	4	3	1	2
	C	3	6	7	9	10	8	5	4	2	1
	D	4	8	7	10	9	6	5	3	2	1
	E	10	9	7	8	6	5	4	3	1	2
	F	2	6	7	9	10	8	5	4	3	1
	G	1	6	8	9	10	5	7	4	3	2

Number of quadrats $= 10$
Number of species/variables $= 7$

In this example, the aim is to produce an ordination diagram for quadrats or samples. This is also known as a normal or R analysis. The opposite of this is known as a Q analysis or inverse analysis and results in the species ordination (if the data are species) or an ordination of the environmental/biotic factors (if the data are environmental/biotic variables).

In the application of PCA to vegetation or environmental/biotic data, initially the data may be standardised. This may be done for either or both species/variables and quadrats. The commonest means of standardisation is zero mean, unit variance. As the name suggests, each species/variable or quadrat is rescaled with a mean of zero and a variance of 1.0. If it is carried out for species/variables, then each species/variable score is expressed in units of standard deviations away from a mean of 0. The general formula is:

$$\frac{SS_i = (S_i - \bar{S})}{\sigma S} = \frac{\text{Each species/variable score} - \text{mean species/variable score}}{\text{Standard deviation of the species/variable}} \qquad (6.7)$$

where

$SS_i =$ the standardised score of species/variable S in quadrat i
$S_i =$ the original scores of species/variable S in quadrat i
$\bar{S} =$ the mean score of species/variable S across all quadrats
$\sigma S =$ the standard deviation of species/variable S across all quadrats

Taking species/variable A in the above example, its mean across the 10 quadrats is 5.5 and the standard deviation is 3.03. The original score of 1 in quadrat 1 becomes -1.49, while the score of 9 in quadrat 4 becomes $+1.16$. All other scores on all other species/variables can be standardised in exactly the same way, using the mean and standard deviation for each species/variable in turn. If the data are not standardised, then the analysis will be biased towards those species/variables with the highest variances, which will in turn reflect the range of measurement values for a particular species/variable. If the correlation coefficient is used as a measure of similarity, then standardisation is automatic. Quadrats or stands can be standardised in exactly the same way, except that the original scores are related to the mean and standard deviation of each quadrat. It is clearly possible to standardise both species/variables and then quadrats and this is known as double standardisation.

The similarity or correlation matrix The next stage of the R analysis is to calculate a similarity or correlation matrix between all species/variables. The most commonly used coefficient is the product-moment correlation coefficient. For the above example, the product-moment correlation matrix is:

Species/variable

	A	B	C	D	E	F	G
A	1.00						
B	0.35	1.00					
C	0.95	0.58	1.00				
D	0.83	0.79	0.93	1.00			
E	0.20	0.96	0.47	0.67	1.00		
F	0.98	0.49	0.99	0.90	0.36	1.00	
G	0.93	0.42	0.88	0.85	0.26	0.90	1.00
	A	B	C	D	E	F	G

Species/variable

The cosine of a right-angled triangle equals the correlation coefficient   Fundamental to the geometrical explanation of PCA is the idea that the correlation coefficient can be expressed as the cosine of a right-angled triangle.

In a right-angled triangle, the cosine of an angle is defined as the ratio of the length of the side adjacent to the angle, to the length of the hypotenuse. In a triangle with sides of lengths of the proportions 3:4:5, the cosine of α is $4/5 = 0.8$, which in cosine tables gives an angle of 37°.

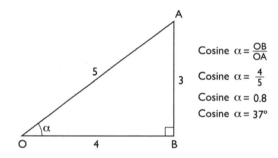

$$\text{Cosine } \alpha = \frac{OB}{OA}$$

$$\text{Cosine } \alpha = \frac{4}{5}$$

$$\text{Cosine } \alpha = 0.8$$

$$\text{Cosine } \alpha = 37°$$

The next step is to consider point O as a hinge, with side OA being lowered onto OB. The result is that the angle α becomes smaller and smaller and the two sides OA and OB become more and more similar in length, provided that the right angle OBA is maintained, until eventually, when $\alpha = 0°$, OA and OB are identical.

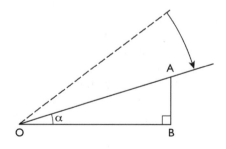

Again, side OA can be moved away from OB until OA is almost vertical. If the perpendicular from A to B is constructed, then the distance becomes very small, until when OA is at right-angles to OB, the distance OB = zero; the ratio OA/OB = 0.0 and the angle $\alpha = 90°$ (cosine $90° = 0.0$).

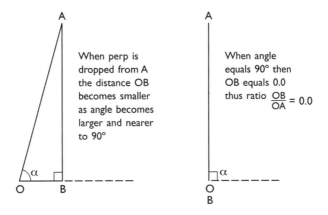

If OA is kept swinging around pivot O, the angle α then becomes greater than 90° and the cosine becomes negative, eventually approaching -1.0 when the angle $\alpha = 180°$.

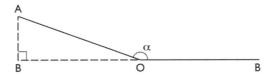

This idea can now be related to correlations between species or environmental/biotic variables. According to the above, the cosine, expressed as a ratio between two sides of a right-angled triangle, varies from $+1.0$ through zero to -1.0, as does the correlation coefficient. It thus becomes possible to represent correlations between species/variables geometrically.

Each species or variable in the analysis can be considered a vector. A vector is a line that possesses properties of both length and direction.

A vector

The length of a vector representing a species/variable is directly related to its variance, but since the first step of PCA is usually to standardise all the species/variables to a variance of 1.0, vectors representing all species/variables in the analysis become the same length, that is 1.0.

Thus if there are two species/variables A and B, these can be expressed as two vectors of the same length. Also, using the cosine principle, the angle between the two vectors at their origin can be taken as an expression of their correlation.

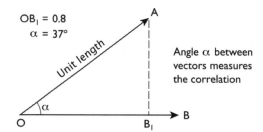

If the perpendicular is dropped from the tip of vector A to intersect vector B, then depending on the angle α, the length of O to the point of intersection alters (B_1). Thus if $OA = 1.0$ (standardised) and angle $\alpha = 37°$, OB_1 must be 0.8 (cosine $37° = 0.8$).

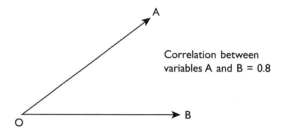

If the correlation between species/variables A and B, represented by vectors A and B, is 0.8, then this is shown geometrically by positioning the vectors at an angle of $37°$ to each other.

This logic can then be followed through for all other correlations between any other pair of species/variables represented as vectors. If the correlation between two species/variables is 0.0, the vectors must be at an angle of $90°$ to each other.

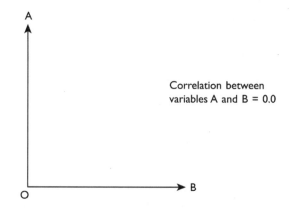

If the correlation between two species/variables is 1.0, then the two vectors will be on top of each other and the angle between them will be zero ($0°$).

If the correlation between two species/variables is −0.8, then the angle between the two vectors will be 143°.

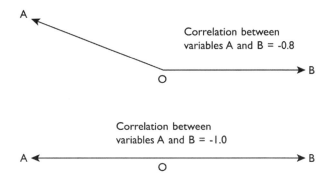

If two species/variables are perfectly negatively correlated (−1.0), then the two vectors will point in opposite directions and the angle between them will be 180°.

This idea is now used to represent the intercorrelations between the seven species/variables as shown on p. 198. Each species/variable is represented as a unit vector and the angles between them correspond to the degree of correlation. Strictly, these should be represented in seven dimensions – one for each species/variable – but obviously this is impossible to visualise. Thus for purposes of explanation, the correlations have been carefully chosen, so as to be depicted on the two-dimensional plane of the page.

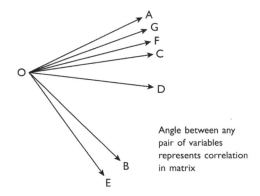

The principal axis Having obtained the geometric picture of the correlations between all seven species/variables, the next stage is to collapse all the species/variables into one or two major trends or components. On the geometric diagram, those species/variables that are very much intercorrelated – A, G, C and F – tend to thrust together in one particular direction. The aim of component analysis is to find this overall directional thrust of the vectors by passing a line or axis through their common origin, enabling each vector representing a species/variable to make a right-angled projection onto this axis, as in the following diagram. This line or axis is known as the principal axis. Since the vectors are of unit length, their projections onto the axis are no more than the cosines of the angles between the vectors and the principal axis.

The important questions are: how is this principal axis located, and how does one know when it shows the direction of strongest thrust of all the vectors?

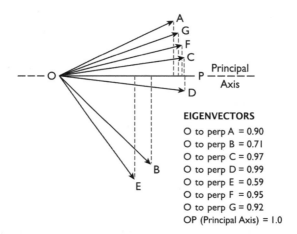

EIGENVECTORS

O to perp A = 0.90
O to perp B = 0.71
O to perp C = 0.97
O to perp D = 0.99
O to perp E = 0.59
O to perp F = 0.95
O to perp G = 0.92
OP (Principal Axis) = 1.0

In mathematical terms, the criteria can be very complex, but in geometric terms, the criteria are relatively simple. The principal axis is swung slowly around the origin (O) and at each instant, the lengths on the principal axis OP from O to the perpendiculars from the tips of all the vectors on to the principal axis are measured, as in the above diagram. These projections are actually the eigenvectors or loadings of all the species onto the principal axis and these can be seen to be equivalent to the degree of correlation of each species/variable with the principal axis.

The position of the principal axis at the point that represents the maximum direction of thrust of all the vectors is found by taking the eigenvectors or loadings for any position of the principal axis, squaring them and then summing the squared values. The optimum position for the principal axis is found when the sum of the squared values is at a maximum. This value, which is the sum of the squared correlations between all species/variables and the axis, is the eigenvalue of that principal axis or first component. In practice, mathematically, there are rapid ways of finding the eigenvalues of any correlation matrix. The optimum position of the principal axis for the example is shown in the above diagram.

In the example, the eigenvector scores at the shown position of the principal axis are:

Species/variable	Eigenvector
A	0.90
B	0.71
C	0.97
D	0.99
E	0.59
F	0.95
G	0.92

The eigenvalue is thus: $(0.90)^2 + (0.71)^2 + (0.97)^2 + (0.99)^2 + (0.59)^2 + (0.95)^2 + (0.92)^2 = 5.33$.

The eigenvalue represents the highest possible degree of correlation of all the species/variables with the principal axis, and thus is a measure of the amount of variation in the data-set accounted for by the first axis. The principal axis represents the first component, which has varying contributions from each of the original species/variables, and those species/variables, whose vectors are closest to the principal axis, are the most important in attempting to explain or interpret the first axis.

Derivation of a second axis Earlier, it was stated that in the simplest models of PCA, the components extracted were always orthogonal (at right angles) to each other, which means that they are uncorrelated. In vector representation, if two species/variables are completely uncorrelated, that is the correlation = 0.0, then the vectors representing those species will be at right angles to each other (90°). By the same logic, if one of the basic tenets of PCA is orthogonality, then the second axis or component must be at right angles to the first. This is shown in the following diagram.

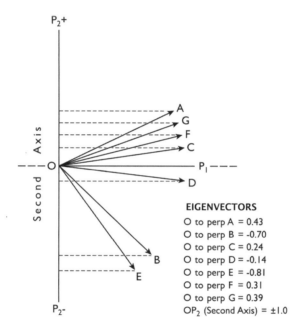

Once the second axis is in position, then the eigenvectors and the eigenvalue for that axis can be derived in exactly the same manner as for the first axis. The projections of the tips of the vectors representing each species/variable onto the second axis at right angles give the eigenvectors for the second axis, and the sum of the squared eigenvector scores gives the eigenvalue. Thus:

Species/variable	Eigenvector
A	0.43
B	−0.70
C	0.24
D	−0.14
E	−0.81
F	0.31
G	0.39

The eigenvalue is thus: $(0.43)^2 + (-0.70)^2 + (0.24)^2 + (-0.14)^2 + (-0.81)^2 + (0.31)^2 + (0.39)^2 = 1.66$.

Note that the second eigenvalue is much lower than the first. One of the features of PCA is that axes are extracted in descending order of importance in terms of their contribution to the total variation in the data-set. Also note that those species/variables that had high eigenvector scores on the first axis tend to have lower eigenvector scores on the second.

Subsequent axes can be extracted in exactly the same manner, by erecting further axes at right-angles to each other. Again a set of eigenvector scores can be determined, and the sum of squared eigenvector scores will give the eigenvalue, although this value will be lower than on the first or second axis. Mathematically, the process of extracting further axes can go on until the number of axes or components equals the number of species or variables, but most higher axes are trivial and contribute little to explanation. For small data-sets, 2–3 axes will summarise most of the variation in the data, and with larger sets 4–5 will suffice. As a guideline, sometimes axes are only extracted up to the point where the eigenvalue associated with a particular axis or component has a value of ≥ 1.0. An eigenvalue of less than 1.0 means that the axis is contributing less to the overall explanation of the variability in the data than any one of the original species or variables.

Calculation of component scores

So far, the analysis has produced two components, each of which has a set of eigenvectors or loadings and a corresponding eigenvalue, which can be seen as summarising the amount of overall variation in the data accounted for by each component. The final stage of analysis is the calculation of the component scores for each quadrat/sample in the original matrix. It is these scores that make up the new matrix of quadrats/samples × components (Figure 6.16).

To obtain the component scores for each of the original ten cases, each of the original values for each species or variable within a quadrat is multiplied by the eigenvector score for that species/variable on a given component. These values are then summed for each quadrat to give the component scores for the quadrats. Thus:

				Quadrats							Eigenvectors	
Species or variables	1	2	3	4	5	6	7	8	9	10	Component I	Component II
A	1	5	7	9	10	8	6	4	3	2	0.90	0.43
B	8	10	7	9	6	5	4	3	1	2	0.71	−0.70
C	3	6	7	9	10	8	5	4	2	1	0.97	0.24
D	4	8	7	10	9	6	5	3	2	1	0.99	−0.14
E	10	9	7	8	6	5	4	3	1	2	0.59	−0.81
F	2	6	7	9	10	8	5	4	3	1	0.95	0.31
G	1	6	8	9	10	5	7	4	3	2	0.92	0.39

To obtain the component score for quadrat 1 on component I:

$$\text{Score} = (1 \times 0.90) + (8 \times 0.71) + (3 \times 0.97) + (4 \times 0.99) + (10 \times 0.59) + (2 \times 0.95)$$
$$+ (1 \times 0.92) = 22.17.$$

To obtain the component score for quadrat 10 on component II:

$$\text{Score} = (2 \times 0.43) + (2 \times -0.70) + (1 \times 0.24) + (1 \times -0.14) + (2 \times -0.81) + (1 \times 0.31)$$
$$+ (2 \times 0.39) = -0.97.$$

This process is carried out for every quadrat on every component extracted. In this example, where only two components have been extracted, this gives a new matrix of 10 quadrats × 2 components as below:

```
Components matrix
                                       Quadrats
Components    1      2      3      4      5      6      7      8      9      10

     I      22.17  41.80  42.90  54.40  53.80  39.50  31.40  21.70  13.40   9.10
     II    -12.10  -7.60  -2.00  -1.80  -3.40   1.40   1.30   0.50   2.10  -0.97
```

Thus the original matrix of ten quadrats and seven species/variables is reduced to ten quadrats with two orthogonal (uncorrelated) components, with a high proportion of the total variation explained. The exact amount explained by each component can be calculated as follows below. The procedure is best explained by taking an extreme example. The highest possible explanation of any one component in PCA would be if 100% of the variation was explained by one axis. For this to occur, all the species/variables in the analysis would have to be perfectly correlated with each other – that is, all the correlations in the matrix would be 1.0 and all the species/variables would be identical in their distribution in the quadrats. In vector form, this would be represented as below, with all the vectors exactly on top of each other lying along the principal axis.

The eigenvector scores would thus also all be 1.0, and the eigenvalue, which is the sum of the squared eigenvector scores, would be equal to the number of species/variables, which in the example would be 7.0. That is:

$$(1.0)^2 + (1.0)^2 + (1.0)^2 + (1.0)^2 + (1.0)^2 + (1.0)^2 + (1.0)^2 = 7.0$$

7 is the number of species/variables.

However, in a typical data-set, where species or variables are intercorrelated to varying degrees, the angles between the vectors representing the species/variables start to appear, as in the previous geometrical diagrams, and thus the eigenvector scores fall below 1.0.

Since the eigenvector scores fall below 1.0, so the eigenvalue, which is the sum of the squared eigenvector scores, must also fall. It follows that the eigenvalue for any component is an exact measure of the proportion of the total variation in the data explained by that component and can be expressed as a fraction of the eigenvalue over the number of species/variables in the analysis.

In the example, the percentage of the total variation explained by the first component is:

$$\frac{5.33 \text{ (the eigenvalue)}}{7.00 \text{ (the number of species/variables)}} \times 100 = 76.20\%$$

The percentage explanation of the second axis is:

$$\frac{1.66 \text{ (the eigenvalue)}}{7.00 \text{ (the number of species/variables)}} \times 100 = 23.65\%$$

The cumulative explanation of the two components is:

$$76.20 + 23.65 = 99.85\%$$

Thus all except for 0.15% of the variation in the original matrix of ten quadrats and seven species/variables is explained by the two components and there would be little value in extracting any further components or axes, although, in theory, this could be calculated up to the seventh.

Production of the final quadrat ordination plot

The quadrat ordination plot is derived by taking the component scores for the quadrats on the two axes and plotting them as a graph, as below. This plot can then be interpreted in exactly the same way as for any other ordination method, based on the principle that the distance between any two points representing quadrats on the graph is an approximation to their similarity of species composition. Thus quadrats 4 and 5 are similar to each other, as are 9 and 10, but 4 and 5 are both very different from 9 and 10.

Clearly, this example has involved considerable simplification. Nevertheless, the basic principles of all applications of PCA are derived from the ideas presented here. For the mathematician, the method is more clearly explained by matrix algebra. Good examples using the matrix algebra approach are in Orlóci (1978), Digby and Kempton (1987) and Causton (1988), Jongman *et al.* (1995), Legendre and Legendre (1998), Waite (2000) and Borcard *et al.* (2011).

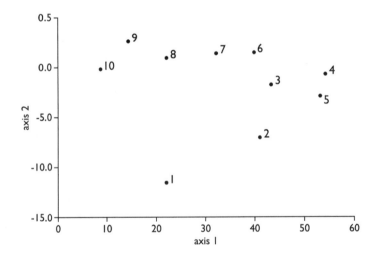

Data centring

A further important distinction in PCA is whether a centred or non-centred analysis is performed (Orlóci, 1967a; Gauch, 1982b; Causton, 1988; Jongman *et al.*, 1995; Legendre and Legendre,

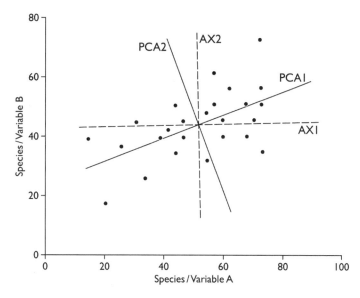

Figure 6.17 Data centring and derivation of principal components (adapted from Gauch, 1982b; redrawn with kind permission of Cambridge University Press).

1998). The idea of data centring is best explained by taking a simple case of two species or variables. When plotted jointly, they give the swarm of points on the graph in Figure 6.17. There are three different sets of coordinates for these points. Firstly, there are the original axes of the graph marked species/variable A and species/variable B. Secondly, there is a pair of axes which are placed on the centroid for the two species/variables, where the centroid is simply the average of the scores for each of the two species/variables. These are labelled AX1 and AX2. This shifting of the coordinate system for the swarm of points to the centroid is known as centring. Thirdly, there is another pair of axes PCA1 and PCA2, which represent the rotation of the first or principal axis around the centroid according to the principles of PCA, so that the projection distances from the points to the axis are minimised, and the axis lies along the direction that summarises maximum variance, as explained above. The second PCA axis in Figure 6.17 is defined a being at right-angles or orthogonal to the first. Clearly, data centring, before the locating of axes, will alter the final position of those axes when compared with a situation where the data have not been centred. Ordinations of non-centred data tend to produce a trivial general first component or axis. Differences between centred and non-centred analyses are usually not great, unless there are major discontinuities in the data.

As an example of this, three PCA quadrat ordinations of the Garraf data-set (Table 4.3) are presented in Figure 6.18. In Figure 6.18a, the data have not been centred, while in Figure 6.18b, they have. In this case, the effects of centring on this data-set are minimal. However, if the data are both centred and standardised, quite striking differences are observable between the plots. Thus standardisation of data is much more likely to have a significant impact on the ordination. In most ecological applications, data are analysed using centred and standardised PCA, and if a variance–covariance matrix is used, data are centred in the process. If the correlation coefficient is used, data are both centred and standardised. The arguments for and against centring and

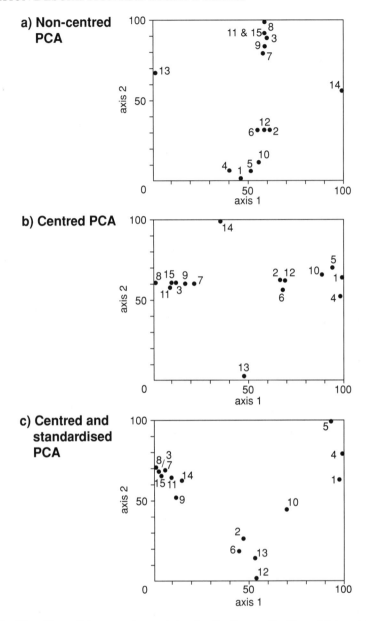

Figure 6.18 The effects of data centring and standardisation on the Garraf data-set (Table 4.3).

standardisation are given by Dagnelie (1960, 1978), Noy-Meir (1973), Noy-Meir *et al.* (1975), Orlóci (1967a, 1978), Greig-Smith (1980, 1983), Hinch and Somers (1987) and Causton (1988).

R and Q ordinations

Some confusion has arisen in the literature over this terminology, which is normally only applied to the use of PCA for ordination. An R ordination is the same as a quadrat ordination, where the component scores derived from a correlation or similarity matrix calculated between species is used to give ordination plots for quadrats. However, a further aspect of PCA is that the eigenvector scores (which represent the weightings of each species on each component) can be plotted to give a species ordination as well. With the R approach, the underlying geometric model is thus one of quadrats in terms of species space. It is clearly possible to carry out PCA on the matrix inversely, so that the component scores would represent species and the eigenvector scores on each component would be used to give an ordination of quadrats (Figure 6.10). This is known as the Q approach. In this case, the underlying geometric model is of species defined in quadrat space. If the same centring and standardisation is applied, then an R analysis will give the same result as a Q analysis.

The biplot method

The fact that both component scores and eigenvectors scores from either an R or a Q analysis can give ordination information on both quadrats and species has led to the development of the biplot method of displaying results. The method was first proposed by Gabriel (1971, 1981) and Bradu and Gabriel (1978). Their use in PCA for floristic ordinations is explained by ter Braak (1983) and ter Braak (1987). Both quadrat scores and species are plotted on the same graph but using different scales, because the scaling of quadrat (component) scores (in an R analysis) will be different for that for species represented by the eigenvector scores. In Figure 6.19a, following a centred PCA on the dune meadow data (Jongman *et al.*, 1995), the quadrats have been plotted in relation to the first two components in a standard quadrat plot. However, superimposed on this, with the axes aligned and suitably scaled, is the plot of species eigenvector scores (loadings) on the first two components. Arrows are then drawn from the joint centred ordination axes (0,0) to the points representing each species. The direction of the arrow indicates the direction in which the abundance of a species increases most rapidly. The length of the arrow shows the rate of change in abundance in that direction. Thus a long arrow indicates gradual rate of change in abundance while a short arrow represents very rapid change. Figure 6.19b shows how this can be used to interpret the abundance of one species in relation to the quadrat plot. Those quadrats containing *Agrostis stolonifera* have been extracted and lines have been constructed from each quadrat point to intersect the oblique arrow or axis representing *Agrostis stolonifera* on the plot. The sequence of quadrats from the top of the arrow through the origin to the tail is an approximation to the rank order of abundance of the species in the quadrats. Thus in Figure 6.19b, the abundance of *Agrostis stolonifera* will be greatest in quadrat 16, second highest at quadrat 13 and lowest at quadrat 6, which is at the other end of the axis. All quadrats to the right of the origin will have abundance above the overall mean of the species, while those to the left will be below.

Distortion effects in using PCA as a method of ordination on species data

It is important first to stress that there is no question that PCA is entirely appropriate for the analysis of environmental/biotic data, where relationships between variables can be assumed to be linear and where there are few or no absences in the data matrix. In contrast, there has been extensive discussion over the use of PCA on species data. The reasons for the doubts concerning its use on species data lie firstly in the fact that the assumed underlying species–environment response model for PCA is linear (Figure 6.5a,b), whereas over environmental/biotic gradients of any length, a Gaussian unimodal, bell-shaped model is usually considered to be more appropriate (Figure 6.5c,d) (Austin, 1999a, 2002, 2005).

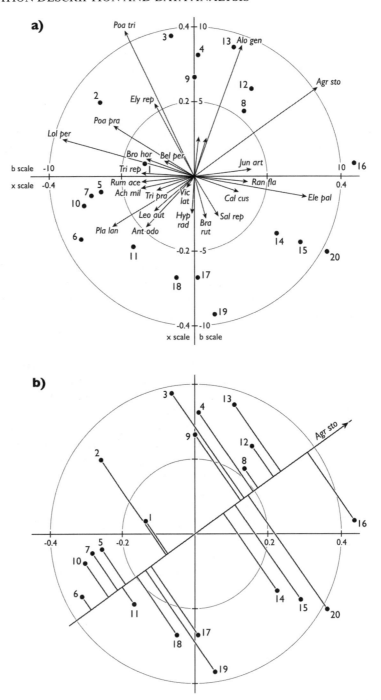

Figure 6.19 Biplot from principal component analysis of the dune meadow data (Jongman *et al.*, 1995). (a) Species are represented by arrows; quadrats by dots; the b scale applies to species and the x scale to quadrats/samples. Species close to the origin are not plotted. (b) Interpretation of the biplot in (a) for *Agrostis stolonifera* (ter Braak, 1995; redrawn with kind permission of Cambridge University Press).

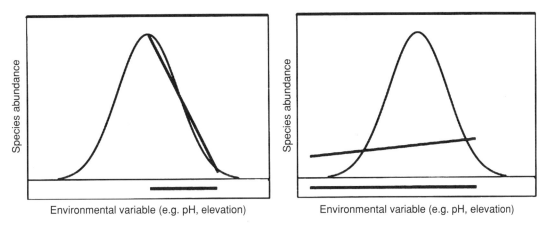

Figure 6.20   (a) Short linear species responses may occur over short distances on rising or falling limbs of unimodal response curves; (b) the discrepancy between linear and unimodal response models over longer gradients. Redrawn and adapted from Lepš and Šmilauer (2003), with kind permission of Cambridge University Press.

Some ecologists (Legendre and Gallagher, 2001; Lepš and Šmilauer, 2003) have argued that it is permissible to use PCA on data that have been collected over short gradients, where species can be assumed to show linear responses to environmental factors over part of their overall unimodal distribution, as in Figure 6.20a. However, this assumes that all species in that data-set exhibit this property for all environmental/biotic factors and gradients, which seems unlikely in most situations. If PCA is used on data containing species that have unimodal response curves (and they are very likely to over gradients of any reasonable length), then there is likely to be a major difference between the assumed linear response and the actual unimodal response, as is shown in Figure 6.20b. This problem gets worse as the length of the environmental/biotic gradient and thus beta diversity increases. Where unimodal species response curves are found, then either correspondence analysis-based methods (DCA, CCA) or non-metric multidimensional scaling (NMS) should be used, as described later in this chapter.

The second reason why the use of PCA on species data is often considered inappropriate is the sparse nature of most species data matrices. With vegetation data, it is commonplace for more than 80% of cells in the matrix to be absences or zero abundances. Use of PCA with either a variance–covariance matrix or a matrix of correlation coefficients is unwise in these situations because the methods assume linearity between species based on continuous (interval or ratio) data (Austin, 1999a, 2005).

The 'horseshoe' distortion effect in PCA

The consequent distortion of the positions of points on ordination plots when using PCA has become known as the 'horseshoe effect' after the original research comparing distortion in different ordination methods using artificial data by Gauch *et al.* (1981). Henderson and Seaby (2008) have shown neatly how easy it is to test methods on the very small simulated data-set in Table 6.5.

With the exception of the slight differences in the first and last samples, the samples should be distributed in order in a linear fashion parallel to the first axis. However, when analysed using PCA,

Table 6.5 Artificial data designed to demonstrate distortion effects in different methods of ordination (Henderson and Seaby, 2008). Reproduced with kind permission of Pisces Conservation Ltd, Lymington, UK.

	Sp1	Sp2	Sp3	Sp4	Sp5	Sp6	Sp7	Sp8	Sp9	Sp10
S1	10	5	0	0	0	0	0	0	0	0
S2	0	10	5	0	0	0	0	0	0	0
S3	0	5	10	5	0	0	0	0	0	0
S4	0	0	5	10	5	0	0	0	0	0
S5	0	0	0	5	10	5	0	0	0	0
S6	0	0	0	0	5	10	5	0	0	0
S7	0	0	0	0	0	5	10	5	0	0
S8	0	0	0	0	0	0	5	10	5	0
S9	0	0	0	0	0	0	0	5	10	0
S10	0	0	0	0	0	0	0	0	5	10

the linear sequence becomes completely distorted into the 'horseshoe' shape with the result that samples 1 and 10, which should be at either end of the gradient and have no species in common, are actually placed next to each other (Figure 6.21). The reason behind this is that PCA treats joint absences as indicative of a positive relationship between species. Obviously this is a grossly simplified and extreme example, but analysis of artificial data-sets such as this has raised questions

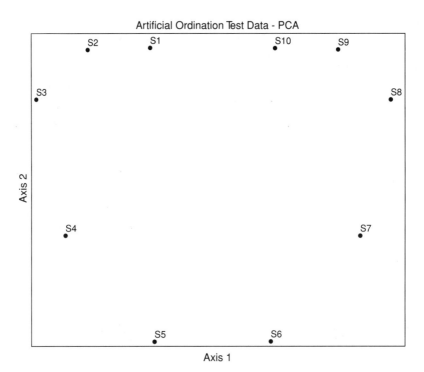

Figure 6.21 The 'horseshoe' effect in the two-dimensional ordination plot resulting from PCA of the data in Table 6.5. The analysis was run using the *PC-ORD* package of McCune and Mefford (1999, 2010). Original work of author.

as to the inevitable severity of distortion with real world data-sets (Gauch and Whittaker, 1972a,b; Gauch *et al.*, 1981; Minchin, 1987a,b; McCune, 1994, 1997).

The Hellinger transformation

Legendre and Gallagher (2001) proposed the use of the Hellinger transformation to overcome distortion problems in PCA. The transformation involves relativising or standardising a data matrix by quadrat/sample totals and then taking the square root of each element in the matrix. If a matrix transformed in this manner is then analysed using Euclidean distances, the result will be a matrix of Hellinger distances. However, Minchin and Rennie (2010) tested PCA on Hellinger-transformed data for 20 simulated data-sets with differing beta diversities and concluded that PCA still performed poorly in virtually all situations. Other standardisations have been also recommended, such as that by Branko and Ranka (1994), but have not been widely applied.

Principal coordinates analysis (PCoA)

Principal coordinates analysis was introduced by Gower (1966), who demonstrated that the PCA method could be used legitimately on a matrix of Euclidean distance coefficients (Chapter 4) calculated between samples. After the distances have been calculated, they are squared and then divided by 2 to transform them into a measure of similarity. An ordination is then obtained by extracting eigenvectors and eigenvalues from that matrix. The method is described in detail in Legendre and Legendre (1998), who stressed that the method can be used with any distance measure. As stated in Chapter 4, however, Euclidean distance measures are not suited to the sparse matrices typical of species data, and thus the method has rarely been used on vegetation data. Like PCA, PCoA is best suited to the analysis of matrices of environmental/biotic data, particularly where variables are of differing data types, including ranked and multistate data. Here Gower's general coefficient (Gower, 1971, 1985, 1987) (Chapter 4) is particularly useful.

To conclude, PCA represents a very complex method of ordination with a range of variations within it. PCA is now widely acknowledged as having serious limitations as a method for the ordination of floristic data. However, PCA is still extremely important as a method of summarising variation in environmental/biotic data (Zitko, 1994; Joliffe, 2002; McGarigal *et al.*, 2000; Wackernagel, 2003). Provided the variables are standardised, usually to zero mean/unit variance, the results of PCA or PCoA may be extremely valuable when interpreted alongside a floristic ordination of data from the same quadrats, carried out by some other ordination method. An excellent example of this is shown in the case study of research by Goldsmith (1973a,b) described below.

Principal curves (PCs)

This approach to ordination is attributed to Hastie and Stuetzle (1989), Tibshirani (1992) and De'ath (1999), and is based on the idea of minimising squared distances to curves, rather than the straight lines that are the case with principal component analysis (PCA). Given the non-linear models of species response to environmental/biotic factors, this technique was shown by De'ath (1999) to be effective for one-dimensional ordinations. As a starting point, a set of ordination scores is presented and a line curving through multidimensional space is fitted so that the distances from the data points to the line are minimised. This is repeated iteratively, using a smoothing algorithm. The method has rarely been used elsewhere, partly because there are various subjective decisions over where the initial line should be located, the choice of smoothing algorithm, and the fact that it has only been shown to work in one dimension thus far. Nevertheless, example applications are described from aquatic ecology and pollution studies by van den Brink and ter Braak (1997, 1998).

Non-metric multidimensional scaling (NMS)

Non-metric multidimensional scaling (NMS) was first used as an ordination method in plant ecology by Anderson (1971), and developed further by Austin (1976), Fasham (1977), Prentice (1977, 1980), Kenkel and Orlóci (1986) and Carroll (1987). NMS is really a set of related techniques that use the rank order information in a matrix of dissimilarities between quadrats or species. The earliest and most frequently used method is that of Shepard (1962) and Kruskal (1964a,b), known just as multidimensional scaling (MDS). A key aspect of NMS is that a range of (dis)similarity indices may be used, including those non-metric ones that are appropriate for species data, such as the Bray and Curtis/Steinhaus (Sørensen/Czekanowski) coefficient (Chapter 4). When these are used, the method becomes non-metric multidimensional scaling (NMS). In NMS, points representing quadrats or species are positioned within a few dimensions or ordination axes known as k-space (usually 1–6 dimensions), so that the distances between the points representing quadrats or species in the ordination k-space have the same rank order as the inter-point dissimilarities in the dissimilarity matrix calculated between all pairs of quadrats or all pairs of species. Conceptually on the ordination plot, it is best to think initially of two dimensions, but k can be any number of dimensions (usually up to six).

Clarke (1993), Cox and Cox (1996), Legendre and Legendre (1998) and McCune and Grace (2002) present many arguments in favour of NMS, and they believe that it is the 'best' indirect ordination method. As a result, it has become increasingly used over the past 20 years. Its main advantages are, firstly, that it does not assume linear relationships between species; secondly, the method uses ranked distances that linearise the relationship between distances measured in species space and distances in environmental space, which overcomes the 'zero-truncation' or 'sparse matrix' problem, typical of species data (Chapters 2 and 4); and thirdly, any distance measure can be used.

Its disadvantage is that it is much more complex both conceptually and also in its application. However, this should no longer put off researchers from using it more widely. In earlier NMS analyses, the number of ordination axes had to be specified at the outset, and depending on whether a quadrat or a species ordination was being computed, an ordination in $k = 1$–6 dimensions produced by another method was required to be supplied as input, along with a matrix of dissimilarity coefficients between the quadrats or the species. In more recent versions, repeated random k-space ordinations are generated and repeated runs are made to find the optimal solution. For any given input or random ordination, NMS then modifies the k-space ordination, so that the rank orders of the inter-point distances between the quadrats or the species are as close as possible to the rank orders of the equivalent values between quadrats or species in the dissimilarity matrix. The measure of how good a fit or match occurs between the two is called the stress, and can be expressed as a single value ≥ 0. If there is a perfect match, then stress $= 0$. The aim of the method is to find the dimensionality that reduces stress as much as possible. The stress function is normally drawn as a diagram, known as a Shepard diagram, with the dissimilarity values in the floristic data plotted in rank order against the distances on the ordination plot. If there is a good match and stress is low, the points will lie on a steadily increasing line or curve. The more the deviations from this smooth line or curve, the higher the stress.

The method

The best description of this complex method is presented in McCune and Grace (2002), and the following is a simplified explanation from that source.

(1) The values in the dissimilarity matrix are ranked.
(2) Following standardisation/normalisation, the elements of the matrix (D) of inter-point (Euclidean) distances on the ordination plot (for k dimensions) are ranked in the same order as those for the dissimilarity matrix.
(3) The degree of fit between the rank order of the coefficients in the dissimilarity matrix and the rank order of inter-point distances on the ordination plot, known as the stress, is calculated.
(4) To achieve this, it is easiest to imagine that a graph is produced of the values for between-sample dissimilarities in rank order against the inter-point k-space ordination distances in the same rank order. If there is a perfect fit between the rank order of the coefficients in the dissimilarity matrix and the rank order of inter-point distances on the ordination plot, a property known as monotonicity is achieved. Monotonicity of a series is when successive values either increase or remain the same but never decrease. Absence of stress would mean that the rank order of dissimilarity values would match the rank order of inter-point distances, and if the ranks were plotted on a graph, they would be on a straight or curved line, monotonicity would have been achieved and stress would equal zero.

    The graph for distance in three-dimensional ordination space against dissimilarity for the 25 quadrats of the Gutter Tor, Dartmoor, data is shown in Figure 6.22. With 25 samples (points) there are $[n \times (n - 1)]/2$ dissimilarity values and inter-point distances $= 300$ of each.

(5) With real data, as in Figure 6.22, the points on the graph of dissimilarity values and ordination inter-point distances move away from this ideal monotonic state, and the less good the matches for a given k-space ordination, the greater the lack of fit or stress.

    The amount of movement required to achieve monotonicity is represented on the graph by the difference between a point d and its position when moved horizontally to achieve monotonicity $\hat{d}$. This difference is shown in Figure 6.23. Stress is the sum of the squared differences between these values for all samples.

$$S_r = \sum_{i=1}^{n-1} \sum_{j=1+1}^{n} (d_{ij} - \hat{d}_i)^2$$

where S_r is the raw stress, n is the number of samples (points), d_{ij} is the inter-point distance in k-dimensional space in rank order of the dissimilarity matrix, and $\hat{d}_{ij}$ is the distance that d_{ij} has to be moved horizontally to achieve monotonicity. The closer the points lie to the monotonic line, the better the fit and the lower the stress.

(6) Raw stress then must be standardised (S_s), usually by the formula:

$$S_s = S_r / \sum_{i=1}^{n-1} \sum_{j=i+1}^{n} d_{ij}$$

and this value is rescaled further as S_R which is the square root of the scaled stress S_S multiplied by 100, which gives values between zero and 100.

(7) Next, stress is minimised by altering the arrangement of the sample points in k-space, which is most frequently performed using a method known as 'the steepest descent algorithm'. McCune and Grace (2002) provide an excellent 'landscape analogy' for this iterative process. Various

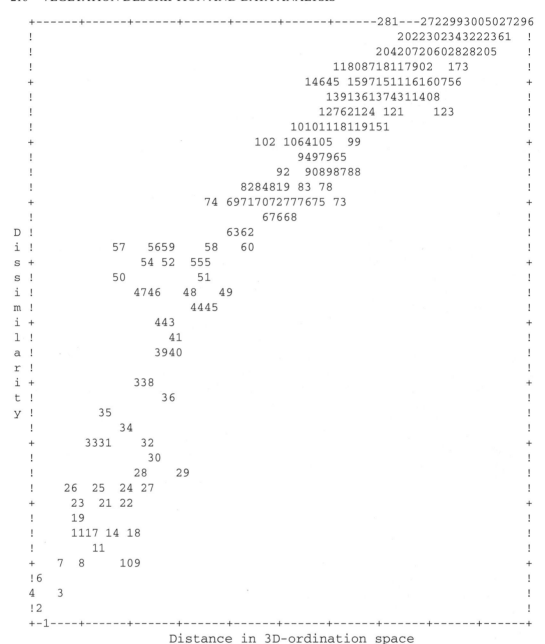

Figure 6.22 Graph of distance in three-dimensional ordination space against dissimilarity for the 25 quadrats of the Gutter Tor, Dartmoor, data. Points are labelled with the rank of dissimilarity. Many points lie on top of each other. Generated using the NMS program within the *PC-ORD* computer software (McCune and Mefford, 1999, 2010). Original work of author.

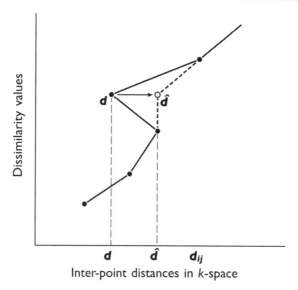

Figure 6.23   A plot of distance expressed as original dissimilarity against distances in k-dimensional ordi-nation space, to illustrate the moving of a point to achieve monotonicity in NMS. Adapted from McCune and Grace (2002), with kind permission of MjM Software Design, and Bruce McCune.

parameters have to be set – the starting configuration (the p-dimensional landscape), the 'step length', the number of iterations (the number of 'steps'), and care has to be taken because of the risk of 'instability' in the process.

> Imagine that you are placed at a random location in a p-dimensional landscape (your 'starting configuration'). Your task is to find the lowest spot (with the minimum stress) in the landscape as efficiently as possible. Because you can see only those parts of the landscape near you, you don't have a perfect knowledge of where the lowest point is located. You proceed by using your hyperspace clinometer, a device for measuring slope and you choose to go in the direction of steepest descent (= the gradient vector). After a prescribed distance (= the step length), you stop and re-assess the direction of steepest descent. The steeper the slope, the larger the step you take. You repeat this a maximum number of times (= number of iterations). Usually you will find yourself in the bottom of a valley well before you reach the maximum number of steps, so that the curve of elevation vs. iteration flattens out. The worriers amongst you will be concerned that there are other still lower places in the landscape, perhaps in a different drainage basin or network (i.e. that you are in a local minimum). Worriers should be dropped by helicopter into several random locations where they can repeat the process until they are convinced that they have found the lowest point (= the global minimum). (McCune and Grace, 2002: p. 128–9).

These various parameters have to be set either by the researcher or, in McCune and Mefford's (1999, 2010) computer program for NMS, can be set automatically in what is described as 'autopilot' mode. Three levels of strictness of criteria are available.

NMS starting configurations

NMS requires a starting ordination plot configuration to be supplied by the user. This can be the axis scores from any pre-existing ordination (Bray and Curtis or DCA). The alternative to this is to supply initial configurations derived at random, that is, ordination axis scores for 1–6 dimensions derived at random. However, if this is done, it is very unlikely the optimal configuration will appear from the first one supplied. It thus becomes necessary to supply a large number of random configurations and then choose the best result from all the random starting configurations for each dimension. Under the 'autopilot' mode in *PC-ORD* (McCune and Mefford, 2010), a minimum of ten different configurations for up to three dimensions are generated and analysed, and for a 'thorough' analysis, 250 runs for up to 6 dimensions are generated and analysed.

The 'best' result is the analysis with the lowest stress for each of the k dimensions. McCune and Grace (2002) state that, following extensive tests, using a large number of random initial configurations is better and gives lower final stress values than using the ordination plot coordinates from a pre-existing ordination, such as Bray and Curtis – this avoids the problem of 'local minima'.

Also Clarke (1993) gives a 'rule of thumb' for 'goodness' of stress values:

<5 An excellent representation with no prospect of misinterpretation – however, rarely achieved.

5–10 A good ordination with no real risk of drawing false conclusions.

10–20 Can still correspond to a usable picture, although values towards 20 indicate a potential to mislead.

>20 Likely to yield a plot that is dangerous to interpret.

Most ecological analyses give values of between 10 and 20, and anything between 10 and 15 should be seen as good. However, once values rise to towards 20, results should be treated with some caution. Clarke also warns against putting too much faith in these guidelines, since stress is likely to increase with increase in sample size, a point also well demonstrated by McCune and Grace (2002).

The ordination plot resulting from NMS analysis of the Gutter Tor, Dartmoor, data is presented in Figure 6.24. This analysis was run using the *PC-ORD* software (McCune and Grace, 1999, 2010) and selecting the 'thorough autopilot' option, which allows for 250 runs with randomly generated ordination coordinate input for six dimensions. A three-dimensional solution is recommended. Thus plots for the first three axes are worth study. In Figure 6.24, the plots for axis 1 v. axis 2 and axis 2 v. axis 3 are presented, and it is interesting to compare these with that derived from detrended correspondence analysis (DCA) of the same data shown in Figure 6.9.

An important aspect of NMS is that, unlike all other methods, axes are not necessarily extracted in descending order of importance in terms of variance explained. Nor is the first axis of a 2-D solution the same as the first axis of a 3-D solution, so that for a given number of dimensions, the solution for any one axis is unique.

Scree plots to decide the appropriate number of dimensions

Selection of the appropriate number of dimensions for a particular analysis is normally done through a scree plot of stress against the number of dimensions. The right number of dimensions or axes is where the plot levels off, bearing in mind Clarke's guidelines on stress above. In the analysis of the Gutter Tor, Dartmoor, data (Figure 6.25; Table 6.6), a three-dimensional solution is recommended and it can be seen that the scree plot levels off after three axes, when the stress value is around 10. A two-dimensional solution would give a stress value of just below 20.

a)

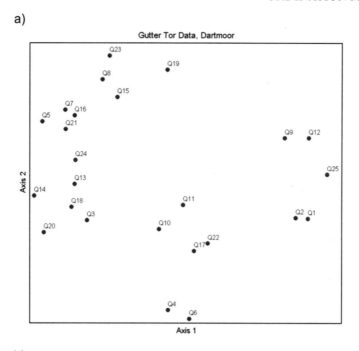

b)

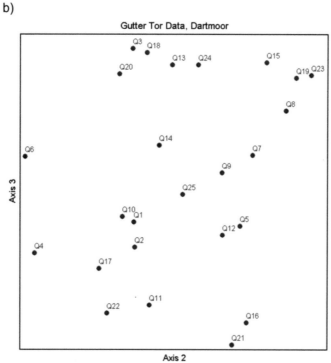

Figure 6.24   The quadrat/sample ordination plots for the first three axes of the NMS analysis of the Gutter Tor, Dartmoor, data (Table 4.4), plotted as (a) axis 1 v. axis 2; (b) axis 2 v. axis 3; (c) three-dimensional plot. Generated using the NMS program within the *PC-ORD* computer software (McCune and Mefford, 1999, 2010). Original work of author.

c)

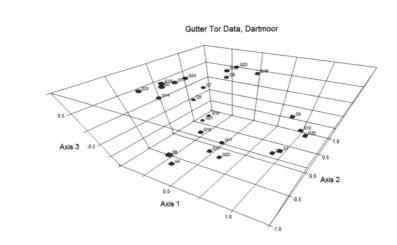

Figure 6.24 (*Continued*).

A Monte Carlo test on the probability of stress having been derived by chance

A Monte Carlo test is available for NMS in *PC-ORD*. This means that the original data are randomised and all the calculations are then redone a good number of times (minimum 25, ideally 100–250+). This is time-consuming in terms of computing. In the results of the number of runs,

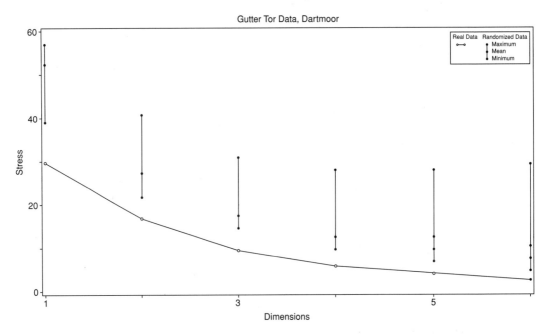

Figure 6.25 The scree plot for NMS analysis of the Gutter Tor, Dartmoor, data, derived using the NMS program from the *PC-ORD* computer software (McCune and Mefford, 1999, 2010). Original work of author.

Table 6.6   The Monte Carlo test for stress in relation to dimensionality (number of axes) for the Gutter Tor, Dartmoor, data. Original work of author.

Axes	Stress in real data 250 runs			Stress in randomised data Monte Carlo test 249 runs			p
	Minimum	Mean	Maximum	Minimum	Mean	Maximum	
1	29.642	47.532	55.381	39.105	52.333	56.994	0.0040
2	16.832	19.494	38.569	21.770	27.391	40.772	0.0040
3	9.468	10.452	28.846	14.651	17.578	30.809	0.0040
4	5.818	6.166	23.147	9.658	12.773	28.049	0.0040
5	4.027	4.391	19.518	6.899	9.598	27.982	0.0040
6	2.637	4.141	22.858	4.723	7.707	29.432	0.0040

p = proportion of randomised runs with stress ≤ observed stress
p = (1 + no. permutations ≤ observed)/(1 + no. permutations)

the proportion of times the real results produce lower scores for stress than the randomised data is a measure of the 'goodness' of the result. In Figure 6.25, for each of the six dimensions, the stress values for the real data lie well below the mean, maximum and minimum for the 249 randomised analyses. Thus the real stress values are unlikely to have been generated by chance. McCune and Grace (2002), however, emphasise that the Monte Carlo test is likely to be less trustworthy in the selection of dimensionality when there are outliers in the data (see below), one or two highly abundant species in the data-set, if there are fewer than ten samples, and if the data-set is particularly sparse with a very high number of zeros.

Instability in the analysis

Stress should fall sharply with the number of iterations (steps) in the analysis and then level off to a relatively stable figure. In some analyses, the curve on a graph of stress value against iteration will follow that pattern but then in later iterations may suddenly perform erratically. When this occurs, the solution is said to be unstable and results should be treated with caution. The graph for the 3-D solution of the Gutter Tor, Dartmoor, data is shown in Figure 6.26 and demonstrates some instability before stabilising after around 50 iterations.

Distortion and performance of NMS compared with other ordination methods

Various authors have tested NMS against other indirect ordination methods using artificial data (e.g. Austin, 1976; Fasham, 1977; Prentice, 1977, 1980; Oksanen, 1983; Minchin, 1987a,b; McCune and Grace, 2002). Some (Oksanen, 1983; Minchin, 1987a,b; McCune and Grace, 2002) have claimed superior results to other methods, while others have found little improvement in using NMS over other methods (Gauch et al., 1981). Applying several runs of NMS to the artificial data of Henderson and Seaby (2008) (Table 6.5) gives the ordination plots shown in Figure 6.27. With no distortion, the points should be plotted in order in linear fashion. Although the points can be seen to follow in sequence in each case, there is clearly considerable distortion present, but it is

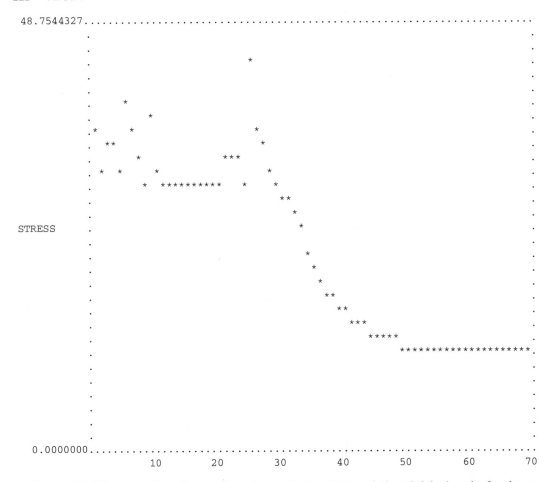

Figure 6.26 The stress v. iteration (step) number for the NMS 3-D solution. Original work of author.

difficult to determine how much of this is due to the small and contrived nature of the data-set, as opposed to the limitations of the method.

Careful reading of the literature seems to imply that in most cases, NMS is an optimal method for indirect ordination, but this conclusion is not universal. Nevertheless, NMS is generally good at recovering gradients of high beta (between-habitat) diversity.

A former problem with NMS was that the computational procedures are very complex. Computation time increases with the square, cube or more of the number of samples or species. This is no longer a problem with modern computers and packages, although a thorough analysis of a large data-set may still take several hours. Also in all methods, quadrat and species ordinations must be analysed separately and the matrix transposed.

The complexity of the methods means that careful attention must be paid to understanding its workings and limitations. The calculation process is iterative and convergence to the best solution does not always occur, (finding 'local' rather than 'global' stress), with several different solutions being possible for some data-sets, depending upon the initial ordination input and the selection of

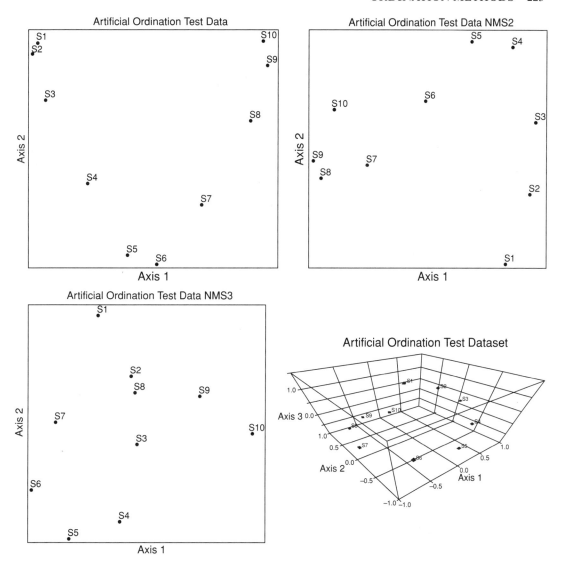

Figure 6.27   Three repeated NMS analyses of the artificial data of Henderson and Seaby (2008) (Table 6.5), using the 'autopilot' approach available in the NMS program in the *PC-ORD* computer software (McCune and Mefford, 1999, 2010). In three dimensions, they plot as a spiral. Original work of author.

key parameters. It is always essential that the method is applied a number of times to any data-set and results compared. Despite these various problems, the method has many strong proponents. As with all methods of ordination, the concept of 'user satisfaction' with the interpretability of the results is extremely important.

In addition to the case study by Robbins and Matthews (2009, 2010) at the end of this chapter, good examples of the use of NMS in vegetation science are to be found in Tong (1989), Dargie (1984), Neitlich and McCune (1997), Peterson and McCune (2001) and Waichler *et al.* (2001).

Correspondence analysis and reciprocal averaging (CA/RA)

Developments in ordination techniques after 1970 were centred principally on correspondence analysis (CA), also known as reciprocal averaging (RA), which is related to the method of weighted averaging devised by Whittaker (1967). The technique was first proposed by statisticians as long ago as 1935 (Hirschfeld, 1935; Fisher, 1940) and was initially applied to problems of attitude scaling in the social sciences. In ecology, it was first used by Roux and Roux (1967), Benzécri (1969, 1973), Hatheway (1971) and Guinochet (1973). Hill (1973b, 1974) and Gauch *et al.* (1977) demonstrated its potential application as an ordination method in plant ecology. The method has been widely applied across all the sciences and has been refined and improved (Greenacre, 1981, 1984; Digby and Kempton, 1987; Legendre and Legendre, 1987, 1998; Ludwig and Reynolds, 1988; Jongman *et al.*, 1995).

As a method, correspondence analysis is very important because it underpinned significant subsequent developments in ordination techniques, and is also at the heart of Two-Way Indicator Species Analysis (TWINSPAN), which has become a widely used method for numerical classification of vegetation data (Chapter 8).

One of the attractions of correspondence analysis is that in the hand-worked version for one axis, it is relatively simple to explain and understand. Another is that the calculation of the quadrat ordination is related to the calculation of the species ordination. The method of weighted averages (p. 176) is applied to a data matrix so that quadrat scores are derived from species scores and weightings, and conversely, species scores can similarly be derived from quadrat scores and weightings. These are carried out successively using an iterative procedure and stabilise out to give a set of scores for species, which give axes for a species ordination and a set of scores for quadrats, which give axes for a quadrat ordination. The starting point is to give weights to the species in a matrix on an arbitrary scale from 0–100 based on their assumed position along a primary environmental gradient. Equally, weights could be given to each quadrat depending upon the level of an important environmental factor. Whether these weights are correct or not is immaterial, since any values can be given across the range of the scale 0–100. However, the nearer the values are to the final solution, the less computation is required.

The method is described below in its simplest form as an iterative two-way averaging process; hence the name 'reciprocal averaging' given to it by Hill (1973b). For this reason, this is the name used in the description below. It can also be computed using matrix algebra and eigenanalysis and as such is related closely to the method of principal component analysis (PCA) described earlier in this chapter. When computed in this manner it is more usual to refer to the method as correspondence analysis.

The method

The method is best explained in relation to an example. For this purpose, the New Jersey salt marsh data coded in presence/absence form are taken (Table 4.2).

(1) Firstly, the row and column totals of the matrix are calculated. With presence/absence data, this gives column Q (Table 6.7), which is the number of quadrats in which each species occurs, and row S, the number of species in each quadrat.

Derivation of the first axis

(2) The next stage is to allocate weights to the 12 species. Here it is assumed there is no information on species tolerances in relation to a primary gradient. Thus for the 12 species, weights are

allocated across the range 0–100 (Table 6.7 – column W). Thus species 1 *Atriplex patula* (*var. hastata*) is given a score of 0, species 2 (*Distichlis spicata*) 9.09, species 3 (*Iva frutescens*) 18.18, species 4 (*Juncus gerardii*) 27.27, and so on up to species 12 (*Suaeda maritima*) with a score of 100.0.

(3) The process of reciprocal averaging then commences. The procedure is iterative in that the averaging process is repeated numerous times to generate a set of quadrat scores, and then those quadrat scores are used to generate an improved set of species scores. On the first iteration, the aim is to derive row Q1 which is the first set of quadrat scores. The value for the first quadrat (42.42) is calculated as follows. Taking the species scores for the first quadrat, each is taken in turn and multiplied by its weighting in column W. Those values are then summed and averaged. Thus:

$$(1 \times 0.00) + (0 \times 9.09) + (0 \times 18.18) + (0 \times 27.27) + (0 \times 36.36) + (1 \times 45.45)$$

$$+(0 \times 54.55) + (0 \times 63.64) + (0 \times 72.73) + (1 \times 81.82) + (0 \times 90.91) + (0 \times 100.0)$$

$$= 127.27$$

This value is then divided by the species total for quadrat 1 (3) = 127.27/3 = 42.42. The remaining values in row Q1 of Table 6.7 are calculated in exactly the same way.

(4) The averaging process is then applied in reverse to give a new set of scores for the species (column S1), using the values just calculated for row Q1. In Table 6.7, the revised value for *Atriplex patula* (*var. hastata*) is calculated by multiplying the species score by the new quadrat scores in row Q1, summing these results and then averaging them. Thus for *Atriplex patula* (*var. hastata*):

$$(1 \times 42.42) + (1 \times 34.09) + (1 \times 32.73) + (1 \times 48.49) + (1 \times 48.49) + (1 \times 36.36)$$

$$+(1 \times 36.36) + (0 \times 47.27) + (1 \times 44.16) + (0 \times 43.94) + (1 \times 41.56) + (1 \times 43.64)$$

$$= 408.30$$

This score is then divided by the quadrat total for species 1 (10) = 408.3 = 40.83.

Scores for the other species in column S1 are calculated in the same manner.

(5) For convenience in computation and to prevent working with very small numbers, the species scores are then rescaled to the range 0–100. The highest value in column S1, 48.49 for *Suaeda maritima*, is rescaled to 100, while the lowest, 35.15 for *Juncus gerardii*, is given a value of 0.0. The range of values (48.49 − 35.15 = 13.34) is made equal to 100 and the remaining scores are rescaled accordingly:

$$S1(Sc) = \frac{(\text{Species value} - \text{lowest species value})}{\text{Range of species values}} \times 100.0$$

where S1(Sc) is the rescaled species value.

Thus for *Atriplex patula* (var. hastata):

$$S1(Sc) = \frac{(40.83 - 35.15)}{13.34} \times 100.0 = 42.58$$

The other values are similarly rescaled to give column S1(Sc) of Table 6.7.

(6) The process of reciprocal averaging from stages 3 to 5 above is then repeated to give row Q2 for the second estimate of quadrat scores and column S2 for the second estimate of species scores, which are rescaled from 0–100 in column S2(Sc) (Table 6.7).

Table 6.7 Calculations for the first axis of a reciprocal averaging (correspondence analysis) of the New Jersey salt marsh data (Table 4.2). Original work of author.

	Quadrats												Quadrat totals (Q)
	1	2	3	4	5	6	7	8	9	10	11	12	
Species													
1. *Atriplex patuila*	1	1	1	1	1	1	1	0	1	0	1	1	10
2. *Distichlis spicata*	0	1	1	1	1	1	1	1	1	1	1	0	10
3. *Iva frutescens*	0	0	0	0	0	0	1	1	1	1	1	1	6
4. *Juncus gerardii*	0	0	1	0	0	1	1	0	0	0	0	0	3
5. *Phragmites communis*	0	0	0	0	0	0	0	1	1	1	1	1	5
6. *Salicornia europaea*	1	1	1	1	1	0	1	0	0	1	0	0	7
7. *Salicornia virginica*	0	0	0	1	1	0	0	0	0	0	0	0	2
8. *Scirpus olneyi*	0	0	0	0	0	1	1	0	0	0	1	0	3
9. *Solidago sempervirens*	0	0	0	0	0	0	0	0	1	1	1	1	4
10. *Spartina alterniflora*	1	1	1	1	1	1	0	1	1	1	0	0	9
11. *Spartina patens*	0	0	0	0	0	0	1	1	1	0	1	1	5
12. *Suaeda maritima*	0	0	0	1	1	0	0	0	0	0	0	0	2
Species totals (S)	3	4	5	6	6	5	7	5	7	6	7	5	
Q1	42.42	34.09	32.71	48.49	48.49	36.36	36.36	47.27	44.16	43.94	41.56	43.64	
Q2	45.75	45.92	36.74	63.95	63.95	32.48	38.26	55.67	54.60	54.52	50.42	56.89	
Q3	49.41	49.34	39.48	66.23	66.23	32.73	38.98	55.64	55.67	56.43	50.58	57.74	
Q14	53.98	51.08	50.02	67.39	67.39	41.92	32.66	20.65	21.29	27.37	17.11	10.85	
Q15	54.22	51.31	50.32	67.54	67.54	42.27	32.94	20.77	21.42	27.50	17.27	10.93	
Q20	54.54	51.62	50.74	67.75	67.75	42.75	33.33	20.95	21.59	27.67	17.49	11.05	
Q20(Sc)	76.70	71.55	70.00	100.00	100.00	55.34	39.29	17.46	18.59	29.31	11.36	0.00	

(Sc); Axis rescaled from 0–100 at the end of each iteration

(7) Reciprocal averaging is then repeated until the amount of change that occurs in the species and quadrat scores is minimal. By the 15th iteration, the scores are relatively stable (Table 6.7). The first species *Atriplex patula* (*var. hastata*) has a scaled score of 46.05, in column S15(Sc). By the 21st iteration, this is only changed to 46.40, in column S21(Sc). The decision as to when to stop the iterations depends on the accuracy required and whether or not the analysis is being done by computer. Clearly the process of averaging is extremely tedious and the only reason for carrying out a small analysis by hand would be to demonstrate the workings of the method. In computer programs that use the averaging process, the average change in species scores between each iteration is calculated and when it falls below a critical value, iteration of the averages is terminated. Obviously, many more iterations can be calculated by computer.

(8) An estimate of the eigenvalue for the first axis is obtained by taking the range of the unscaled scores on the final iteration (column S21) (max. = 67.75; min. = 19.45; range = 48.3) and expressing this as a proportion of the range of the scaled values for the previous iteration (S20), which obviously had a range of 100. This estimate is thus 0.48 and is an approximation to the eigenvalue for the first axis of the ordination. It may be seen as a measure of the proportion of the total variation in the data explained by the axis in the same way as for principal component analysis.

Table 6.7 (*Continued*)

W	S1	S1 (Sc)	S2	S2 (Sc)	S3	S3 .... (Sc)	S15	S15 (Sc)	S16	S16 .... (Sc)	S21	S21 (Sc)
0.00	40.83	42.58	48.90	46.47	50.64	46.57....	41.37	46.05	41.57	46.20....	41.86	46.40
9.09	41.35	46.44	49.65	49.16	51.13	48.22....	39.69	42.57	39.89	42.71....	40.16	42.89
18.18	42.86	57.52	51.73	56.54	52.50	52.94....	21.65	5.18	21.80	5.23....	22.01	5.30
27.27	35.15	0.00	35.83	0.00	37.06	0.00....	41.53	46.40	41.84	46.76....	42.28	47.26
36.36	44.12	67.21	54.42	66.12	55.21	62.22...	19.45	0.62	19.56	0.62....	19.75	0.62
45.45	40.93	43.34	49.87	49.93	52.30	52.24....	49.99	63.92	50.19	64.06....	50.49	64.26
54.55	48.49	100.00	63.95	100.00	66.23	100.00....	67.39	100.00	67.54	100.00....	67.75	100.00
63.64	38.10	22.08	40,39	16.22	40.76	12.69....	30.56	23.66	30.82	23.93....	31.19	24.31
72.73	43.32	61.28	54.11	65.01	55.10	61.84....	19.15	0.00	19.27	0.00....	19.45	0.00
81.82	42.00	51.32	50.40	51.81	52.35	52.41....	44.57	52.69	44.76	52.81....	45.03	52.98
90.91	42.60	55.84	51.17	54.55	51.72	50.25....	20.51	2.81	20.66	2.88....	20.88	2.96
100.00	48.49	100.00	63.95	100.00	66.23	100.00....	67.39	100.00	67.54	100.00....	67.75	100.00

(9) The scores in column S21(Sc) (Table 6.7) represent the first axis of the species ordination, while the rescaled values of row Q20, row Q20(Sc) represent the first axis of the quadrat ordination. These are plotted as the first axes of the species and quadrat ordination plots in Figure 6.28, and may be interpreted in the standard way described earlier in this chapter.

Derivation of the second axis

(10) Hill (1973b), in an Appendix to his paper, gave a simple example of how a second axis can be extracted. The same reciprocal averaging process was repeated but with a new set of initial scores. He recommended a set of starting values for species scores near to the end of the first axis calculations, for example, the scores in column S16(Sc). At the end of each iteration, when the new set of species scores have been calculated, the set of first axis species scores have to be multiplied by an integer and then subtracted from the new species scores. If this correction is not made, the first axis will slowly re-establish itself.

In the above example, the second axis scores for the 12 species and the 12 quadrats, calculated using the computer program of Kent (1977) and after 18 iterations, are shown in Table 6.8. The eigenvalue estimate for this second axis is 0.29.

(11) A third axis and subsequent axes may be extracted in a similar manner.

(12) The two axis species and quadrat ordination plots are presented in Figure 6.28. These plots are interpreted in the usual manner as described earlier in this chapter.

All computer programs for RA and CA use matrix algebra to provide a more efficient and accurate solution, and the matrix algebra approach is described in the Appendix to Hill's (1973b) paper, as well as in Ludwig and Reynolds (1988), Jongman *et al.* (1995) and Legendre and Legendre (1998). When calculated in this way, the method is usually known as correspondence analysis (CA). As with PCA, the eigenvalues decrease in size and importance with successive axes, but the cumulative sum of the eigenvalues is not the same as the sum of the variances of either species or quadrats. Hence the eigenvalues in CA are less important than they are in PCA. The individual elements of the eigenvectors represent the axis scores that are used for the construction of ordination

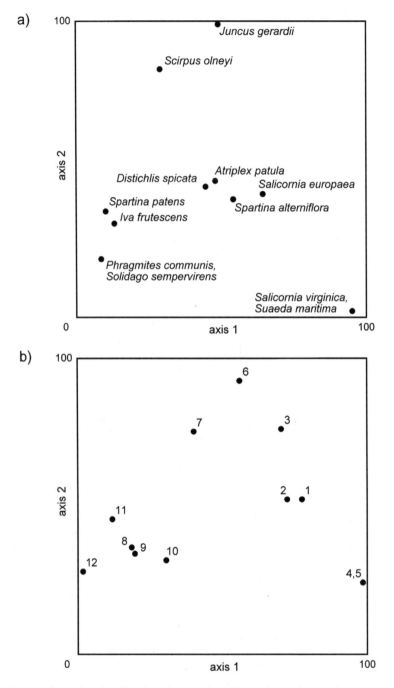

Figure 6.28 The two-dimensional ordination plots produced by reciprocal averaging/correspondence analysis of the New Jersey salt marsh presence–absence data (Table 4.2). (a) Species ordination; (b) quadrat ordination. Original work of author.

Table 6.8 Species and quadrat scores for the second axis of reciprocal averaging/correspondence analysis of the New Jersey salt marsh data, Table 4.2). Original work of author.

Species			Quadrats	
1.	*Atriplex patula*	46.85	1.	43.31
2.	*Distichlis spicata*	44.70	2.	43.65
3.	*Iva frutescens*	32.52	3.	54.92
4.	*Juncus gerardii*	100.00	4.	29.11
5.	*Phragmites communis*	20.84	5.	29.11
6.	*Salicornia europaea*	42.52	6.	63.12
7.	*Salicornia virginica*	0.00	7.	55.13
8.	*Scirpus olneyi*	83.49	8.	34.89
9.	*Soligado sempervirens*	21.00	9.	34.62
10.	*Spartina alterniflora*	40.56	10.	33.69
11.	*Spartina patens*	35.82	11.	40.75
12.	*Suaeda maritima*	0.00	12.	31.41

plots. Hill (1973b) pointed out that CA/RA is very similar to non-centred PCA standardised by species.

As with PCA, CA/RA can be described in geometric terms. Like PCA, CA/RA aims to find new axes that summarise the variation within a multidimensional cloud of points representing distances between quadrats or species, reducing the dimensionality in the process. CA/RA, however, uses chi-squared distances, an origin at the centroid (the centre of gravity of the swarm of points making up the cloud), and has both quadrat and species weights proportional to the quadrat or species totals (double transformation by totals). Thus the methods are similar but differ in the way that the points are projected onto the axes.

One advantage of CA/RA was the simultaneous analysis of species and quadrats and the relationship between the species ordination plot and the quadrat ordination plot. Plotting of the distribution of scores of individual species in each quadrat (Table 4.2) on the quadrat ordination plot (Figure 6.28b) and comparison of the resulting quadrat plot with the position of that species on the species ordination plot (Figure 6.28a) will show this relationship. The position of the species on the species ordination plot will be at the centre of gravity or the 'average' position of the distribution of the same species on the quadrat ordination plot.

Gauch (1982b) stated that CA/RA was usually best when analysing long community or environmental gradients with medium-high beta diversity, and since a majority of analyses are of this type, it was initially considered an optimal ordination method. However, for relatively homogeneous data-sets with short gradients, he accepted that other methods, such as PCA, would sometimes be better.

CA/RA has been introduced here using presence–absence data. It is equally applicable to quantitative data with the species or quadrat weightings being multiplied by the abundance data rather than the 1/0 of the qualitative data.

Problems with correspondence analysis and reciprocal averaging

Although Hill (1973b) and Gauch *et al.* (1977) claimed that CA/RA was generally superior to all other ordination techniques existing at that time, particularly for data collected across long environmental gradients, two major problems soon emerged in its application.

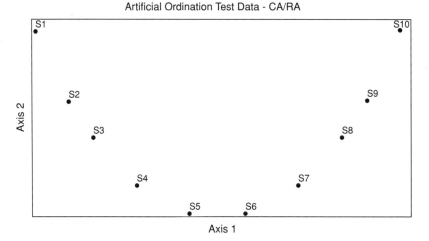

Figure 6.29 The 'arch effect' (inverted) in CA/RA demonstrated by analysis of the artificial data of Henderson and Seaby (2008) (Table 6.5) using the *PC-ORD* computer software (McCune and Mefford, 1999, 2010). The data should be plotted linearly in order, parallel to the first axis. Original work of author.

The 'arch effect'

This is well demonstrated in the ordination plot of the New Jersey salt marsh data (Figure 6.28), in the analysis of the artificial data of Henderson and Seaby (2008) (Table 6.5; Figure 6.29), where the arch is inverted, and in the case study of the Narrator Catchment, Dartmoor, described at the end of this chapter (Figure 6.49). The arch is produced when the first two axes are plotted as in Figure 6.28 and reflects the fact that the second axis may be simply a quadratic distortion of the first axis. This problem continues into higher dimensions – the third axis may have a cubic distortion, the fourth a quartic, and so on. However, because most ordination plots use only the first two axes, it was the quadratic distortion of the second axis that caused most difficulty. As a result, important secondary gradients in the data might not emerge until higher axes because the eigenvalue of the quadratic distortion could be greater than that of the actual secondary gradient, particularly when the secondary gradient was less than about half the extent of the primary gradient (Gauch, 1982b). Interpretation thus became problematic, and it was difficult to predict which axes carried ecologically meaningful information. Where a clear arch occurs and there is a strong single underlying gradient, it can be beneficial to interpret the ordination by examining the distribution of points along the length of the arch. This will usually clarify the primary gradient of the first axis. However, the second axis should then be ignored, and it may then be useful to examine the third and fourth axes for meaningful environmental/biotic relationships.

The axis compression effect

This second problem is demonstrated in Figures 6.30 and 6.31. Figure 6.30 shows the results of a one-axis CA/RA ordination of a set of artificial data made up of 18 quadrats and 23 species (Hill and Gauch, 1980). Both the quadrat and species single axis ordinations are plotted together, and in theory, because of the properties of the artificial data, the points in the grid should all be equally spaced in relation to both the quadrat and the species axes. In practice, this does not occur

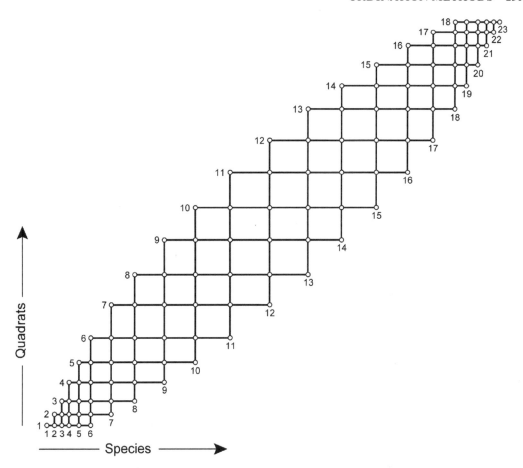

Figure 6.30 The arrangement of an artificial and regular data structure containing 18 quadrats in rows and 23 species in columns, spaced according to the first axis of a correspondence analysis ordination. The presence of a species in a sample is indicated by a dot. In theory, these quadrats and species should be spaced evenly, but the effect of correspondence analysis is to compress the ends of the axes relative to the middle (Hill and Gauch, 1980; reproduced with kind permission of Kluwer Academic Publishers).

and points nearer to the ends of the axes are compressed, while those nearer the centre are more spread out. This is related to the arch effect, and the combined arch and compression effects for the single axis quadrat ordination of the artificial data are shown in Figure 6.31.

Detrended correspondence analysis (DCA)

Detrended correspondence analysis (DCA) and the associated FORTRAN computer program *DECORANA* were devised by Hill (1979a) and Hill and Gauch (1980) in order to solve the two problems of CA/RA – the 'arch effect', and compression of points at the ends of the first axis (Figures 6.30 and 6.31). The 'arch effect' occurred because, although the second and higher axes of CA/RA were uncorrelated, they were not independent of each other. The first two axes were uncorrelated, or orthogonal, because the positive correlation on one side of the arch was matched

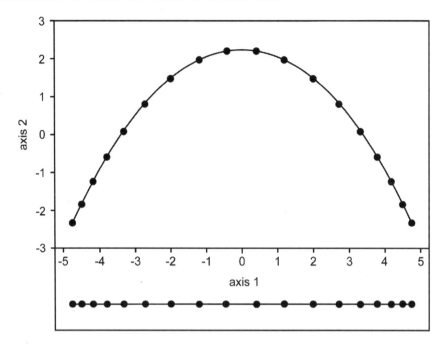

Figure 6.31   The two major faults of correspondence analysis: the arch effect and distortion of the second axis, and the compression of the ends of the first axis relative to the middle. These are the same data as for the quadrat ordination in Figure 6.30, with 18 equally spaced points in a straight line (Hill and Gauch, 1980; reproduced with kind permission of Kluwer Academic Publishers).

by the negative correlation of the other side of the arch as in Figure 6.31. The arch was a result of the quadratic relationship between the first and second axes and was thus rarely a reflection of the ecological content of the data, but instead was a consequence of the mathematics behind the method.

Removal of the 'arch effect' by DCA involved what was known as 'detrending'. The first axis was divided into a number of segments, and within each segment the second axis scores were recalculated so that they had an average of zero, as in Figure 6.32. When this was done for all segments, it meant that all second axis scores were expressed as deviations from a mean of zero. In the original computer program *DECORANA*, the first axis was divided into many segments and the averaging was achieved through a sophisticated running averages procedure. Figure 6.32 shows how this reduced the arch effect and straightened out the arch trend in relation to the first and second axes.

Detrending was applied to the quadrat scores at the end of each iteration of the correspondence analysis, except that once convergence was achieved, the final quadrat scores were derived from weighted averages of the species scores without detrending. For the third axis, quadrat scores were detrended in relation to both first and second axes and the pattern was similar for higher axes. In reality, very few examples exist of DCA results based on higher (>2) ordination axes.

The second problem of CA/RA was the compression of points at the ends of the first axis relative to the middle (Figure 6.31). This difficulty was overcome again by segmenting the axis and expanding those segments at the end and contracting those in the middle, so that the species turnover (the arrival and departure of species along the first axis gradient) occurred at as uniform a rate as

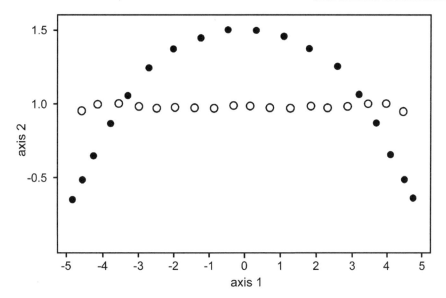

Figure 6.32 A simplified representation of the method for detrending used in detrended correspondence analysis. For explanations, see text. Quadrat scores before detrending are shown as •, after detrending as ○. Redrawn from Hill and Gauch (1980), with kind permission of Kluwer Academic Publishers.

possible. Hence equal distances on the quadrat ordination axes correspond to equal differences in species composition. The rescaling of axis segments was achieved by expanding or contracting small segments of the species ordination, while trying to equalise the average within-segment dispersion of the species scores at all points along the quadrat ordination axis (Figure 6.33).

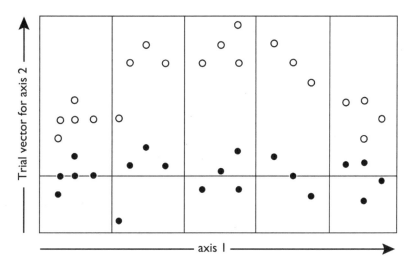

Figure 6.33 Within-segment standard deviation of species scores in relation to position along the first (quadrat) axis in detrended correspondence analysis (•), compared with non-detrended correspondence analysis (○). Redrawn from Hill and Gauch (1980), with kind permission of Kluwer Academic Publishers.

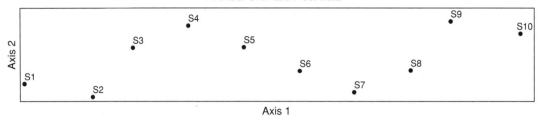

Figure 6.34 The removal of the 'arch effect' (inverted) in CA/RA by detrending using DCA, demonstrated by analysis of the artificial data of Henderson and Seaby (2008) (Table 6.5) using the *PC-ORD* computer software (McCune and Mefford, 1999, 2010). The arch effect is reduced, although distortion is not entirely eliminated, and the points are located in order, parallel to the first axis with the samples evenly spaced along the first axis. Original work of author.

Thus the species ordinations were adjusted so that the quadrat scores were the weighted mean values of the scores of the species that occur within them (Hill and Gauch, 1980; Gauch, 1982b).

Another important feature of DCA was that, in the rescaling process, the axes were scaled in units of the average standard deviation of species turnover (SD) (Gauch, 1982b). Along a gradient, a species appears, rises to its mode and then disappears over a distance of almost 4 SD, and similarly a complete turnover in the species composition of samples occurs in about 4 SD. A change of 50% in the composition of a quadrat (called a half-change) occurs in about 1 SD or slightly more. DCA scales axes in these SD units. Thus, in DCA, axes can be of variable length and, unlike CA/RA, are not scaled into an arbitrary range of 0–100 according to the size of the eigenvalues.

Criticisms of DCA as an ordination method

The detrending in DCA is generally effective in removing the 'arch effect' in relation to plots of first and second axes (Figure 6.34).

Gauch (1982b) concluded that DCA gives 'results at least as good as and usually superior to other ordination techniques' (p. 159). However, Hill and Gauch (1980) did acknowledge that there were still problems, although these tended to be forgotten by many researchers, given the enthusiasm with which DCA was initially received. Outliers (individual quadrats or species which are well separated from the rest of the points on an ordination diagram and hence are very different in species composition or distribution) and discontinuities (gaps in species or quadrat distribution along axes) both present problems. Outliers are best dealt with by removal from the data-set, and most computer programs have an option for this. Large discontinuities mean that the widths of gaps in the gradient have to be estimated and this can lead to inaccuracies.

Since the 1980s, particularly with the widespread availability of the computer program *DECORANA*, DCA has achieved considerable prominence and has become widely used (Kent and Ballard, 1988; Kent, 2006; Von Wehrden *et al.*, 2009). However, its application has not been without criticism (Ejrnaes, 2000). Dargie (1986) noted that a certain amount of distortion in DCA ordination is attributable to variations in species richness (alpha diversity) along gradients and axes in addition to the arch and compression effects. DCA does not necessarily correct for this. He concluded that the literature has emphasised variations in beta diversity (between-habitat diversity) as a significant source of distortion, but species richness (within-habitat diversity) is probably also important because it varies systematically along most successional and environmental gradients.

Wartenberg *et al.* (1987: p. 438) commented on the manner in which DCA has been adopted rather uncritically by plant ecologists, and they argued that there is 'no empirical justification for the method, since the DCA model is not consistent with the structure of the data... and ... there is no theoretical justification for the method, since DCA is, as Hill and Gauch (1980) point out, an *ad hoc* adjustment of CA/RA.' In relation to the 'arch effect', the position of points on the first axis is not changed, and all DCA does is to flatten the arch on the original plot. They continued '...This deception does not enhance our understanding of the data or help to identify the cause of the observed distortion' (p. 435). Concerning the end of axis compression problem, they also question the assumption that rates of species turnover are constant or even along gradients and axes and that all species can be treated equally. Also, Wartenberg *et al.* (1987) and Minchin (1987a) pointed out that the detrending and flattening of the arch may result in loss of ecological information if some of the arch form represents a real pattern in the original data.

Both Legendre and Legendre (1998) and McCune and Grace (2002) provided summaries of many further criticisms that have been made of DCA, and argued strongly against its use. They also cited the work of Tausch *et al.* (1995) and Podani (1996), who found that the axis scores in DCA were influenced by the input order of the data, particularly on higher axes. Oksanen and Minchin (1997) attributed the instability to both lax criteria in the stability thresholds of the solutions, and an error in the smoothing algorithm of the original *DECORANA* program. Both of these were subsequently corrected and virtually all present-day versions of the program incorporate those corrections.

The method of detrending has also been criticised and refined (Greenacre, 1984; Oksanen, 1988; ter Braak, 1988a,b; Knox, 1989). The original detrending using segments and subtracting of a moving average was shown to be unstable in some situations and with certain types of data. Ter Braak (1988a,b) provided an alternative approach called polynomial detrending, where axis scores were replaced by residuals from a multiple regression on polynomial functions (up to a specified degree) of the axes already obtained. This represented an alternative for the reduction of the arch effect, but did not assist with rescaling to reduce end of axis compression. Knox (1989), however, questioned whether polynomial detrending represented any real improvement, and Jackson and Somers (1991) showed that the number of segments selected for the detrending procedure gave varying results on second and higher axes. They stated 'Although detrended solutions may be interpretative (an attribute that encourages the use of DCA), the arrived at solution may be only one of many possible results... we caution against the acceptance of DCA as an ecological panacea... we have found that the choice of axis segmentation may substantially affect the interpretation and hence the utility of higher dimensions generated with DCA' (p. 711).

In summary, despite its limitations, DCA still remains one of the most widely used methods for indirect ordination. Various researchers have argued that the method still produces clearly interpretable ordinations in many instances (e.g. Peet *et al.*, 1988). The key point in these complex discussions is that interpretation of results from DCA is best carried out with some knowledge of its limitations and in comparison with other ordinations of the same data, such as non-metric multidimensional scaling (NMS) (see below).

For comparative purposes, the ordination plot for the first two axes of the DCA of the New Jersey salt marsh data is presented in Figure 6.35. When compared with the original CA/RA plot (Figure 6.28b), the effects of detrending and rescaling of the first axis to avoid compression are clearly seen.

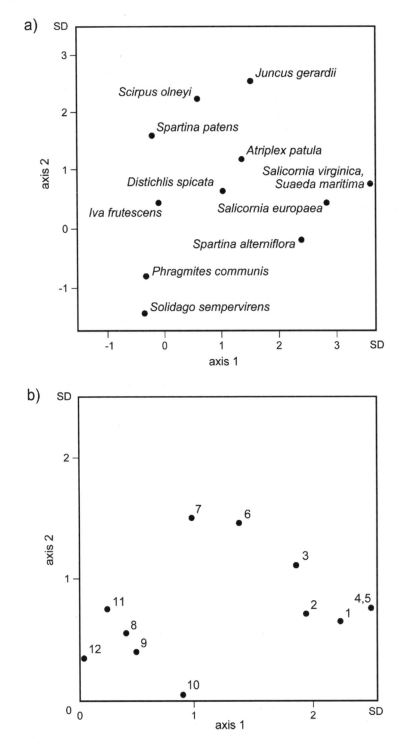

Figure 6.35 Ordination plots for the first two axes of the DCA of the New Jersey salt marsh presence–absence data (Table 4.2). Original work of author.

Canonical correspondence analysis (CCA)

Canonical correspondence analysis (CCA) was developed by ter Braak (1985, 1986a, 1987, 1988a,b,c, 1994, 1995). The application of the technique has been greatly aided by the availability of the associated *CANOCO* computer program (ter Braak, 1988a,b; Lepš and Šmilauer, 2003), although CCA is now also available in a range of other computer software (Chapter 9). CCA is different to all the ordination methods discussed thus far. Ultimately, ordination is an exercise in examining relationships between species distributions and their associated environmental/biotic factors and gradients. Of particular relevance to this is the application of methods of correlation and regression (ter Braak and Prentice, 1988) (Chapter 5). All the ordination methods discussed so far in this chapter, along with Bray and Curtis (polar) ordination and principal component analysis, have this goal, but they are all indirect, in that the analysis is performed on the species data alone first and then interpretation of underlying environmental/biotic gradients is made by superimposing environmental/biotic data on the ordination plots and looking for patterns and correlations. Some analyses may go as far as correlation and regression of quadrat axis scores with environmental factors, but for various reasons this is not always satisfactory.

CCA differs from this classical indirect approach because it incorporates the correlation and regression between floristic data and environmental factors within the ordination analysis itself. Thus the input to CCA usually consists of not just a data matrix of quadrats × species, but also a second paired data matrix of quadrats × environmental/biotic factors. Using multivariate analysis and particularly techniques of multiple regression (Chapter 5, Case Study), together with correspondence analysis or reciprocal averaging (CA/RA), an integrated ordination of species with associated environmental/biotic data is obtained. In view of this, CCA is best defined as a method of direct ordination, with the resulting ordination being a product of the variability of the environmental/biotic data as well as the variability of the species data. It also follows from this that CCA may only be performed effectively if a good set of paired environmental/biotic data has been collected for the samples or quadrats in the analysis.

This approach of using both species and environmental/biotic data in the actual ordination process is known as canonical analysis. The resulting ordination diagram thus not only expresses patterns of variation in floristic composition but also demonstrates the principal relationships between the species and each of the environmental/biotic variables.

The exact process by which CCA works is very complex and is explained in detail in ter Braak (1986a, 1987, 1994), Legendre and Legendre (1998), and Lepš and Šmilauer (2003). The method uses multiple regression to select the linear combination of environmental variables that explains most of the variation in the species scores on each axis. Using the iterative approach of correspondence analysis or reciprocal averaging (CA/RA), within each iteration or averaging cycle a multiple regression is carried out between the quadrat ordination scores for an axis (dependent variable) and various combinations of the environmental/biotic variables (the independent variables). The CA iteration process can also be carried out by eigenanalysis, as was the case with basic CA/RA. The calculated best-fit values for quadrats from the regression for the combination of environmental/biotic variables are then taken as an improved estimate of those quadrat ordination axis scores. Often a 'stepwise' multiple regression procedure is employed that selects the combination of independent environmental/biotic variables that gives the highest overall explained variance in the original quadrat axis scores. CA iteration then continues, with another multiple regression being performed to improve fit on the next iteration, and so on. The scores eventually stabilise in the same manner as for CA/RA. Blanchet *et al.* (2008) offer valuable perspectives on the use of forward selection methods in multiple regression, and hence CCA and related methods (see RDA below).

CCA thus becomes a restricted correspondence analysis. Ter Braak and Prentice (1988) call this a 'constrained ordination' – the quadrat axis scores are constrained by the environmental/biotic variables. These restrictions are reduced as more environmental/biotic variables are included in the analysis, since more of the variation in the quadrat scores is likely to be accounted for. The 'arch effect' described for CA/RA can also occur in CCA (ter Braak, 1986a, 1987) and the method of detrending used in detrended correspondence analysis (DCA) (Hill and Gauch, 1980) can be applied to remove it (giving DCCA). In the computer program *CANOCO* (ter Braak, 1988a,c; Lepš and Šmilauer, 2003), options are available for detrending by segments, as in the original DCA program *DECORANA* (Hill, 1979a) or by polynomials (ter Braak, 1988a,c). However, ter Braak (1987) pointed out that the arch can be removed 'more elegantly' (p. 139) by dropping superfluous environmental/biotic variables. The environmental/biotic variables most likely to be superfluous are those most highly correlated with the arch axis (second axis). CCA with such variables removed should not need detrending.

An important part of the output from CCA, in addition to standard plots for species and quadrat ordinations, is the use of biplots of the species ordination diagram together with the environmental factors (Gabriel, 1971; ter Braak, 1986a, 1987; Lepš and Šmilauer, 2003). The principles of biplots were introduced in the section of this chapter on PCA (p. 209). They greatly enhance the interpretation of environmental/biotic gradients and, in particular, allow individual species to be related to all major environmental/biotic factors.

As an example, the Gutter Tor data from Dartmoor, southwest England (Table 4.4), have been analysed by CCA within the *PC-ORD* computer program (McCune and Mefford, 1999, 2010). In addition to the species data, environmental/biotic data were available on soil/peat pH, soil moisture content, soil/peat depth and animal grazing intensity as measured by faecal units.

In Figure 6.36, the quadrat–environment biplot is presented. The points represent individual species and an arrow representing each environmental/biotic variable is plotted, pointing in the direction of maximum change of the environmental variable across the diagram. The length of the arrow is proportional to the magnitude of change in that direction, and for interpretation purposes, each arrow can also be extended backwards through the central origin. Those environmental/biotic factors with long arrows are more important in influencing community variation than those with short arrows. The angles between the arrows representing environmental/biotic variables are indicative of the degree of correlation between those variables, in the same manner as explained for correlations between variables in the explanation of PCA (pp. 198–206). Arrows on top of each other (angle $= 0°$) have a correlation of $+1.0$, those at right angles ($90°$) have a correlation of 0.0, while those arrows pointing in opposite directions have a correlation of -1.0. Arrows with intermediate angles correspond to the equivalent positive or negative correlation. In the same manner, the angle between the arrow representing each environmental/biotic variable and each ordination axis indicates the correlation between that variable and the ordination axes.

A point corresponding to an individual species can be related to each arrow representing an environmental factor by drawing a perpendicular from the line of the arrow up to the point representing the species. The order in which the points project onto the arrow from the tip of the arrow downwards through the origin is an indication of the position of the species in relation to the environmental factor. Species with their perpendicular projections near to or beyond the tip of the arrow will be strongly positively correlated with and influenced by the arrow. Those at the opposite end will be less strongly affected (ter Braak, 1987).

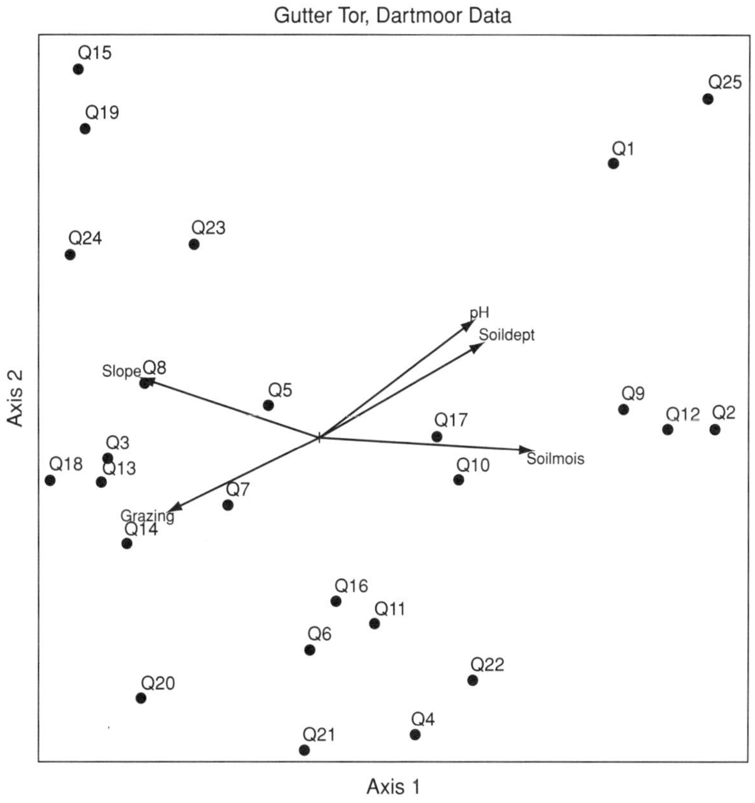

Figure 6.36 The quadrat/sample-environmental biplot from canonical correspondence analysis (CCA) of the Gutter Tor, Dartmoor, data, derived using the CCA program from the *PC-ORD* computer software (McCune and Mefford, 1999, 2010). Original work of author.

The CCA species biplot for the Dartmoor data (Figure 6.37) shows the species and community distribution very clearly, with various groupings of species emerging. Species of the wet bogs and flushes (*Drosera rotundifolia*, *Narthecium ossifragum*, *Juncus effusus*, *Sphagnum* spp., *Trichophorum cespitosum* and *Carex nigra*) are shown to the right of the plot. Heath species are found in the lower centre (*Calluna vulgaris*, *Erica tetralix*, *Vaccinium myrtillus*, *Cladonia portentosa*, *Agrostis curtisii*). Finally, the species of the improved acidic grasslands (*Agrostis capillaris*, *Festuca ovina*, *Trifolium repens*, *Danthonia decumbens*) are located towards the left of the diagram.

These groupings and the individual species are clearly shown in relation to the arrows representing environmental factors and gradients. The wet bogs and flush species on the right of Figure 6.37 are shown to have higher pH, due to being flushed with rainwater, deeper soil/peat depths, and high soil moisture with low slope angles. Equally, the grazing axis is most strongly influencing the improved acidic pastures to the lower left. The effects of slope are again well illustrated, with *Pteridium aquilinum*, *Galium saxatile* and *Viola riviniana* all found in well-drained soils with steeper slopes, which is well known to correspond to their environmental preference.

The position of each environmental arrow with respect to each axis indicates how closely correlated the axis is with that factor. Thus soil moisture is most highly correlated with the first

Gutter Tor, Dartmoor Data

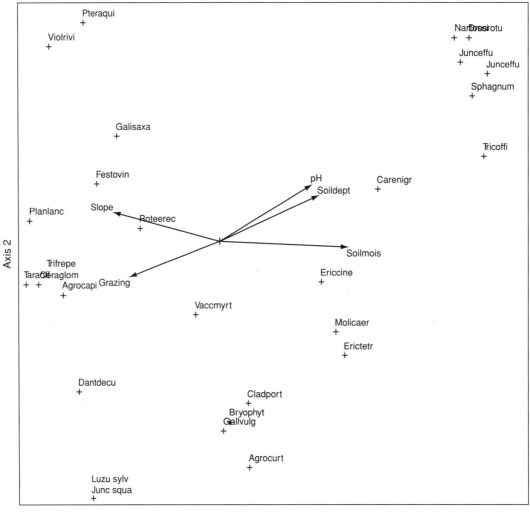

Figure 6.37   The species-environmental biplot from canonical correspondence analysis (CCA) of the Gutter Tor, Dartmoor, data, derived using the CCA program from the *PC-ORD* computer software (McCune and Mefford, 1999, 2010). Original work of author.

axis, and the CCA results confirm this correlation as 0.977 (Table 6.9). The other factors are correlated with both first and second axes, although the highest correlations are with the first axis in each case. The eigenvalue for the first axis of the quadrat/environment biplot is 0.837 and the second 0.493, representing 21.2% and 12.5% of the total variance respectively. Thus the first two axes explain 33.7% of the variance in the species/environment data.

Figures for the correlation (Pearson and Kendall's tau) between the quadrat/sample ordination scores, which are the LC scores (linear combinations of the environmental/biotic variables), and the total variation attributed to all the environmental/biotic variables are also presented in

Table 6.9 Canonical correspondence analysis (CCA) of the Gutter Tor, Dartmoor, data (Table 4.4). Summary statistics, eigenvalues, percentage variances explained and correlations of quadrat/sample ordination scores on the first two axes are given, with the environmental variables (LC scores – linear combinations of environmental/biotic variables). Correlations are the 'intraset' correlations of ter Braak (1988a). Original work of author.

	Axis 1	Axis 2	Axis 3
Eigenvalue	0.837	0.493	0.429
% variance explained	21.2	12.5	10.9
Cumulative % explained	21.2	33.7	44.5
Pearson correlation, Spp-Envt	0.983	0.861	0.836
Kendall (rank) corr., Spp-Envt	0.847	0.673	0.600
Correlations			
Soil/peat depth	0.751	0.480	−0.156
Slope angle	−0.810	0.304	0.156
pH	0.704	0.591	−0.331
Soil moisture	0.977	−0.061	−0.197
Grazing	−0.691	−0.377	−0.580

Table 6.9. These are, perhaps not surprisingly, always very high, and McCune and Grace (2002) argue that the statistics are misleading in that they are often interpreted as the amount of variation in the community data that is explained by the environmental variables. This is not the case.

Lastly, the arrow for any single environmental/biotic variable may be taken and extended right across the plot for either quadrats/samples or species. Figure 6.38 shows this for the soil moisture

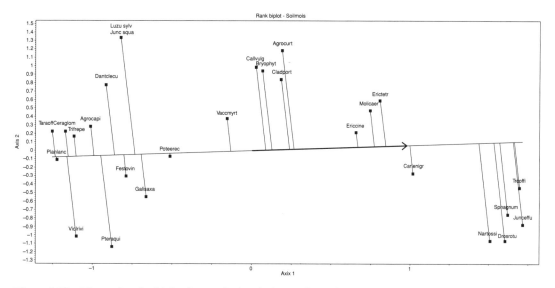

Figure 6.38 The rank order biplot for species in relation to the environmental factor of soil moisture from the CCA ordination of the Gutter Tor, Dartmoor, data. The second axis is inverted (a mirror-image) compared with the species biplot produced by the *PC-ORD* software in Figure 6.37. However, the relative inter-point distances are identical. Analysis performed using the Ecom 2 computer program of Pisces Conservation Ltd. (2008a). Original work of author.

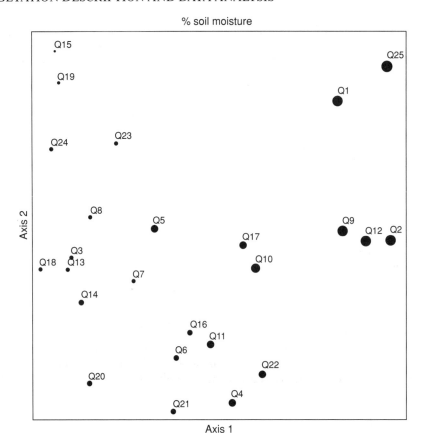

Figure 6.39 Percentage soil moisture values for the 25 quadrats of the Gutter Tor, Dartmoor, data-set overlaid on the CCA quadrat/sample ordination plot (Figure 6.36) using proportional circles and derived using the CCA program from the *PC-ORD* computer software (McCune and Mefford, 1999, 2010). Original work of author.

arrow on the Gutter Tor species biplot of Figure 6.37. For each species point, a perpendicular is dropped to the line of the extended arrow. The order in which the perpendiculars fall along the line correspond to the relative tolerance of each species along the soil moisture gradient. In Figure 6.38, those species at the right end are tolerant of wet conditions, while those at the left end are found in dry conditions. This procedure may be followed for any significant environmental/biotic factor arrow on either the species or quadrat biplot. In the case of quadrats, the quadrats are ranked along the gradient in the same manner as for the species.

In addition to the quadrat- or species-environment biplots, the standard quadrat or sample ordination plot can be produced in the usual way, and individual environmental/biotic variables or species abundances can be overlaid as proportional symbols (Figure 6.39). In some packages, the environmental data may also be plotted as centroids (averages) of their distribution in the quadrats in relation to the ordination axes.

It is important always to check that the ordination scores used in CCA are the LC scores (ter Braak, 1994), which are those derived from the multiple regression of the quadrat ordination axis scores with the environmental/biotic variables. In early versions of the *CANOCO* program, the WA scores were used, which were those resulting from the weighted averages procedure within the correspondence analysis part of the method. In some data-sets, these scores can be quite different (McCune and Grace, 2002). Linked to this, two sets of figures for correlations of environmental/biotic variables with the quadrat ordination axes are produced in some packages. It is important that those correlations derived from the LC scores, known as the 'intraset' correlations, are used (ter Braak, 1988a). McCune and Grace (2002) stressed that, although the correlations are a useful indicator of the relative importance of the environmental variables on each ordination axis, they should not be seen as an independent measure of the degree of relationship between the environmental/biotic variables and the variability of the vegetation, as expressed in the ordination axes.

A Monte Carlo significance test

In CCA, a Monte Carlo test is used to assess the significance of the environmental/biotic variables in explaining the variation in the ordination axis scores. The first test relates to the eigenvalues, which represent the amount of variability in the data summarised by each axis. However, the size of the eigenvalues on their own do not represent the significance of the result because the eigenvalues provide no information as to whether the variability explained by the environmental/biotic variables is greater than would be expected by chance (Økland, 1999). A Monte Carlo randomisation test (Manly, 1997; Smith, 1998) is used to overcome this problem. The original data are shuffled at random to produce new data-sets a large number of times (up to 1000) and the whole CCA is re-run on each new data-set, generating new eigenvalues for each axis. The proportion of times that the eigenvalue for the many runs of the shuffled data is greater than or equal to that for the real data-set gives the *p* value for significance (Table 6.10). A similar test is also performed on the figure for the species-environment correlation, but as discussed above, there are questions about the validity of these correlations. In the case of the Gutter Tor data, the eigenvalue for the first axis is clearly highly significant. There are problems over the calculation of significance for the second and third axes, however.

Table 6.10 Monte Carlo randomisation test results for the CCA of the Gutter Tor, Dartmoor, data. Original work of author.

Axis	Real data	Randomised data, Monte Carlo test, 998 runs			
		Mean	Minimum	Maximum	p
Eigenvalue					
1	0.837	0.363	0.164	0.633	0.001
2	0.493	0.222	0.089	0.437	
3	0.429	0.130	0.051	0.297	
Species-env. correlation					
1	0.983	0.733	0.525	0.889	0.001
2	0.861	0.628	0.433	0.825	
3	0.836	0.540	0.351	0.755	

p = proportion of randomised runs with either the eigenvalue or the species-environment correlation greater than or equal to the observed eigenvalue or species-environment correlation; i.e. $p = (1 + \text{no. permutations} \geq \text{observed})/(1 + \text{no. permutations})$. p is not reported for axes 2 and 3 because using a simple randomisation test for these axes may bias the p values.

Criticisms of CCA and cautions on its use

McCune and Grace (2002) are foremost in presenting criticisms and discussing the limitations of CCA. They stress that the method has been very widely applied (see Birks *et al.*, 1996), often uncritically. Their main criticism is expressed in the following quote:

> Multivariate analysis is a way of getting messages from a high-dimensional world that we cannot see directly. Using CCA is like getting these messages through a narrow mail slot in a door. The edges of the slot are defined by the measured environmental variables. Messages that do not fit the slot are either deformed to push them through or just left outside. (McCune and Grace, 2002: p. 164)

The key point is that CCA has multiple linear regression at its core, and thus all the limitations of multiple regression apply (Chapter 5). Principally, these are that in the regression, only linear relationships between the quadrat/sample ordination axis scores (dependent variable) and the environmental/biotic variables (independent variables) are examined.

As emphasised in Chapter 5, multiple regression should really be seen as a modelling technique, whereby careful selection and combination of independent variables is performed. In CCA, as with multiple regression, as the number of independent or environmental/biotic variables increases compared with the number of quadrats/samples, the results become increasingly questionable and, in the case of CCA, the species-environmental relationships may appear to be strong, when in fact strong relationships can be found even with randomly generated variables (McCune and Grace, 2002). This problem also affects the interpretation of any correlation coefficients (Pearson product-moment, Kendall's tau or Spearman rank) that may be calculated between quadrat/sample ordination axis scores and environmental/biotic variables.

A further problem with CCA is that the calculation of the eigenvalues and the percentage variances explained by each ordination axis and in total is open to interpretation and varies between one computer package and another. The reasons behind this are explained in McCune and Grace (2002).

Further issues linked to the assumptions of the multiple regression model relate to the problem of multicollinearity. If there are significant intercorrelations between the independent variables, the results become unreliable (Graham, 2003). Prior to a CCA analysis, the intercorrelations between all the independent environmental/biotic variables should always be examined (Table 6.11), and where two variables are highly correlated, one should be removed. These problems can also be explored by computing successive multiple regressions with each of the environmental/biotic variables as the dependent variable and the other variables as independent variables. Any variables with high r^2 values should be examined carefully and have their inclusion evaluated. Some programs, for example CCA in the *CANOCO* package, also calculate a value for the variance inflation

Table 6.11   The intercorrelations (Pearson) between the environmental/biotic variables in the Gutter Tor, Dartmoor, data-set. Original work of British Ecological Society/Wiley-Blackwell.

	Soildept	Slope	pH	Soilmois	Grazing
Soildept	1.000	−0.557	0.833	0.746	−0.527
Slope	−0.557	1.000	−0.503	−0.815	0.381
pH	0.833	−0.503	1.000	0.710	−0.548
Soilmois	0.746	−0.815	0.710	1.000	−0.533
Grazing	−0.527	0.381	−0.548	−0.533	1.000

factor (VIF) (p. 169) for each independent or environmental/biotic variable. Any variable with a VIF of greater than 1 should be considered for possible removal from the data-set.

From the above discussion, the key point to realise is that, being based on multiple regression, use of CCA should be seen as a modelling exercise, whereby a carefully planned series of exploratory and final analyses is made.

Partial canonical correspondence analysis

CCA also has the facility to carry out 'partial' analyses of species-environmental relationships, where certain environmental/biotic 'covariables' can be controlled for and eliminated from the ordination. Such an analysis examines the residual variation in the species data and its relationship with any specific environmental variables that may be of interest. An example of the application of this approach is presented in the case study by Cushman and Wallin (2002) at the end of this chapter.

One of the key assumptions of the multiple regression model that underpins CCA is that variables are not subject to spatial autocorrelation (Chapter 3), but it is likely that, in many applications of CCA, environmental/biotic variables, as well as the ordination axes themselves, may be spatially autocorrelated. Thus, Type I errors relating to the significance levels in the regression (rejecting the null hypothesis when it should actually be accepted) may occur. This situation will be exacerbated if there is additional collinearity or intercorrelation among the environmental/biotic variables (Graham, 2003). A number of ecologists and statisticians have examined this problem (Burrough, 1995). Borcard *et al.* (1992), Legendre (1993), Palmer (1993) and Borcard and Legendre (1994) first suggested a solution involving the use of CCA within the computer program *CANOCO*, which allows the calculation of partial canonical correspondence analysis (ter Braak, 1986a, 1987, 1988a,b,c; Økland and Eilertsen, 1994; Méot *et al.*, 1998; Lepš and Šmilauer, 2003; Økland, 2003; ter Braak and Šmilauer, 2003), and they have demonstrated, with varying degrees of complexity, how the explained and unexplained variance in the regression can be partitioned so that the relative contributions attributable to spatial variability, as opposed to the environmental/biotic variables, can be assessed.

CCA has now been used very widely by ecologists and vegetation scientists, but examples of the application of such variance partitioning to remove spatial autocorrelation in vegetation science are still comparatively few (Pinel-Alloul *et al.*, 1995; Anderson and Gribble, 1998; Tuittila *et al.*, 2007). Qinghong and Bråkenhielm (1995) and Cushman and Wallin (2002) (see case study) provide examples of its application within plant ecology. Further research is thus still required into the importance of such spatial-environmental interactions in ordination analyses.

Various new developments in this area are of interest (Kent, 2006; Kent *et al.*, 2006). The first is the multiplicative habitat modelling package for non-parametric regression developed by McCune (2004, 2006) and programmed in the *Hyperniche* computer package by McCune and Mefford (2009). Although as yet only used on comparatively few vegetation data-sets (e.g. Berryman and McCune, 2006; Lintz *et al.*, 2011), this non-parametric regression package should allow for more effective analysis of the non-linear and often Gaussian-shaped response models that are characteristic of most species–environment data (Figure 6.40). The techniques should also allow for removal of spatial autocorrelation followed by non-parametric analysis of residuals. A particularly important feature of the package is the possibility of taking the scores for any single ordination axis as the response variable in the regression and to then regress them non-parametrically against the environmental/biotic variables using a wide variety of different response models.

Askernish Outer Hebrides - NMS ordination axes

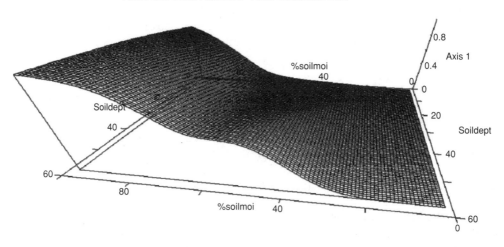

Askernish Outer Hebrides - NMS ordination axes

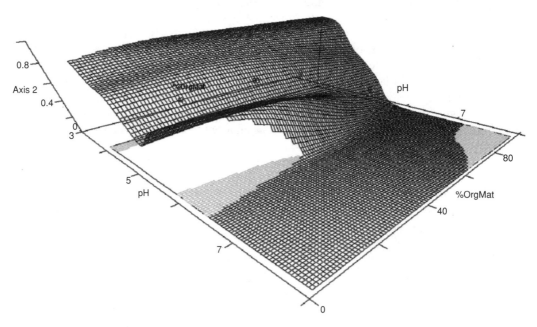

Figure 6.40 Modelling of two NMS ordination axes as response surfaces against environmental variables using the *Hyperniche* multiplicative habitat modelling package for non-parametric regression developed by McCune (2004, 2006) and programmed in the *Hyperniche* computer package by McCune and Mefford (2009). The data are 217 quadrat samples across the machair sand-dune–inland transition on South Uist, Outer Hebrides, Scotland (Kent *et al.*, 1997). The modelling is described more fully in Chapter 8. Original work of author.

In another development, Borcard and Legendre (2002) and Borcard *et al.* (2004) have proposed a new approach aimed at identifying spatial patterns across the entire range of spatial scales that may exist within a data-set. The method, 'principal coordinates of neighbour matrices' (PCNM), involves the calculation of the principal coordinates of a matrix of geographical neighbours among a set of spatially distributed samples. Using four example data-sets, they argue that PCNM analysis is preferable to classical regression or CCA because the environmental relationships could be linked to one or several spatial scales. As Borcard *et al.* (2004: p. 1831) stated: 'This often represents crucial information to interpret the results of a study, because the scale of observation strongly influences the perception of relationships amongst variables'. The method is explained in detail in Borcard and Legendre (2002) and creates a set of spatial explanatory variables that have structure at the full range of scales in any data-set, and assesses to which of these new variables the species data may be responding. Variance partitioning among both spatial and environmental variables may be employed within this framework. Further, more advanced developments are in progress, linked to Moran's eigenvector maps (Dray *et al.*, 2006), but are too complex to be summarised here.

Wagner (2003, 2004) demonstrated how CA and CCA can be partitioned by distance (indirect and direct multiscale ordination) and integrated with geostatistics using various forms of variogram analysis. Despite these advances, the validity of several of these variance partitioning approaches has, however, been questioned (Gilbert and Bennett, 2010).

Hypothesis testing using CCA

Canonical correspondence analysis and the *CANOCO* program can also be used in experimental community ecology for hypothesis-testing – deductive rather than inductive analysis (Chapter 1) because the Monte Carlo significance test is available to test for the effects of specific environmental variables after the influence of other variables has been removed or 'controlled'. In this way, the method can be used to analyse data from randomised block experiments or data from a number of different locations. Also it is possible to restrict the permutations to those among samples-within-blocks or samples-within-locations. Again, this represents more complex and advanced analysis. Good examples of this approach are described in Wassen *et al.* (1990), Pyšek and Lepš (1991) and Lepš and Šmilauer (2003).

Redundancy analysis (RDA), distance-based redundancy analysis (dbRDA) and canonical analysis of principal coordinates

Redundancy analysis (RDA) (Rao, 1964; van den Wollenberg, 1977; Gittins, 1979; ter Braak, 1995; Legendre and Legendre, 1998; Borcard *et al.*, 2011) involves calculating principal component analysis (PCA), rather than CCA, on a species data matrix constrained by performing multiple regression with a matrix of paired environmental/biotic variables. In concept, the method is related to canonical correspondence analysis (CCA), and partial RDA can also be performed in the same manner as for CCA described above. A full explanation of RDA is given in Legendre and Legendre (1998) and Borcard *et al.* (2011), but when applied to sparse species data, the method suffers from the same limitations as PCA in that it firstly assumes linear relationships between species responses and environmental gradients, whereas in reality many species show more unimodal response curves (Chapter 2), and secondly, the use of correlation coefficients is unsuited to sparse data. The method is thus only suited to non-sparse data-sets along short linear species response gradients, and not to data-sets showing high beta diversity.

In an attempt to overcome this problem, Legendre and Anderson (1999), McArdle and Anderson (2001), Anderson and Willis (2003) and Millar *et al.* (2005) introduced 'distance-based' RDA (dbRDA) and canonical analysis of principal coordinates, whereby a choice of distance coefficients may be used in combination with principal coordinates analysis (PCoA), rather than principal components analysis (PCA).

Co-inertia analysis (CoIA)

Dolédec and Chessel (1994), Franquet *et al.* (1995) and Dray *et al.* (2004) introduced this method as a variant of CCA. PCA is first carried out on the species and environmental/biotic data matrices, and new axes are then calculated to maximise the covariance between the scores for quadrats/samples in the two sets of resulting ordinations. Once again, in using PCA, the linearity problem remains and there are also issues in cross-multiplication of the two matrices (McCune and Grace, 2002). It has rarely been used in plant ecology.

USE OF ORDINATION METHODS TO ANALYSE SUCCESSIONAL OR TIME-SERIES DATA

Any of the ordination methods described in this chapter may be used to analyse successional or time-series data, whereby repeat measurements of presence/absence or abundance data are recorded from the same quadrats or sample locations at successive points in time. Each time a set of quadrats or samples is recorded it can be seen as a 'time-slice', as depicted in Figure 6.41.

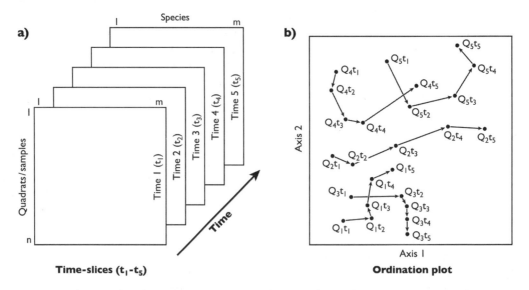

Figure 6.41   The use of ordination to analyse successional or time-series data. Each repeated set of measurements in the same set of quadrats/samples produces a quadrat/samples × species presence/abundance 'time-slice'. (a) Five 'time-slices' are shown. (b) On the associated ordination plot, each quadrat thus occurs five times and the successive points in time are joined by arrows to give a 'trajectory analysis'. Only the trajectories of the first five quadrats are shown. Original work of author.

These 'time-slices' can then be amalgamated on a spreadsheet into one large data matrix and any method of ordination applied. The result is a sample/quadrat ordination plot, where the same sample/quadrat appears several times, and following the sequence of repeated points in time, a series of lines or arrows may be drawn connecting them. The distance a point moves between any two points in time is indicative of the amount of change in species composition over that time period. This approach is sometimes called 'trajectory analysis', since it follows the trajectory of change in species composition through time and in many situations, and this will represent the operation of vegetation successional processes. In other situations, it could indicate the effects of management action for conservation or response to climate change. Some computer packages, for example *PC-ORD* (McCune and Mefford, 1999, 2010), offer the facility to plot the lines directly on the computer ordination plot.

PROBLEMS IN THE APPLICATION OF ORDINATION METHODS

The nature of the species-response model and conflict with the assumptions of linear relationships between species

This issue has already been raised several times. The prevailing model of species response to an environmental factor is the unimodal, Gaussian or bell-shaped response curve (Figures 2.4 and 2.5), and most researchers using ordination still accept this as being the most realistic model (Whittaker, 1953, 1975; Gauch and Whittaker, 1972a,b; ter Braak and Prentice, 1988). Clear understanding of the idea of unimodal species response curves has been confused because sometimes a position on an environmental gradient has been equated with actual physical location on a transect. It is very important to appreciate that environmental gradients are an abstract concept representing a theoretical distribution of a species, and do not necessarily imply an underlying spatial relationship (Austin, 1985, 2005).

Testing for the existence of unimodal response curves in the real world has been carried out by various authors (Huisman *et al.*, 1993; Bio *et al.*, 1998; Bio, 2000; Oksanen and Minchin, 2002; Heikkinen and Mäkipää, 2010) but most notably by Austin (1980, 1985, 1987, 1999a, 2005), Austin and Austin (1980), Austin *et al.* (1984) and Austin and Smith (1989). The results show that, although unimodal bell-shaped species response curves are found, they are not universal, they may be bimodal and are often positively skewed (Figures 2.5 and 2.12) (Austin, 1990; Austin and Gaywood, 1994). Patterns of species response curves along environmental gradients are seen to be highly variable and are greatly influenced by species richness (beta diversity) and the overall length of the gradients. Austin (2005) concluded that an adequate general model of species response still does not exist (Chapter 2).

All methods of ordination make assumptions about the nature of the data. Some, such as PCA and RDA, assume that species are linearly related to each other and to environmental/biotic gradients and are thus based on the linear model. This problem was discussed earlier in this chapter (Figures 6.5a–d and 6.20). The idealised linear relationship between two species and an environmental gradient is shown in Figure 6.5a, resulting in the linear relationship between the species shown in Figure 6.5b. If, however, the unimodal or bell-shaped species response model is accepted (Figure 6.5c), then the joint relationship between two species along an environmental gradient produces the highly distorted curve of Figure 6.5d. Clearly most species are not linearly related, and Austin's work indicates that many species do not have perfect unimodal response curves either. Thus there is still a serious problem of the inadequacy of the existing underlying model of species response to environment and the way in which ordination methods treat data from the real world (Austin, 2005).

The consequence of this is that, when applied to real world data, all methods of ordination cause distortion of the true relationships between species and samples and associated environmental data because of the disparities between the ecological and mathematical models. The key question is, how serious is this distortion and does it matter? As described above, it is possible to examine the form of species response curves in real world data-sets using options in various software packages such as *CANOCO* (ter Braak and Šmilauer, 2003) and *Hyperniche* (McCune and Mefford, 2009), but it remains difficult to decide objectively how far species in a given data-set conform to either the linear or the unimodal model, or even to neither of the models.

The range and choice of ordination techniques and the search for a 'best' method given the problems of 'distortion'

The range and choice of methods for ordination remains a substantial problem for both researchers and students (Kent and Ballard, 1988; von Wehrden *et al.*, 2009). Even now, there is no absolute agreement over which is the 'best' method giving least distortion of the data. A clear consensus over which method should be recommended for general use has never emerged, and a kind of pluralistic approach prevails (von Wehrden *et al.*, 2009). As each technique or group of techniques has evolved, it has usually been assumed to represent the best available. Unfortunately, subsequent evaluation and testing through either real world application or using simulated data has virtually always led to a reappraisal of methods. Thus exactly why a particular researcher chooses a certain method or combination of methods for a given set of data is often not very clear. Choice is often based on the availability of computer programs or the current feeling on the 'goodness' of certain methods by different groups of researchers based on recent articles or textbooks or their own particular experiences with their own data-sets.

Numerous authors have compared different ordination methods on both simulated data and on real world data-sets, and in the process have tried to evaluate the degree of distortion inherent in different methods. The original research in this area by Gauch and Whittaker (1972a,b, 1976) is worth summarising. They introduced the idea of simulation of species response to environmental/biotic factors and gradients and the construction of artificial data-sets based on the coenocline or simulated gradient of community composition. From such an artificial gradient, data-sets of known properties can be produced. Figure 6.42 shows a simulated coenocline with 17 species and 13 quadrats (Causton, 1988). The 17 species are each shown with a bell-shaped response curve and the 13 quadrats have been sampled at 13 equally spaced points along the coenocline.

As species come and go along the coenocline, species turnover occurs. The amount of species turnover along the coenocline is the same at the beta diversity. Gauch and Whittaker (1972a,b, 1976) devised a measure of this beta diversity by defining units of 'half-change' along the gradient. A half-change is defined as the distance or separation along the coenocline at which the similarity between samples or quadrats is 50%. In Figure 6.42 the number of half-changes from left to right is shown, based on the Steinhaus (Sørensen/Czekanowski) coefficient (Chapter 4). A single gradient coenocline can be extended to two gradients producing a coenoplane. Again these have been used to test for distortion in different ordination methods.

Using the coenocline principle, Gauch and Whittaker (1972a,b) and Gauch *et al.* (1977) compared the distortion attributable to various ordination methods. The results for correspondence analysis/reciprocal averaging (CA/RA) and two variants of principal component analysis (PCA) are shown in Figure 6.43. In all cases, the line representing quadrats or samples along the

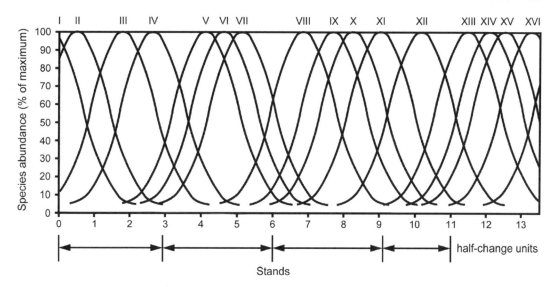

Figure 6.42   A simulated coenocline with 17 species and 13 quadrats or samples. Single half-changes are shown along the coenocline, calculated on the basis of the Steinhaus (Sørensen/Czekanowski) similarity coefficient. Redrawn from Causton (1988), with kind permission of Chapman and Hall/Unwin Hyman. Reproduced with permission from Chapman and Hall/Unwin Hyman.

coenocline should be straight and parallel to the first axis, but it becomes distorted into the now familiar 'arch' of CA/RA and 'horseshoe' of PCA (see also Figures 6.21 and 6.29). The results of these and other comparisons indicated that detrended correspondence analysis (DCA) shows least distortion (see also Figures 6.34, 6.49 and 6.50), followed by polar ordination (PO), then CA/RA, and finally PCA. The results for PO were a surprise, and as an ordination method it still has its supporters (Beals, 1984; Causton, 1988; McCune and Beals, 1993; McCune and Grace,

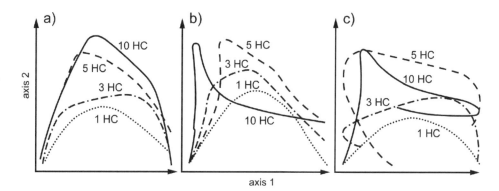

Figure 6.43   The 'arch' and 'horseshoe' effects demonstrated by analysis of simulated coenocline data: (a) correspondence analysis/reciprocal averaging; (b) centred and species-standardised principal component analysis; (c) centred and non-standardised principal component analysis. Sample ordinations for four levels of beta diversity (1, 3, 5, 10 half-changes) are shown. In all cases the result should be a horizontal straight line. Redrawn from Gauch *et al.* (1977), with kind permission of the British Ecological Society and Wiley-Blackwell.

2002; Huerta-Martínez *et al.*, 2004). Further refinements for simulation of community data were described by Minchin (1987b), and additional examples of testing a range of ordination methods using coenoplanes can be seen in McCune and Grace (2002) and Hirst and Jackson (2007). Confusing results are reported from these sources, particularly with respect to the performance of non-metric multidimensional scaling (NMS). McCune and Grace (2002), Clarke (1993), Clarke and Warwick (2001) and Clarke and Gorley (2006) are all very strong proponents of NMS as the best method for direct ordination, but Hirst and Jackson (2007) and to a lesser extent Henderson and Seaby (2008) (Figure 6.27) are less certain about the lack of distortion in NMS.

A further consideration is that a good performance using simulated data does not automatically mean that a method always works best with different real world data-sets. Numerous other researchers have compared different combinations of ordination methods on their own real world data (Austin, 1976; Fasham, 1977; Whittaker and Gauch, 1978; Gauch et al., 1977, 1981; Oksanen, 1983; Brown *et al.*, 1984; Ezcurra, 1987; Minchin, 1987a). No completely consistent results have emerged, and recommendations have depended very much upon the properties of the particular data-sets analysed and the combinations of methods applied. Thus no consensus over a 'best' method has emerged. At the present time non-metric multidimensional scaling (NMS) probably remains the most recommended method for indirect ordinations, while detrended correspondence analysis (DCA) still has its supporters, despite the serious and valid criticisms of the artificial nature of the detrending process.

Canonical correspondence analysis (CCA) and to a lesser extent redundancy analysis (RDA) are widely accepted as 'best' methods for direct ordinations, but they each have their problems, as discussed above. CCA can be a substantial improvement if a good set of environmental/biotic data can be supplied with the species data and a carefully considered modelling approach is adopted, bearing in mind the limitations of the linear multiple regression model. However, a great deal of effort in environmental/biotic measurement is necessary in order to obtain such good quality data. Soil variables are clearly often very important, but such measures, particularly of soil chemistry, can be very time-consuming for large numbers of samples and are frequently intercorrelated. Biotic factors such as grazing, burning and human impact often need to be measured, but obtaining reliable and consistent data on these variables is difficult.

In summary, where a good set of environmental/biotic data are available in addition to the species data, then a comparative analysis of NMS and CCA is often recommended, particularly where beta diversity is high and there are long gradients. This corresponds to the 'pluralistic' approach that tends to be adopted with ordination analysis today (von Wehrden *et al.*, 2009) and which is greatly facilitated by the availability of most methods for comparison in the majority of computer packages (Chapter 9). Where good quality environmental/biotic data are not available, NMS is the probably the best and most appropriate choice, but again it is usually both easy and interesting to run a comparative analysis with DCA on the same data-set, bearing in mind its limitations. The merits of using either PCA or RDA and their variants on short linear gradients still appear to be a matter of debate.

The problem of outliers

Outliers will affect an ordination regardless of which technique is applied. An outlier is a quadrat which is very different from the others in a data-set in terms of species composition, or a species which is very different in its distribution in a set of quadrats. Thus when a quadrat ordination is performed, the resulting ordination plot shows the outlying quadrat or quadrats at one extreme, and the rest of the quadrats grouped tightly together at the other extreme (Figure 6.44a), since

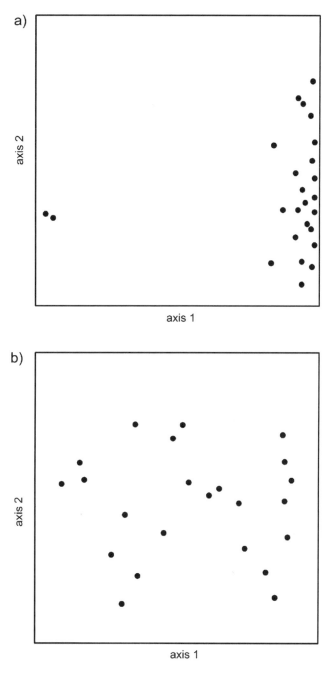

Figure 6.44   (a) The effect of 'outliers' on a two-dimensional ordination; (b) the plot with the two outlying quadrats in plot (a) removed from the data-set. The remaining points are spread out and the relationships between them are demonstrated much more clearly. Original work of author.

the majority of quadrats are all much more similar to each other than to the outlying quadrat or quadrats. By definition, an outlying quadrat usually contains several species that only occur in the outlying quadrat and nowhere else in the data. A similar situation can occur in a species ordination.

The solution is to remove the offending quadrat(s) and to reanalyse the data. This has the effect of spreading out the remaining quadrats and displaying the similarity among them much more clearly (Figure 6.44b). In the process, unusual and outlying species are generally also removed. Virtually all computer packages such as *PC-ORD* and *CANOCO* have options that allow for the deletion of quadrats while the program is running, so that data files do not have to be edited to remove quadrats and or species. Removal of a quadrat does not mean that it can then be conveniently forgotten. In interpretation, it should be explained why the quadrat or quadrats were outliers, and the particular ecological conditions that caused the problem to occur should be investigated.

The problems of hypothesis generation and over-interpretation

At the start of Chapter 6, ordination techniques were described as a group of methods for hypothesis generation. Once an ordination has been run, study of ordination plots and species/quadrat–environment relationships should enable patterns in the data to be seen and new ideas to be generated. However, Kent and Ballard (1988) showed that very few examples of published research clearly progress from the description of the results to the generation of hypotheses, leading to further new research involving the actual testing of the hypothesis. One notable exception to this is the work of Goldsmith (1973a,b) described at the end of this chapter. For some researchers, ordination still becomes an end in itself, whereas in many situations, it should really just be a beginning – a method of exploratory data analysis leading to the formulation of new ideas and new research based on hypothesis testing, experimentation and modelling.

A related problem is the tendency to over-interpret ordination results and to try to read too much into them. A student or researcher, seeing ordination as an end in itself, will be tempted to take those results as final and conclusive, rather than accepting that further data collection and experimentation centred on a newly formulated hypothesis is probably necessary.

Ordination and multivariate analysis as a panacea

Over many years, various authors have raised many questions about the application and teaching of multivariate analysis generally and ordination in particular (Gower, 1987; Gittins *et al.*, 1987; James and McCullogh, 1990; McCune and Grace, 2002; Austin, 2005). The problems identified in the previous section are again highlighted. James and McCullogh (1990) struck a cautionary note when they stated (p. 158–9):

> Ecologists and systematists need multivariate analysis to study the joint relationships of variables. That the methods are primarily descriptive in nature is not necessarily a disadvantage... we are forced to agree in part with the criticism that multivariate methods have opened a Pandora's box. The problem is at least partly attributable to a history of cavalier applications and interpretations. We do not think that the methods are a panacea for data analysis, but we believe that sensitive applications combined with focus on natural biological units, modelling and an experimental approach to the analysis of causes would be a step forward.

This seems an appropriate caution with which to conclude this chapter. Few vegetation scientists who have used ordination methods could say that this criticism, at least in part, did not apply to them and some of their work!

Case studies

The vegetation and environmental controls of sea cliffs at South Stack, Anglesey – principal component analysis as an ordination technique (Goldsmith, 1973a,b)

The vegetation of sea cliffs has rarely been studied in detail, principally because of the obvious problems of working in such a difficult environment. Goldsmith (1973a,b) carried out a study of sea cliff communities at South Stack on Holy Island, Anglesey, Wales, with the aim of generating and testing hypotheses concerning relationships between sea cliff plant communities, distributions of individual species and causal environmental factors.

The collection of vegetation data on sea cliffs posed a number of problems, particularly over access to sites and sampling, bearing in mind the crenulations of the coastline and the near-vertical nature of many cliffs. As a result, a stratified sampling system was adopted. The range of exposures, aspects and slopes of the cliffs were determined and then vegetation composition was sampled within each of these types. Quadrats were taken where there was a visually homogeneous stand of vegetation within a certain combination of aspect, slope and geology.

The geology of the cliffs comprises Precambrian rocks derived from grits, sandstones and shales, over which both podsolised and brown earth soils had formed. A quadrat or sample area with a size of 3 m × 3 m was used for vegetation description, and data from 65 stands or quadrats were collected. In each 3 m × 3 m area, percentage cover was assessed using 200 random points and a sampling pin with a diameter of 1.5 mm, and shoot frequency was measured in 50 randomly placed 10 cm × 10 cm subquadrats within each 3 m × 3 m area for all flowering plants. In order to determine the major environmental gradients controlling sea cliff communities and to enable variation in environmental factors to be synthesised, a total of 18 environmental factors were measured in each quadrat (Table 6.12).

For data analysis, both ordination and numerical classification methods were used. Of primary interest here is the use of principal component analysis (PCA) on the floristic data to give both species and quadrat ordinations, but also of importance is the further but separate application of PCA to the environmental data. Thus variation in plant response was identified by one ordination, while variation in environmental factors and

Table 6.12   Environmental factors recorded in each 3 m × 3 m sample stand of sea cliff vegetation on Anglesey, Wales (Goldsmith, 1973a,b). Reproduced with kind permission of the British Ecological Society and Wiley-Blackwell.

pH	Total bases
Conductivity	Phosphate (PO_4)
% soil moisture	Conductivity
Soil water-holding capacity	Height above sea
% organic matter	Distance from sea
Soil depth	General aspect
Sodium (Na)	Specific aspect
Potassium (K)	Slope
Calcium (Ca)	Potassium/calcium (K/Ca)

gradients was characterised by another. This corresponds to both the indirect or-dination approach (left pathway) and the synthesis of environmental factors (right pathway of Figure 6.1). The extent to which the two ordinations were related would show how closely the variation in floristics correlated with variation in environmental controls.

Quadrat ordination of floristic data

The ordination diagrams of the quadrat ordination of the 65 quadrats, using the frequency data standardised by quadrat, are shown in Figures 6.45 and 6.46. The weighted simi-larity coefficient (Orlóci, 1966) was used as the coefficient of similarity. The gradients of

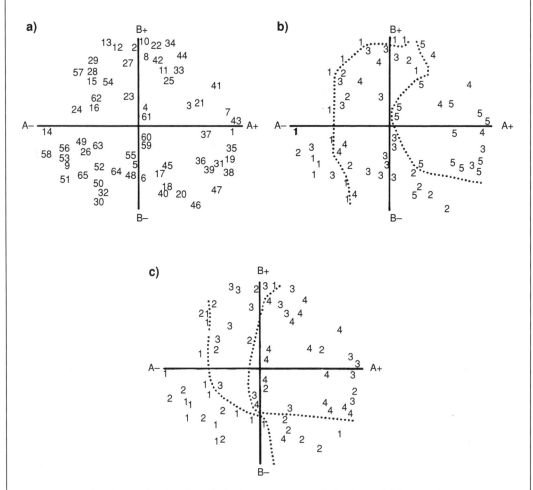

Figure 6.45   Quadrat ordination for principal component analysis of sea cliff data on Anglesey, Wales: (a) quadrat plot; (b) species richness in quadrats plotted as quintiles (1 = low; 5 = high); (c) total plant cover plotted as quartiles (1 = low; 4 = high). From Goldsmith (1973a), redrawn with kind permission of the British Ecological Society and Wiley-Blackwell.

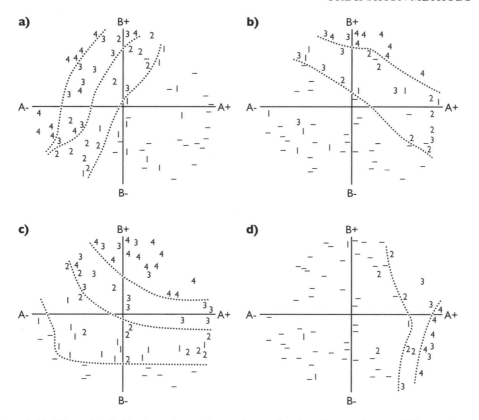

Figure 6.46 Plots of individual species on the quadrat ordination diagram for sea cliff vegetation on Anglesey, Wales. Values are species frequencies in quartiles (1 = low; 4 = high). (a) *Armeria maritima*, (b) *Plantago maritima*, (c) *Festuca rubra*, (d) *Calluna vulgaris*. From Goldsmith (1973a), redrawn with kind permission of the British Ecological Society and Wiley-Blackwell.

environmental factors on the ordination diagram are shown in the first plot. Note that axes are centred on zero, since in PCA both positive and negative scores occur on each axis.

The first plot (Figure 6.45a) shows the quadrat distribution in relation to the first two axes. In Figures 6.45b, the species numbers in each quadrat have been plotted as quintiles, while in Figure 6.45c, total plant cover is plotted as quartiles. A clear gradient of both species numbers and cover is shown on both diagrams, with low values in quadrats to the left of the first (horizontal) axis and higher values to the right. Quadrats on the extreme left were typical of exposed cliff areas, while those on the extreme right corresponded to cliff-top grassland and heath communities. Further evidence of this was obtained by plotting the abundance of certain species on the quadrat ordination diagram (Figure 6.46). The salt-tolerant species *Armeria maritima* occurs with high frequencies on the left-hand side of the plot, particularly in the sector A−B+, and is virtually absent on the right-hand side of the plot. *Festuca rubra* and *Plantago maritima* have distributions centred on the sector A+B+ and these quadrats are typical of the sea cliff grasslands found in more sheltered sites on cliff tops and slightly inland. *Calluna vulgaris*, however,

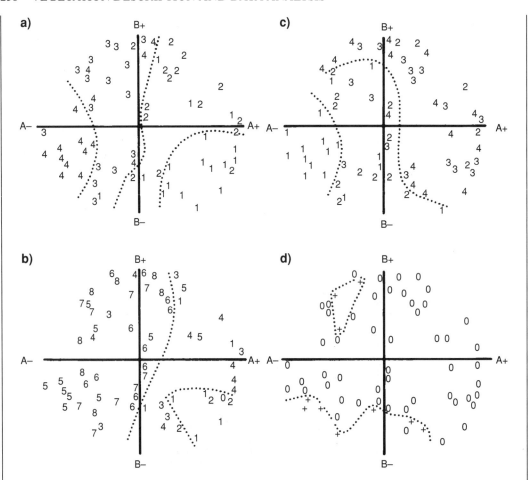

Figure 6.47 Plots of environmental data superimposed upon the quadrat ordination diagram for sea cliff vegetation on Anglesey, Wales: (a) soil conductivity/salinity as quartiles (1 = low; 4 = high); (b) aspect (SSW = 9; SW,S = 8; WSW,SSE = 7; W,SE = 6; WNW,ESE = 5; NW,E = 4; NNW,ENE = 3; N,NE = 2; NNE = 1; flat = 0); (c) soil moisture as quartiles (1 = low; 4 = high); (d) bird influence. From Goldsmith (1973a), redrawn with kind permission of the British Ecological Society and Wiley-Blackwell.

occurs predominantly in sector A+B− and is representative of several dominant heath species again found in more sheltered sites on the cliff top or inland.

Plots of environmental data superimposed on the quadrat ordination are shown in Figure 6.47. The plots for conductivity (a) and soil moisture (c) fit in well with the observations on species richness and individual species distributions, with quadrats exposed to high salt spray concentrations on the left and low salinity concentrations on the right, where the more sheltered inland grassland and heathland species occur. The aspect plot (b) is more complicated, but indicates a visual correlation of quadrats exposed to the prevailing southwest winds with high salinity. Soil moisture shows a reverse trend to salinity/conductivity and is also related to the overall amount of plant and soil cover as well as evaporation and exposure. Lastly, a plot for bird influence is presented (d) where no definite trend is observable, although, as a biotic variable, it is known that sea birds contribute large amounts of phosphates to soils through their droppings and may damage vegetation by trampling and collection of nesting material.

Ordination of environmental factors

The same 65 quadrats were then ordinated using the 18 environmental variables of Table 6.12. The resulting quadrat ordination diagram for the first two components is presented in Figure 6.48a. Note that when compared with the previous figures, axes A and B are plotted differently, but again are centred on zero. Plotting of selected environmental factors shows the first axis to be characterised by salinity (Figure 6.48b) and the second by the distribution and amount of organic matter (Figure 6.48c). Examination of Figures 6.45a and 6.48a shows some similarity of distribution of the quadrats with respect to each other. However, there are also clear differences. One of the reasons for this is that species distribution and response are not solely determined by environmental controls. Other factors, such as competition and plant species strategies, have to be considered as well.

Hypothesis generation and testing

On the basis of this evidence, various hypotheses were erected concerning plant species distributions on the sea cliffs of Anglesey in relation to environmental controls. Low species cover and numbers and the presence of maritime species such as *Armeria maritima* (sea pink) were thought to be correlated with high salinity, as reflected in the conductivity, exposure and soil moisture data. However, salinity is highest when soils dry out, and the absence of other species was hypothesised as being the result of high total salts plus the desiccation factor. The effect of salinity on *Armeria* was thus an indirect one, in that *Armeria* survives not only because of its tolerance of high salinity, but also because other species, which otherwise would compete with it, are precluded by the severe environmental conditions. Thus it is principally the absence of competition that allows *Armeria* to flourish.

These ideas were put forward as hypotheses which were then tested in more detailed experiments. Firstly, the pattern of deposition of salt spray, which determines the salinity levels in the soil, was studied, with the hypothesis that spray deposition would be a function of distance from the sea. Plastic beakers of non-saline soil were placed at carefully chosen locations around the cliffs and left for periods of up to six months. Conductivity analysis of the samples after this time showed that variations in the amount of salt deposited were correlated principally with height and distance from the sea, and secondly with local topographic features which provided increased shelter or exposure. No direct variation in response to a transect up a cliff and away from the sea was observed, however. Instead, variations in aspect leading to differences in evaporation were found to be of significance. This was further reflected in temporal variations between summer and winter, where southerly aspects experienced much higher evaporation rates in summer.

A second experiment investigated the hypothesis that species such as *Armeria* were found in exposed saline sites, not because the salinity was favourable to them but rather that the salinity prevented the establishment of competitive species found further

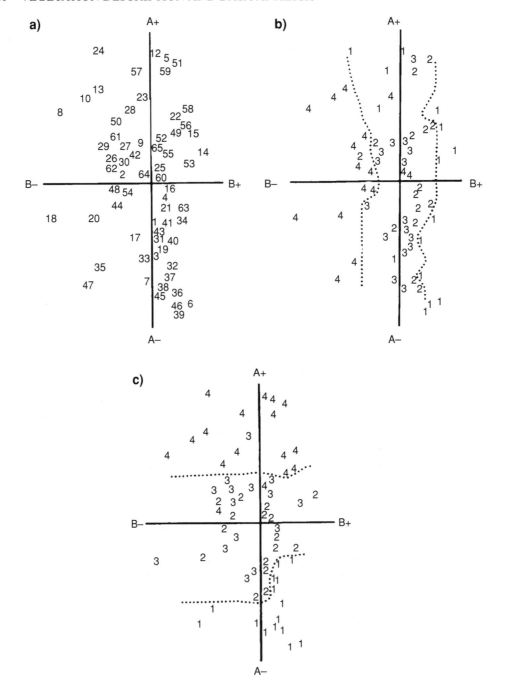

Figure 6.48   Ordination plot of the sea cliff quadrats using PCA on the 18 environmental variables of Table 6.12: (a) quadrat plot; (b) distribution of humus determined by loss on ignition, as quartiles (1 = low; 4 = high); (c) soil conductivity/salinity as quartiles (1 = low; 4 = high). From Goldsmith (1973a), redrawn with kind permission of the British Ecological Society and Wiley-Blackwell.

inland. The maritime species *Armeria* and the inland cliff species *Festuca rubra* (red fescue) were grown separately in pots and half were watered with fresh and half with salt water. Also they were grown together in a replacement series experiment (de Wit, 1960), which was also subject to a fresh and saline watering regime. In such a series, two species are grown together along a gradient of inverse proportion to each other with one species gradually replacing the other, while density of plants remains the same. The results showed that when grown alone under either treatment, there was no difference between yields of the two species. However, in the replacement series, *Armeria* had the competitive advantage in the sea-watered pots, whereas *Festuca* had the advantage in the fresh-watered pots. It was concluded that maritime species do not require saline conditions but are salt-tolerant and flourish on exposed cliffs, largely because of the absence of other species that would normally out-compete them. This lack of competitive ability is due to their slow growth rate and semi-prostrate form. On exposed sea cliffs, competition from inland species is limited by high soil salinity and damage to leaves by salt spray during gales.

Although this research was completed over 40 years ago, it still remains an excellent example of the application and interpretation of ordination methods. Interestingly, the ordination method employed was PCA, which, as discussed above, has been increasingly criticised as a method of ordination because of distortion effects. This study, nevertheless, demonstrates that highly interpretable results can be obtained using PCA. This is almost certainly because the main salinity, soil moisture and organic matter gradients inland, away from the sea, were comparatively short, resulting in a more linear species response, better suited to the underlying linear PCA model (Figure 6.20).

The use of quadrat ordinations to display data on environmental limiting factors and individual species distributions is clearly demonstrated. Most importantly, this research also shows the value of ordination as a means of hypothesis generation, with the hypotheses being tested by more detailed experimentation afterwards. A major criticism of the application of ordination methods in plant ecology over the past 40 years has been that comparatively few researchers have gone beyond the stage of description of vegetation variation and environmental gradients to generation of hypotheses which they have subsequently tested (Kent and Ballard, 1988; von Wehrden *et al*., 2009). Finally, Goldsmith's work demonstrates the care that is necessary in the interpretation of causal relationships between species and environmental factors. In the case of *Armeria*, the deterministic explanation that it is found by the sea because of its ability to tolerate salt spray has been shown to be too simplistic, and a more detailed understanding of competitive interaction between *Armeria* and other species is required.

Correspondence analysis/reciprocal averaging (CA/RA) and detrended correspondence analysis (DCA) as ordination methods used to analyse species–environment relationships in the Narrator Catchment, Dartmoor (Kent and Wathern, 1980)

A large number of experimental catchments have been set up in Britain and elsewhere to enable detailed studies of the movement of water through the land phase of the hydrological cycle. One such catchment was set up in the Narrator Brook, on the southwest margins of Dartmoor, southwest England in 1973. Research concentrated on the effects of different vegetation types and land use management practices on both the quantity and quality or chemistry of water moving through the watershed. The features of plants that are significant in determining the rate of movement of water through vegetation are their vertical structure and physiognomy together with the floristic composition. Thus in the initial stages of the research, a survey of the nature and extent of the plant communities was required with a detailed knowledge of the species assemblages present.

A total of 162 1 m × 1 m quadrats were collected from within the 4.35 km² of the catchment and analysed in various ways to establish the various plant communities and their environmental controls. Ordination methods were used to show the ecological relationships between the 162 quadrats and the 82 species. In the original paper (Kent and Wathern, 1980), correspondence analysis/reciprocal averaging (CA/RA) was applied to the data, giving the quadrat ordination diagram shown in Figure 6.49. This diagram is an excellent example of the 'arch effect' attributable to CA/RA. Nevertheless, it was possible to make a very clear interpretation of the data, particularly by following the trend of the arch. In addition, rank correlation was used to relate quadrat ordination axis

a)

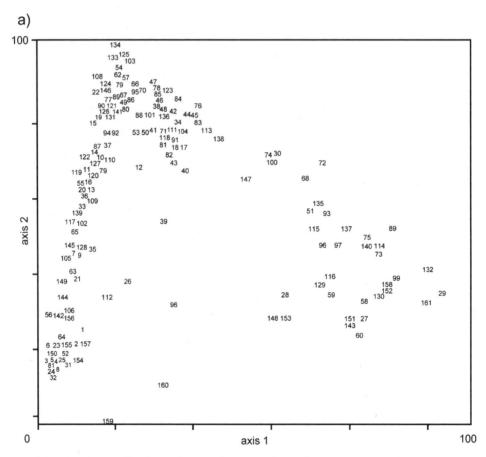

Figure 6.49 Quadrat ordination of vegetation data from the Narrator Catchment, Dartmoor, England, using correspondence analysis/reciprocal averaging. From Kent and Wathern (1980), reproduced with kind permission of Kluwer Academic Publishers.

scores to environmental data. Significant correlations emerged between the first axis and soil moisture content, peat/soil depth, slope angle and peat/soil pH. Thus the primary gradient was inferred as being drainage quality and related variables. The second axis showed significant correlations with altitude, *Calluna vulgaris* (heather) age and pH. Taken together, these were interpreted as being a reflection of biotic factors and land use management practices such as burning and grazing.

At the time of the original data analysis (1979), detrended correspondence analysis (DCA) was not yet available. Thus when the original computer program for DCA (*DECO-RANA*) was released, the data were re-analysed in order to see the effects of detrending on the ordination. The quadrat ordination diagram for DCA is presented in Figure 6.50. Comparison of Figures 6.49 and 6.50 clearly shows how the detrending process flattens the arch. However, even after detrending, a form of arch structure still remains, suggesting that some of the original 'arch effect' was attributable to the inherent properties of the data, as well as to the use of CA/RA.

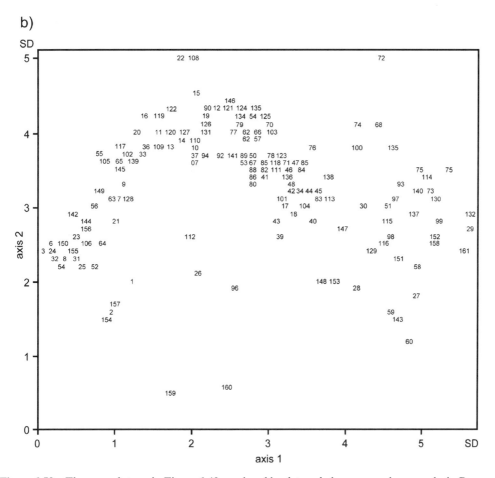

Figure 6.50 The same data as in Figure 6.49, analysed by detrended correspondence analysis. Reproduced with permission from John Wiley & Sons, Inc.

In terms of interpretation, the DCA ordination probably makes very little difference, since the relative positions of points on the first axis are always identical, although there is a significant change in the distribution of points on the second axis. Thus although DCA clearly gives an 'improved' ordination, the extent of that improvement is hard to assess accurately. However, significantly higher rank correlations were found between the second axis and the associated environmental variables of altitude, *Calluna vulgaris* age and pH. Thus the interpretation of both ordination axes remained much the same.

Non-metric multidimensional scaling (NMS) and the ordination of pioneer vegetation data on glacier forelands in southern Norway (Robbins and Matthews, 2009, 2010 – Chapters 1, Chapter 3 and 8)

This research was introduced as a case study in Chapter 1 and the field methods were presented as a case study in Chapter 3. Following field survey, a total of 42 glacier foreland sites were available for data analysis, containing 71 species. Non-metric multidimensional scaling ordination was then employed to examine the variability of pioneer species composition across the 42 glacier foreland sites (Figure 1.2; Plate 1.2). The NMS program based on the Steinhaus (Sørensen/Czekanowski) similarity coefficient in the *PC-ORD* computer package was used with the 'autopilot' option on 'thorough', and a Monte Carlo randomisation test was calculated on the stress scores (McCune and Grace, 2002; McCune and Mefford, 1999, 2010). The stress figure was 13.7 for a 3-dimensional solution and 19.7 for a 2-dimensional solution. Although the 2-dimensional figure was close to the upper limit of 20 in the Clarke (1993) guidelines for 'goodness of stress', the 2-dimensional solution was taken because the cumulative r^2 or percentage variance explained by the ordination was 0.76 for the 2-dimensional ordination but only increased to 0.81 in the 3-dimensional ordination. Instability was also higher in the 2-dimensional solution, but was still within acceptable limits.

The resulting site/sample ordination plot was presented in Figure 1.3. The data show an absence of any clear group structure and demonstrate that the 42 sites are largely distributed as a continuum. Three sites also occur as 'outliers'. These data were also analysed using numerical classification, as described in the case studies in Chapter 8, and Figure 1.3 shows the overlay of the four group cluster levels superimposed on the 2-dimensional ordination diagram. A key finding of environmental correlations with the ordination axes was that both altitude and distance eastwards were highly significant, and that altitude and distance were themselves highly correlated.

In a second paper, Robbins and Matthews (2010) again used NMS ordination on the same data, but this time the data for each site were separated into four successional stages: pioneer; up to 70 years; up to 250 years; and mature (>250 years). Data from all four successional stages were available at 39 of the glacier foreland sites, and these were analysed together in one NMS ordination (156 samples) using the same settings as in the previous analysis. Again the 2-dimensional solution was selected, with a stress score of 20, and the 2-dimensional solution was rotated to maximise correlation between axis 1 and successional stage ($r = 0.79$; $p < 0.01$).

The first analysis had demonstrated the importance of altitudinal zonation in determining overall vegetation species composition. Thus for interpretational purposes, the 39 sites were divided into four groups on the basis of their mean altitude; <1000 m;

1100–1480 m; 1500–1600 m and 1620–1860 m. The rotated NMS ordination diagram was then plotted as four separate graphs (Figure 6.51a–d), corresponding to each of the four altitudinal zones. The membership of the four successional stages was then plotted with four centroids surrounded by confidence ellipses, based on two standard deviations, each of which enclose 95% of samples in a particular successional group. Arrows were also plotted linking the centroids of each stage to show the relative movement of successional stages along Axis 1 (Figure 6.51).

Examination of the plot and associated species data showed the pattern of vegetation development in relation to altitude (see also Figure 8.16):

(a) <1000 m – early development of birch woodland occurs within 70 years.
(b) 1100–1600 m – vegetation dominated by *Poa alpina* (alpine meadow grass) and *Oxyria digyna* (mountain sorrel) or *Salix herbacea* (dwarf/snowbed willow) and *Luzula arcuata* (curved woodrush).
(c) >1600 m – *Salix herbacea* (dwarf/snowbed willow) and *Luzula arcuata* (curved woodrush) or *Juncus trifidus* (three-leaved rush) and *Hieraceum alpinum* (alpine hawkweed), with the latter becoming more frequent by the mature stage. At sites above 1600 m, the *Salix-Luzula* community is not found until after 250 years and pioneer vegetation persists until well after this length of time.

This research clearly demonstrates the value of NMS as an ordination technique and also the manner in which it can be combined with other approaches, such as numerical classification (Chapter 8), to give a thorough and informative analysis of successional processes in pioneer glacier foreland vegetation.

Canonical correspondence analysis (CCA) and the ordination of vegetation data from the evergreen broad-leaved forests of Eastern China (Wang et al., 2007)

The importance of this world vegetation type and its rarity in China was introduced as a case study in Chapter 1. In order to inform conservation activity and to define plant community types and their underlying environmental/biotic controls, data on all vascular plants were collected from 199 10 × 10 m plots using the Braun-Blanquet cover abundance scale. A total of 237 plant species were found, demonstrating the considerable diversity of these forests. Data on ten environmental/biotic factors and successional stage were also recorded at each quadrat (Table 6.13). In order to explore species–environmental/biotic gradients, CCA was used for ordination and analyses were run in both the *CANOCO* (ter Braak, 1986a, 1987, 1988a,b, 1995) and *PC-ORD* (McCune and Mefford, 1999, 2010) computer packages.

In terms of multiple regression modelling, after the initial run with ten environmental/biotic variables, four variables, all related to human and biotic impact, were removed, since they had minimal explanatory power. The summary statistics for the CCA ordination are presented in Table 6.13.

The Monte Carlo tests for both the eigenvalue and the species-environment correlation on the first axis were highly significant ($p = 0.001$). The first ordination axis was highly correlated with distance from the nearest mature forest stand, slope angle and altitude. Thus across all forest types, there is an altitudinal-slope gradient from flatter habitat at low altitude to steeper slopes at higher altitude, with the mature EBLF found on the

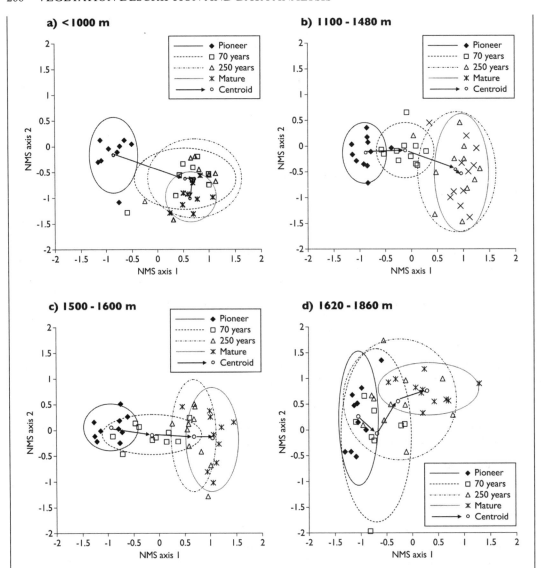

Figure 6.51   Non-metric multidimensional scaling ordination of vegetation data from four successional stages (a–d) within 39 glacier foreland sites within four altitudinal zones. Centroids of each successional stage are plotted, with arrows connecting the four stages. Ellipses indicate ± 2 standard deviations (95% of samples) from each stage centroid. From Robbins and Matthews (2010), reproduced with kind permission of Jane Robbins, John Matthews and the Institute of Arctic and Alpine Research, Boulder, Colorado.

Table 6.13 Canonical correspondence analysis (CCA) of the evergreen broad-leaved forest data from eastern China using six environmental/biotic/successional variables. Summary statistics, eigenvalues, percentage variances explained and correlations of quadrats/sample ordination scores on the first two axes with the environmental variables (LC scores) are given. Correlations are the 'intraset' correlations of ter Braak (1988a). Original work of Elsevier.

	Axis 1	Axis 2	Axis 3
Eigenvalue	0.311	0.201	0.146
% variance explained	2.9	1.9	1.4
Cumulative % explained – species	2.9	4.8	6.2
Cumulative % explained – Spp-Envt	36.7	60.5	77.7
Pearson correlation, Spp-Envt	0.836	0.773	0.798
Kendall (rank) corr., Spp-Envt	0.622	0.501	0.502
Correlations			
1. Altitude	0.630	0.098	0.769
2. Slope	0.737	0.259	−0.117
3. Asptrans	0.258	0.010	0.003
4. Succstag	−0.228	−0.909	−0.029
5. DistNatV	−0.892	0.272	0.055
6. Soil Dep	−0.153	−0.468	0.142

Variables: 1. Altitude (m). 2. Slope angle (degrees). 3. Aspect – linearised to north–south range using the formula cosine $(180° − x) + 1.1$, where x is the aspect in degrees from north. Sites with no slope and hence no aspect were given a score of 1. 4. Successional stage 1–7, based on the model of Song and Wang (1995). 5. Distance of each sample point from the nearest stand of mature natural EBLF vegetation (GIS). 6. Soil depth (cm). Also measured but not included because of minimal contribution to explained variance in initial ten-variable analysis: distance to nearest road; distance to nearest village; past land use; present land use (both on impact scale 1 = natural vegetation; 2 = plantation; 3 = fuel forest; 4 = burned within past three years; 5 = abandoned cultivated land).

highest steeper slopes. The second axis was very highly correlated with successional stage, which is linked to disturbance and also soil depth. In Figure 6.52, the five main physiognomic types of forest have been plotted onto the quadrat ordination diagram, and the link between physiognomic development and successional development is displayed vertically in the diagram and correlated with the second axis.

From these results, the need for a fuller understanding of forest regeneration and succession was realised, and in particular the importance of resprouting mechanisms in EBLF tree species. Based on the ordination results and also on associated analysis of community structure using numerical classification (TWINSPAN) (Chapter 8), a model of succession involving paired pathways of resprouting and recruitment of trees from seed was derived (Figure 6.53). Further research is underway to examine the roles of resprouting and seedbanks in the restoration and conservation of the evergreen broad-leaved forests.

Use of partial canonical correspondence analysis in the study of the composition of forest communities in the Sikote-alin Mountains of the Russian Far East (Cushman and Wallin, 2002)

Cushman and Wallin (2002) examined the composition of forest communities in the Sikhote-alin Mountains of the Russian Far East. The forests are some of the most

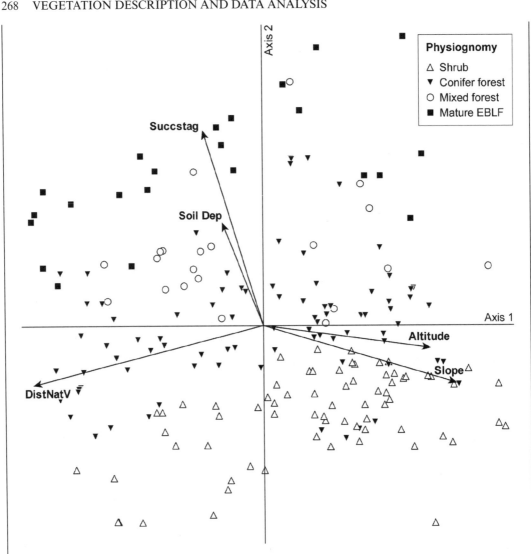

Figure 6.52 Quadrat–environmental biplot (axes 1 and 2) from canonical correspondence analysis ordination of the degraded evergreen broad-leaved forest vegetation from the Tiantong and Dongqian Lake area of eastern China (eigenvalues: axis 1: 0.311; axis 2: 0.201). Membership of five physiognomic groups of forest types is superimposed (Wang *et al.*, 2007). Reproduced with kind permission of Elsevier.

diverse and productive forests within the enormous area of the Russian taiga or boreal coniferous forest biome. The area of study in the Sikhote-alin Mountain range lies to the north of Vladivostock near to the far eastern coast opposite Japan. The undisturbed primary forests are home to some of the most endangered animals in Russia, such as the Siberian tiger. The forests are under threat from extensive uncontrolled and catastrophic wildfires and logging. The aim of the research was to examine the relationships between

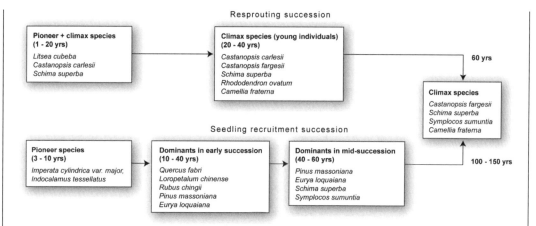

Figure 6.53 The relationship between the resprouting and seedling recruitment pathways in EBLF succession in the Tiantong/Dongqian region of eastern China (Wang *et al.*, 2007). Reproduced with kind permission of Elsevier.

patterns of forest composition and environmental and particularly biotic factors linked to disturbance.

Tree species composition was recorded on 75 0.1 ha circular plots in seven watersheds during summer 1996. Plots within each watershed were located subjectively but stratified according to major differences in topographic position. Plots were selected for homogeneity and were all closed-canopy stands. In each plot, the presence/absence of 14 genera of tree types were recorded, apart from *Populus tremula*, which was separated from all other species of poplar, and the diameter breast high (dbh) of all stems over 10 cm dbh was measured in 10 cm size classes. Nine environmental variables were recorded at each plot, as well as information on its geographical position (Table 6.14).

Table 6.14 The 12 variables measured in each stand in the forest communities in the Sikhote-alin Mountains (Cushman and Wallin, 2002). Reproduced with kind permission of Elsevier.

Type	Variable	Definition
Environmental	Elevation	Metres above sea level
	Slope	sin (slope in degrees)
	Aspect	−cos (angle in degrees − 35)
	Solar influx	tan (slope in degrees × aspect)
	Solar heat	Solar influx/elevation
	Valley	Binary, 1 if in valley
	Hillside	Binary, 1 if on hillside
	Ridge	Binary, 1 if on ridge
Fire	Fire	Binary, 1 if evidence of stand replacing fire
Space	X	UTM X coordinate, standardised to zero mean and unit standard deviation
	Y	UTM Y coordinate, standardised to zero mean and unit standard deviation
	XY	Product of X and Y

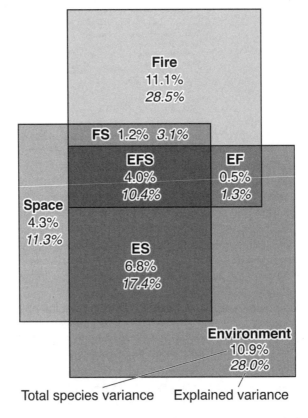

Figure 6.54   A Venn diagram demonstrating the partitioning of the explained variance of the three groups of variables in determining forest species composition in the Sikhote-alin Mountains of the Russian Far East (Cushman and Wallin, 2002). Seven components are identified: (1) partial effects of the environment; (2) partial effects of fire disturbance; (3) partial effects of space; (4) partial joint effects of environment and space (ES); (5) partial joint effects of environment and fire disturbance (EF); (6) partial joint effects of fire and space (FS); (7) joint effects of environment, fire disturbance and space (EFS). The area of each component is proportional to the variance it explains. For each component, the first figure is the percentage of total species variance and the second figure is the percentage of explained variance of the joint species/environment relationship explained by that component. Redrawn and reproduced with kind permission of Elsevier.

Using partial CCA ordination within the *CANOCO* computer package, they examined three groups of variables – environmental factors, fire disturbance history and spatial influences on species composition. Figure 6.54 shows the partitioning of the explained variance into seven components based on the original three groups of variables, and Table 6.15 lists the 12 CCA ordinations performed to examine the various components of the explained variance in terms of environmental variables, fire and space. Given a total variance explained by all three groups of variables of 38.8%, the unique effects of the spatial variables in terms of the total explained variance were only 4.3%, the environment–space interaction accounted for an additional 6.8%, together with a further 4.0% for the environment–fire–space interaction, denoting a degree of collinearity

Table 6.15 The 12 CCA ordinations used to decompose the explained variance in the tree communities of the Sikhote-alin Mountains in relation to environmental, fire, space and their interactions. Also included is the full null model (13), which included all independent variables (Cushman and Wallin, 2002). Reproduced with kind permission of Elsevier.

Analysis number	Independent variables[a]	Co-variables[b]	Sum of canonical eigenvalues[c]	% of species[d]	Significance of axis 1[e]	Significance of all axes[e]
1	Space	None	0.375	16.4	<0.002	<0.002
2	Space	Env	0.128	5.6	0.12	0.30
3	Space	Fire	0.255	11.2	<0.002	<0.002
4	Space	Env and Fire	0.100	4.4	0.07	0.09
5	Fire	None	0.386	16.9	<0.002	<0.002
6	Fire	Env	0.283	12.4	<0.002	<0.002
7	Fire	Space	0.266	11.6	<0.002	<0.002
8	Fire	Env and Space	0.254	11.1	<0.002	<0.002
9	Env	None	0.507	22.2	<0.002	<0.002
10	Env	Space	0.262	11.5	0.19	0.04
11	Env	Fire	0.404	17.7	0.006	0.006
12	Env	Space and Fire	0.249	10.9	0.126	0.180
13	Env-Space-Fire		0.890	39.1	<0.002	<0.002

[a] The independent variable set is the variables used as predictor variables in the CCA model.
[b] The covariable set is the variables whose influence on the community is factored out in the partial CCA models.
[c] The sum of canonical eigenvalues equals the total variance explained by each CCA model.
[d] The percentage of species variance records the proportion of the variance in the species data-set that is explained by each CCA model.
[e] The significance of axis 1 and all axes were quantified using Monte Carlo permutation tests.

between space and elevation, which was the primary environmental gradient in the ordination.

The variance decomposition summarised in Figure 6.54 and Table 6.15 indicated that both environmental and disturbance history vary together along spatial gradients. Over time, the increasingly intensive disturbance regime (fire and logging) since 1900 could be seen as a wave, moving across the study area from east to west away from the coast from low to high elevation. Use of partial CCA in this manner is very important and is increasingly being used to 'partial out' the spatial autocorrelation component of variation in both species and environmental/biotic variables. However, although such results are valuable, they do not assess the contributions of individual variables and the precise effects of collinearity and spatial autocorrelation combined, bearing in mind the comments of Lennon (2000) and Legendre *et al.* (2002). The approach has also been questioned by Gilbert and Bennett (2010).

Chapter 7

Phytosociology and the Zürich-Montpellier (Braun-Blanquet) school of subjective classification

INTRODUCTION

The concept of the plant community was discussed at length in Chapter 2. The next two chapters are concerned with methods for recognising and defining plant communities. This process is known as phytosociology – phyto means plant, and sociology means the study of communities or groupings (of plant species). The practice of phytosociology implies that workers must agree to a large extent with the ideas of Clements (1916) and his belief that distinct assemblages of plant species, which repeat themselves over space, can be identified and termed plant communities. In contrast, Gleason (1917, 1926), with his individualistic view, did not believe in identifiable communities; instead, plant species were seen as being distributed as a series of continua, responding to different environmental gradients. Supporters of the Gleasonian view therefore reject the whole basis of phytosociology. Many researchers, however, tend towards the Clementsian view and the climax pattern ideas of Whittaker (1953).

CLASSIFICATION

All methods for recognising and defining plant communities are methods of classification. The aim of classification is to group together a set of individuals (quadrats or vegetation samples) on the basis of their attributes (floristic or plant species composition). The end product of a classification should be a set of groups derived from the individuals, where, ideally, every individual within each group is more similar to the other individuals in that group than to any individuals in any other group. In practice, this ideal is rarely achieved, particularly in phytosociology. Nevertheless, this remains the ultimate aim of the process of classification. The groups derived from a set of individual quadrats through classification on the basis of their floristic content are usually taken as the plant communities of the area under study.

Vegetation Description and Data Analysis: A Practical Approach, Second Edition. Martin Kent.
© 2012 John Wiley & Sons, Ltd. Published 2012 by John Wiley & Sons, Ltd.

The methods for carrying out classification are many and varied. Earliest methods, such as those of the Braun-Blanquet school, were based on sorting of floristic data tables by hand, and have often been described as 'subjective'. After the advent of computers in the late 1950s and early 1960s, various numerical methods based on mathematics and statistics were devised, and these are described as 'objective', although the word 'objective' needs to be defined and used with care. This point is discussed further at the start of the next chapter. Although the distinction between 'subjective' and 'objective' methods is made, the difference relates more to the whole approach, including both purpose and methods of data collection, as well as the methods of classification (Westfall *et al.*, 1997). Thus the 'subjective' methods are dealt with separately in this chapter, and the 'objective' techniques are covered next.

A final introductory point is that all the methods of classification and related computer software discussed in the next three chapters are applied to floristic data, that is, data collected on species presence/absence and/or abundance.

PHYTOSOCIOLOGY USING SUBJECTIVE CLASSIFICATION

There is a very long history and tradition to the methods of subjective classification. This history is described in detail in Whittaker (1962), Shimwell (1971), Werger (1974a,b), van der Maarel (1975), Westhoff and van der Maarel (1978) and Dierschke (1994). The methods have been developed primarily in Europe, and Shimwell (1971) identified four major groups or schools of method which evolved during the period 1900–1960.

(a) **The Zürich-Montpellier School**. This school was established by Professor Braun-Blanquet in 1928, when he published his book *Pflanzensoziologie*, which was translated into English in 1932. This text includes a classification of the vegetation of the French Mediterranean and the central Alps, and immediately shows the regional nature of the whole approach. Subsequently Professor Braun-Blanquet's ideas were developed further by Professor Rheinhold Tüxen in Germany, and together Braun-Blanquet and Tüxen carried on the development of the classification system well into the 1960s (Podani, 2006).

(b) **The Uppsala School**. This was based in Scandinavia and its origins can be traced back to the work of von Post (1862). However, the most significant developments of the School came with Professor du Rietz's (1921) paper on method in phytosociology. The centre of research in Uppsala generated many students and followers (du Rietz, 1942a,b), who together produced many papers on the plant communities of Sweden (Lawesson *et al.*, 1997).

(c) **The Raunkaier (Danish) School**. Raunkaier is most famous for his work on vegetation life-form (1934, 1937), but he also developed a method for describing vegetation based on floristics and for tabulating vegetation samples into community types (Raunkaier, 1928). Subsequently, however, his methods have been little used.

(d) **'Hybrid Schools'**. Several researchers developed their own subjective methodologies based primarily on the methodology of the Zürich-Montpellier and Uppsala schools. Most notable among these was the 'British' School of Poore (1955a,b,c, 1956) and the work of Poore and McVean (1957) and McVean and Ratcliffe (1962) on the vegetation of the Highlands of Scotland. A further example was that of Rieley and Page (1990), who produced a phytosociological account of British vegetation.

Two points need to be made about the traditions of subjective classification. Firstly, the methods have been devised and applied largely in Europe and Scandinavia (Mucina, 1997a,b), although some attempts have been made to apply them in North America (see the case studies). Outside of these areas and particularly in the tropics, they have been very limited in their application.

Secondly, although the various schools did have differences in initial approach, most methodology has now converged on the technique of the Zürich-Montpellier School and Braun-Blanquet. For this reason, the method that is usually described is that of the Zürich-Montpellier School, and it is this that is included here. However, there are problems in that the method is not very well covered in the literature, and a number of aspects of the method remain unclear, even to experienced workers (Ewald, 2003).

THE METHOD OF THE ZÜRICH-MONTPELLIER SCHOOL (BRAUN-BLANQUET)

The following description of the method is derived from several sources, notably Braun-Blanquet (1928, 1932/1951), Poore (1955a,b,c, 1956), Ellenberg (1956), Becking (1957), Pawlowski (1966), Shimwell (1971), Werger (1974b), Mueller-Dombois and Ellenberg (1974), van der Maarel (1975), Westhoff and van der Maarel (1978) and Mucina (1997b).

The ultimate purpose of the whole methodology of Braun-Blanquet was to construct a global classification of plant communities. The method is based on several fundamental concepts and assumptions.

The relevé (or aufnahme)

This is a vegetation sample or stand, equivalent in terms of vegetation description to a quadrat. The most important point is that the location of all relevés is entirely non-random. The site for vegetation description is thus deliberately and carefully selected as a representative area of a particular vegetation type. This presupposes that the worker has a very thorough knowledge of the vegetation in the region under study and already has a clear, if subjective, impression of the major vegetation types. The samples are thus selected to be representative of those types. This means that the methodology is really only capable of being used by those who have had a long and intimate experience of the vegetation.

Homogeneity

A further complication is that the relevé or sample should be uniform and homogeneous. This means that the particular assemblage of species which are believed to be representative of the community type being described should exist over a sizeable local area, without any detailed variations within it. Thus local micro-environmental and micro-habitat variations should be either avoided or ignored. The existence of such uniform or homogeneous plots in all vegetation types is highly questionable, particularly if there are mosaics within the vegetation.

Minimal area

The method also states that the relevé or quadrat must be large enough for a representative sample of uniform vegetation to be taken. This will vary according to the life-form and physiognomy of the dominant vegetation type, and the number of species which are found in the relevé as the size of the plot increases. To determine the relevé size, the method of minimal area is used (Chapter 3).

This is a graph of species numbers against increase in relevé or quadrat size, also known as the species/area curve. The curve is derived by taking a small quadrat size and counting the number of species. Quadrat size is then doubled, and the number of species found is recorded again. The process is repeated, with the quadrat size being progressively doubled and species numbers being counted. At some point, depending on the diversity and physiognomy of the vegetation, the graph will level off. The point at which this occurs is known as the minimal area, and the quadrat size at that point is the smallest area that will adequately describe the vegetation. An example of the graph is shown in Figure 3.7. In practice, considerable difficulty often occurs in determining the exact point on the graph at which the 'break' occurs and the line for species numbers against area levels off. Cain (1932, 1934b) described a means of 'recognising' this break point, but later he admitted that it depended on the ratios of the axes used for the graph. The subject of minimal area curves was reviewed by Hopkins (1955, 1957) and Dietvorst et al. (1982). They stressed the difficulties of defining the break point, and Dietvorst et al. (1982) devised a method based on the calculation of similarity between plots in a series of nested samples of increasing size. The method is, however, described as tedious, although it does overcome some of the problems of subjectivity. Chytrý and Otýpková (2003) reviewed plot sizes employed in various European phytosociological surveys.

Concepts of minimal area are also related to homogeneity. A smooth minimal area curve, levelling off beyond a break point, will only occur if the vegetation is homogeneous. If the doubling of quadrat size brings the relevé into an adjacent local area of different vegetation, the curve may level off but then start to rise again. This indicates that the sample is not homogeneous. Also, in highly diverse tropical environments, particularly rainforests, minimal areas may be impossible to define, or could be represented by areas of several square kilometres.

The association

This is the basic unit of the classification system, corresponding to the level of the plant community. An association is a plant community type, found by grouping together various sample relevés that have a number of species in common. An example of such a set of relevés, constituting an association from McVean and Ratcliffe's survey of the vegetation of the Scottish Highlands (1962), is presented in Table 7.1. More recent discussion on the nature of the association concept is found in Willner (2006) and Jennings et al. (2009).

Abstract and concrete communities

An important feature of Braun-Blanquet associations is that they are 'abstract' in type. Thus in the McVean and Ratcliffe (1962) classification of the plant communities of the Scottish Highlands, the various communities found in the region are presented with lists of typical relevés (Table 7.1). The complete set of such tables, showing the various communities and associations, represents an inventory of the range of vegetation types in the Scottish Highlands. Apart from the few example relevés listed in each table, there is, however, no indication of the detailed location and distribution of the various community types. The associations are thus 'abstract' and are simply described as occurring 'somewhere' at various or numerous locations in the Scottish Highlands. Researchers who had described the plant communities and vegetation types at a particular location in Scotland could then look up McVean and Ratcliffe's abstract classification to see where their results 'fitted' within the overall structure of Highland communities.

The alternative to the idea of the 'abstract' community is the 'concrete' community. If the vegetation of one area of the Scottish Highlands was described in detail and perhaps mapped,

Table 7.1 An association table from McVean and Ratcliffe's classification of Scottish Highland vegetation (1962) (reproduced with kind permission of English Nature).

	1	2	3	4	5	6	7	8	9
Relevé reference number	M55	M55	R36	M58	M57	M58	M58	M58	M58
Map reference	8652 1999	8645 2006	8645 2004	8312 2225	8898 2463	8075 2899	8078 2984	8075 2930	8042 2883
Altitude (feet)	150	150	150	600	700	850	1300	1000	1300
Aspect (degrees)	325	360	360	360	360	—	—	—	—
Slope (degrees)	5	3	5	20	8	0	0	0	0
Cover (%)	100	100	100	100	100	100	100	100	100
Height (cm)	60	60	65	60	30	—	—	—	—
Plot area (cm^2)	16	16	4	8	16	4	4	4	4
Trees, shrubs and dwarf shrubs									
Betula pubescens	+	+							
Calluna vulgaris	6	7	8	7	7	4	5	7	8
Empetrum nigrum		3		1	2			+	3
Erica cinera				+					
Erica tetralix	1	3		1		+			
Ilex aquifolium		+							
Pinus sylvestris	1	+		1				1	
Sorbus aucuparia	2	1			1				
Vaccinium myrtillus	6	5	4	4	6	7	7	5	6
Vaccinium vitis-idaea	3	3	3	3	3	3	4	6	3
Ferns and fern allies									
Blechnum spicant				2	1		1		1
Pteridium aquilinum	1	2	1	2			2		
Grasses, sedges and other monocotyledons									
Deschampsia flexuosa	2	1	1	3	1	3	3	3	3
Listera cordata	+			2				1	1
Luzula multiflora						+			
Dicotyledon forbs									
Melampyrum pratense	2	1							
Oxalis acetosella							1		
Bryophytes									
Aulacomnium palustre				1	3				2
Dicranum majus	3	3	1	3	2		1		+
Dicranum scoparium		1	1			2		3	
Hylocomium splendens	3	5	7	4	3	4	8	9	5
Hylocomium umbratum			2						
Hypnum cupressiforme			2	1			2	1	
Plagiothecium undulatum	2	2	2	3	2	3	2	+	2
Pleurozium schreberi			3	2		3	2	4	2
Polytrichum commune							+		
Polytrichum formosum									3
Ptilium crista-castrensis	1	2	3	4	+	9	3	4	4
Rhytidiadelphus loreus	3	1	3		2		4	2	2
Sphagnum girgensohnii		+			5	+			+
Sphagnum nemoreum	7	8	7		8		2		2
Sphagnum palustre	+								

(*Continued*)

Table 7.1   (*Continued*)

Releve reference number	1 M55	2 M55	3 R36	4 M58	5 M57	6 M58	7 M5K	8 M58	9 M58
Sphagnum quinquefarium	7		7				2	+	7
Sphagnum russowii				8					
Thuidium tamariscinum	3	+	2		2				
Liverworts									
Anastrepta orcadensis				3					
Calypogeia trichomanis				3	3		2		3
Cephalozia bicuspidata					3				
Cephaloziella sp.				3					
Frullanfia tamarisci	+		2						
Herberta hutchinsiae		+							
Leptoscyphus taylori	1			4					
Lophocolea bidentata				1		2			
Lophozia floerkii				2	3				
Lophozia obtusa							3		
Lophozia ventricosa				2				3	2
Mastigophora woodsii	+								
Plagiochila asplenoides							2		
Scapania gracilis			2						
Lichens									
Cladonia carneola							1		
Ctadonia coccifera								2	
Cladonia cornuta							1	2	
Cladonia floerkeana								1	
Cladonia furcata					1				
Cladonia gracilis					1				
Cladonia impexa					1			1	
Cladonia pyxidata		+				1	1		
Cladonia rangiferina agg.								1	
Cladonia squamosa						1		2	
Peltigera horizontalis					1				
Number of species (60)	23	23	19	29	20	15	23	22	20

Localities	1–3	Coille na Glas Leitre, Loch Maree, Wester Ross
	4	Mullardoch, Glen Cannich, Inverness
	5	Amat Wood, Bonar Bridge, Wester Ross
	6	Loch an Eilean, Rothiemurchus, Inverness
	7	Glenmore, Inverness
	8	Iron Bridge, Rothiemurchus, Inverness
	9	Invereshie, Inverness

then the communities would be termed 'concrete' because they are precisely shown in terms of location (Legg, 1992).

Tabular comparison and sorting of relevés

The final associations, which represent groups of similar relevés, are derived by a subjective process of tabular sorting and rearrangement of both relevés and species. The exact method varies and

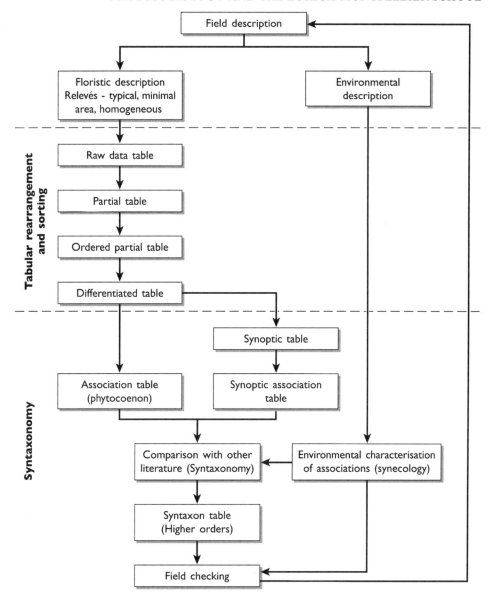

Figure 7.1   Flowchart of stages in the subjective classification of relevés using the Braun-Blanquet method (adapted from Westhoff and van der Maarel, 1978).

is not well described. Thus difficulties often occur, even when the approach is being applied by experienced workers. Generally, sorting involves the following stages, and a diagrammatic summary of the whole process is shown in Figure 7.1.

(1) *Compilation of the raw data table.* This comprises a set of relevés from the region under study. The data for these relevés will have been collected following minimal area analysis in 'representative' samples of homogeneous vegetation. The data will usually be based on the

Braun-Blanquet or Domin cover abundance scales (Table 3.4). To illustrate the application of the method, the Gutter Tor data, from Dartmoor (Table 4.4), transformed into Braun-Blanquet cover values, is presented in Table 7.2. It should be remembered that this is a small and simplified data-set. In practice, all plant groups are normally included.

(2) *Calculation of the constancy of each species*. This is simply the number of relevés within which each species occurs. The species in the raw data table are then rearranged on the basis of constancy, from high to low (Table 7.3). The purpose of this is to assist in the identification of differential species at the next stage. Dengler *et al.* (2009) provide warnings on the effects of differing quadrat or plot sizes on constancy, particularly when several data-sets are combined.

(3) *Finding good differential species*. These will be species of medium to low constancy that tend to occur together in a series of quadrats, and thus can be used to characterise groups. In Table 7.3, examples of sets of good differential species have been underlined using different symbols for different groups. There may be as few as two or as many as six or eight differential species characterising a group.

(4) *Drawing-up of partial or extract tables*. These show groups of quadrats characterised by sets of differential species. For the Dartmoor data, this is completed in two stages. Firstly, in Table 7.4, the quadrats of Table 7.3 have been rearranged to place relevés containing similar differential species into groups. Secondly, the species are then also rearranged, so that the differential species that characterise each group are placed next to each other (Table 7.5). The result is an ordered partial table. This last stage has the effect of concentrating entries in the data matrix down the diagonal. The top of the table containing the differential species is also separated from the rest.

(5) *Secondary sorting* of relevés within the groups of the partial table, so that the most similar quadrats are placed next to each other. This has been done in Table 7.6. The process of secondary sorting will involve species that are not differential species, known as the companion species. The resulting table is known as the differentiated table. The various groupings of relevés then emerge. Occasionally, a relevé or sample may not fit in any of the established groupings. Quadrat 10 of the Dartmoor data is a good example. This relevé must be discarded.

(6) Each of these groups would then be characterised as an *association or plant community*. In the Zürich-Montpellier method, there is a system of nomenclature using the names of the characterising species and suffixes to denote the association. Thus in the first group of Table 7.6, the characterising species names would be run together to give the title Sphagno-Junceto-Caricetum.

Fidelity

These characterising species are also described as having degrees of fidelity. Thus Braun-Blanquet (1951) defines five degrees of fidelity:

Fidelity 5 – exclusive species, completely or almost completely confined to one community.
Fidelity 4 – selective species, found most frequently in a certain community but also rarely in other communities.
Fidelity 3 – preferential species, present in several communities more or less abundantly but predominantly in one certain community, and there with a greater deal of vigour.
Fidelity 2 – indifferent species, without a definite affinity for any particular community.
Fidelity 1 – accidentals, species that are rare and accidental intruders from another community or relics of a preceding community.

Table 7.2 Quadrat data from Gutter Tor (Table 4.4) transformed ro Braun-Blanquet cover values (Table 3.4) to provide a raw table (C = constancy). Original work of author.

Species	1	2	3	4	5	6	7	8	9	10	11	12	13	14	15	16	17	18	19	20	21	22	23	24	25	C
Agrostis capillaris			5				1						5	5				5		5				5		7
Agrostis curtisii				5	2	5														2						4
Bryophytes					1	1	2							1		1					1	1				7
Calluna vulgaris				2	3					2	2										5				1	6
Carex nigra	2	2					1		1	2	1		+			5			1						2	10
Cerastium glomeratum																+		1								2
Cladonia portentosa				1					2	2			1	2		2					2	2				8
Danthonia decumbens																				1					1	2
Drosera rotundifolia	1									1			1													3
Erica cinerea				1																						1
Erica tetralix										2	2	2					2					2				5
Festuca ovina			2		3	1	3	3					3	2	3	1		2	2	2			3	3		14
Gelium saxatile			2		1	1	3	3	1			1	2	2	2		1	2	2				3	2		15
Juncus effusus	1	2						2		2															2	5
Juncus squarrosus																				1						1
Luzula sylvatica																				1						1
Molinia caerulea	3	3		3		2				2	4						5					5			2	9
Narthecium ossifragum		2										2													1	3
Plantago lanceolata																		1						1		2
Potentilla erecta			2	1		1		2	1	2	1		2	1	1		2	1	1					1		14
Pteridium aquilinum								5	4						4				5							4
Sphagnum sp.	5	5										4											5		5	5
Taraxacum officinale																		+								1
Trichophorum cespitosum	1			1								2													1	4
Trifolium repens													1						1							2
Vaccinium myrtillus			3				2	2		2				2		1	1					2	1		1	10
Viola riviniana	1												1											1		3

Quadrats

Table 7.3 Raw table with species rearranged by constancy (C) and with groups of matching species underlined. Original work of author.

Species																Quadrats										
	1	2	3	4	5	6	7	8	9	10	11	12	13	14	15	16	17	18	19	20	21	22	23	24	25	C
Galium saxatile		**2**	**2**		1	1	3	3	1	2		1	**2**	2	2		1	**2**	2	2			3	**2**		15
Festuca ovina		**2**	**2**		3		3	3					**3**	2	3	1		**2**	2	2	1		3	**3**		14
Potentilla erecta			**2**	1		1		2			2		**2**	1	1		2	**1**	1					**2**		14
Carex nigra	**2**	**2**							**1**	2	1	**1**				+	1		1						**2**	10
Vaccinium myrtillus					3		2	2		2	2			2		1	1				2	1				10
Molinia caerulea		3		2		2	2			2	4						5					5			2	9
Cladonia partentosa				1					2	2	1			2		2					2	2				8
Agrostis capillaris			**5**						1				**5**	5				**5**		5				**5**		7
Bryophytes				2	2		2							1		1				5	1					7
Calluna vulgaris				2	3											5					5				1	6
Erica tetralix				1							2	2					2					2				5
Juncus effusus	**1**	**2**							**1**			**1**													**2**	5
Sphagnum sp.	**5**	**5**							**4**			**4**													**5**	5
Agrostis curtisii				5	5	5		5		2										2						4
Pteridium aquilinum								5							4				5				5			4
Trichophorum cespitosum		**1**							**1**			**2**													**1**	4
Drosera rotundifolia	1											1													1	3
Narthecium ossifragum	2											2													1	3
Viola riviniana													1					1	1					1		3
Danthonia decumbens													1							1						2
Cerastium glomeratum													+					1								2
Plantago lanceolata																		1						1		2
Trifolium repens																		1		1						2
Erica cinerea										1																1
Juncus squarrosus																				1						1
Luzula sylvatica			1																	1						1
Taraxacum officinale																			+							1

Table 7.4   The rearrangement of quadrats as part of construction of a partial table from the raw table of the Gutter Tor, Dartmoor data (Table 7.3). Original work of author.

Species	Quadrats																								
	1	2	9	12	25	11	17	22	8	15	19	23	3	13	14	18	20	24	16	21	4	6	5	7	10
Galium saxatile		1	1	1					3	2	2	3	2	2	2	2	2	2	2			1	1	3	2
Festuca ovina		2							3	3	2	3	2	3		2	2	3					3	3	
Potentilla erecta			1			1	2		2	1	1		2	2	1	1		2	1	1	1	1			2
Carex nigra	2	2	1	1	2	1	2														1				2
Vaccinium myrtillus	2	3		2	2	2	1	1	2						2				+	2					2
Molinia caerulea	3	3			2	4	5	5	2		1								1	2	3			2	2
Cladonia portentosa			2	2		1		2							2				2	1	3	2	3	2	2
Agrostis capillaris			1									5	5	5	5	5	5	5			1				
Bryophytes								1							1				1	5	2	1	1		
Calluna vulgaris				2	1	2	2												5	5	2		3	2	
Erica tetralix	1	2	1	2	1	2	2								1						1				
Juncus effusus	1		1	1	2	2	2	2																	
Sphagnum sp.	5	5	4	4	5			2																	
Agrostis curtisii																	2				5	5			
Pteridium aquilinum									5	4	5	5													
Trichophorum cespitosum		1	1	2	1																				
Drosera rotundifolia	1			1	1																				
Narthecium ossifragum	2			2	1																				
Viola riviniana											1			1			1	1							
Danthonia decumbens														1		1	1								
Cerastium glomeratum														+											
Plantago lanceolata																1		1							
Trifolium repens													1			1									
Erica cinerea																									
Juncus squarrosus																	1								1
Luzula sylvatica																	1								
Taraxacum officinale																+									

Table 7.5 Further rearrangement of species to give a partial table of the Gutter Tor vegetation data. Original work of author.

Quadrats

Groups

Species	I					II			III				IV						V		VI		VII		
	1	2	9	12	25	11	17	22	8	15	19	23	3	13	14	18	20	24	16	21	4	6	5	7	10
Differential species																									
Carex nigra	2	1	1	1	2	1																			2
Juncus effusus	1	2	1	1	2																				
Sphagnum sp.	5	5	4	4	5																				
Trichophorum cespitosum	1	1	1	2	1														+						
Drosera rotundifolia	1			1	1																				
Narthecium ossifragum	2		2		1																				
Molinia caerulea	3	3			2	4	5	5	2		1				2				1	2					2
Vaccinium myrtillus			2			2	1	1											1	1	3	2	3	2	2
Erica tetralix						2	2	2													1		3	3	2
Pteridium aquilinum																									
Festuca ovina									5	4	5	5													
Galium saxatile						1		1	3	3	2	3									1	1	3		2
Potentilla erecta						1	2		3	3	2	2									1	1	1		2
Agrostis capillaris									2	2	1	1	2						5	5	2	1	3		
Calluna vulgaris						2	1						2	3	2	2	2	3	5	2	5	5	3		2
Cladonia portentosa								2					2	2	1	1	2	2			2	1		2	2
Agrostis curtisii													2	2	1	1			1		5	5			
Bryophytes								1					5	5	5	5	5	5							
Companion species																									
Viola riviniana											1						1	1							
Danthonia decumbens																1	1								
Cerastium glomeratum														+	1										
Plantago lanceolata													1	1	1										
Trifolium repens													1		1										
Erica cinerea																									
Juncus squarrosus																									
Luzula sylvatica																	1	1							
Taraxacum officinale																+									1

Table 7.6 The final differentiated table following secondary sorting within groups. Original work of author.

Species	Quadrats																								
	12	25	1	2	9	11	22	17	8	15	19	23	3	13	18	24	14	20	16	21	4	6	5	7	10
Associations	I					II			III				IV						V		VI		VII		
Differential species																									
Carex nigra	1	2	2	2	1	1	1												+						2
Juncus effusus	2	1	2	2	1																				
Sphagnum sp.	4	5	5	5	4																				
Trichophorum cespitosum	2	1	1	1	1																				
Drosera rotundifolia	1	1		1	1																				
Narthecium ossifragum	2	1	2																						
Molinia caerulea	2	3	3	3		4	5	5	2								2		+						
Vaccinium myrtillus						2	1	1									2		1	2	3	2	3	2	2
Erica tetralix	2					2	2	2											1	1	1	1			2
Pteridium aquilinum												1													
Festuca ovina			1	1	1			1	5	4	5	5	2	3	3	2	2	2	1	1	1	1	3	3	
Galium saxatile			1	1	1	1		1	3	3	2	3	2	2	2	3	2	2			1	1	1	3	
Potentilla erecta					1	1		2	3	3	2	2	2	2	2	2	1	1							
Agrostis capillaris	1								2	1	2	1	5	5	5	5	5	5							
Calluna vulgaris				2	2	2											2	2	5	5	2	2	3		2
Cladonia portentosa			2	1	2	1	2									2			5	2	5	5	1	2	2
Agrostis curtisii				2	2													2	1		5	5	1	2	
Bryophytes						1							1	1	1	1		1							
Companion species																									
Viola riviniana												1	1	1				1							
Danthonia decumbens													1	1	1										
Cerastium glomeratum													+	+											
Plantago lanceolata														1	1	1									
Trifolium repens													1		1	1	1	1							
Erica cinerea																							1		
Juncus squarrosus																	1	1							
Luzula sylvatica																		1							
Taraxacum officinale															+										

The concept of fidelity has been much criticised and misunderstood. Poore (1955a) pointed out that the degree of fidelity of a particular species can only be fully assessed when all the vegetation of a region has been described. It is thus very much a geographically defined concept. However, the concept must also depend on the size of the geographical region used to define fidelity. Also fidelity has been confused with constancy of species within groups or associations. A species with a high constancy in an association (i.e. occurs in every, or almost every, relevé of the association group) may not necessarily have the highest degree of fidelity. As a small example, in the Dartmoor data, *Juncus effusus* is entirely faithful to group I and is 100% constant because it occurs in all five quadrats of the group. However, *Galium saxatile* is a constant species of associations III, IV and V, but is not faithful to any one of those groups. Even in the Dartmoor example, the apparently high fidelity of *Juncus effusus* to group I of the data-set does not mean that it automatically has high fidelity when compared with other community types from other studies that may have been completed on Dartmoor and surrounding regions. The same principle applies to much larger-scale studies. Again, the key point is that fidelity is entirely dependent upon the size of the vegetation region that is being described, and the extent to which that description is complete.

Further discussion of the problems of fidelity are presented in Poore (1955a), Moore (1962), Shimwell (1971), Werger (1974b), Mueller-Dombois and Ellenberg (1974), Westhoff and van der Maarel (1978), Kershaw and Looney (1985), Mucina (1997b) and Bruelheide (2000).

Synoptic tables

Once associations have been defined and recognised, a synoptic table may be produced for each association. Each community type is represented by a column in which each characterising species of each association is indicated as a percentage or class value. An example of such a table is shown in Table 7.7, which contains data from a survey of all the rich-fen systems of England and Wales (Wheeler, 1980). In the first of several papers, a total of 22 communities or associations were identified and the synoptic table shows percentage constancy values on a 5-point scale.

Higher orders of classification

In the Braun-Blanquet system, the level of the association is fundamental and represents the basic unit of vegetation description, equivalent to the plant community. However, under the system, higher and lower orders can be recognised within an overall floristic association system, as shown in Table 7.8.

A grouping of two or more associations which have their major species in common and which differ only in detail can be combined to give an alliance. Alliances can similarly be grouped at a higher level into orders, and orders into classes. Equally, the association can be subdivided into sub-associations; sub-associations can be divided into variants, and within these various facies can be recognised. An example of how this hierarchical structure works is shown in Table 7.9, which is again taken from Wheeler's work on rich-fen systems in England and Wales (1980). In this manner, the whole hierarchy of vegetation units in a region can be described and their relationships to each other displayed. To assist with this process over large areas, an international code of botanical nomenclature has been established, based on the idea of syntaxonomy, which is a set of rules governing the naming of communities and hierarchies under the Braun-Blanquet system (Moravec, 1971). Table 7.9 is an example of a syntaxon table. The higher order classification of syntaxonomy for the British Isles is presented in Shimwell (1971: Appendix III).

Table 7.7 A synoptic table of common species in the 22 rich-fen communities of England and Wales (Wheeler, 1980); the digits are constancy values: 1 = 5–20%; 2 = 21–40%; 3 = 41–60%; 4 = 61–80%; 5 = 81–100%; the occurrence of species with very low constancy is not shown (reproduced with kind permission of Wiley-Blackwell and Dr B.D. Wheeler).

	1	2	3	4	5	6	7	8	9	10	11	12	13	14	15	16	17	18	19	20	21	22
Phragmites communis	5	5	5	4	3	5	5	5	5	3	5	1	2	3	2	2	4	5	3	2	4	4
Galium palustre	5	5	2	5	5	5	3	5	2	5	1	1	4		2			3	2	5	3	4
Mentha aquatica	2	3	1	4	3	4	2	5	3	5	5	1	4	2	2	1			1	3	3	4
Acrocladium cuspidatum	1	1	1	2	4	5	3	5	2	5	5	4	5	5	5	4		5	1	5	5	5
Angelica sylvestris		2			4	3	5		3	5	5	1	3	3	5	1			1	3	1	3
Eupatorium cannabinum	1	3		4	2	4	4	4	5	2	5		2	4		2	1	5	2	1	1	3
Lythrum salicaria		3		4	3	3		4	3	2	4		3	1			1	3			3	3
Juncus subnodulosus		2		4	2	1	1	4	5	2	5		4	5		1	1	4				
Carex rostrata				2	5					5	2	1	1							4		2
Eriophorum angustifolium		1			5					5	3	3	1							2		
Equisetum fluviatile		1		4						5	3		2		1	1				4		2
Menyanthes trifoliata		3		5	5	2		2		5	4	2	2		1					4		1
Potentilla palustris				5	5	2		3		5			2		1		1	3	2	3	1	3
Cladium mariscus		5			1			4	5	1	2		1	1		2	2				1	1
Ranunculus lingua	2	1	5		3	3		1		2			1							1		1
Typha angustifolia	3	1	5			2		1										2				
Sium latifolium		1	5					1														
Cicuta virosa		1	5					1	1													
Carex pseudocyperus	1	1	5		1			1														2
Carex lasiocarpa		2						2		4		1	1							2		
Carex diandra								1		5	1		1									
Acrocladium giganteum								1		5	2		1							2		
Carex elata				5	1		1	4	1	1	2		1					1	2		2	2
Carex paniculata						5	3	3		3	2		2		1					2	2	5
Campylium stellatum						2	1	3	5	4	5	4	3	3	2							
Valeriana officinalis								3	2	2	4		2	1	2	1		3		4	3	4
Filipendula ulmaria						2	4	3	2	5	3	1	5	2	4	5	1	3	1	5	3	2
Lysimachia vulgaris						4	2	4	1				2				1	3	1		2	3
Peucedanum palustre						4		5					1				2	5	3		1	1
Thelypteris palustris						4		4					1				1	5	3		2	2
Calamagrostis canescens						1	1	4					1				1	5	2			2
Epilobium hirsutum						1	4	1					2			5						1
Thalictrum flavum							2	1					2	1		1						
Carex panicea							3	2	5	5	5	4	5	4						1		
Carex nigra							1	1	5	5	5	2	3	5						2		
Carex lepidocarpa								1		3	5	5	1									
Pedicularis palustris								2		3	5	4	2	1								
Schoenus nigricans								2	3	1	5	4	2	1								
Epipactis palustris								2	1	1	5		2	2								
Anagallis tenella										1	4		2	1								
Eriophorum latifolium										1	5	4	1									
Drepanocladus revolvens										3	5	4	1									
Pinguicula vulgaris											5	5	1									
Carex dioica										1	3	4										
Molinia caerulea							2	5	3	5	5	3	5	5						3	3	1
Cirsium dissectum							1	4		3		1	5									
Potentilla erecta							1	5	2	5	4	3	5	4						2	2	

(Continued)

Table 7.7   (*Continued*)

	1	2	3	4	5	6	7	8	9	10	11	12	13	14	15	16	17	18	19	20	21	22
Succisa pratensis								1	3	3	5	5	4	5	4				2	2		
Cirsium palustre						1	2	3	2	3	5	3	5	4	3	3		3		4	3	5
Vicia cracca								1	1	3	1	4		5	4	3				2		
Galium uliginosum								1		2	3	3	1	4	5	2	1			2	3	1
Valeriana dioica								1	2	2	4	3	5	4	4	4	1	1		4	2	3
Caltha palustris		2		1		2	2	3		5	3	2	4		3	1				5	1	4
Iris pseudacorus						1	3	3					3					2	1		4	4
Holcus lanatus										2	3	1	5	3		1				3	2	1
Ranunculus acris										2	2	1	5	2	3					4	2	2
Poa trivialis											1	1	5		1					5	2	3
Carex acutiformis											1		3		2					1	3	4
Scrophularia aquatica						2					1		3									2
Epilobium parviflorum											1		3									
Hypericum tetrapterum											2		3	1		1						
Cerastium holosteoides												1	4	2								
Rumex acetosa													4	4		1						
Sanguisorba officinalis										2			1		5					2		
Geum rivale										1					3					3		
Climacium dendroides										2		1	1		5					4		
Crepis paludosa										1					4					5		
Myrica gale									3		1						5	3	5	2		1
Alnus glutinosa						4	2	3			2		1	1			4	3	1	2	2	5
Salix cinerea						4	3	3		3	2		2	1	1	1		3	3	5	5	5
Betula pubescens							1	2		2	3		1	1	2		1	5	5	4		3
Mnium undulatum										1	2		3		1					3	3	3

Key to community-types: 1, *Scirpo-Phragmitetum*; 2, *Cladietum marisci*; 3, *Cicuto-Phragmitetum*; 4, *Caricetum elatae*; 5, *Potentillo-Caricetum*; 6, *Caricetum-paniculatae*; 7, *Angelico-Phragmitetum*; 8, *Peucedano-Phragmitetum*; 9, *Cladio-Molinietum*; 10, *Acrocladio-Caricetum*; 11, *Schoeno-Juncetum*; 12, *Pinguiculo-Caricetum*; 13, Fen meadow communities; 14, *Cirsio-Molinietum*; 15, *Carex nigra-Sanguisorba officinalis* community; 16, *Epilobium hirsutum-Filipendula ulmaria* community; 17, *Myricetum gale*; 18, *Betuio-Dryopteridetum cristatae*; 19, *Betulo-Myricetum*; 20, *Crepido-Salicetum*; 21, *Salix cinerea* carr; 22, *Osmundo-Alnetum*.

Table 7.8   The hierarchical classification units of the Braun-Blanquet system. Original work of author.

Rank	Suffix	Example
Class	-etea	Molinio-Arrhenatheretea
Order	-etalia	Molinietalia
Alliance	-ion	Junco-Molinion
Association	-etum	Cirsio-Molinietum
Sub-association	-etosum	Cirsio-Molinietum caricetosum
Variant	–	Specific names used
Facies	–	Specific names used

Table 7.9 A synopsis of the plant communities of the rich-fen systems of England and Wales (Wheeler, 1980) (reproduced with kind permission of Wiley-Blackwell and Dr B.D. Wheeler).

PHRAGMITETEA
  PHRAGMITETALIA
    PHRAGMITION (COMMUNIS)
      Scirpo-Phragmitetum W. Koch 1926
        Phragmites-dominated swamp of pools and wet places in fens.
      Cladietum marisci Zobrist 1935 em. Pfeiffer 1961
        Cladium-dominated swamp of pools and wet places in fens.
      Cicuto-Phragmitetum Wheeler 1978
        Semi-floating fen dominated by *Phragmites communis*, with much *Carex pseudocyperus, Cicuta virosa, Ranunculus lingua* and *Sium latifolium.*
  MAGNOCARICETALIA
    MAGNOCARICION
      Caricetum elatae Koch 1926
        Carex elata-dominated swamp of pools and wet fens.
      Caricetum paniculatae Wangerin 1916
        Carex paniculata-dominated swamp of pools and wet fens.
      Potentillo-Caricetum rostratae ass. nov. prov.
        Wet sedge fen, often dominated by *Carex rostrata* or *Juncus effusus* (sometimes *Carex nigra* or *Phragmites communis*), usually with much *Eriophorum angustifolium* and *Potentilla palustris.*
      Angelico-Phragmitetum ass. nov. prov.
        Tall fen dominated usually by *Phragmites communis* or sometimes *Carex paniculata*, with a range of tall herbaceous dicotyledons. Widespread.
      Peucedano-Phragmitetum Wheeler 1978
        Tall fen dominated by *Phragmites communis, Cladium mariscus* or *Calamagrostis canescens*, with *Carex elata, Juncus subnodulosus, Lysimachia vulgaris, Peucedanum palustre, Thelypteris palustris*, etc. Mainly in the Norfolk Broads.
      Phragmites communis-dominated communities
        Species-poor vegetation characterized by dominant *Phragmites.*
      Cladium mariscus-dominated communities
        Species-poor vegetation characterized by dominant *Cladium.*
      Glyceria maxima-dominated communities
        Species-poor vegetation characterized by dominant *Glyceria maxima.*
      Cladio-Molinietum ass. nov.
        A mixed-sedge community, generally species-poor, with much *Cladium mariscus* and *Molinia caerulea.* East Anglia and Anglesey.
PARVOCARICETEA
  TOFIELDIETALIA
    CARICION DAVALLIANAE
      Schoeno-Juncetutn subnodulosi Allorge 1922
        Sedge communities of calcareous fens. *Schoenus nigricans* and/or *Juncus subnodulosus* dominate, with a wide range of associates including *Anagallis tenella, Caraex lepidocarpa, Epipactis palustris, Eriophorum latifolium*, etc. Mainly in southern Britain.
      Pinguiculo-Caricetum dioicae Jones 1973 em. Wheeler 1975
        Communities of calcareous sotigenous mires on peat or mineral gleys, often displaying a sedge sward of *Carex dioica, C. lepidocarpa. C. nigra, C. panicea.* and *Eriophorum latifolium.* Mainly, but not exclusively, in N. England, both lowland and upland.
      Acrocladio-Caricetum diandrae (Koch 1926) nom. nov.
        Communities of very wet calcareous fens, usually topogenous, *Carex diandra, C. lasiocarpa* and *Acrocladium giganteum* are usually abundant and characteristic, often with much *Carex rostrata, Eriophorum angustifolium, Menyanthes trifoliata* and *Potentilla palustris.*

(Continued)

Table 7.9   (*Continued*)

MOLINIO-ARRHENATHERETEA
  MOLINIETALIA
    CALTHION PALUSTRIS
      Fen-meadow communities
        Rush- or sedge-dominated communities, usually in spring fens or wet grassland.
        Juncus subnodulosus, J. articulatus and *J. acutiflorus* may all be important.
        Important sedges include *Carex acutiformis* and *C. disticha.* A very variable group.
    JUNCO (SUBULIFLORI)-MOLINION
      Cirsio-Molinietum Sissingh et De Vries 1942
        Grassland dominated usually by *Molinia caerulea,* with *Carex hostiana, C. panicea,*
        C. pulicaris, Cirsium dissectum, Gymnadenia conopsea. Potentilla erecta, Succisa pratensis
        (*Juncus subnodulosus*).
      Carex nigra-Sanguisorba officinalis community Proctor 1974
        Herb-rich vegetation normally dominated by *Molinia caerulea* or *Carex nigra,* with
        Crepis paludosa, Epilobium palustre, Geum rivale, Sanguisorba officinalis, Climacium dendroides.
      Molinia-Myrica community
        Species-poor vegetation dominated by *Molinia caerulea* and *Myrica gale.*
      Molinia sociation
    FILIPENDULION
      Epilobium hirsutum-Filipendula ulmaria community
        Vegetation dominated by tall herbs, normally *Epilobium hirsutum* and/or
        Filipendula ulmaria. Often species-poor.
      Phragmites-Urtica dioica community
        Species-poor *Phragmites*-dominated vegetation with much *Urtica dioica.*
FRANGULETEA
  SALICETALIA AURITAE
    SALICION CINEREAE
      Myricetum gale (Gadeceau 1909) Jonas 1935
        Low scrub dominated by *Myrica gale*
      Betulo-Dryopteridetum cristatae Wheeler 1975
        Acidophilous open birch scrub with much *Sphagnum,* and with *Dryopteris cristata, D.*
        carthusiana, Calamagrostis canescens, Cladium mariscus, Peucedanum palustre (Pyrola
        rotundifolio). Confined to Broadland.
      Betulo-Myricetum ass. nov.
        Birch scrub with much *Myrica gale.*
      Crepido -Solicetum pentandrae Wheeler 1975
        Salix pentandra- or *Salix cinerea*-dominated carr, with *Betula pubescens, Equisetum fluviatile,*
        Geum rivale and *Crepis paludosa.* N. Britain.
      Salix cinerea carr.
        Specios-poor *Salix cinerea*-dominated vegetation.
      Frangula alnus carr
        Low dense scrub dominated by *Frangula alnus.*
      Rhamnus catharticus carr
        Species-poor carr dominated by *Rhamnus catharticus.*
ALNEIEA GLUTINOSAE
  ALNETALIA GLUTINOSAE
    ALNION GLUTINOSAE
      Osmundo-Alneturn Klötzli 1970
        Alder carr vegetation, typically with *Carex acutiformis, C. paniculata, Eupatorium cannabinum,*
        Iris pseudacorus, Lythrum salicaria, Solanum dulcamara and *Urtica dioica.* widespread in
        lowland fens.

COMPUTERISED METHODS OF TABULAR REARRANGEMENT

The subjective nature of the process of tabular rearrangement has been reduced to some extent by the writing of various computer programs to carry out tabular rearrangement. Early versions were produced by Benninghoff and Southworth (1964), Moore G.W. *et al.* (1967), Moore J.J. *et al.* (1970), van der Maarel *et al.* (1978) and Westfall *et al.* (1982). Some of these programs are based on simple matching coefficients, while others are related to the more objective methods of numerical classification, particularly similarity analysis, described in the next chapter. Probably the most important computer program for the analysis of phytosociological data at present is the *JUICE* package (Tichý, 2002; Tichý and Holt, 2006), which is discussed in greater detail in Chapter 9. This program is specifically designed for the compilation, editing, classification and analysis of large phytosociological tables. The latest version can handle up to 1 million relevés, caters for relevés collected in different data-sets with varying formats, performs classification using both similarity analysis and Two-Way Indicator Species Analysis (TWINSPAN) (Chapter 9), computes fidelity measures and average Ellenberg values, facilitates the preparation of synoptic tables and has wide-ranging options for table output formatting. Bruelheide and Jandt (1997) and Bruelheide and Chytrý (2000) discussed issues related to the handling of very large phytosociological data-sets.

However, it is important to realise that even when the more 'objective' methods of classification, such as similarity analysis and TWINSPAN, are used, there is still a high degree of subjectivity in the overall approach, particularly with regard to sampling and the selection of typical or representative relevés by the Braun-Blanquet method, and also in the selection and designation of final groupings or associations. Numerical approaches have also been applied to the process of syntaxonomy and are reviewed by Mucina and van der Maarel (1989) and Fischer and Bemmerlein (1989).

ENVIRONMENTAL DATA

Virtually all tabular rearrangement and interpretation of association and group structure is carried out on the basis of floristic data. However, it is usual to collect a certain amount of environmental data on each relevé, for example height, aspect, soil type and drainage quality. Although these are not used in the actual process of classification, it is clear that the associations that result are also characterised by certain environmental conditions. Braun-Blanquet himself introduced the concept of the synecosystem, whereby environmental factors are reflected in the vegetation and *vice versa*. Final description and discussion of associations will thus often refer to their site character and the local environment. Van der Maarel (1993) also described relationships between floristically-defined phytosociological groups and Ellenberg indicator values (Chapter 3).

DISCUSSION OF THE ZÜRICH-MONTPELLIER SYSTEM

Even though their application has been increasingly questioned over the past 50 years, the techniques of the Zürich-Montpellier School and Braun-Blanquet remain very important in vegetation science (Moravec, 1992; Ewald, 2003). The main reason for this is that they provide the methodology for the major classification system of European vegetation types as well as some other parts of the world (Mucina, 1997a,b). The approach is, however, not without its problems. Egler (1954) presented one of the most eloquent criticisms, claiming that the method was oversimplified and represented the forcing of a weak methodology onto a much more complex real world. A major reason for Egler's comments was that he was a follower of the Gleasonian individualistic view of the plant community. Even he, however, admitted that much valuable work had been completed by Braun-Blanquet and his colleagues in Europe. Nevertheless, valid criticism of the method exists and centres on the following points:

(a) The subjectivity of the whole methodology, particularly the methods of field sampling. The selection of 'typical' or 'representative' relevés is often said to be highly biased and does assume a substantial knowledge of the vegetation prior to any attempt at description. Also, non-homogeneous and transitional or ecotone areas between typical and representative samples are not recorded under this method, yet are still clearly plant communities. The question may thus be asked: when does a transitional community between two major community types become a new community in its own right? Part of the answer to this problem was provided by Poore (1955a,b,c, 1956) with his concept of the nodum. Within any region, there is a set of major plant communities or associations which are distinctive and characteristic and can be recognised consistently in the field. However, between these major types, there are zones of transition and ecotones. Thus each of the major plant communities can be described as a nodum and every region has a set of noda. However, certain transitional types do exist between the noda and should be recognised as such.

(b) The concept of 'abstract' communities. This has proved a confusing concept, particularly for students and inexperienced researchers.

(c) The process of tabular rearrangement. The exact methodology for carrying this out varies from one worker to another, although the principles are generally agreed. Development of computerised methods has helped with the practical aspects of relevé sorting.

(d) The discarding of relevés that do not fit any of the associations that have been defined from a set of data has also been questioned. The reason for doing so lies under (a) above, in that such a relevé must have been badly chosen at the stage of field description. However, this inevitably raises questions of circular arguments along the lines that the communities must already have been defined in the worker's mind before data collection started, and data that do not fit the preconceived ideas of community structure may be discarded, when in reality they represent an association not previously recognised and thus not sampled. The answer to this point lies in the necessity of the researcher having a very thorough knowledge of the vegetation before work commences.

(e) Terminology – much strange and perhaps unnecessary terminology is used at all stages. Also, different schools have invented various new terms, each with slightly different meanings, a process which has not been helped by the use of various European languages.

(f) Perhaps most important of all is the criticism that the whole methodology is not well described in the literature, has many variations and has an air of mystique about it which is difficult to dispel. This is not really surprising when the whole approach can only be applied by very experienced workers in the field. This is unfortunate and makes the topic a difficult one for students and young ecologists.

ONGOING PHYTOSOCIOLOGICAL RESEARCH

On the positive side, the system undoubtedly works, and a very detailed and accurate classification of most of the vegetation of Europe and Scandinavia has been achieved (Mucina *et al.*, 1993, 2000; Rodwell *et al.*, 1995, 1997; Mucina, 1997a,b; Lawesson *et al.*, 1997; Ewald, 2003; Podani, 2006). In addition to the work of McVean and Ratcliffe (1962) and Wheeler (1980) in Britain, good examples of the application of the method can be found in Shimwell (1971), Pawlowski

(1966) and Yeo *et al.* (1998), and numerous further examples in the journal *Phytocoenologia* and some issues of the journals *Folia Geobotanica*, *Plant Ecology* (formerly *Vegetatio*) and *Community Ecology* (formerly *Coenoses*). Phytosociology is also important as a basis for vegetation mapping (see the case studies from Newfoundland and Minnesota below), and vegetation mapping has benefitted greatly from significant advances in both remote sensing (Alexander and Millington, 2000; Aplin, 2005; McDermid *et al.*, 2005) and geographical information systems (GIS) (Burrough and McDonnell, 1998; Johnston, 1998; Wadsworth and Treweek, 1999; Longley *et al.*, 2010).

Using the numerical methods described in Chapter 8 to assist with relevé sorting, the National Vegetation Classification (NVC) has been produced for the whole of mainland Britain (not including Northern Ireland), and is also described as a case study in Chapters 3 and 8. Slow but steady progress is also being achieved in the US on a National Vegetation Classification (USNVC) (Sánchez-Mata, 2003; Jennings *et al.*, 2002, 2008, 2009), and the USNVC is presented as a case study below. Canada is also working on a similar classification linked to the USNVC (Ponomarenko and Alvo, 2000). Despite this, many contemporary ecologists nevertheless see phytosociology as a purely descriptive activity, pursued largely for its own sake and often devoid of scientific rigour (Ewald, 2003).

Case studies

The phytosociology of the heathlands of Newfoundland – Meades (1983)

As part of a symposium volume on the biogeography and ecology of Newfoundland (South, 1983), Meades (1983) presented a full phytosociological account of the heathland vegetation of the island. The greater part of the island is made up of forests of balsam fir (*Abies balsamea*) and black spruce (*Picea mariana*) (56%), and wetland (24%). The remaining 20% is heathland distributed within five general groups as shown in Figure 7.2. Meades aimed to define these heath communities in much greater detail using the methods of subjective classification. Given the typical vegetation types of Newfoundland, he used terminology that was closer to that of the Uppsala School and du Rietz (1942a,b). Although field methods are not described in detail, apart from a short Appendix, the methods were very similar to those of the Zürich-Montpellier School described in this chapter, including the use of Braun-Blanquet cover scales for recording.

At the start of the survey, the major community types were already known and Meades was able to relate them to three basic phytosociological alliances of heathlands proposed by du Rietz in 1942:

(1) EMPETRION – representing exposed alpine and subalpine heaths dominated by *Empetrum hermaphroditum*, a very close relative of *Empetrum nigrum* (crowberry). On Newfoundland, this is represented by the very similar species *Empetrum eamesii*. This alliance includes the alpine heath, *Empetrum* heathland and moss heath on Figure 7.2.

(2) MYRTILLON – these are the less exposed heaths below the tree-line, where there is adequate snow accumulation in winter. The heaths are dominated by *Vaccinium myrtillus* (bilberry), a species restricted to the northwest part of the American continent. This is the Kalmia heath on Figure 7.2.

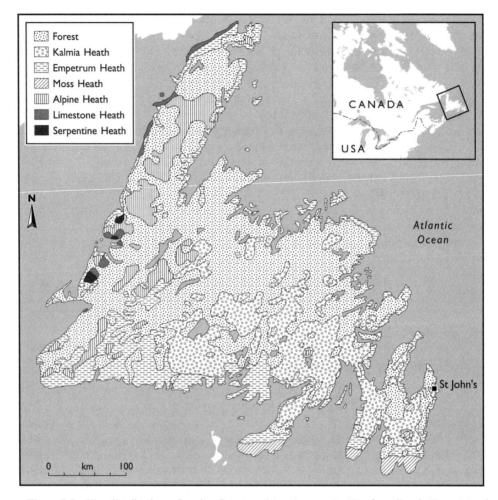

Figure 7.2 The distribution of major forest and heath types in Newfoundland. From Meades (1983), redrawn with kind permission of Junk, The Hague.

(3) DRYADION – this alliance is the species-rich vegetation of neutral and basic soils dominated by *Dryas octopetala* (Mountain avens). These are the limestone and serpentine heaths on Figure 7.2.

As an example, part of the differentiated table for the limestone and serpentine heaths is presented in Table 7.10. The first point to note is the double score for each species. The first value is for abundance using a slightly modified scale:

1: cover less than 5%
2: cover 5–25%
3: cover 26–50%
4: cover 51–75%
5: cover 76–100%.

Table 7.10 Part of the differentiated table of acidic dwarf shrub heaths in Newfoundland (Meades, 1983). Reproduced with kind permission of Junk, The Hague.

	Group A							Group Re										Group E										Group K				Group V			
Altitude	70	70	240	150	120	30	75	30	75	120	170	30	240	120	30	150	30	90	170	90	120	75	60	180	180	230	180	60	75	90	150	150	150	15	150
Slope	0	0	20	5	0	5	10	30	30	30	0	30	30	20	0	0	5	0	20	30	10	20	20	40	30	5	40	0	5	0	50	0	30	0	30
Aspect	–	–	N	S	–	SE	NE	S	NW	W	W	N	N	W	–	–	S	–	E	E	NE	N	SE	E	W	NE	W	–	W	E	E	S	–	E	–
Site	71	72	71	71	71	71	71	71	71	71	71	71	71	71	71	71	71	71	71	71	71	71	71	71	71	71	71	71	71	71	71	71	71	71	71
Number	137	138	74	20	42	18A	5B	18B	1A	42P	45B	90	49J	19	55	15	8D	24C	46C	47B	5C	52B	41C	1A	5	65B	69C	41D	35D	51C	26C	37A	38G	39A	36C
Differential Species of the Association																																			
Arctostaphylos alpina	.1	.1	1.2	1.2																															
Loiseleuria procumbens	.1	1.2	1.2	1.2																															
Diapensia lapponica	.1	2.2	1.2	1.2																															
Cetraria nivalis	.1	1.2	1.2	1.2																															
Ochrolechia frigida	.2	.2	.2	.2			.1																												
Rhacomitrium lanuginosum	.2	1.2	2.2	2.2		4.4	4.4	4.4	4.4		2.2		2.2																						
Cladonia boryii	1.2	1.2	1.2	2.2		1.1	2.2	2.2	1.1				.2	1.2																					
Sphaerophorus globosus	1.1	1.1	2.2	1.2		.1	1.2						.2																						
Cetraria islandica	.1	.1	1.2	1.2			1.2										1.2																		
Vaccinium uliginosum	2.2	2.2	.2	.1		2.1	2.1					1.1	1.1																						
Empetrum eamesii	2.2	2.2	2.2	1.2		1.2						1.1	1.1																						
Potentilla tridentata	2.1	1.1	1.1	1.1		1.1	1.1	1.1	1.1		1.1	1.1	1.1	.1		2.1																1.1	1.1	1.1	1.1
Deschampsia flexuosa	1.2	1.2	1.2	1.2		1.2	1.1	1.2	1.2		1.2	1.2	1.2	1.1		1.2																2.2	2.2	1.2	1.2
Calamagrostis pickeringii			1.2	.2		1.1	2.1	1.1	2.1	1.1	1.1	1.2					1.1									1.1						1.2	1.1		
Juniperus communis	1.2	.2		2.2		1.2	1.2	2.2																										2.2	1.2
Prenanthes trifoliolata	1.1	1.1		.1		1.1																		.1	1.1									
Solidago uliginosa	.1			.1		.1							1.1																						
Rhododendron canadense													1.1					2.1	2.1	1.1	3.1	.1	2.1		1.1							2.1			
Viburnum cassinoides																			.1	2.2		2.1	1.1	1.1	1.1						1.1				
Nemopanthus mucronata													1.1									1.1	1.1	.1	.1						1.1				
Amelanchier bartramiana													.1					.1	.1			2.1	2.1			.1	1.1				1.1				
Luzula campestris																		.1		.1		.1	2.1									1.2	1.2	1.2	1.2
Lycopodium obscurum																																.1	.1		1.1
Taraxacum officinale																																.1	.1	.1	.1
Achillea millefolium																																.1	.1	.1	
Hieracium murorum																																.1			
Dicranum fulvum																																2.2	2.2		
Anthoxanthum odoratum																																1.1	1.1		
Solidago rugosa																																1.1	1.1		.1

(Continued)

Table 7.10 (*Continued*)

Differential Species of the Sub-Associations	Db	Dj	-	Em	Es	Et	Km	Ks	Ka	Kt	
Betula pumila	1.1 1.1										
Salix uva-ursii	2.2 1.2										
Polygonum viviparum	1.1 1.1										
Leontodon autumnalis	1.1 1.1										
Platismatia glauca		1.2 1.2									
Cetraria cucullata		.2 .2									
Juncus trifidus		1.2 1.2									
Lycopodium selago		.1 .1	.1	.1							
Myrica gale				2.1 2.1	2.1 2.1 .1		2.1 2.1	1.1 1.1 1.1 2.1			
Chamaedaphne calyculata				1.1		1.1	1.2	1.1 1.1 1.1	1.1		
Clintonia borealis				1.1			.1		1.1 1.1	.1	
Smilacina trifoliata					1.1		.1		1.1 .1	.1	
Coptis groenlandica			1.1				1.1	1.1 1.1			.1 1.1
Osmunda cinnamomea				1.1	.1 1.2		1.2	2.2			.1
Sphagnum nemoreum					3.3 3.3 3.2			3.3 4.4 5.5			
Mylia anomala					.1 .1 .1			.1 .1			
Ptilium crista-castrensis					2.2 2.2 1.2			2.2 2.2	2.2 2.2		1.2
Vaccinium oxycoccus					1.2 .2 1.2	2.2		.1 1.2 1.2	1.2	1.2	
Sarracenia purpurea					1.1	.2		.1			
Drosera rotundifolia					.1 .1			.1			
Alnus crispa								3.1	3.1 2.1 3.1 3.1		
Pteridium aquilinum								1.1 2.1 2.1 1.1	.1	2.1	
Betula papyrifera								1.1 1.1	.1		
Cyripedium acaule								1.2 2.2			
Dicranum undulatum									1.1 1.1		
Solidago macrophylla									1.1 2.1		
Sorbus americana								.1 1.1 2.2			2.1
Polytrichum commune											1.1

Companion Species	A				Re				E									
Vaccinium angustifolium	.1	1.1	1.1	1.1	1.1	1.1	.1	1.1	1.1	1.1	1.1	1.1	1.1	2.1	1.1	1.1	2.1	2.1
Kalmia angustifolia	.1		.1	1.1	1.1		1.2	1.1	.1	1.1	2.2	2.1	2.1	1.1	1.1	2.1		1.1
Ledum groenlandicum			.1	1.1	.1	.1	1.1		.1	1.1	.1	1.1	1.1	2.1		1.1	.1	2.1
Cladonia arbuscula	1.2	.1	1.2	1.2		.2	1.2			2.2	2.2	.2		2.2	2.2	1.2	2.1	
Cladonia mitis	1.2	.2	1.2	1.2		.2	2.2	2.2		2.2	1.2	1.2	1.2	1.2	2.2	2.2	3.2	2.2
Cladonia rangiferina	2.2	1.2	1.2	1.2	.1	.2	2.2			3.3	2.2	2.2	2.2	2.2	2.2	1.2	2.2	2.2
Cladonia alpestris										2.2	.2	1.2		1.2	1.2	2.2	.2	2.2
Cornus canadensis	1.1	1.1	1.1	1.1	.1	1.1			1.1	1.1	.1	1.1		.1		.1	1.1	1.1
Empetrum nigrum	2.2	2.2	1.2	1.2	1.2	1.2	.2		3.3	3.3	2.3	2.2	2.3	.2	3.3	3.3	4.4	4.4
Vaccinium vitis-idaea	1.1	1.1	2.2	2.2	2.2	2.2	2.2	2.2	2.2	2.2		1.2	1.2	2.2	2.2	1.2	2.2	2.2
Pleurozium schreberi	.1			.2								3.3	2.2	1.2			2.2	1.2
Hylocomium splendens											2.2	2.2		2.2	2.2			
Dicranum scoparium										2.2				1.2	2.2	2.2		
Maianthemum canadense				.1				1.1	1.1	.1					.1			
Trientalis borealis	.1	.1			.1	.1			.1		.1	.1					1.1	
Linnaea borealis							1.2			1.2					1.2		.2	
Pyrus floribunda							1.1		1.1									1.1
Dicranum fuscescens				1.2				2.2										
Cornicularia aculeata	.2	.2														1.2		
Cladonia uncialis	.2	.2			1.2	1.2		1.2						.2				
Cladonia terrae-novae					1.2				.2				1.2		2.2			
Cladonia elongata											1.2	.2					.2	
Kalmia polifolia											1.1				.1	1.1		
Hypogymnia physodes	.2	.2		1.2													.2	
Spiraea latifolia																.1		
Polytrichum juniperinum		.1		.1								.1						
Aster novi-belgii																		
Gaultheria hispidula														1.2				
Larix laircina			.1	.1														
Hypnum imponens	.1																	
Sanguisorba canadensis															1.1			
Lycopodium clavatum													1.1					.1

Companion Species	K														V			
Vaccinium angustifolium		1.1	1.1	1.1	.1	1.1	1.1	1.1	1.1	1.1	1.1	2.1	2.1	1.1	4.1	3.1	2.1	3.1
Kalmia angustifolia	3.1	1.1	3.1	1.1	2.1	2.1	4.1	4.1	1.1	2.1	4.1	3.1	4.1	4.1	1.2	1.1	1.1	1.1
Ledum groenlandicum	1.1	1.1			1.1	.1	2.1	1.1		2.1	2.1		2.1	1.1			.1	1.1
Cladonia arbuscula	.2	2.2	2.2		1.2		.2	.2			2.1	2.2	2.2	1.2			2.2	
Cladonia mitis	1.2	1.2	1.2		.2						1.2	1.2	1.2	1.2				
Cladonia rangiferina		2.2	3.3	1.2	1.2		.2	.2										
Cladonia alpestris		2.2	2.2		.2						2.2		2.2					
Cornus canadensis		.1	1.1	.1	1.1	.1	1.1					1.1	.1		1.1	1.1	1.1	.1
Empetrum nigrum				.1	.1	1.1					1.1						2.2	2.2
Vaccinium vitis-idaea		.1		1.1	.1						.1		1.1			2.2	2.2	
Pleurozium schreberi			2.2	2.2	2.2	1.2	.2	1.2			4.4	3.3	2.2					2.2
Hylocomium splendens				3.3			2.2	.2				2.2						
Dicranum scoparium	2.2	2.2		2.2	2.2		2.1	1.2	1.2		2.2				1.2		3.3	2.2
Maianthemum canadense		.1							1.1			1.1				1.1	1.1	
Trientalis borealis							1.1	1.1						.1	.1		.1	
Linnaea borealis	1.1											2.2						1.2
Pyrus floribunda	1.1			1.1					1.1	2.1								
Dicranum fuscescens							1.2	1.2	1.2					1.2				
Cornicularia aculeata																		
Cladonia uncialis	.2	1.2																
Cladonia terrae-novae																		
Cladonia elongata																		
Kalmia polifolia				1.1	1.1						.1							
Hypogymnia physodes			.1		.2													
Spiraea latifolia																	1.1	1.1
Polytrichum juniperinum	1.1	.1	1.1			1.1						.1	1.1		1.1			
Aster novi-belgii		.1															.1	
Gaultheria hispidula	1.2																	1.2
Larix laircina																		
Hypnum imponens																		.2
Sanguisorba canadensis				.1														
Lycopodium clavatum				1.1														1.2

Additional Species:
72137: *Campanula rotundifolia* (.1)
71074: *Picea glauca f. parva* (.1); *Stereocaulon paschale* (1.2); *Cladonia coccifera* (.1)
71018A: *Pinus sylvestris* (.1)
71041D: *Abies balsamea* (.1)
71035D *Rubus pubescens* (.1); *Lycopodium annotinum* (.1)
71052B *Geocaulon lividum* (.1)
71069C: *Prunus pennsylvanica* (1.1)
71036C: *Anaphalis margaretacea* (.1)
71052B: *Sphagnum papillosum* (1.2); *Aulacomnium palustre* (.1)
70137A: *Fragaria virgianum* (.1); *Lonicera villosa* (1.1); *Rosa nitida* (.1)
71036: *Cladonia cristatella* (.1); *Cladonia pyxidata* (.1)
71046: *Cladonia squamosa* (.2)
71055: *Cladonia coccifera* (.1)

The second digit in the table is the Braun-Blanquet index of sociability:

1: growing singly
2: in small tufts or tussocks
3: in large patches
4: carpet-forming, carpet covers at least half plot
5: forming an almost continuous carpet.

Note that in Table 7.10 the differential species are clearly delimited at the top of each group and the companion species are listed below. Overall, the following list of communities was recognised:

Heath communities of acid soils

(1) DIAPENSIO-ARCTOSTAPHYLETUM ALPINAE (alpine heath association)
    Named after *Diapensia lapponica* and *Arctostaphylos alpina*

    Sub-associations: i) JUNCETOSUM (*Juncus trifidus*)
                        ii) BETULETUM (*Betula pumila*)
(2) EMPETRO-RHACOMITRIETUM LANUGINOSAE (moss heath association)
    Named after *Empetrum eamesii* and *Racomitreum lanuginosum*
(3) EMPETRETUM (*Empetrum* heaths)
    Named after *Empetrum eamesii*

    Sub-associations: i) EMPETRETUM TYPICUM (*Empetrum eamesii*)
                        ii) EMPETRETUM MYRICETOSUM (*Myrica gale*)
                        iii) EMPETRETUM SPHAGNETOSUM (*Sphagnum nemorum*)
(4) KALMIETUM (*Kalmia* heaths)
    Named after *Kalmia angustifolia*

    Sub-associations: i) KALMIETUM TYPICUM (*Kalmia angustifolia*)
                        ii) KALMIETUM MYRICETOSUM (*Myrica gale*)
                        iii) KALMIETUM SPHAGNETOSUM (*Sphagnum nemorum*)
                        iv) KALMIETUM ALNETOSUM (*Alnus crispa*)
(5) VACCINIETUM ANGUSTIFOLII (blueberry heaths)
    Named after *Vaccinium angustifolium*

Heath communities of limestone soils

(1) EMPETRETUM (*Empetrum* heaths on limestone)
    Named after *Empetrum eamesii*

    Sub-associations: i) EMPETRETUM-SALICETOSUMCORDIFOLIAE (*Salix cordifolia*)
                        ii) EMPETRETUM-SALICETOSUM RETICIULATAE (*Salix reticulata*)
(2) HERACULETUM-SANGUISORBETUM CANADENSE (forb-dominated association)
    Named after *Heracleum maximum* and *Sanguisorba canadensis*

(3) POTENTILLETUM (shrubby cinquefoil association)
Named after *Potentilla fruticosa*

Sub-associations: i) POTENTILLETUM-DRYADETOSUM INTEGRIFOLIAE (*Dryas integrifoliae*)
ii) POTENTILLETUM-JUNCETOSUM ALPINAE (*Juncus alpinus*)

Heath communities of serpentine soils
Sub-associations: i) LYCNETUM TYPICUM (*Lychnis alpina var. americana*)
ii) LYCNETUM ADIANTETOSUM (*Adiantum pedatum*)

Following this categorisation of the association, the origins and the successional and climax status of the heathlands are discussed. This study is an excellent example of the application of the methods of subjective classification and is unusual in being from North America. Finally, it is worth noting that no higher classification was attempted by Meades. In the early 1980s, when this was written, he stated 'the existing Zürich-Montpellier classification of heath ... will have to undergo considerable revision before the North American heaths can be incorporated into it'. This again demonstrates the emphasis of the whole approach on European vegetation.

Phytosociological studies in the Hoyfjellet, Southern Norway (Coker, 1988)

The purpose of this research was to carry out phytosociological studies of an area across the tree-line in southern Norway and to map and investigate the distribution of the major plant noda (syntaxa). A further aim was to evaluate the efficiency of various methods of multivariate analysis when applied to phytosociological work. The multivariate methods described in Chapters 6 and 8 were used to assist with tabular sorting and the classification of quadrats or relevé samples, leading to an interpretation in the conventional phytosociological terms described earlier in this chapter. Finally, the associations and phytosociological groups that were identified were correlated with the major environmental and temporal features of the montane environment using ordination methods.

Method

In order to describe the spatial variation in floristics within the 8 km^2 area across the tree line, a stratified random sampling design was adopted. For each relevé or quadrat, a 2 m × 2 m area was taken, with species abundance measured using subjective estimates of percentage cover. All groups of plants, apart from algae and fungi, were recorded in a total of 500 relevés. Within these, 425 species were found.

Eighty relevés selected from the original 500 were also investigated for environmental variables. Soil samples were taken for pH, organic matter content, total nitrogen and exchangeable phosphate and calcium. Slope, aspect, and an estimate of the length of snow lie were also recorded. Aspect data were not included, owing to problems of transformation. It was realised subsequently that angular transformation would have been possible (Batschelet, 1981).

Analysis

Hand-sorting of the 500 relevés was regarded as totally impracticable and thus numerical methods were adopted. A number of different techniques for both classification (Chapter 8) and ordination (Chapter 6) were applied and evaluated. An early version of the *MULVA* phytosociological analysis program (Wildi and Orlóci, 1996; Wildi, 2010) and Two-Way Indicator Species Analysis (TWINSPAN) (Hill, 1979b) were used for classification, while detrended correspondence analysis (DCA) (Hill, 1979a) and canonical correspondence analysis (CCA) (ter Braak, 1988a,b) were used for ordination.

Classification

The *MULVA* program produced interesting results and showed the promise of this numerical development of the traditional approach to multivariate analysis (Wildi and Orlóci, 1996). However, better results were obtained from TWINSPAN, particularly when rare species (those occurring in only one relevé) were removed from the data-set. This reduced the species total from 425 to 336.

The TWINSPAN groups were therefore interpreted, giving the following syntaxa (alliances and associations) using nomenclature provided in Dahl (1956, 1985) and Økland and Bendiksen, 1985). Alliance names are in bold:

Arctostaphyleto-Cetrarion nivalis
Cetrarietum nivalis typicum trifidetosum
Alectorieto-Arctostaphyetum uvae-ursi
Potentilleto-Festucetum ovinae

Phyllodoco-Vaccinion myrtilli
Phyllodoco-Vaccinetum myrtilli dicranetosum

Oxycocco-Empetrion hermaphroditi
Chamaemoreto-Sphagnetum acutifolii
Betuleto-Sphagnetum fusci

Lactucion
Geranieto-Betuletum
Corno-Betuletum
Rumiceto-Salicetum lapponae

Aconition septentrionalis
Acconitum septentrionalis ass.

Nardeto-Caricion bigelowii
Hylocomieto-Betuletum nanae juniperetosum

Kobresio-Dryadion
Dryas-Antennaria alpina (nodum)

Cratoneureto-Saxifragion aizoidis
Cratoneureto-Saxifragion aizoidis
Cariceto-Saxifragetum aizoidis (?)

Cassiopeto-Salicion herbaceae
Dicranetum starkei
Luzulo-Cesietum
Lophozieto-Salicetum herbacae typicum conostometosum

Ranunculeto-Oxyrion digynae
Alchemilletum alpinae
Polygoneto-Salicetum herbaceae

Sphagneto-Tomenthypnion
Carex atrofusca ass.
Carex dioica-Tomenthypnum nitens ass.
Filipendula-Mnium ass.
Salix glauca-Paludella-Sphagnum warnstorfianum ass.

Stygio-Caricion limosae
Scorpidieto-Caricetum limosae

Mniobryo-Epilobion hornemannii
Mniobryo-Epilobietum hornemanni
Philonoto-Saxifragetum stellaris

Ordination

These groupings were further confirmed by the ordination analyses, with CCA giving much better results than DCA. The major use of ordination, however, was on the subset of 80 quadrats for which environmental data had been collected. The CCA results showed that duration of snow-lie was the most significant environmental factor related to the first axis, while calcium concentration and pH were of significance on the second. Surprisingly, phosphate levels, slope and organic content appeared to be of little importance.

Examination of CCA biplots for both species and quadrat ordinations revealed species groups that were strongly correlated with snow patches and calcium-rich soils. There were also distinctive groups for calcareous woodland and mire communities. Although soil moisture values were not determined, there appeared to be evidence of a wet–dry gradient corresponding to the second axis. At one end, mire species and those typical of irrigated sites formed a well-defined group, while at the other, there was a clear group of lichen–heath species, characterising dry sites. The relationship of this potential moisture gradient to the pH and calcium gradients was investigated further and it was concluded that the wetter sites were also the most calcareous, while the drier were the least calcareous.

This research is important in demonstrating the way in which methods of data analysis can be used to assist with classical phytosociology, particularly where there is a very large set of data. The relationships between classification and ordination methods are also shown, with classification assisting primarily with the phytosociology, but ordination proving invaluable in the exploration of environmental characteristics of the phytosociological groups and in determining the positions with respect to the primary environmental gradients. A final point is that the associations recognised above were mapped within the study area to show their biogeographical and spatial relationships.

The US National Vegetation Classification (USNVC) (Grossman et al., 1998; Faber-Langendoen et al., 2009; Jennings et al., 2008, 2009)

Until around 1990, perhaps surprisingly, there had been no attempt to develop a classification of the vegetation types of the US. However, following the first overall description of the vegetation of North America by Barbour and Billings in 1988, later revised (2000), the first steps were taken to provide a framework for the establishment of federal standards for systematic vegetation classification, resulting in the first Federal Geographic Data Committee (FGDC) standard for vegetation classification produced in 1997 (Grossman *et al.*, 1998) and revised in 2008 (Table 7.11) (Peet, 2008; Faber-Langendoen *et al.*, 2009).

Table 7.11 A summary of the USNVC revised hierarchy and criteria for natural vegetation (Faber-Langendoen *et al.*, 2009: 89). Reproduced with kind permission of the Ecological Society of America.

Hierarchy level	Criteria
Upper	Physiognomy plays a predominant role
1.1 Formation Class	Broad combination of general dominant growth forms that are adapted to basic temperature (energy budget), moisture and substrate/aquatic conditions
1.2 Formation subclass	Combinations of general dominant and diagnostic growth forms that reflect global macroclimatic factors driven primarily by latitude and continental position, or that reflect overriding substrate/aquatic conditions
1.3 Formation	Combinations of dominant and diagnostic growth forms that reflect global macroclimatic factors as modified by altitude, seasonality of precipitation, substrates and hydrological conditions
Middle	Floristics and physiognomy play dominant roles
1.4 Division	Combinations of dominant and diagnostic growth forms and a broad set of diagnostic plant species that reflect biogeographical differences in composition and continental differences in mesoclimate, geology, substrates, hydrology and disturbance regimes.
1.5 Macrogroup	Combinations of moderate sets of diagnostic plant species and diagnostic growth forms, that reflect biogeographical differences in composition and subcontinental to regional differences in mesoclimate, geology, substrates, hydrology and disturbance regimes
1.6 Group	Combinations of relatively narrow sets of diagnostic plant species (including dominants and co-dominants), broadly similar composition, and diagnostic growth forms, that reflect regional mesoclimate, geology, substrates, hydrology and disturbance regimes
Lower	Floristics plays a predominant role
1.7 Alliance	Diagnostic species, including some from the dominant growth form or layer, and moderately similar composition that reflect regional to subregional climate, substrates, hydrology, moisture/nutrient factors, and disturbance regimes
1.8 Association	Diagnostic species, usually from multiple growth forms or layers, and more narrowly similar composition that reflect topo-edaphic climate, substrates, hydrology, and disturbance regimes

The table shows clearly the focus on establishing the correct top-level structures for the syntaxonomy of North American vegetation. The definition of terms of the revised

2008 hierarchy levels, using both physiognomic and floristic criteria, are also shown in Table 7.11.

Table 7.12   The guiding principles of the US National Vegetation Classification Standard (Federal Geographic Data Committee, 1997/2008; Jennings *et al.*, 2008, 2009: p. 176). Reproduced with kind permission of the Ecological Society of America.

- The classification is applicable over extensive areas.
- The vegetation classification standard is compatible, wherever possible, with other Earth cover classification standards.
- The classification will avoid developing conflicting concepts and methods through cooperative development with the widest possible range of individuals and institutions.
- Application of the classification must be repeatable and consistent.
- When possible, the classification standard will use common terminology (i.e. terms should be understandable and jargon should be avoided).
- For classification and mapping purposes, the classification categories were designed to be mutually exclusive and additive to 100% of an area when mapped within any of the classification's hierarchical levels (Division, Order, Class, Subclass, Formation, Alliance or Association). Guidelines have been developed for those instances where placement of a floristic unit into a single physiognomic classification category is not clear. Additional guidelines will be developed as other such instances occur.
- The classification standard will be dynamic, allowing for refinement as additional information becomes available.
- The NVCS is of existing, not potential, vegetation and is based upon vegetation condition at the optimal time during the growing season. Vegetation types are defined on the basis of inherent attributes and characteristics of the vegetation structure, growth form and cover.
- The NVCS is hierarchical (i.e. aggregatable) to contain a small number of generalised categories at the higher level and an increasingly large number of more detailed categories at the lower levels. The categories are intended to be useful at a range of scales.
- Upper levels of the NVCS are based primarily on physiognomy (life form, cover, structure, leaf type) of the vegetation (not individual species). Life forms (e.g. herb, shrub, or tree) in the dominant or uppermost stratum will predominate in classification of the vegetation type. Climate and other environmental variables are used to help organise the standard, but physiognomy is the driving factor.
- Lower levels of the NVCS are based on actual floristic (vegetation) composition. The data used to describe Alliance and Association types must be collected in the field using standard and documented sampling methods. The Alliance and Association units are derived from these field data. These floristically-based classes will be nested under the physiognomic classes of the hierarchy.

Jennings *et al.* (2009) presented the guiding principles of the USNVC Classification Standard (Table 7.12) and described the standard procedures for field survey and the processes of classification, which may both be either subjective, in the manner of Braun-Blanquet, or objective, with the assistance of the numerical classification methods described in Chapter 8. The importance of the core levels of the association and the alliance are stressed, with the association being defined as '. . . a vegetation classification unit defined on the basis of a characteristic range of species composition, diagnostic species, species occurrence, habitat conditions and physiognomy' and an alliance defined as '. . . a vegetation classification unit containing one or more associations, and defined by a characteristic range of species composition, habitat conditions, physiognomy, and diagnostic species, typically at least one of which is found in the uppermost or dominant stratum of the vegetation' (Jennings *et al.*, 2009: p. 177–8). Examples of USNVC Association and Alliance names are given in Table 7.13.

Table 7.13   Examples of USNVC association and alliance names (Jennings *et al.*, 2009: p. 191). Reproduced with kind permission of the Ecological Society of America.

Examples of association names
- *Schizachyrium scoparium-(Aristida* spp.) Herbaceous Vegetation
- *Abies lasiocarpa/ Vaccinium scoparium* Forest
- *Metopium toxiferum-Eugenia foetida-Krugiodendron ferreum-Swietenia mahagoni/ Capparis flexuosa* Forest
- *Rhododendron carolinianum* Shrubland
- *Quercus macrocarpa-(Quercus alba-Quercus velutina)/Andropogon gerardii* Wooded Herbaceous Vegetation

Examples of alliance names
- *Pseudotsuga menziesii* Forest Alliance
- *Fagus grandifolia-Magnolia grandiflora* Forest Alliance
- *Pinus virginiana-Quercus (coccinea, prinus)* Forest Alliance
- *Juniperus virginiana-(Fraxinus Americana, Ostrya virginiana)* Woodland Alliance
- *Pinus palustris/ Quercus* spp. Woodland Alliance
- *Artemisia tridentata* ssp. *wyomingensis* Shrubland Alliance
- *Andropogon gerardii-(Calamagrostis canadensis, Panicum virgatum)* Herbaceous Alliance

---

Standardised procedures are also established for data storage and databases, peer review of classification structures once complete, and presentation of reports.

Now that this over-arching structure is in place, the aim is that phytosociological studies from across the whole range of North American habitats may be located within it in a systematic and standardised fashion. An excellent example is provided by Faber-Langendoen *et al.* (2007), who classified and mapped the vegetation of Voyageurs National Park in Minnesota according to the principles of the USNVC.

The park, which occupies 54,243 ha of northern Minnnesota in the sub-boreal region of North America, was established in 1975. From previous surveys, a preliminary list of 37 associations was compiled in 1996, followed by field description of 259 classification plots. Plot sizes were 20 m × 20 m for forests and woodlands and 10 m × 10 m for shrublands, herbaceous and non-vascular vegetation. Abundance was recorded on a 7-point Braun-Blanquet scale (Table 3.4). A wide variety of environmental data were collected and sample locations were recorded by global positioning systems.

Data analysis involved several of the methods of classification and ordination described in Chapters 6 and 8, including detrended correspondence analysis (DCA), non-metric multidimensional scaling (NMS) and flexible beta cluster analysis, using the *PC-ORD* computer software (McCune and Mefford, 1999, 2010) (Chapter 9). Assignment of plots to associations was made on the basis of both ordination and classification results in relation to the preliminary survey associations, with plots from each association being combined to give a vegetation type summary or set of association tables. Colour aerial photography was available from 1995–96 and was then used for mapping of both associations and alliances. An important feature of this was 'ground-truth' checking and mapping accuracy assessment. In 1997–98, mapped polygons for each association type were selected at random, allowing for some logistical constraints, and

1251 accuracy test plots were recorded. These were then cross-checked against map polygons resulting in an overall thematic map accuracy of 82.4%.

This case study is important as an example of the type of survey that has been and still is being completed within the overall structure of the USNVC. Particularly significant is the link between phytosociology, used to derive associations and alliances, and the mapping of those associations and alliances using aerial photography and geographical information systems (GIS) (Burrough and McDonnell, 1998; Johnston, 1998; Wadsworth and Treweek, 1999; Longley *et al.*, 2010). Ground-truthing and the need for assessment of mapping accuracy is another key feature of the research.

Chapter 8

Numerical classification, cluster analysis and phytosociology

INTRODUCTION

The principles of classification and their relevance to phytosociology using tabular rearrangement were introduced in Chapter 7. This chapter is concerned with objective or numerical methods for classification. The goals of classification using numerical methods are the same as for tabular rearrangement: the grouping of a set of individuals (quadrats or vegetation samples) into classes on the basis of their attributes (floristic composition). Ideally each group should contain quadrats with very similar species composition. These groups or classes are then interpreted and used to define a set of plant communities for the area under study.

The field of numerical classification is a very broad one and the techniques described here have not simply been used for phytosociology within plant ecology but have been applied across the whole range of sciences from biology (taxonomy), geology, geography, chemistry, medicine and astronomy to psychology, sociology, archaeology and history (Cormack, 1971; Williams, 1971; Anderberg, 1973; Sneath and Sokal, 1973; Sokal, 1974; Clifford and Stephenson, 1975; Gillison and Anderson, 1981; Dunn and Everitt, 1982; Gordon, 1987, 1996; Gower, 1988; Kaufman and Rousseeuw, 1990; van Tongeren, 1995; Everitt *et al.* 2001; Fielding, 2007). The methods are also widely described as techniques of cluster analysis, based on the concept of grouping together points representing individuals with similar characteristics in mathematical space.

Methods of numerical classification, like methods of ordination, are techniques for data reduction and data exploration. They are used to look for pattern and order in a set of data. The search for order is frequently an end in itself, with classification and the production of a set of groups being the only concern. However, as with ordination, the methods can be used for hypothesis generation, leading to further, more detailed research. The development of such methods has been closely linked to the advent of computers and their increasing power and sophistication. Classification is a tedious process, and the subjective methods of tabular arrangement described in Chapter 7 are only suitable for relatively small sets of data. The numerical techniques now available can handle hundreds and even thousands of samples or species in one analysis.

Numerical methods of classification are defined as objective only in the sense of repeatability. When used for phytosociology, a method for numerical classification represents a set of rules

Vegetation Description and Data Analysis: A Practical Approach, Second Edition. Martin Kent.
© 2012 John Wiley & Sons, Ltd. Published 2012 by John Wiley & Sons, Ltd.

governing the process of grouping individuals or quadrats together. Thus for one set of data, any researcher using the same numerical method should obtain the same result. The element of subjectivity in the classification process is removed. However, as will be seen later, different numerical methods give varying results for the same set of data, and these variations are dependent on the mathematical properties of each technique. Thus although any one numerical method is objective in the sense of repeatability for one set of data, there is no unique solution or single classification of a set of data. As with ordination methods, the 'best' classification is one that enables a clear ecological interpretation to be made. The idea of user satisfaction is very important, and although the classification process can be described as objective, the interpretation still remains subjective. This point is discussed very effectively by Goodall (1978).

NATURAL AND FORCED CLASSIFICATIONS OF VEGETATION DATA

When classification is applied to a set of vegetation data, ideally the 'natural' group structure within the data should be found. Different data-sets will contain individuals or quadrats that have varying intensities of group structure. Ordination diagrams (Chapter 6) are a good way of illustrating this point. In Figure 8.1a, the points representing quadrats on the ordination diagram show a distinct clustering. Remembering the fundamental principle of ordination diagrams, that an increase in distance between any two points is a reflection of decrease in their degree of similarity of species composition, then a set of points close together, separated from another set, as in Figure 8.1a, represents a clear group structure. Classification of these data would produce well-defined groups and these natural groupings would be found relatively easily.

However, if classification is applied to a set of data represented on an ordination diagram such as Figure 8.1b, then it is unlikely that any meaningful groupings will be found, since the vegetation data are distributed as a continuum (Chapter 2). Nevertheless, it is still possible to apply methods of numerical classification to these data. If this is done, the points will be forced into groups and quite arbitrary boundaries would be drawn between the points. It is for this reason that some

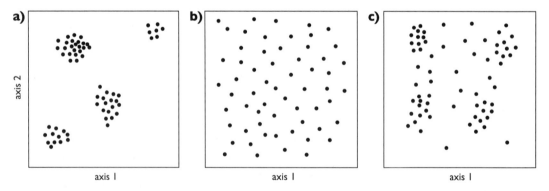

Figure 8.1   Principles of grouping or clustering shown on a two-dimensional ordination diagram: (a) clear group structure with tight clusters – numerical classification should find these easily; (b) a continuum of points with no group structure – numerical classification would arbitrarily partition the continuum; (c) the more common situation between (a) and (b), where some group structure exists but there are transitional points between groups – if numerical classification is applied, it will find the 'cores' of the groups and transitional points between groups will be allocated to the nearest group, depending on which method of numerical classification is applied. Original work of author.

ecologists, particularly in America, believe that phytosociology and classification of vegetation samples is inherently wrong, since if vegetation is distributed as a continuum, then forcing a classification upon it will produce misleading results. McIntosh (1967b) discussed this at length.

In reality, most vegetation data produce ordination diagrams that lie between these two extremes, as in Figure 8.1c. Here there are clusters of points that act as cores for groups and definitely indicate some form of group structure, which may be taken as representing the major plant communities. However, there are also some points that are transitional between groups and have properties (floristic composition) common to more than one group. There is a debate as to how to interpret this situation. Some subjective classifiers (Chapter 7) would tend to reject or ignore the transitional points, thus artificially strengthening the group structure. Other ecologists would regard these points as interesting and possibly representing samples from ecotone or transitional areas between major plant communities. However, numerical classification of the data portrayed in Figure 8.1c would cause all points to be allocated to their nearest group. At the interpretation stage, the decision would have to be made as to whether to disregard intermediate points between groups, or to identify them as transitional types, worthy of particular attention.

The important conclusion to this discussion is that when numerical classification is applied to a set of vegetation data, the researcher must believe that some group structure is present in the data and that reasonably distinct community types exist. Otherwise, the classification will only arbitrarily partition a continuum. The differences of viewpoint between Clements and Gleason on the nature of the plant community, described in Chapter 2, are worth remembering here.

This section also serves as an introduction to the relationships between classification and ordination. Increasingly, the two groups of methods are being used together on the same set of data in order to examine patterns and to search for group structure (Kent and Ballard, 1988; Kent, 2006). This is known as complementary analysis.

CHARACTERISTICS OF METHODS OF NUMERICAL CLASSIFICATION

A set of general principles that apply to the various methods of numerical classification has evolved (Whittaker, 1978b; Bridge, 1993; Milligan, 1996):

(a) *Techniques are based on hierarchical principles*

Most techniques are hierarchical in nature. This means that the results can be portrayed as a dendrogram (tree or linkage diagram). The reason why hierarchical methods prevail is that such a dendrogram shows different levels of similarity or dissimilarity very clearly, and the different levels displayed in the dendrogram are often very helpful when it comes to making ecological interpretations. An alternative method for the presentation of hierarchical data is the icicle plot, but these have never been widely used in plant ecology. The principles of their construction are described in Kruskal and Landwehr (1983). Non-hierarchical methods, based on partitioning of multidimensional space, such as k-means clustering and fuzzy clustering, do exist (Grabmeier and Rudolph, 2002) but have not been widely applied in vegetation science. These are nevertheless covered briefly towards the end of this chapter.

(b) *Divisive or agglomerative*

Divisive methods start with the total population of individuals and progressively divide them into smaller and smaller groups. Division ceases either when each group is represented by a single individual, or when some form of predetermined stopping rule is applied to halt division. Agglomerative methods start with each individual and join individuals and then individuals and groups together into larger and larger groups, until all the individuals are in one big group. Again, decisions are usually made to halt the process of agglomeration before this point when an interpretable set of groups has emerged.

(c) *Quantitative or qualitative data*

Methods of classification may be applied to either quantitative (interval or ratio) or qualitative (binary) data. Most methods will accept either type of data and the decision as to whether to use quantitative data or not depends on the research objectives and the type of problem being analysed. One method, Two-Way INdicator SPecies ANalysis (TWINSPAN), uses the idea of the pseudospecies, whereby the presence of a species at different predetermined levels of abundance is used (Hill *et al.*, 1975; Hill, 1977, 1979b). Thus in TWINSPAN the percentage cover scale is often divided into six using five cut levels. Thus the first pseudospecies may be 1–2% cover of the species, 3–5% the second pseudospecies, 6–10% the third, 11–20% the fourth, 21–50% the fifth, and over 50% the sixth. These six levels of abundance of a species are then used in presence/absence form to make the classification. Pseudospecies cut levels can also be applied to any other abundance scales.

(d) *Equal emphasis of species*

Most analyses assume that all species present are given equal importance in the analysis. It is possible, however, to downweight rarer species or to increase the importance of common or dominant species in a classification.

(e) *Normal and inverse analysis*

A normal analysis is where samples or quadrats are classified into groups on the basis of species composition. Inverse analysis is when groupings of species are produced on the basis of their distribution in a series of samples or quadrats.

(f) *Single or joint (two-way) analysis*

Many methods analyse quadrats or samples separately from species (separate normal and inverse analyses). More recent methods, notably TWINSPAN, carry out a joint classification of quadrats and species simultaneously. Considerable advantages are claimed for this approach (Hill, 1979b; Gauch, 1982b). In the most recent version of the *PC-ORD* package (McCune and Mefford, 1999, 2010), however, the dendrograms from separate analyses of samples and species can also be combined to provide a two-way table, even though there is no direct mathematical relationship between the two, as in Figure 8.6.

(g) *Robustness*

The idea of robustness is that the effectiveness of a method of classification should not be dependent on properties of a particular set of data and a technique should perform well in most applications. Methods vary greatly in their robustness and thus the extent to which they have found widespread use.

HIERARCHICAL CLASSIFICATION

Hierarchical methods are best discussed under the subheadings of agglomerative and divisive. The following discussion assumes that a normal analysis is being performed (grouping of quadrats or samples on the basis of species composition).

AGGLOMERATIVE TECHNIQUES

Agglomerative methods proceed from individual samples or quadrats and progressively combine them in terms of their similarity until all the quadrats are in one group. The use of the terms

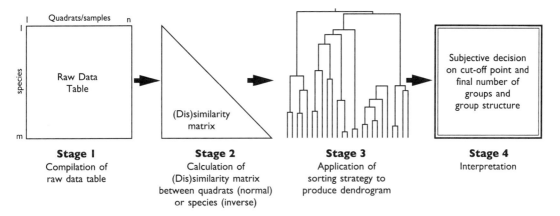

Figure 8.2 Stages in agglomerative classification using similarity analysis. Original work of author.

similarity and dissimilarity and the idea of similarity and dissimilarity coefficients was introduced in Chapter 4. Similarity coefficients measure how alike any two quadrats are in terms of species composition. Dissimilarity coefficients assess how unlike any two quadrats are in species composition. Dissimilarity is thus the complement of similarity and the two are very closely related.

Similarity analysis

As shown in Figure 8.2, most methods contain four stages, as follows.

Stage 1: The raw data matrix

Data are first arranged in a raw data table as described in Chapter 4. In numerical classification, data are then sometimes standardised before analysis. Ideas of standardisation (also known as relativisation) were introduced in Chapter 4. The following standardisations are possible:

(a) *Standardisation to sample/quadrat total* – Species scores within a quadrat are summed and each abundance is divided by the total. The effect of this is to correct for the overall abundance or 'size' of the sample (total cover or number of individuals). This method is sensitive to species richness.

(b) *Standardisation to species total* – In this case, abundances are totalled over all the quadrats and then individual scores are divided by that total. This tends to overweight rare species and downweight common species. It is not widely used.

(c) *Standardisation to sample/quadrat maximum* – Here, all species scores are divided by the maximum attained by any species in the sample or quadrat.

(d) *Standardisation to species maximum* – In a similar way to (c), all quadrat scores for a species are divided by the maximum attained by the species across all the quadrats. This tends to weight less abundant species more equally. The data are also less dependent upon the type of abundance measure used (% cover, frequency, density or biomass).

(e) *Standardisation by species sums of squares* – The sum of the squared abundances for each species then equals one.

(f) *Standardisation to zero mean and unit variance and vector length* – The difference between each species abundance and the species' mean is divided by the standard deviation of the species abundances. Species means are then 0 and standard deviations are 1. This standardisation is not compatible with the Steinhaus (Czekanowski/Sørensen/Bray and Curtis) (dis)similarity coefficient.

Stage 2: The calculation of a similarity or dissimilarity matrix

The degree of matching between each pair of quadrats is then calculated on the basis of a similarity or dissimilarity coefficient. Various (dis)similarity coefficients were introduced in Chapter 4, notably the Steinhaus (Sørensen/Czekanowski/Bray and Curtis) and Euclidean distance coefficients. Many others also exist, and new and improved ones are still being devised (Crawford and Wishart, 1967; Noest and van der Maarel, 1989). In most computer packages for numerical classification, such as *PC-ORD* (McCune and Mefford, 1999, 2010), *Primer-E* (Clarke and Gorley, 2006) or *JUICE* (Tichý, 2002), for example, a number of different coefficients for continuous variables are presented. Nevertheless, for species data, the Steinhaus coefficient and its related variations are the most widely used, while for environmental and non-species data, Euclidean distance, which is a dissimilarity measure, is usually the most appropriate. When calculated between all pairs of samples or quadrats, a (dis)similarity matrix is produced. An important point about dissimilarity coefficients is that they are often equated with distances between points in mathematical space. Thus with Euclidean distance, the higher the value, the greater the distance between two points or between points and groups. Further descriptions of other (dis)similarity coefficients are given in Legendre and Legendre (1998), Waite (2000), Everitt *et al.* (2001), McCune and Grace (2002) and Wildi (2010).

Stage 3: The application of a sorting strategy

A sorting strategy is a set of rules by which a set of samples or quadrats are progressively allocated to groups on the basis of the information in the (dis)similarity matrix. The ideas of iteration and fusion are important features of such methods. Most sorting strategies are iterative, in that a series of passes is made through the similarity matrix, looking for the pair of individuals, the individual and group, or pair of groups that are most similar to each other. This will be the pair with either the highest similarity value or the lowest dissimilarity value. When the most similar pair is found, the pair is fused into the same group. Then the next most similar pair is found and so on. This iterative process of successive grouping is known as fusion. Each fusion that occurs, whether between individuals, individuals and groups or between groups, decreases the number of groups by one.

Combinatorial versus non-combinatorial strategies   The idea of combinatorial and non-combinatorial sorting strategies was introduced by Lance and Williams (1966, 1967) and relates to how the (dis)similarity matrix is reduced during sorting. In a dissimilarity matrix of order $n \times n$ (n is the number of samples), if two individuals or groups G_p and G_q with n_p and n_q elements respectively have the smallest squared dissimilarity (d_{pq}^2) in the dissimilarity matrix, they will be the next to be joined or fused into a group. Joining of two groups is expressed as $G_p \cup G_q = G_r$. If the squared dissimilarity between G_r and all other groups, for example G_i, is calculated, before fusion, values for $d_{pr}^2, d_{qr}^2, d_{pq}^2, n_p, n_q$ and n_r are known. If d_{ir}^2 can be calculated using these values,

Table 8.1   Combinatorial coefficients for selected sorting strategies (Lance and Williams, 1967). Reproduced with permission from John Wiley & Sons, Inc.

	Coefficients			
Sorting strategy	α_p	α_q	β	γ
Nearest neighbour	0.5	0.5	0	−0.5
Furthest neighbour	0.5	0.5	0	0.5
Group average	n_p/n_r	n_q/n_r	0	0
Centroid	n_p/n_r	n_q/n_r	$-\alpha_p\alpha_p$	0
Ward's method (minimum variance)	$\dfrac{n_i + n_p}{n_i + n_r}$	$\dfrac{n_i + n_q}{n_i + n_r}$	$\dfrac{-n_i}{n_i + n_r}$	0
Flexible beta	$(1 - \beta)/2$	$(1 - \beta)/2$	β	0

n_p = number of elements in G_p; n_q = number of elements in G_q; n_r = number of elements in $G_r = G_p \cup G_q$; n_i = number of elements in G_i; $i = 1, n$ except $i \neq p$ and $i \neq q$.

then the method is said to be combinatorial and, most importantly, the original data do not need to be kept after the first set of values have been computed.

The general combinatorial equation to summarise this is:

$$d_{ir}^2 = \alpha_p d_{ip}^2 + \alpha_q d_{iq}^2 + \beta d_{pq}^2 + \gamma \left| d_{ip}^2 - d_{iq}^2 \right| \tag{8.1}$$

The combinatorial coefficients for the various different sorting strategies described below are presented in Table 8.1. These values are best imagined as weights that define how distances from the pre-fusion group are fused into a new set of distance scores for the new post-fusion group.

If a strategy is non-combinatorial, it means that the dissimilarity values must be calculated from the original data after each fusion and cannot be calculated from the previous ones, which is much more inefficient.

Space-conserving or space-distorting strategies   When a (dis)similarity matrix is calculated, distances in space are dependent upon the properties of the particular coefficient or resemblance measure employed. During fusion, the varying approaches of different strategies to measuring the distances between individuals and groups may alter the original properties of the space. Some methods may cause the space around individuals and groups to be expanded or contracted, and are thus known as space-distorting. If space contracts, the property of chaining may result, whereby a group tends to move near to some or all of the remaining individuals and the likelihood of an individual joining an existing group, rather than forming a new group, is increased. The outcome is a poor group structure, where large numbers of individuals successively join the same group in the manner of a chain. When space expands, sorting strategies tend to produce new groups, because groups seem to move further apart and individuals not yet in a group are more likely to start new groups.

A number of different sorting strategies have been developed under the general heading of similarity analysis. Those most frequently used in vegetation analysis are discussed below.

Single linkage clustering (minimum or nearest neighbour method) (Sneath and Sokal, 1973)   The method looks for the most similar pair of quadrats in the (dis)similarity matrix in order to form groups. In geometric terms, this means that quadrats and groups are joined or fused by looking for the nearest individual or member of another group (Figure 8.3). Single linkage clustering

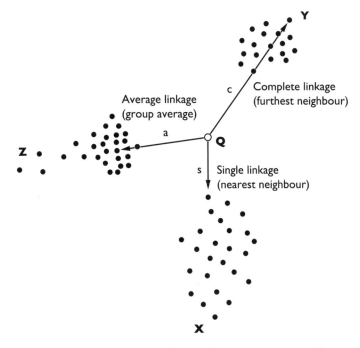

Figure 8.3   A graphical illustration of three different sorting strategies or clustering methods. Q represents a quadrat to be assigned to one of three groups, X, Y and Z. The shortest distance from Q to the nearest neighbour (single linkage) = s; furthest neighbour (complete linkage) = c; and average linkage (group average) = a. Original work of author.

tends to produce 'straggly' clusters and is not usually that effective for phytosociological work. Nevertheless, it is more frequently used in taxonomic research.

Complete linkage clustering (maximum or furthest neighbour method) (Sneath and Sokal, 1973) This is the opposite of single linkage clustering. The distances between individuals and groups are now defined as the distance between their most remote pair of individuals (Figure 8.3). The method starts in the same way as single linkage, by looking for the most similar pair, but then further fusions depend on finding the minimum distance between the furthest points in existing groups (not the nearest, as with single linkage).

Average linkage (group average) (Sokal and Michener, 1958) In this case, the process of fusion is based on the minimum average distance between individuals and groups (Figure 8.3). The most commonly used is the unweighted pair-groups method using arithmetic averages (UPGMA).

Centroid sorting (Sokal and Michener, 1958) The method works by finding the most similar pair of quadrats in the (dis)similarity matrix and then fusing them by averaging of their attributes. The (dis)similarity between these fused quadrats and the rest of the data is then recalculated. The process is then repeated with the most similar pair of individuals, individual and group, or group and group, the fusion on each iteration being carried out in the same manner, by averaging of their attributes. The method is different from average linkage because in the latter case it is the (dis)similarity values of all individuals in a group which are averaged, rather than the attributes upon fusion of individuals and groups. Centroid sorting has a number of problems attached to it, notably reversals, where after one dendrogram fusion, the next takes place at a lower level than the original when normally it should be higher (Clifford and Stephenson, 1975).

Minimum variance or error sums of squares clustering (Ward's Method) (Ward, 1963)   This is similar to average linkage clustering and is based on the idea that at each stage of the analysis, the loss of information on fusing quadrats into groups can be measured by the sum of squared deviations of every quadrat from the mean of the group to which it belongs. On each iteration, every possible pair of quadrats and groups is taken and those two individuals or groups whose fusion results in the lowest increase in the error sum of squares (or the variance) are combined.

Flexible clustering (flexible beta) (Lance and Williams, 1967)   Flexibility refers to the flexibility in the combinatorial Equation (8.1). The method is defined by the four constants:

$$\alpha_p + \alpha_q + \beta = 1; \quad \alpha_p = \alpha_q; \quad \beta > 1; \quad \gamma = 0$$

where the value of β is provided by the user. 'Flexible beta' is flexible because the user controls the space-distorting properties. When β is near to $+1$, it is space-contracting and severe chaining will occur, while as the value approaches 0 and becomes negative, the method becomes increasingly space-expanding and individuals are more closely grouped. McCune and Grace (2002) provided some example dendrograms from repeated clustering of the same data using the flexible approach with different values of β. Lance and Williams (1967) stated that a β value of -0.25 is produces results very similar to Ward's method.

Although these are the six most commonly used methods of similarity analysis (Williams *et al.*, 1966), others exist, notably median cluster analysis and McQuitty's method, but they have rarely been used in vegetation data analysis. As described above, all of these methods have been shown to be related through the combinatorial approach (Lance and Williams, 1966, 1967) and can be very efficiently programmed as computer algorithms.

Information analysis (Williams et al., 1966)   This could be described as another type of similarity analysis, but deserves specific mention because it is based on the idea of information content as an approach to similarity, and was devised and used specifically for phytosociological purposes. The method is based on the information statistic (I). A group of quadrats contains a certain amount of information, and the greater the differences between the quadrats making up a group, the greater the information content and the higher the value of I. The information statistic (I) for a group is calculated as:

$$I = \sum_{j=1}^{i} [n \log n - a_j \log a_j - (n - a_j) \log(n - a_j)] \tag{8.2}$$

where n is the number of quadrats, j is the species, and a_j is the number of quadrats in the group containing the jth species. The data are in presence/absence form.

Individual quadrats have an information content of zero. When two quadrats are combined, the more closely the species composition of the two quadrats match, the lower the I values of the combined two quadrats. When two groups of quadrats are combined, the information content of the new group is equal to the information content of the first group plus the information content of the second, plus an increment in information content caused by the differences in species composition between the two groups. Thus:

$$I_{\text{group 1}} + I_{\text{group 2}} + I_{\text{heterogeneity}} = I_{\text{new group}}$$

Those two groups resulting in the least increase in information content (that is, which are most similar) are combined. At the start, those two quadrats forming a group with the lowest value of I are found and combined. Worked examples of information analysis are given in Poole (1974) and Causton (1988). Orlóci (1972) presented alternative information functions.

In this way, fusions are progressively made and a dendrogram is constructed. The method was particularly favoured by Australian ecologists (the Canberra School) in the 1960s and 1970s. Elsewhere it has never been widely used, but it nevertheless provides an interesting alternative to similarity analysis.

Stage 4: Interpretation using dendrograms

With agglomerative methods of similarity analysis and information analysis, a complete dendrogram is usually drawn. Usually dendrograms are scaled using a simple distance function, the distance being the dissimilarity or similarity score for each fusion of individuals or groups. Many computer packages scale dendrograms using the objective function of Wishart (1969), which measures the information lost at each fusion in the hierarchy. The functional scaling is well explained in McCune and Grace (2002). Some packages also offer the option of plotting the dendrogram on a log scale, as opposed to an arithmetic scale, which expands the detail in the lower section of a dendrogram.

The choice as to the most meaningful number of final groups, which are then taken to represent plant communities, is a subjective decision, which relies on the ecological knowledge and experience of the user. One commonly applied method is to draw a line through the higher levels of the dendrogram at a suitable point and to interpret the groups that result. While this is a sensible starting point for interpretation, there is no reason why all groups should be taken from this one arbitrary level. If, on interpretation, a large group looks as though it can be subdivided on phytosociological and ecological grounds, then it should be. Equally, two groups that do not appear to be clearly distinct could be recombined. The longer the 'stems' on a dendrogram, the more well-defined the group at the foot of the stem will be. The key point is that, at this interpretative stage, the knowledge of the researcher is paramount and phytosociological and ecological sense should prevail.

Similarity and information analysis of the Gutter Tor data (Table 4.4)

In Figure 8.4, the Gutter Tor data-set from Dartmoor has been analysed by five different methods of similarity analysis and by information analysis. The resulting dendrograms show a high degree of similarity between the various methods and sorting strategies. The following groups emerge with a high degree of consistency:

Quadrats 1, 2, 9, 25 and 12	Quadrats 16 and 21
Quadrats 3, 18, 13, 20, 24 and 14	Quadrats 11, 17 and 22
Quadrats 8, 23, 19 and 15	Quadrats 4, 6 and 10 (5 and 7)

Greater differences occur between the information analysis and the various similarity analyses, since the information analysis was performed only on presence/absence data. Nevertheless, the

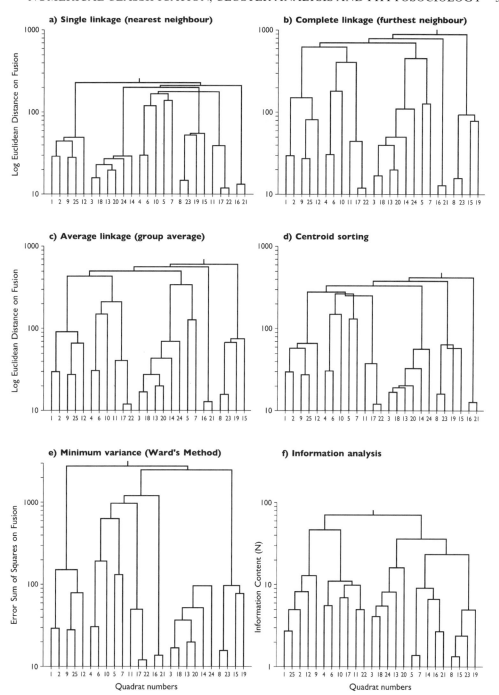

Figure 8.4   Dendrograms produced by five methods of similarity analysis and information analysis of the Gutter Tor, Dartmoor, data (Table 4.4). (a) Single linkage (nearest neighbour); (b) Complete linkage (furthest neighbour); (c) average linkage (group average); (d) centroid sorting; (e) minimum variance clustering (Ward's method); (f) information analysis. a–d were calculated with the Steinhaus (Czekanowski/Sørensen/Bray and Curtis) (dis)similarity coefficient. Original work of author.

general patterns still remain. Another noticeable feature is the occurrence of 'reversals' in the centroid sorting dendrogram (Figure 8.4d). This is a well-known failing of the method.

Comparison of the groups from the various methods of similarity analysis of the Dartmoor data (Figure 8.4) and the subjective classification of Table 7.6 shows that very similar groups emerge from within this small set of data. The reason for the consistency between the different methods and the subjective classification is that the Gutter Tor data show a relatively clear group structure. It is likely that comparison of different methods on a data-set that was closer to a continuum would result in much more variation between methods. This demonstrates the point about the subjectivity of the methodology. Even though the methods can be described as objective, the classification depends upon the properties of the particular technique applied and the resulting interpretation is still subjective, relying on the ecological knowledge and experience of the user, as well as an understanding of the advantages and disadvantages of the chosen method.

Inverse analysis

The six dendrograms of Figure 8.4 were all normal analyses or quadrat/sample classifications, where the quadrats were sorted into groups on the basis of their species content. Inverse analysis or species classification involves turning the data matrix around, so that species are sorted into groups on the basis of their distribution across the set of quadrats. McCune and Mefford (1999, 2010) recommend standardisation or relativisation of species in inverse analyses because, without standardisation, common species tend to separate into different groups from rare species, and standardisation helps to reduce, although not eliminate, that effect.

In Figure 8.5, the dendrogram produced by inverse analysis of the Gutter Tor data using group average clustering on the Steinhaus (Czekanowski/Sørensen/Bray and Curtis) (dis)similarity coefficient is shown. Although inverse analysis was not described in the section on subjective classification in Chapter 7, the groupings of species in Table 7.6 are confirmed by the similarity analysis.

Two-way similarity analysis

When both normal and inverse analyses are performed on the same data, it is possible for the two resulting dendrograms to be plotted jointly to give a 'two-way' dendrogram plot, with the original species data rearranged on the basis of both sample and species dendrogram sequences (McCune and Mefford, 1999, 2010). Using the routines available in their *PC-ORD* package, the plot for the Gutter Tor data is presented in Figure 8.6. This type of diagram enables relationships between sample and species groups to be more easily identified. The *PC-ORD* package (McCune and Mefford, 1999, 2010) also enables editing of the table to highlight blocks of species and samples and to allow for rotation of groups on the dendrogram. McCune and Grace (2002) stressed that dendrograms are best imagined like a child's mobile, so that groups on the end of dendrogram stems may be free to rotate.

DIVISIVE TECHNIQUES

Divisive methods of classification proceed by taking all samples or quadrats in a data-set and dividing them into two groups, those two groups into four, four into eight, and so on.

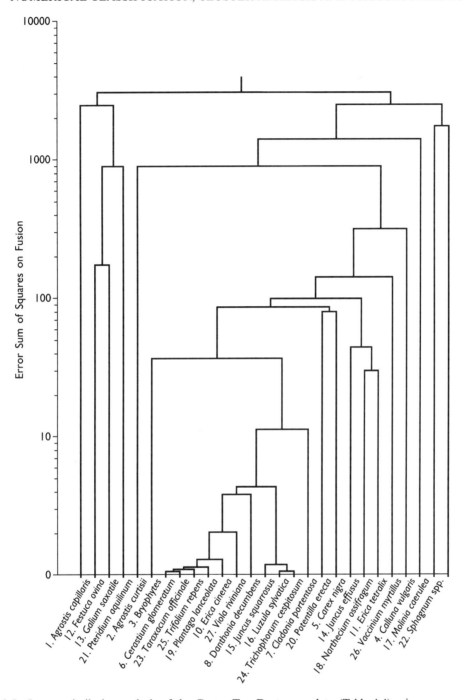

Figure 8.5 Inverse similarity analysis of the Gutter Tor, Dartmoor, data (Table 4.4) using group average similarity analysis with the Steinhaus (Czekanowski/Sørensen/Bray and Curtis) (dis)similarity coefficient. Original work of author.

Gutter Tor, Dartmoor Data - Group Average Similarity Analysis

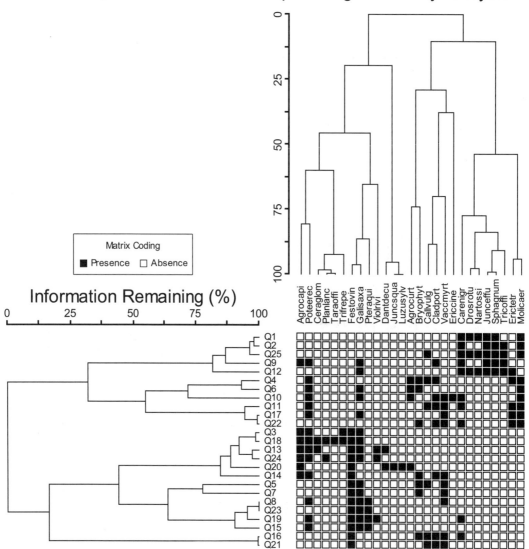

Figure 8.6 Two-way similarity analysis of the Gutter Tor, Dartmoor, data, using group average clustering with the Steinhaus (Czekanowski/Sørensen/Bray and Curtis) (dis)similarity coefficient, using the *PC-ORD* package (McCune and Mefford, 1999, 2010). Solid squares represent species occurrences in the reordered data matrix. Original work of author.

Two-Way INdicator SPecies ANalysis (TWINSPAN)

Over the past 30 years, Two-Way Indicator Species Analysis (TWINSPAN) (Hill, 1979b; Gauch and Whittaker, 1981) has been one of the most widely used techniques for numerical classification, but in recent years has also become one of the most criticised. It was preceded by several

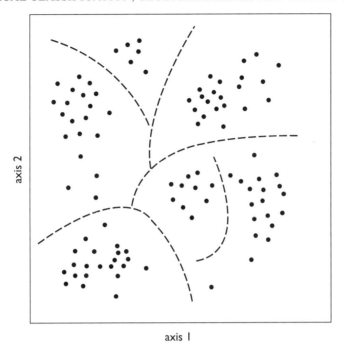

axis 2

axis 1

Figure 8.7 Subjective partitioning of a two-dimensional ordination plot. Original work of author.

developments that were similar in concept, notably AXOR (Lambert *et al.*, 1973), and POLY-DIV and REMUL (Lance and Williams, 1975; Williams, 1976a,b). TWINSPAN was originally described as indicator species analysis in 1975 (Hill *et al.*, 1975). Hill (1979b) himself described this terminology as unfortunate and confusing, and suggested that perhaps the best name would have been dichotomised ordination analysis, since the method is based on progressive refinement of a single axis ordination from reciprocal averaging or correspondence analysis (Chapter 6) and the notion of partitioning ordination space (Roux and Roux, 1967).

The concept of partitioning ordination space may be explained as follows. On a 1-, 2- or 3-dimensional quadrat ordination plot, lines or partitions can be drawn through those areas of the plot with the lowest density or absence of points (Figure 8.7). This can be done subjectively, as in Figure 8.7, and if the clustering of points is marked, a satisfactory classification can be found. Thus conceptually, divisive methods of classification are best envisaged as trying to place divisions through those least dense areas of points representing quadrats in multidimensional space. In practice, however, they are rather more complex than this.

The TWINSPAN method – introductory principles

Pseudospecies In phytosociology, the idea of differential species in defining plant communities is very important (Chapter 7). The differential species are used to make a division or a dichotomy and thus to separate one group of quadrats from another, with one set of differential species characterising one group and another set the other. However, the idea behind differential species is based on presence/absence, rather than abundance, and is thus a qualitative rather than a quantitative concept. Hill *et al.* (1975) devised the pseudospecies concept as a way of adapting

Table 8.2   The TWINSPAN sorted table for the Garraf data (Table 4.3) (see text for explanation). Reproduced with permission from John Wiley & Sons, Inc.

```
                        1 11    11 1
                      145206249373581

  2   Bracramo    3-4------------   0000
  6   Ericmult    433-3----------   0000
  8   Lavaangu    111------------   0000
 15   Salvsp.     133------------   0000
  4   Chamhumi    11-----2-------   0001
 12   Pistlent    323-4----------   0001
  5   Cortsell    ---12242-------   0010
  7   Euphsp.     22-11-22-------   0010
 10   Philangu    23233314-------   0010
 11   Pinuhale    33322233-------   0010
 13   Quercocc    5644454511211-1   0011
 16   Sedumsp.    ---1-1112------   01
  1   BareRock    343334356666666   10
 14   Rosmoffi    32311-213221232   10
  3   Cerasili    --------122-1--   11
  9   Oleaeuro    --------2313344   11
 17   Smilaspe    -------2-122231   11

                  000000001111111
                  000111110111111
                   00111 000011
```

abundance data to a qualitative equivalent which can then be used in the process of making a division. Thus any species abundance scale is partitioned into a series of pseudospecies, each of which can then be used in the process of making a dichotomy.

As an example, if percentage cover data were being analysed, these could be redefined as five pseudospecies, and a typical species, for example, *Agrostis capillaris*, could be recoded as:

Agrostis capillaris 1 – cover up to 2%
Agrostis capillaris 2 – cover 3–5%
Agrostis capillaris 3 – cover 6–10%
Agrostis capillaris 4 – cover 11–20%
Agrostis capillaris 5 – cover over 20%

An important principle is that these classes are non-exclusive and cumulative; if only 2% *Agrostis capillaris* occurred in a quadrat it would be coded as *Agrostis capillaris* 1, but if it occurred in another quadrat with 60% cover, all five pseudospecies would be present (*Agrostis capillaris* 1–5).

TWINSPAN output   Although the detail of the method is complex, one of the most attractive features of the technique is the computer-generated two-way table which is produced at the end of a TWINSPAN analysis. The TWINSPAN table for the garigue and maquis data from Garraf in northeast Spain (Table 4.3; Plate 4.1) is shown in Table 8.2. This represents a sorted two-way table of the original data matrix of 15 quadrats × 17 species. Quadrats have been sorted column-wise and species row-wise. Various features of this table need explanation:

(a) Quadrat numbers are written vertically across the top of the table above each column. Thus the first column is quadrat 1, the second quadrat 4, the third quadrat 5, and so on.

(b) Species names and numbers are written down the left-hand side with the species number as originally coded on data entry, followed by two blocks of four letters corresponding to abbreviations of the Latin binomial. Thus the first row is species 2 *Brachypodium ramosum*, which is abbreviated to Bracramo.

(c) The main block of the table presents the sorted data. Both quadrats and species have been sorted so that they form groups of both quadrats and species. The effect of this is to concentrate entries down the diagonal of the table from top left to bottom right. Species that do not fit easily into the overall diagonal trend may appear at the top or bottom of the table.

(d) The numbers from 1–6 within the table correspond to the scores for pseudospecies described above. In this instance, with percentage cover data, the scale is:

- species absent
1 less than 2% cover
2 2–5% cover
3 6–10% cover
4 11–20% cover
5 21–50% cover
6 51–100% cover

(e) At the bottom of the table is the dichotomised key for quadrats, which shows both the group structure and the sequence of divisions. Taking the first line, the quadrats are split into the two groups indicated by the row of eight zeros (first group) and the continuation of the row as seven 1s (second group). This corresponds to the first division of the quadrats into two groups of eight and seven. Each of these two groups is then split again as indicated by the second row of zeros and 1s at the foot of the table. Thus the eight zeros of the first row are split in the second row into three zeros followed by five 1s. Similarly, the seven ones of the first row are split into one zero and six ones. Thus the second row summarises the second level of division, with four groups of three, five, one and six quadrats. The third row summarises the division of each of those four groups into eight, and so on. In a standard analysis, division is not continued any further if there are four or fewer quadrats in a group. Thus only two groups (Group 2 with five quadrats and Group 4 with six quadrats) are split any further, as shown in the third row at the bottom of Table 8.2.

(f) On the right-hand side of the table is the dichotomised key for the species. The principles are exactly the same as for quadrats, starting with the first division into two groups shown by the first column of zeros and 1s. These two groups are then split into four in the second column and so on. Once again, in a standard analysis, a group is not divided further if it contains four or fewer species.

Interpretation   Interpretation of the table is subjective. For the quadrat classification, the level of subdivision at which interpretation is made depends on the size of the data-set and the number of levels in the classification. In the case of the small set of data from Garraf (Table 8.2), only the second and third levels may be taken. Not all groups have to be taken from one level. If further subdivisions of a large group make more ecological sense, then it should be subdivided by looking at the next level. If two subdivisions of a small group seem to produce artificial subgroups, then it may make sense to amalgamate them again. Similar principles apply to the species classification using the columns on the right-hand side.

The following groups emerge from the quadrat classification of the Garraf data using the second level:

Group A: Quadrats 1, 4 and 5
Group B: Quadrats 2, 10, 6, 12 and 14 (which could be subdivided further at level 3 into 2, 10 and 6, 12, 14)
Group C: Quadrat 9
Group D: Quadrats 3, 7, 8, 11, 13 and 15 (which could be subdivided further at level 3 into 3, 7, 13, 15 and 8, 11)

Interestingly, in Table 4.3, data on the aspect of the sites are presented as a north/south division:

Sites with a north-facing aspect = 1 2 4 5 6 10 12 14
Sites with a south-facing aspect = 3 7 8 9 11 13 15

On the first division at level 1, this gives a clear separation on the basis of aspect, with all eight north-facing quadrats of the first side of the division separated from the remaining seven, which are south-facing (Table 8.2).

TWINSPAN – the classification method

The detailed workings of TWINSPAN are very complex. Nevertheless, it is possible to present a reasonably simplified explanation. The following is based on Hill *et al.* (1975), Hill (1979b), Causton (1988) and Jongman *et al.* (1995).

Making a dichotomy   A key concept of TWINSPAN and phytosociology is that for each division of a set of quadrats, a dichotomy can be made with a group of quadrats on one side characterised by one set of differential species, and a second group on the other side characterised by a second set of differential species. Ideally, species will belong to one side of the dichotomy or the other. In practice, of course, this never occurs. If it did, classification would be easy and use of numerical methods hardly necessary.

This general principle of division is applied in a series of levels, starting with the whole set of quadrats or species, dividing them into two groups, each of those into two to give four groups, each of the four into two to give eight groups, and so on. The process of dividing a group at any level is carried out in exactly the same manner as follows.

Derivation of pseudospecies   The first stage of the analysis is the conversion of the data into pseudospecies. The concept of pseudospecies has already been introduced. Even with a small set of data, the production of pseudospecies results in a large table. Part of the pseudospecies table for the Garraf data (the first 6 and the last – the 17th – species across all 15 quadrats) is presented in Table 8.3. In the original data (Table 4.3), the data were coded as a six-point Domin cover scale. For the TWINSPAN analysis, the six-point scale was used to define six pseudospecies as shown in Table 8.3. Thus in quadrats 1 and 2, the first three pseudospecies of bare ground are present, while in quadrat 3, all six are present, and so on. Full coding of these data gives a total of 57 pseudospecies across the 17 species, and it is these that are used in presence/absence form in the TWINSPAN analysis.

Table 8.3 Part of the pseudospecies table for the Garraf data (Table 4.3) – the first six and last (17th) species are shown. There are up to six pseudospecies for each original species. Reproduced with permission from John Wiley & Sons, Inc.

Species		Quadrat number															Number of pseudospecies	Cumulative pseudospecies
		1	2	3	4	5	6	7	8	9	10	11	12	13	14	15		
Bare rock	1	1	1	1	1	1	1	1	1	1	1	1	1	1	1	1	6	1
	2	1	1	1	1	1	1	1	1	1	1	1	1	1	1	1		2
	3	1	1	1	1	1	1	1	1	1	1	1	1	1	1	1		3
	4				1			1	1	1	1	1	1	1	1	1		4
	5							1	1	1		1		1	1	1		5
1	6							1	1	1		1		1		1		6
Brachypodium ramosum	1	1		1		1											4	7
	2	1		1		1												8
	3	1		1		1												9
2	4					1												10
	5																	
	6																	
Ceratonia siliqua	1			1				1		1						1	2	11
	2			1				1										12
3	3																	
	4																	
	5																	
	6																	
Chamaerops humilis	1	1			1										1		2	13
	2														1			14
4	3																	
	4																	
	5																	
	6																	

(Continued)

Table 8.3 (*Continued*)

Species		1	2	3	4	5	6	7	8	9	10	11	12	13	14	15	Number of pseudospecies	Cumulative pseudospecies
														Quadrat number				
Cortaderia selloana	1		1				1				1		1		1		4	15
	2						1				1		1		1	1		16
	3												1					17
5	4												1					18
	5																	
	6																	
Erica multiflora	1	1			1	1					1						4	19
	2	1			1	1					1							20
	3	1			1	1					1							21
6	4	1																22
	5																	
	6																	
Smilax asper	1			1				1	1			1	1	1	1	1	3	55
	2							1	1				1	1	1	1		56
	3								1									57
17	4																	
	5																	
	6																	

Total number of species 17
Total number of pseudospecies 57

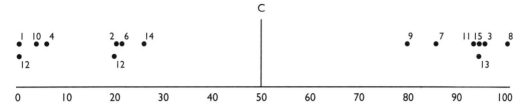

Figure 8.8   The primary ordination axis for the data from Garraf (Table 4.3), showing the positions of the 15 quadrats. The location of the centroid (the mean quadrat score C on the ordination axis) is shown. Original work of author.

The primary ordination   This is carried out on the original raw species data using reciprocal averaging or correspondence analysis (Chapter 6). Only the first axis ordination is used, and this primary ordination for the Garraf data is shown in Figure 8.8. The 15 quadrats are shown together with the centroid or mean of their scores. The quadrats can be divided into two groups on either side of the centroid – those on the left are known as the negative group, and those on the right the positive group. It is important to stress that these do not refer to the presence or absence of any species but are merely a convenient terminology for a dichotomy or a division. In the more sophisticated version of the program produced by Hill in 1979, rarer pseudospecies are downweighted in the reciprocal averaging ordination. This is done to avoid sets of similar rare species splitting off small groups of quadrats at an early stage of the classification.

Identify preferential pseudospecies to one side or other of the dichotomy to give indicator species
When the distribution of pseudospecies in each of the two groups is examined, some pseudospecies are found to be only on one side of the division while others are found only on the opposite side. Such pseudospecies, which occur exclusively on either side of the dichotomy, would be good indicator species.

   Thinking of the pseudospecies data in presence/absence form, most pseudospecies will tend to be present more on one side of the division than on the other. The main exception to this would, of course, be pseudospecies that occur in every quadrat. It thus becomes possible to work out a form of indicator value. For the jth pseudospecies, an indicator value (I) can be calculated as:

$$I_j = \frac{nj^+}{n_+} - \frac{nj^-}{n_-} \tag{8.3}$$

where n_+ is the total number of quadrats on the positive side of the division, and n_- is the total number of quadrats on the minus side. nj^+ is the number of quadrats on the positive side which have the jth pseudospecies and nj^- is the number of quadrats on the negative side which have the jth pseudospecies. Thus when a pseudospecies occurs in every quadrat on the positive side but in none on the negative side, $I_j = 1$, and when a pseudospecies occurs in all quadrats on the negative side and in none of the positive, $I_j = -1$. Such pseudospecies are called perfect indicators. If a pseudospecies occurs in every quadrat, I_j will equal 0.

   In the Garraf data, *Pinus halepensis*1 and *Pinus halepensis*2 (Aleppo pine) occur in all quadrats on the negative side and in none on the positive side, while *Olea europaea*1 (Olive) occurs in no quadrats on the negative side and all quadrats on the positive side. Thus *Pinus halepensis*2, as the highest pseudospecies value for *Pinus,* would be given a score of −1, while *Olea europaea*1 would be given a score of +1. A pseudospecies occurring in all quadrats in the data (e.g. Bare ground3) would obtain a score of 0 (zero).

The basic rule with multiple pseudospecies for each species is that the pseudospecies with the highest indicator value counts as the overall indicator value for the species, e.g. with *Pinus halepensis*, the second pseudospecies (*Pinus halepensis*2) has a score of -1 while the third only has a score of -0.625 (5/8− and 0/7+). *Pinus halepensis*2 is thus a perfect indicator, while *Pinus halepensis*3 is not. Thus the final rule is that only one actual species can be an indicator species, and this will be the pseudospecies number with the highest indicator value for a species.

In the TWINSPAN program, up to five indicator pseudospecies and hence species may be used to make a division. These would be the five highest indicator values regardless of whether they were positive or negative.

The refined ordination   Each quadrat is then allocated an indicator score by adding $+1$ for each positive indicator and -1 for each negative indicator that it contains. In the Garraf data, only one indicator is chosen: *Olea europaea*1, which is a positive indicator ($+$). In this very simple situation, only *Olea europaea*1 is chosen because it is a perfect indicator. *Pinus halepensis*1, *Pinus halepensis*2, *Phillyrea angustifolia*1 and *Quercus coccifera*4 could also have been chosen, since they are also perfect indicators, although negative. However, there is no need to include them, since they would give exactly the same result. In the refined ordination, each quadrat containing *Olea europaea*1 is thus given a score of $+1$, while those without *Olea europaea*1 are given a score of 0. This divides the quadrats into the two groups with eight on the negative side and seven on the positive.

In a more complex data-set, where there are several indicator pseudospecies, which will usually not be perfect indicators but just pseudospecies with high indicator values, each is assigned the value of $+1$ or -1, depending on whether they have positive or negative values of I. Then each quadrat is given an indicator score based on the number of indicator pseudospecies that it contains. As an example, where there are five indicator pseudospecies, and three are positive and two are negative, then if a particular quadrat contains two of the positive indicators and two of the negative, then the indicator score is $+2 - 2 = 0$.

If a maximum of five indicator pseudospecies are to be used, then the range of indicator scores is shown as below:

Positive and negative indicators		Range of indicator scores					
5 @ −1	0 @ +1	−5	−4	−3	−2	−1	0
4 @ −1	1 @ +1	−4	−3	−2	−1	0	1
3 @ −1	2 @ +1	−3	−2	−1	0	1	2
2 @ −1	3 @ +1	−2	−1	0	1	2	3
1 @ −1	4 @ +1	−1	0	1	2	3	4
0 @ −1	5 @ +1	0	1	2	3	4	5

To make a division, an indicator threshold needs to be set. If, with five indicator pseudospecies, there are three in the positive and two in the negative, then in the above table, a threshold of 0 will place quadrats with scores of -2, -1 and 0 in the negative group and 1, 2 and 3 in the positive. The final selection of the indicator threshold is determined by finding the position where the quadrats on the negative and positive sides, as defined by the indicator scores, agree most closely with the distribution of the quadrats on either side of the division on the original primary ordination. Thus each of the five indicator thresholds is tested in turn, and the one that gives the least discrepancy between their location on the original primary ordination and the new refined ordination is the one that is chosen.

In the small set of Garraf data, the situation is extremely simplified. Only one species *Olea europaea*1(+) is selected as an indicator. Thus any quadrat containing *Olea europaea*1 has a sum of +1 and a quadrat without it has a score of 0. Here there is complete agreement between the division on the original primary ordination and the refined ordination, since *Olea europaea*1 is a perfect indicator.

In the TWINSPAN program, which Hill produced in 1979, this process of refinement of the ordination is modified and is achieved by a transfer or iterative relocation algorithm (Gower, 1974) with a discriminant function used to make the transfers. The discriminant function is derived by adding together two ordinations. The first ordination is obtained by adding together the 'preference scores' of the commoner species. A preference score of +1 is given to each pseudospecies that is at least three times more frequent on the positive side of the division than on the negative side and *vice versa*, with scores of −1 being given to a pseudospecies that is three times more frequent on the negative side than the positive. Other common pseudospecies are scaled within these limits. The second ordination is derived by taking a mean of the preference scores for all pseudospecies in each quadrat. The two resulting ordinations are each scaled to an absolute maximum value of 1 and added together. The first ordination will tend to be dominant at the lower levels of the hierarchy, and the second at the higher levels. Hill (1979b) stated that, in practice, this process polarises the ordination quite strongly and reduces the number of borderline cases.

Misclassifications, borderline cases and the 'zone of indifference'   In the Garraf data, the dichotomy is very clear-cut and there are no quadrats near to the central divide or axis centroid of the primary ordination axis (Figure 8.8). However, in more complex data-sets, a number of quadrats will be close to or at the divide. In several cases, the location of the quadrat may be in the negative group on the primary ordination and the positive group on the refined ordination (using the indicator scores), and the possibility of misclassification occurs. In the original method, the assumption is made that the position of quadrats on the primary ordination is in general correct, since it is based on the complete floristic composition of the data-set. In contrast, the refined ordination, being based on only 1–5 indicator pseudospecies, is less reliable. However, the purpose of the classification method is to allocate quadrats to groups on the basis of the refined ordination and the indicator scores, so the indicator score for a quadrat is preferentially used to show where a quadrat should be, while bearing in mind that the position of the quadrat on the primary ordination was different.

In order to deal with this problem, it is important to realise that most misclassifications are probably only borderline, in that their primary ordination scores differ only marginally from their refined ordination scores. The exact side of the division to which they are allocated is really a matter of indifference. Thus a narrow zone of indifference is defined on the original primary ordination, where all quadrats are classified according to their indicator score whatever their primary ordination score.

To define the zone, the primary ordination axis is divided into segments. These segments are calculated such that a critical zone is recognised around the centroid. Thus using the quadrat scores (x) centred on their mean ($xmean$), the zone is defined as extending from $[xmean - xmin]/5$ below the mean to $[xmax - xmean]/5$ above (Figure 8.9a). This zone, which represents a fifth of the length of the original ordination, is then further subdivided into eight segments (Figure 8.9b). The zone of indifference is then set to half the size of the critical zone (that is, four segments). This zone is then gradually moved across the area of the critical zone, that is, sections

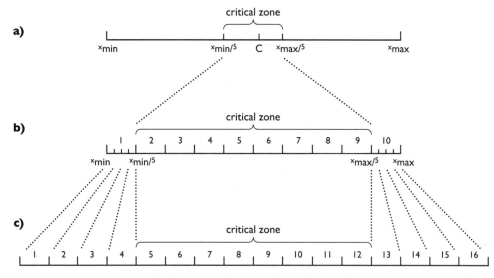

Figure 8.9 The critical zone in the refined ordination of TWINSPAN: (a) the definition of the critical zone as lying between [*xmean* – *xmin*]/5 and [*xmean* – *xmax*]/5. The critical zone thus represents a fifth of the length of the ordination axis; (b) subdivision of the critical zone into eight segments. In the TWINSPAN output, the areas on either side of the critical zone are also divided into four segments each, giving 16 segments in all; (c) in the stylised diagram which can be printed as part of TWINSPAN output (Table 8.4), the zones on either side of the critical zone are each subdivided into four segments, which, together with the eight segments of the critical zone, gives 16 segments in all. Original work of author.

2–5, 3–6, 4–7, 5–8, 6–9. In each case, the indicator scores of any quadrats are re-examined and compared for any misclassifications that occur. The final optimal position is that where the fewest misclassifications occur.

The final division   The final point of division is thus moved so that it fits as well as possible with the indicator threshold used to produce the final indicator ordination. A diagram can be drawn comparing the positions of the quadrats on the primary ordination against their positions on the refined ordination. This graph shows the six possible positions where quadrats could be located (Figure 8.10).

(a) Negative group
(b) Borderline negatives – allocated to the negative group because of their indicator scores
(c) Misclassified negatives – these are definitely placed with the negative group but their indicator scores would have placed them in the positive group
(d) Positive group
(e) Borderline positives – allocated to the positive group because of their indicator scores
(f) Misclassified positives – these are definitely placed within the positive group but their indicator scores would have placed them in the negative group.

Ideally, as is the case with the Garraf data, all quadrats would lie within groups (a) and (d). As few quadrats as possible should lie in groups (b), (c), (e) and (f).

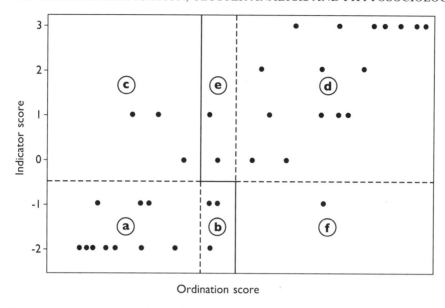

Figure 8.10   A scattergraph showing the relationship between the indicator score and the ordination score for a set of hypothetical data. Regions (a) and (d) represent the negative and positive groups respectively. Regions (c) and (f) are the misclassified negatives and positives, while the borderline quadrats lie within regions (b) and (e). The indicator threshold is set at −1. Redrawn from Hill *et al.* (1975), with kind permission of the British Ecological Society and Wiley-Blackwell.

Preferential pseudospecies   Once the division has been made, the preferential pseudospecies are tabulated. A pseudospecies is regarded as preferential to one side or other of the dichotomy if it is more than twice as likely to occur on one side than the other. Three categories of preferentials are recognised: positive preferentials, negative preferentials and non-preferentials. An important aspect of this is that calculation of degree of preference is dependent upon group size. Thus if 60 quadrats are divided into one group of 10 and another of 50, a pseudospecies that occurs in 8 out of the 10 quadrats in the first group will have a score of 0.8, while if it occurs in 18 out of the 60 in the second, it will only have a score of 0.3. The pseudospecies will therefore be deemed preferential to the smaller group, even though it occurs in many more quadrats in the other larger group.

Printed output for a division   An example of the printed information relating to a division within TWINSPAN is presented in Table 8.4. This is the information produced for the first division of the Garraf data. The first line gives the division number and the number of quadrats plus the group code (*). The second line gives the information from the primary reciprocal averaging ordination. Below this are listed the indicators with their sign. In the Garraf output, only the one indicator (*Olea europaea1*) (+) is listed, together with the maximum indicator score for the negative group (0) and the minimum indicator score for the positive group (1).

If requested, a stylised diagram of the pattern of the final division as shown in Table 8.4 can be printed, showing the relationship between the primary ordination axis (along the x axis) and the indicator ordination (on the y axis). The whole axis is displayed as 16 segments, with the eight-segment critical zone extending from segments 5 to 12. On either side, the positive and negative zones are each subdivided into four equal subzones, to enable the pattern of quadrat distribution to be seen (Table 8.4 and Figure 8.9c). The zone of indifference is shown as lying between segment

Table 8.4 The information printed by TWINSPAN for the first division of the Garraf data. Reproduced with permission from John Wiley & Sons, Inc.

```
Garraf Vegetation Data                    Classification of samples
***********************************************************************

    DIVISION    1  (N=     15)         i.e. group *
    Eigenvalue: 0.5847  at iteration    3
    INDICATORS and their signs:
    Oleaeuro 1(+)
    Maximum indicator score for negative group    0
    Minimum indicator score for positive group    1
    1 2 3 4 5 6** 7 8 9 10** 11 12 13 14 15 16****
    *********************************************
    0 0 0 0 0 0** 0 0 0  0**  0  0  0  0  2  5**1
    *********************************************
    4 1 3 0 0 0** 0 0 0  0**  0  0  0  0  0  0**0

    ITEMS IN NEGATIVE GROUP    2  (N =     8)       i.e. group *0
    QUAD1      QUAD2     QUAD4     QUAD5     QUAD6    QUAD10    QUAD12    QUAD14

    ITEMS IN POSITIVE GROUP    3  (N =     7)       i.e. group *1
    QUAD3      QUAD7     QUAD8     QUAD9     QUAD11   QUAD13    QUAD15

    NEGATIVE PREFERENTIALS
    Bracramo1(2,    0)   Chamhumi1(3,   0)   Cortsell1(5,   0)   Ericmult1(4,   0)
    Euphsp. 1(6,    0)   Lavaangu1(3,   0)   Philangu1(8,   0)   Pinuhale1(8,   0)
    Pistlent1(4,    0)   Salvsp. 1(3,   0)   Sedumsp.1(4,   1)   Bracramo2(2,   0)
    Cortsell2(4,    0)   Ericmult2(4,   0)   Euphsp. 2(4,   0)   Philangu2(7,   0)
    Pinuhale2(8,    0)   Pistlent2(4,   0)   Quercocc2(8,   1)   Salvsp. 2(2,   0)
    Bracramo3(2,    0)   Ericmult3(4,   0)   Philangu3(5,   0)   Pinuhale3(5,   0)
    Pistlent3(3,    0)   Quercocc3(8,   0)   Salvsp. 3(2,   0)   Quercocc4(8,   0)
    Quercocc5(4,    0)

    POSITIVE PREFERENTIALS
    Cerasili1(0,    4)   Oleaeuro1(0,   7)   Smilaspe1(1,   6)   Cerasili2(0,   2)
    Oleaeuro2(0,    6)   Smilaspe2(1,   4)   Oleaeuro3(0,   5)   BareRock4(3,   7)
    Oleaeuro4(0,    2)   BareRock5(1,   7)   BareRock6(0,   7)

    NON-PREFERENTIALS
    BareRock1(8,    7)   Quercocc1(8,   6)   Rosmoffi1(7,   7)   BareRock2(8,   7)
    Rosmoffi2(4,    6)   BareRock3(8,   7)   Rosmoffi3(2,   2)

              -------- E N D   O F   L E V E L   1 --------
```

7 and segment 10. However, given the marked polarisation of the ordination, it is not actually used. The indicator ordination shows the eight quadrats of the negative group with indicator scores of 0 clustered at the left-hand end and the seven quadrats of the positive group with indicator scores of 1 clustered to the right. An important point to realise is that the six zones of the stylised table correspond exactly to the six zones of Figure 8.10.

Below this are listed the quadrats in the positive and negative groups. In a more complex analysis, borderline negatives, misclassified negatives, borderline positives and misclassified positives would all be listed here. Finally, the preferential pseudospecies are listed under the three headings of negative preferentials, positive preferentials and non-preferentials. Each pseudospecies is followed by two numbers in brackets, for example, Bracramo1 (2, 0), which are the number of quadrats on each side of the divide in which the pseudospecies occurs. Thus *Brachypodium ramosum*1 occurs in two quadrats on the negative side and none on the positive. Non-preferentials are shown as occurring in similar proportions of quadrats in the two groups, depending on group size.

Further divisions   Further divisions are made in exactly the same way, so that each of the two groups resulting from the first division is taken and divided in turn to give four groups. These are then each divided again to give eight, and so on. In a standard analysis, division terminates when there are four or fewer quadrats in a group. However, this, along with most other parameters, can be varied at the start of the program.

Production of a quadrat by species two-way table   In order to produce a final table such as that shown in Table 8.2, the divisions must first be ordered and then the species must also be classified. The ordering of the quadrat groups is achieved by comparing the two site groups produced at any level with the quadrat groups at two higher levels in the hierarchy. Thus if the TWINSPAN hierarchy of divisions is numbered as in Figure 8.11, then if the positions of groups 4, 5, 6 and 7 have already been determined, each of their next divisions (8 and 9 for group 4; 10 and 11 for group 5, etc.) can be swivelled and compared with the adjacent groups in the next level of the hierarchy. If 8 and 9 are swivelled and compared with group 5, then if 8 is more similar to 5 than 9, the order of groups 8 and 9 in the final table will be reversed. Again, if group 11 is more similar to group 4 than group 10, and also is less similar to group 3 than group 10, the ordering 11 10 will be used. Similar comparisons are made for all groups at the third level, and switches made if necessary.

Classification of the species is achieved in a similar way to that of quadrats, except that the species classification is based on the fidelity (the degree to which species are confined to particular

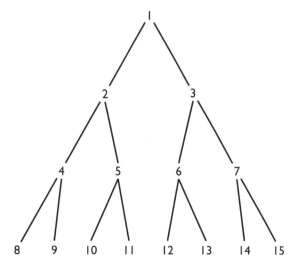

Figure 8.11   The numbering of the hierarchical group structure produced by TWINSPAN. Original work of author.

groups of quadrats) rather than on the raw data. Again a primary species ordination axis is taken out, but species are only placed in positive or negative groups with little information presented on the borderline cases between groups. The final table is then printed as in the example of Table 8.2.

The binary notation from either the bottom or the right-hand side of a TWINSPAN table can be used to produce a dendrogram of either the quadrat or the species classification. These diagrams are not usually scaled but are simply presented as a series of levels showing the pattern of divisions at each level. Although they can be useful, they are not nearly as informative as the TWINSPAN table itself.

This explanation of the workings of TWINSPAN is still fairly complex and is difficult to grasp on a single reading. However, the really attractive feature of the method and the program is that the final table is very easily understood and relatively straightforward. Despite the complexities of the method, the end product can be seen as no more than a sorted two-way table. It is this feature which has made it so simple to use and has encouraged its widespread application.

Criticisms of TWINSPAN as a classification method

Despite its extremely widespread use, not just in vegetation science but in many other areas of ecology, TWINSPAN has been severely criticised by some authors. Foremost among these were McCune and Grace (2002: p. 97), who stated:

> Ecologists should not use *TWINSPAN*, except in the very special case where a two-way ordered table ... is needed for a data set with a simple, one-dimensional underlying structure. Presence of a single strong underlying gradient can be manifest from prior knowledge or it can be detected with non-metric multi-dimensional scaling ... *TWINSPAN* cannot represent complex datasets in its one-dimensional framework (van Groenewoud, 1992; Belbin and McDonald, 1993).

Legendre and Legendre (1998) and McCune and Grace (2002) also criticised the complexity of the method, the artificial nature of the 'pseudospecies' employed and the fact that, apart from the above description in the previous edition of this text, there is no detailed published description of the method. Belbin and McDonald (1993) showed that methods of similarity analysis, notably group average (UPGMA), performed markedly better than TWINSPAN on various trial species data-sets.

Also a computational error was found in the original version of the program by Tausch *et al.* (1995), related to instability in the results, initially when using quantitative data, but subsequently also with presence/absence data, so that the resulting groups varied, depending on the input order of the data. The error was shown to be linked to the use of too lax a criterion for the achievement of stability in the correspondence analysis/reciprocal averaging algorithm (Oksanen and Minchin, 1997). A tolerance level of 0.003 and maximum number of iterations of 5 was replaced by a tolerance level of 0.0000001 and a maximum of 999 iterations. All current versions of the TWINSPAN program now incorporate these corrections. Despite these problems and criticisms, TWINSPAN has been and still is nevertheless very widely used by vegetation scientists and plant ecologists, partly because of the widespread availability of the program, which was re-released in a Windows version by Hill and Šmilauer (2005) (Chapter 9), but also because of the attractiveness of its final two-way sorted table.

Equally as many examples of research using TWINSPAN continue to be published as those using some form of similarity analysis. Presumably the researchers appear to believe that despite its failings, use of TWINSPAN still meets the requirement of 'user satisfaction'. Whether this is because they are too uncritical of the method remains an open question. Roleček *et al.* (2009)

argued that the reservations of some authors against TWINSPAN were seen as advantages by some others, because TWINSPAN does not extract natural groups of the most similar sites. Rather, as a divisive technique, it partitions the dissimilarity space determined by the main gradients in the data, disregarding possible discontinuities. Agglomerative clustering works better where natural clusters are present, while TWINSPAN is better for partitioning data-sets with a weaker group structure, where samples are more evenly distributed across the main environmental gradients. They proposed a modification to the method, involving analysis of the heterogeneity of the clusters prior to each division, with the modified algorithm dividing only the most heterogeneous cluster at each step.

Carleton *et al.* (1996) replaced the correspondence analysis/reciprocal averaging with canonical correspondence analysis in their *COINSPAN* computer program, but perhaps because of the problems with TWINSPAN, it has rarely been used in published research and similar criticisms to those levelled at both TWINSPAN and canonical correspondence analysis can be applied.

NON-HIERARCHICAL CLASSIFICATION

Most numerical classification of vegetation data uses hierarchical methods. However, various non-hierarchical methods exist, although they have only rarely been used in vegetation science. Such techniques are commonly known as partitioning methods that are cluster-based, rather than centring on the individuals or objects (quadrats or species) being classified (Legendre and Legendre, 1998; Everitt *et al.*, 2001; Grabmeier and Rudolph, 2002; Fielding, 2007). In one approach, for example, initial starting points in multidimensional space are taken to 'seed' groups, and other points within a given radius of each group 'seed' are gradually incorporated into each group (Figure 8.12), depending on the method and algorithm used.

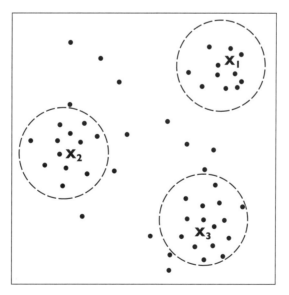

Figure 8.12   The principles of non-hierarchical clustering. X_1, X_2 and X_3 represent selected starting or 'seed' individuals and other individuals within a given radius are taken in to form a group. As the radius expands and circles start to overlap, rules for allocation of individuals to their nearest cluster have to be established or criteria have to be set for reallocation from smaller clusters to larger ones. Original work of author.

k-means clustering

Perhaps the most widely used partitioning method is k-means clustering (MacQueen, 1967; Hartigan and Wong, 1979; Legendre and Legendre, 1998; Fielding, 2007), which has four steps:

(1) Commence by selecting a number (k) of clusters and make an initial partitioning of the data in p-variable (species) space to those k clusters. This may be done at random or based on some 'seed' values.
(2) Reallocate individuals by assigning each individual to its nearer cluster centre.
(3) Recalculate the cluster centres as centroids (mean position of all individuals in each of the k clusters in multivariable space).
(4) Repeat steps 2 and 3 either a fixed number of times or until some criterion of stability is reached, which is usually based on the number of points changing position from one group to another.

The main reason why k-means clustering has seen little application in ecology is because of the requirement of pre-selecting the number of clusters, since this is usually unknown in advance of most analyses, and the fact that distances to cluster means are usually calculated using some form of the Euclidean distance or error sum of squares measure, which is not suited to sparse species data. Nevertheless, example output from k-means clustering of the Gutter Tor, Dartmoor, data with five groups specified at the outset is shown in Figure 8.13. This 'star diagram' shows the five clusters of samples linked to their group centroids and plotted on a two-dimensional ordination diagram derived from non-metric multidimensional scaling (NMS) (Chapter 6).

A variant on the k-means approach is the k-medians approach (PAM – partition around medoids) (Kaufman and Rousseeuw, 1990), which uses medians to define cluster centres instead of means because they are less affected by outliers.

Fuzzy clustering

Techniques of fuzzy classification or clustering, although originally devised some years ago (Bezdek, 1974, 1981; Marsili-Libelli, 1989; Roberts, 1989; Podani, 1990; Equihua, 1991), are becoming more widely used in vegetation science, partly due to their availability with the computer programs available from Pisces Conservation Ltd (2008d). In fuzzy clustering, samples are not allocated to a single group but instead they relate to a membership function that indicates the strength of relationship to all or some of the clusters in the data-set. The notion of a membership function is linked to fuzzy logic, which is an extension of Boolean logic, where ideas of true and false are replaced by that of partial truth (Moraczewski, 1993a,b, 1996). Generally, the number of core groups is assumed to be known in advance of the application of the method, and the membership function for each sample is estimated using some iterative function, similar to the k-means approach above. De Cáceres *et al.* (2010) applied three different approaches to a set of 531 Catalonia pasture relevés with some success, but also found problems and limitations in all cases. Roberts (1986, 1989, 2009) has demonstrated how the analysis of fuzzy sets can also be applied to ordination methods, once more demonstrating the links between numerical classification and ordination methods. Olano *et al.* (1998) provided an example of fuzzy classification linked to constrained ordination, while Banyikwa *et al.* (1990) demonstrated the combined approach of fuzzy set ordination and classification in the Serengeti grasslands of Africa.

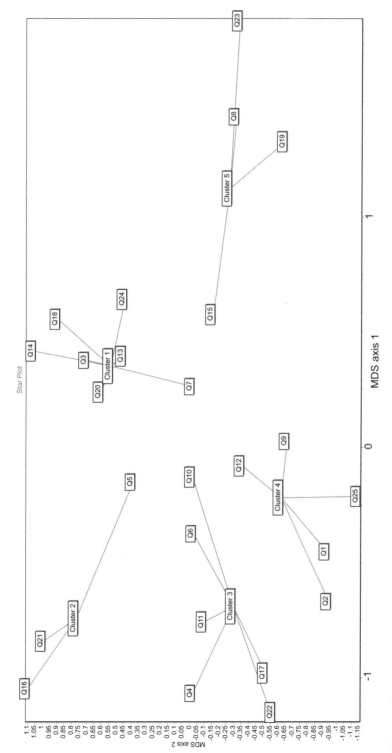

Figure 8.13 A 'star' diagram from *k*-means clustering of the Gutter Tor, Dartmoor, data, with five groups specified at the outset. Analysis performed using the fuzzy classification program of Pisces Conservation Ltd (2008d). Original work of author.

Neural networks

Neural networks are computing algorithms that seek to explore the computational possibilities of large and complex networks of relatively simple elements. The analogy is with the neurons in the brain. They are primarily concerned with pattern recognition and hence their potential value for numerical classification (Ripley, 1996; Callan, 1999; Haykin, 1999; Everitt *et al*., 2001). Most applications have the disadvantage that they assume that the group classes are known in advance (a supervised classification) and are thus better suited to the allocation of new samples to an existing classification, but of greatest relevance to unclassified vegetation data is the unsupervised method of self-organising maps (SOM) (Kohonen, 1982, 1997; Lek and Guégan, 1999; Giraudel and Lek, 2001). Use in vegetation science is at an early stage, but an example of the application of supervised classification and allocation of new quadrats or relevés to an existing classification is presented in Černá and Chytrý (2005). Neural networks are also often better seen as a modelling technique, as exemplified by the research by Hilbert and Ostendorf (2001) into the modelling of changes in vegetation distribution in response to climate change.

RECOMMENDED NUMERICAL CLASSIFICATION ANALYSIS FOR SPECIES AND ENVIRONMENTAL DATA

For interpretational purposes, the 'best' method should be the one that makes most ecological sense. Group average, which is equivalent to the unweighted pair-groups method (UPGMA) using the Steinhaus (Czekanowski/Sørensen/Bray and Curtis) (dis)similarity coefficient is probably regarded as the optimal method of similarity analysis for species data, which are virtually always 'sparse', although, despite its failings, TWINSPAN is still widely used and nevertheless appears to provide interpretable results in many circumstances. For environmental data (interval and ratio data), Ward's minimum variance clustering, which implicitly uses the Euclidean distance coefficient, is recommended, because of its space-conserving properties. These conclusions are strongly supported by other researchers who have compared a range of different methods (Belbin and McDonald, 1993; Cao *et al*., 1997; McCune and Grace, 2002).

Using nine geometric and five non-geometric evaluators (see paper for details), Aho *et al*. (2008) came to similar conclusions when comparing five hierarchical agglomerative methods applied to two real-world data-sets: flexible-β linkage (Lance and Williams, 1966, 1967), Ward's method (Ward, 1963), complete linkage (McQuitty, 1960), group average linkage (UPMGA) (Sokal and Michener, 1958), and single linkage (Sneath and Sokal, 1973); the hierarchical divisive method TWINSPAN (Hill, 1979b); and two non-hierarchical methods: partitioning around medoids (PAM: Kaufman and Rousseeuw, 1990) and *k*-means analysis (MacQueen, 1967; Hartigan and Wong, 1979). The value β = −0.25 was used for flexible-β linkage (Lance and Williams, 1967; Legendre and Legendre, 1998). They concluded that, with the exception of single linkage analysis, all methods produced comparatively similar groups, but within this, flexible-β linkage, group average linkage and Ward's error sum of squares method performed best.

CLASSIFYING LARGE DATA-SETS

Where large sets of floristic data have been collected, it is often necessary to break the analysis down into several stages. Simply entering all data to one similarity analysis or TWINSPAN will not necessarily give satisfactory results. An alternative strategy is to subsample the data to produce an initial classification, and then to allocate the remaining quadrats or samples to those groups.

Gauch (1979, 1980) wrote a program for composite clustering (*COMPCLUS*), which takes any quadrat from a set of data at random as a starting point, and clusters or forms a group of all sites within a specified radius from that site (Figure 8.12). This process is then repeated until all quadrats are accounted for. At a second stage, quadrats from smaller groups or clusters are reallocated to larger groups by selecting a larger radius. Janssen (1975) devised a similar method but started by taking the first quadrat in the data-set. In a computer program called *CLUSLA* (Louppen and van der Maarel, 1979), once a quadrat is found which lies outside of the radius of the first group, that quadrat initiates a new group. Subsequent quadrats are compared with all existing groups and allocated to that which is most similar. In all these methods, there is a strong dependence on the order in which quadrats enter the classification. *FLEXCLUS* (van Tongeren, 1986) is the most advanced of these methods, where a certain number of quadrats are selected at random as initiating points or are chosen by the researcher. All other sites are then allocated to the groups formed around those quadrats. A process of relocation is applied until a certain stability is achieved. A number of variations of these methods exist, but these non-hierarchical methods are not widely used. Digby and Kempton (1987) provided useful further discussion on this topic.

Van der Maarel *et al.* (1987) reviewed the problems of handling large data-sets and suggested a two-step approach to classification based on stratification of the data by either geographical area or major differences in vegetation type. Numerical classification is then performed on each subset. The resulting clusters are then summarised by calculating a 'synoptic cover–abundance value' for each species in each cluster. All clusters are then subjected to further classification using these synoptic scores, and the groups resulting from these are interpreted as community types.

More recently, the computer programs *JUICE* (Tichý, 2002; Tichý and Holt, 2006) and *TURBOVEG* (Hennekens, 1996; Hennekens and Schaminée, 2001) (Chapter 9) have been released, which facilitate the analysis of large phytosociological data-sets. These packages incorporate routines to store and export data in formats that enable the use of a wide range of computer packages for numerical classification within an overall database management system. Up to 1 million relevés can be stored and processed. Since 1994, *JUICE* has been accepted as the standard computer package for the European Vegetation Survey.

TESTING FOR DIFFERENCES BETWEEN GROUPS

There have been important advances in techniques for determining whether groups of samples assigned to classes by appropriate criteria in advance of analysis are significantly different from each other. For various reasons, notably the problems of sparse data matrices and non-linearity between variables (species), when analysing species data, the classical parametric multivariate analysis of variance (MANOVA) methods are inappropriate. The non-parametric 'analysis of similarities' (*ANOSIM*) program (Clarke and Green, 1988; Clarke and Warwick, 1994; Clarke and Gorley, 2006) is becoming more widely used in this context, along with the multi-response permutation procedures (MRPP) method of Mielke *et al.* (1976, 1981) and Mielke and Berry (2001), available in the *PC-ORD* package (McCune and Mefford, 1999, 2010), which performs a similar function. However, it is important to realise that groups derived from cluster similarity analysis or TWINSPAN cannot be tested on the same data using these methods because of the danger of circularity of argument (McCune and Grace, 2002; Clarke and Gorley, 2006).

Indicator species analysis (ISA) (Dufrêne and Legendre, 1997; Legendre and Legendre, 1998)

ISA identifies the characterising species for a set of groups produced from similarity analysis or TWINSPAN, and assesses how well species are separated between groups. Statistical significance of groups is determined by a Monte Carlo randomisation test (as described in Chapter 6 for canonical correspondence analysis (CCA) ordination) with the null hypothesis that the species have no value as indicators. The advantages of ISA is that computation involves use of both the abundance and frequency of species, the Indicator Value (IV) is computed independently for each species in the data-set, and it can also be applied to any *a priori* set of classification groups, as well as computed classifications.

The method involves computation of the Indicator Value (IV) for species i in group j. The index is derived from the product of the relative abundance, also known as the specificity (A_{ij}), and the relative frequency, also known as fidelity (B_{ij}):

$$A_{ij} = \frac{\bar{x}_{ij}}{\sum\limits_{j=1}^{n} \bar{x}_i} \tag{8.4}$$

$$B_{ij} = \frac{n_{ij}}{n_j} \tag{8.5}$$

$$IV_{ij} = A_{ij} \times B_{ij} \times 100 \tag{8.6}$$

where

   $\bar{x}_{ij} =$ the mean cover of species i within group j
   $\sum\limits_{J} \bar{x}_i =$ the sum of the mean cover of species i in all groups
   $n_{ij} =$ the number of samples in group j occupied by by species i
   $n_j =$ the total number of samples in group j.

IV_{ij} is zero when species i is absent from group j, and equals 100 when species i occurs in all samples within group j and is not found in any other groups. Uncommon and rare species have low IV_{ij} scores. Common species have high B_{ij} scores and thus higher IV_{ij} scores. However, they are unlikely to be statistically significant because permutations in the Monte Carlo randomisation test will also give high B_{ij} scores. Those species that occur in all samples within only one group will have the highest and most significant IV_{ij} scores for that group.

The method is explained further in Dufrêne and Legendre (1997), Legendre and Legendre (1998) and McCune and Grace (2002). Further refinements of the approach are described in Bakker (2008).

Comparing dendrograms and group structures produced by different objective classification methods

Methods for comparing dendrograms and group structures have been devised, notably cophenetic correlation (Sneath and Sokal, 1973; Rohlf, 1974), maximisation of between- or within-group variance (Orlóci, 1967b), and prediction of variables from cluster membership (Gower, 1967). Digby and Kempton (1987) described the use of two-way contingency tables to compare sets of groups resulting from different classifications of the same data-set. More recent reviews of methods are presented in Legendre and Legendre (1998), Everitt *et al.* (2001) and Fielding (2007).

In general, the 'goodness' of a classification of a set of vegetation data should be assessed in terms of the ease with which ecological sense can be made of the results. Once again, this involves considerable subjectivity on the part of the researcher and demonstrates the point that the interpretation of a classification remains partly an art, depending on the experience and knowledge of the user.

Characterising plant community types in terms of environmental and biotic factors

Once a set of communities has been identified by classification, they are often further characterised by associated environmental/biotic data. At its simplest, this can involve description of each group in terms of individual environmental/biotic variables, such as pH, soil moisture content, nutrient status, grazing intensity or management regime. Calculation of simple descriptive statistics (Chapter 5), such as the mean, maximum, minimum and standard deviation or their non-parametric equivalents (stem and leaf plots, medians) for continuous data, is performed for the quadrats in each group. For ordinal and nominal data the construction of histograms will assist with interpretation.

More sophisticated methods for interpretation exist, notably the use of simple discriminant functions based on a set of environmental/biotic variables to test for significant separation between groups. Ter Braak (1982, 1986b) wrote a program called *DISCRIM* to do this within the TWINSPAN program, using only those environmental/biotic variables that optimally predict the classification. The TWINSPAN classification can be drawn as a dendrogram, and the most discriminating environmental/biotic variables at each branch are shown.

Where environmental/biotic data are paired with species data, the differences between groups resulting from classification of the species data may be tested using the environmental/biotic data-set and programs for multivariate analysis of variance (MANOVA) or the non-parametric equivalents *ANOSIM* (Clarke and Gorley, 2006) and MRPP (Mielke *et al.*, 1976, 1981; McCune and Grace, 2002) as discussed above.

A further important development in this area is that of classification and regression trees (CART) (Breiman *et al.*, 1984; De'ath and Fabricus, 2000; Vayssieres *et al.*, 2000; Zuur *et al.*, 2007). CART splits a set of samples recursively into subgroups, on the basis of a single response variable at each split that may be either categorical (classification trees) or continuous (regression trees). The subgroups become progressively homogeneous in relation to predefined groups. The tree is derived by splitting the data into two groups, those two into four, and so on, using a simple rule based on a single explanatory variable. Splitting is continued until an overgrown tree is produced, which can then be 'pruned back' to give an optimal result. The result is a tree-like classification based on a dichotomous key. This can then be used to assign unknown samples into the groups. CART is now used in conjunction with geographical information systems (GIS) to produce maps of plant communities, habitats or land cover based on spatially referenced environmental or impact data. Used in this manner, the technique represents a valuable form of ecological response modelling, as demonstrated by Michaelsen *et al.* (1994), who used it to assist with the efficient planning of vegetation surveys and sampling strategies. CART only works effectively with medium to large numbers of samples (200+). An introduction to CART is presented in Urban (2002), and a further example of CART combined with GIS in the area of soil classification is given in Scull *et al.* (2004).

An example of a classification tree is presented in Figure 8.14. The study area on South Uist, in the Outer Hebrides of Scotland, was located across a transition from the machair coastal sand dunes to ca. 50 m above sea-level. The dune communities consist of species-rich grasslands above alkaline shell sand. Inland, upland communities of acidic grassland, heath and bog were present.

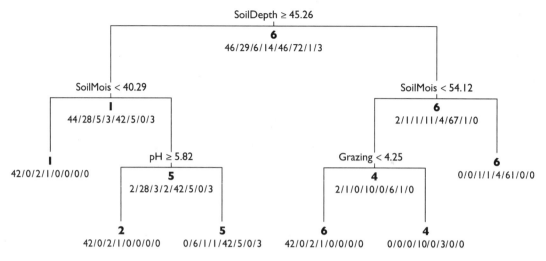

Figure 8.14   The classification tree resulting from CART analysis of the Hebrides species and environmental/biotic data (217 samples × 249 species; Kent *et al.*, 1996). Original work of author.

Between the dunes and the uplands, where some of the shell sand had blown onto peaty soils, intermediate 'blackland' communities occurred (Kent *et al.*, 1996). Sampling was carried out along a rectangular network of samples in a triangular sampling grid (a 'tranome', *sensu* Kent *et al.*, 1997), measuring 2162 m in length (orientated west–east, perpendicular to the shoreline and across the machair–upland transition) and 200 m in width (orientated north–south). A total of 217 samples were positioned ca. 50 m apart within the grid. Using a 1 m × 1 m quadrat, all species of vascular plant, bryophyte and lichen present in the quadrats were identified along with an estimate of percentage cover. In total, 249 species were recorded, along with 12 environmental variables.

The species data were first analysed by similarity analysis (group average with the Steinhaus (Sørensen/Czekanowski/Bray and Curtis) (dis)similarity coefficient. Inspection of the dendrogram indicated eight interpretable groups. Using the R software package, a classification tree analysis was performed with the membership of the eight groups derived from the similarity analysis of the species data as the categorical response variable, along with the 12 environmental variables, resulting in the classification tree shown in Figure 8.14. The samples are divided at each point in the tree on the basis of the variable and value above each division. The distribution of samples in the original eight groups from the species data is shown below each division. The relationship between the original eight species groups and the critical environmental variables and their values at each split in the tree are thus demonstrated. The primary division relates to sand/soil/peat depth, with soil moisture critical at the divisions below this. At the third level, both pH and grazing achieve significance.

The Mantel test

A different approach to the comparison of variability between a set of species data and a set of environmental/biotic variables is the Mantel test, which was originally devised to analyse patterns of disease clustering in cancer research (Mantel, 1967) and was introduced to ecology by Sokal (1979) followed by Legendre and Fortin (1989). The Mantel coefficient analyses a pair of (dis)similarity or distance matrices from two data-sets that have been collected from the same set of sample locations, with the hypothesis that there is a match between the values in one matrix with those in the second. It thus handles multivariate data. In plant ecology, it has usually been calculated between a matrix of (dis)similarities of species composition between samples or quadrats (ecological or floristic distance) and a matrix of (dis)similarities between any environmental/biotic data (environmental/biotic distance) collected from those same samples or quadrats. The Mantel test may be applied to whole data-sets or, perhaps more interestingly, to subsets of samples or quadrats derived from numerical classification or cluster analysis of species data, where the aim is to examine variations in the inter-relationships between species and environmental data in different sample groups.

The coefficient is normally based on calculation of the product-moment correlation coefficient between the paired values in the two matrices, and assumes that only the evidence of a linear relationship between the two is of interest. Non-parametric rank correlation methods may also be used (Dietz, 1983). Assessment of significance, using the Mantel z statistic, involves comparison of the resulting correlation coefficients with correlation coefficients derived following randomisation of values in one of the matrices (a Monte Carlo test) (Fortin and Jacquez, 2000; Fortin and Gurevitch, 2001). An asymptotic t-approximation may also be used (Fortin, 1999). McCune and Grace (2002) provide a detailed introduction to the topic.

Another way in which Mantel tests have been applied is in examining spatial autocorrelation in a set of species or environmental/biotic data collected in a set of quadrats, where the spatial coordinates of the sample location have been recorded using some form of global positioning system (GPS). A third matrix of actual geographical distances between the sampling points (Kent et al., 1997; Fortin, 1999) can then be derived and correlated against either or both the species and environmental/biotic (dis)similarities/distances. A further modification to the method, known as the Mantel correlogram, involves dividing the correlation coefficients into a series of distance categories followed by testing of the values for each distance category against the overall average for autocorrelation in the complete data-set. Positive and negative scores of the Mantel z value then represent autocorrelations greater or less than the overall average autocorrelation for the survey area (Oden, 1984; Oden and Sokal, 1986; Legendre and Fortin, 1989). Koenig and Knops (1998) criticised the value of standard Mantel tests and Mantel correlograms in this context because they tend to show only the predictable pattern of spatial autocorrelation declining with increase in distance between sampling points. Also the result is a single value, which is the average spatial autocorrelation for the whole study area, assuming an isotropic surface.

Numerical classification and ordination – complementary analyses

By now it should be clear than both ordination and classification can be performed on the same data. This is known as complementary analysis (Kent and Ballard, 1988). One of the most effective ways of showing the relationships between the two techniques is to plot the group membership of quadrats resulting from classification onto the quadrat ordination diagram from an ordination of the same data. In Figure 8.15a, the two-dimensional ordination plot for detrended correspondence analysis of the Garraf data is presented. In Figure 8.15b, the TWINSPAN group membership (four groups A–D) of each quadrat has been superimposed. The relationship between the classification and the ordination should be self-evident. When used together with other diagrams for environmental factors, this approach can assist greatly with interpretation. Such plots also give some information on the group structure of the data and the extent to which the quadrats are distributed as a continuum or not. Unfortunately, however, the distortions inherent in most methods of ordination often tend to blur this picture. Another example of this approach is shown in Figure 8.13.

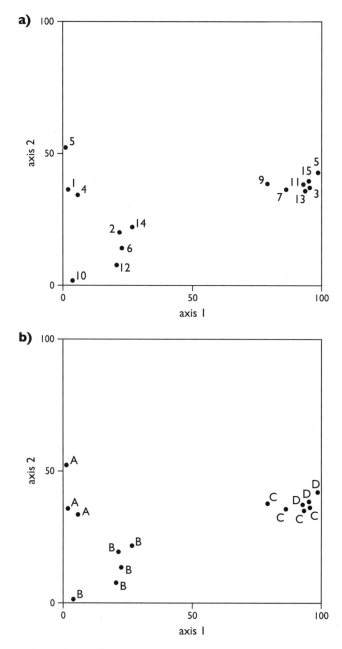

Figure 8.15   (a) Two-axis quadrat ordination plot produced by detrended correspondence analysis of the Garraf data (Table 4.3). (b) TWINSPAN groupings of the Garraf data superimposed on the ordination diagram of Figure 8.14(a). Groups A–D are those defined by TWINSPAN earlier in this chapter. Original work of author.

Numerical classification and hypothesis generation

Just as with ordination, methods for numerical classification can be seen as a means of hypothesis generation. They are techniques for data exploration and data reduction. However, again, as with ordination, there has been a tendency for researchers to carry out classification as an end in itself rather than using the results as a starting point for further work. Where the goal of the research is purely phytosociological, then this is acceptable, but there are many other situations where new research ideas may be generated as a result of an initial descriptive classification.

Numerical classification and subjective classification (the Zürich-Montpellier School)

Often methods of numerical classification may be used within a phytosociological framework, which is essentially that of the Zürich-Montpellier School (Chapter 7). Here the database and numerical methods, such as those incorporated in the *TURBOVEG* (Hennekens and Schaminée, 2001) and *JUICE* packages (Tichý, 2002; Tichý and Holt, 2006) are used to carry out the tabular rearrangement process, rather than doing it by hand and therefore subjectively. One of the positive features of TWINSPAN (Hill, 1979b) is that it produces a table that provides a useful starting point for a full Zürich-Montpellier type exercise in phytosociology. However, it is important to realise that with Zürich-Montpellier, the whole philosophy of approach is different, and at the interpretation stage the primary aim will be to fit any groupings that have been found into the established hierarchy of nomenclature and the existing abstract categorisations of the Zürich-Montpellier system.

Many other situations exist where plant community types may be recognised at the local scale for a whole range of different purposes, such as trying to understand plant–environment relationships, generate and test hypotheses concerning different management regimes for certain types of vegetation, or categorisation of communities for purposes of conservation management. Such work will often contain an element of phytosociology, in that one aim is to define and recognise plant community groupings, but there may be no need to relate them to the Zürich-Montpellier classification groups for Europe or elsewhere. Perhaps this is a suitable point at which to restate the importance of having a clear aim and purpose to any research work in plant ecology.

To conclude this chapter with a view to the future, developments in numerical classification continue, and the potential of Bayesian model-based analysis as an alternative to the classical cluster analysis approach has been demonstrated by ter Braak *et al.* (2003), although a number of problems still exist with the methods. As yet, only Witte *et al.* (2007) have applied the method to vegetation data with some success.

Case studies

Application of similarity analysis to a study of vegetation change on highway verges in southeast Scotland (Ross, 1986)

The vegetation of road verges has always been of interest to plant ecologists (Way, 1969). Ross estimated that by the mid-1980s, the extent of motorway verges in Britain was increasing at the rate of 250 km per year. Verges are important as reservoirs and refuges for local species that invade the original landscaping seed mix. Such species might otherwise be under threat from urbanisation, industrialisation, intensive agriculture and commercial forestry.

Surveys of verges have demonstrated the importance of both salt, which is used for deicing, and lead from vehicle exhausts in affecting plant growth. Chow (1970), Davison (1971), Spencer *et al.* (1988) and Spencer and Port (1988) studied such effects in some detail. As would be expected, both sodium and lead have been demonstrated to decrease as distance from the highway increases. Application of herbicides and periodic cutting for safety reasons are also commonly employed management techniques.

Ross surveyed seven sites along the A90 dual carriageway in East Lothian, Scotland, at two time periods, exactly 10 years and 20 years after the original creation and seeding of the verges. The original seed mixture was as recommended by the then Department of the Environment:

Lolium perenne	Perennial rye grass S23	53.6%
Festuca rubra	Red Fescue S59	17.9%
Cynosurus cristatus	Crested dog's tail	10.7%
Poa pratensis	Smooth-stalked meadow grass	8.9%
Trifolium repens	White clover S100	8.9%

The verges were originally sown in 1964 and surveys were carried out in 1974 and 1984. Sampling was stratified according to the verge type:

Site 1:	flat verge
Sites 2 and 5:	smooth, grassed embankment
Site 3:	rocky embankment
Sites 4 and 6:	disturbed embankment, beside lay-bys
Site 7:	rock cutting

All sites had a flat 2–3 m verge adjacent to the highway. At each site, two transects were established at right angles to the road. Percentage cover was recorded using a modified Domin scale in consecutive 1 m × 1 m quadrats along the transects to the nearest field boundary. A total of 65 quadrats were sampled in 1974, and 90 in 1984.

Numerical classification was applied to the data in order to recognise species groupings and habitat types and to compare the plant communities after 10 and 20 years. Similarity analysis using two different sorting strategies (centroid sorting and single linkage) was applied within the program available in the GENSTAT package (Alvey *et al.*, 1980). The data were analysed in both abundance and presence/absence form. The dendrograms for centroid sorting of the 10- and 20-year presence/absence data are presented in Figure 8.16, with details of the major habitat groups in Table 8.5. For both the 10- and 20-year classifications, four distinct groupings were present:

(1) the flat mown verge association (1–4 m from road) (*Lolium perenne*, *Holcus lanatus* and *Trifolium repens*)
(2) the verges under a tree-shrub canopy – *Fagus/Acer* understorey
(3) rock/scree cuttings
(4) grassed embankments – subdivided into two types at the 10-year period and three types at the 20-year period.

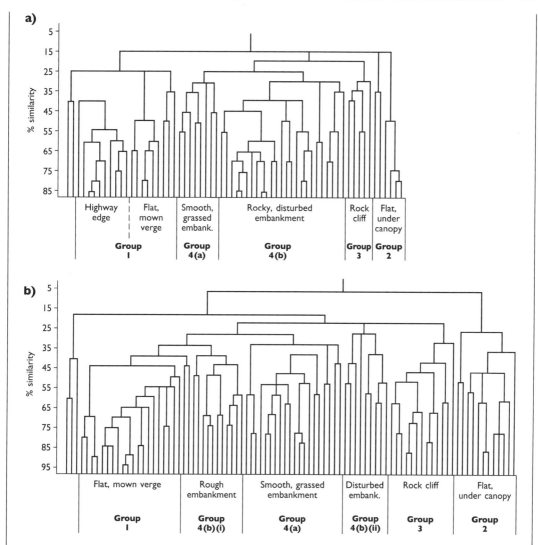

Figure 8.16 Dendrograms from similarity analysis (centroid sorting) of (a) 10-year and (b) 20-year data from road verges in East Lothian, Scotland. Redrawn from Ross (1986), with kind permission of Wiley-Blackwell.

The results demonstrate that the grass species determine all but one of the major groups. In group 2, none of the original seed mix species were present as dominants or frequents at either 10 or 20 years. After 20 years, *Cynosurus cristatus* and *Trifolium repens* were not even associates in any group. *Festuca rubra*, however, became the main dominant in 4 out of the six 1984 groups. Other individual species distributions are discussed in both space and time, followed by an evaluation of the effects of management practices, particularly mowing and the application of herbicides. The analysis showed that the species richness of the verges was higher in 1974 than 1984, reflecting classical successional models, where there is an initial influx of local species followed by competition

Table 8.5 Habitat groups from similarity analysis (centroid sorting) of (A) 10-year and (B) 20-year data on road verges in East Lothian, Scotland (Ross, 1986; reproduced with kind permission of Wiley-Blackwell).

(A) Group No.	No. members, 1974	Habitat	Dominants	Frequents	Associates
				1974 (10-year survey)	
1	16	Mown verge	Lolium perenne Trifolium repens Holcus lanatus	Poa annua	Bromus perennis Plantago major
2	6	Flat under canopy	Aegopodium podagraria Agropyron repens Galium aparine	Ulex europaeus Chamerion ansustifolium	Dactylis glomerata Holcus lanatus
3	5	Rock cliff	Festuca rubra Chamaenerion angustifolium	Cirsium arvense Ulex europaeus	Dactylis glomerata Rubus fruticosus
4(a)	8	Smooth grassed embankment	Festuca rubra Poa pratensis	Holcus lanatus Trifolium repens Tussilago farfara	Dactylis glomerata Trifolium repens Ranunculus acris
4(b)	25	Rocky disturbed embankment	Festuca rubra Holcus lanatus Cirsium arvense	Cynosurus cristatus	Taraxacum officinale

(B) Group No.	No. members, 1984	Habitat	Dominants	Frequents	Associates
				1984 (20-year survey)	
1	21	Mown verge	Lolium perenne Poa pratensis Poa annua Taraxacum officinale	Trifolium repens Festuca rubra Bellis perennis Agrostis stolonifera Dactylis glomerata	Agrostis capillaris Plantago lanceolata Senecio jacobaea Holcus lanatus Pseudoscleropodium purum

(Continued)

Table 8.5   (*Continued*)

(B) Group No.	No. members, 1984	Habitat	Dominants	1984 (20-year survey) Frequents	Associates
2	12	Flat under canopy	*Aegopodium podagraria* *Galium aparine* *Agrostis stolonifera*	*Agrostis capillaris* *Acer pseudoplantanus* *Fagus sylvatica*	*Lolium perenne* *Bromus ramosus*
3	13	Rock cliff	*Festuca rubra* *Chamaenerion angustifolium* *Taraxacum officinale*	*Rumex acetosella* *Dactylis glomerata* *Ulex europaeus*	*Senecio jacobaea* *Trifolium repens*
4(a)	20	Smooth grassed embankment	*Festuca rubra* *Taraxacum officinale* *Rhytidiadelphus squarrosus*	*Lolium perenne* *Agrostis stolonifera* *Senecio jacobaea* *Ulex europaeus* *Pseudoscleropodium purum*	*Poa pratensis* *Holcus lanatus*
4(b) (i)	9	Rough embankment	*Festuca rubra* *Agrostis stolonifera* *Rhytidiadelphus squarrosus* *Eurhynchium striatum* *Pseudoscleropodium purum*	*Dactylis glomerata* *Senecio jacobaea*	*Lophocolea heterophylla* *Lolium perenne* *Poa pratensis*
4(b) (ii)	12	Disturbed embankment	*Festuca rubra* *Agrostis stolonifera* *Bromus mollis*	*Holcus lanatus*	*Lolium perenne* *Taraxicum officinale*

and interaction between them and the sown species and between themselves. Also the grass embankment group appeared to be diversifying in terms of habitat, with three subgroups recognised in 1984 compared with two in 1974.

This research illustrates not only the use of similarity analysis as a means of finding pattern in a set of data. It represents a typical research project in plant ecology, where an interesting new habitat is taken and various ideas in relation to the ecology and management of that habitat are formulated. However, a necessary starting point is the recognition and definition of community types. The work is also interesting in demonstrating how repeat sets of data, in this case at 10-year intervals, can be analysed using numerical classification to enable comparison of species composition of the same quadrats at different points in time.

Finally, the work demonstrates one of the problems of similarity analyses. The results presented are for centroid sorting of presence/absence data, but the data were analysed using both centroid sorting and single linkage analysis on both presence/absence and quantitative data. This highlights the difficulties of choice in using variations of similarity analysis. More recent research would suggest that group average cluster analysis combined with the Steinhaus (Sørensen/Czekanowski/Bray and Curtis) (dis)similarity coefficient would probably be the best choice of similarity analysis. Nevertheless, the extent to which reanalysis using these methods would alter the interpretation of results remains uncertain.

Similarity analysis using the flexible-β sorting strategy and multi-response permutation procedures (MRPP) in the analysis of pioneer vegetation on glacier forelands in southern Norway (Robbins and Matthews, 2009, 2010)

This case study was introduced in Chapter 1 and the fieldwork strategy in Chapter 3. Vegetation data on pioneer vegetation growing on 43 glacier forelands in Norway were collected. A total of 71 higher plant species were recorded in 42 locations, with one location being removed from the analysis because there was no recordable vegetation on the glacier foreland at that site. As described in Chapter 6, ordination analysis was completed using non-metric multidimensional scaling (NMS), and to complement this, with the aim of examining the variability of the pioneer species data, similarity or cluster analysis was applied using the Steinhaus (Sørensen/Czekanowski/Bray and Curtis) (dis)similarity coefficient with the flexible-β sorting strategy with β set to the recommended value of -0.25. Assessment of the resulting dendrogram and the decision on group structure was based on three further approaches:

(1) The use of multiple response permutation procedures (MRPP) (Mielke and Berry, 2001; McCune and Grace, 2002) to calculate and compare within-group distances with those from random permutations of the data.
(2) Indicator species analysis (ISA) (Dufrêne and Legendre, 1997; Legendre and Legendre, 1998), which identifies characterising species for a set of groups, such as those derived from similarity analysis, and computes a measure of how well species are separated between groups. Statistical significance of groups is determined by a Monte Carlo randomisation test (as described in Chapter 6 for canonical correspondence analysis (CCA) ordination) with the null hypothesis that the species have no value as indicators.

(3) Overlaying of cluster group centroids on the NMS ordination diagram (Figure 1.3) with surrounding confidence ellipses at ± 2 standard deviations from the mean, enclosing approximately 95% of samples within each cluster to show the degree of overlap of groups.

As the overlay diagram for four groups in Figure 1.3 demonstrates, there were no clear clusters in the data that would enable a set of classification groups to be derived, and the ellipses overlap quite significantly. Thus the data may be considered to be distributed more as a continuum, and any partitioning would be arbitrary. The MRPP results indicated no differences between groups, and in any case, although helping to confirm lack of group structure, the analysis was not strictly valid, since the MRPP analysis was based on the same data used in the similarity analysis to produce the classification groups.

ISA indicated some significant species differences at the two-group level, but again a clear statistical separation was not indicated. The interpretation of these results concluded that although clear separation was not achieved and the vegetation was distributed as a continuum, two overlapping pioneer subcommunities could be identified. The first was characterised by *Deschampsia alpina* (alpine hair grass) – *Oxyria digyna* (mountain sorrel) with a wide altitudinal and geographical distribution, and a second by *Saxifraga cespitosa* (tufted saxifrage) – *Trisetum spicatum* (spike trisetum) occurring only in the Jotunheimen area above the tree-line. Robbins and Matthews concluded that the glacial forelands of higher altitudes are characterised by species that start to colonise at an earlier stage of succession than at lower altitudes, and that there is a greater separation of species within the pioneer stage up to an altitude of around 1600 m. These findings were taken further in their second paper (2010). Rate of successional advancement was shown to change with altitude: below about 1000 m, in the sub-alpine belt, the transition from pioneer vegetation to birch woodland occurs within 70 years. Above about 1600 m in the high-alpine belt, herbaceous pioneer vegetation can persist indefinitely and little successional change occurs. At intermediate altitudes, the dwarf-shrub and snowbed vegetation types of the low- and mid-alpine belts develop within c. 250 years (Figure 8.17).

The floristic classification of British hedgerows (French and Cummins, 2001)

Hedgerows comprise a significant element of the British agricultural landscape and are important for livestock control and shelter, as boundaries, and for erosion control. Ecologically, they are extremely valuable as wildlife refuges and corridors and as reservoirs of species diversity within the less ecologically-rich agricultural landscape (McCollin *et al.*, 2000). Often hedgerows can be dated back over several centuries (Hooper, 1970; Rackham, 1977). Many hedgerows have been lost to agricultural intensification; for example, Barr *et al.* (1991) estimated a net loss of around 20% in England between 1984 and 1991, and in Wales of 25%. In this period, complete removal of hedgerows accounted for 9500 km per year, but significant additional losses were attributable to lack of management. Interestingly, between 1990 and 1993, the rate of new planting at 4400 km per year exceeded the rate of removal (3600 km/year), reflecting some government initiatives, but unfortunately there was still a net decrease in hedgerow length of 18,000 km per year, once again often due to poor or non-existent management, with

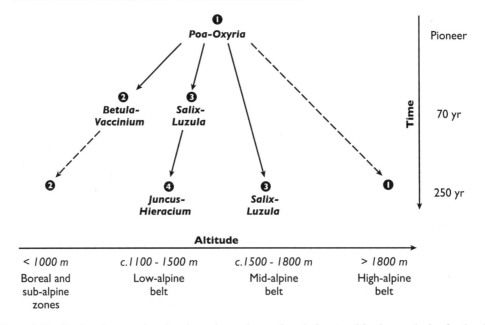

Figure 8.17 Regional successional trajectories and rates in relation to altitude on glacier forelands in south-central Norway. Full arrows indicate directional change; dashed arrows indicate no further directional change. Numerical values for altitude and time are approximate, and low and mid-alpine trajectories are affected by small-scale topographic variations and associated habitat variations. Redrawn with kind permission of the Institute of Arctic and Alpine Research, Boulder Colorado.

hedges being re-described as lines of trees or 'gappy' shrubs. Thus, rather than complete clearance of hedgerows, lack of adequate management is now the main factor behind the decline in the quality of British hedgerows.

French and Cummins (2001) focused on surveys of hedgerows completed throughout Britain in 1978 and 1990 in order to provide information on a national overview of hedgerow quality. As a starting point for the provision of data for conservation planning, the data enabled the researchers to classify and identify the main types of hedgerow.

Data were collected across Britain (but not Northern Ireland), in 1 km squares of the Ordnance Survey national grid and stratified by the 32 land classes derived from the Countryside Surveys of Britain (Bunce *et al.*, 1996). In 1978, 256 squares were sampled, which were resampled in 1990, along with 252 additional squares selected according to the principles of the Countryside Survey outlined in Barr *et al.* (1993). Sampling plots were 10 m × 1 m lengths of hedgerow with all species rooted in the quadrat being recorded, using a visual estimate of percentage cover (to the nearest 5%) for all species with over 5% cover. Interestingly, two observers were used simultaneously as a check on data quality.

In both 1978 and 1990, two hedgerow plots were measured in every square that contained at least 30 m of hedge. In squares with 10–30 m of hedge, only one sample was taken. The selection process was random with controls to ensure spatial separation. There had to be at least 1 m of hedge between the centre of the hedge and any adjacent feature such as a ditch or road. In total, for both years, 1213 plots were recorded.

To meet the aim of categorising the main types of hedgerow in terms of species composition, the data were subjected to numerical classification using Two-Way Indicator Species Analysis (TWINSPAN) (Hill, 1979b). Only data for vascular species, including pteridophytes, were included. Two data-sets were compiled, one for woody species and one for herbaceous species, with some in both categories being grouped into aggregates, giving 55 woody 'species' and 185 herbaceous 'species'. Woody data were classified first, using the percentage cover data, while with the herbaceous data, only presence/absence data were used. The TWINSPAN dendrograms for both classifications are presented in Figure 8.18, and the descriptions of the main higher level classes for the woody components of hedges in Table 8.6 and herbaceous hedge-bottoms in Table 8.7.

Table 8.6 Highest-level hedge classes from TWINSPAN analysis of woody data (French and Cummins, 2001). Reproduced with kind permission of Wiley-Blackwell.

Class	Name	Number of plots	Description
H1	Planted exotics	9	Mostly exotic conifers or *Ligustrum ovalifolium.*
H2	Wild privet	5	*Ligustrum vulgare* always present with at least 20% cover. *Crataegus* spp. and *Acer pseudoplatanus* can reach substantial proportions.
H3	Beech	19	*Fagus sylvatica* dominant. *Crataegus* spp. up to 40% and *Buxus sempervirens* may be present. Other species only incidental.
H4a	Hawthorn-dominant	552	*Crataegus* spp. dominant. Many incidental species but only *Rosa* species, *Sambucus nigra* and *Prunus spinosa* in more than 10% of plots. (NB: by default TWINSPAN includes here a small group of monotypic hedges lacking indicators for any other class).
H4b	Rich-hawthorn	61	Similar to H4a but *Crataegus* species are not always dominant and usually with more woody species present, of which some may have substantial cover.
H4c	Elder-hawthorn	40	*Sambucus nigra* dominant or co-dominant. *Crataegus* or other thorny species frequently present but never dominant.
H5a	Willow	6	*Salix* species generally dominant, *Rosa* species the main subdominant.
H5b	Mixed-hazel	157	*Corylus avellana* and *Crataegus* species present in most plots, but usually no clear dominant. Only H4a, with more than three times the number of plots, contains more incidental species.
H6	Blackthorn	270	*Prunus spinosa* the main species not always dominant. *Crataegus* the main secondary species. Moderate number of incidental species.
H7	Elm	49	*Ulmus* species dominant or co-dominant. Few other species.
H8	Gorse	8	*Ulex europaeus* dominant species. Generally low cover, even of *Ulex.*

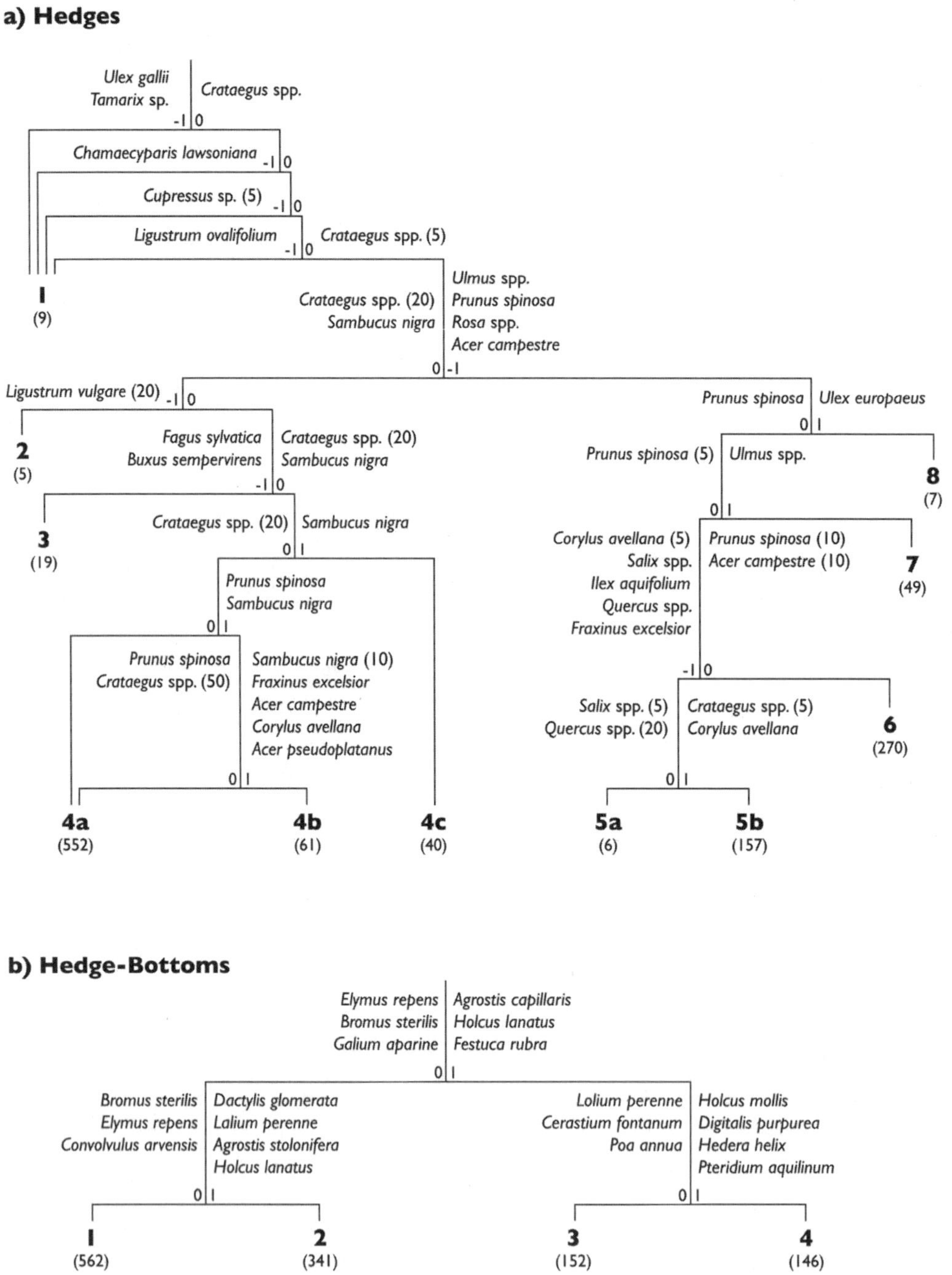

a) Hedges

b) Hedge-Bottoms

Figure 8.18 TWINSPAN dendrograms for classification of British hedgerows. (a) Woody hedges; (b) herbaceous hedge-bottoms. Numbers in brackets after species names are the minimum cover (%) required for the species to be counted when scoring. Redrawn from French and Cummins (2001), with kind permission of Wiley-Blackwell.

Table 8.7   Highest-level hedge-bottom classes from TWINSPAN analysis of herbaceous data (French and Cummins, 2001). Reproduced with kind permission of Wiley-Blackwell.

Class	Name	Number of plots	Description
HB1	Intensive arable	562	Communities of mainly lowland arable areas, subject to intensive management, particularly cultivation and fertiliser input.
HB2	Arable rotational	341	Similar to HB1 but with species more typical of less intensively managed agricultural ground, including rotational systems. This and HB1 both include communities of formerly managed or cultivated ground, now abandoned, neglected or 'derelict'.
HB3	Grassland	162	Completely dominated by grassland vegetation including temporary and permanent pasture and ley silage crops.
HB4	Woodland	146	Communities of woodlands or similar environments.

This classification of hedgerow types was extremely important for planning of conservation and rural policy initiatives. Subsequent testing of the detailed keys in the field showed that they were easy to use and could be used in rapid environmental assessments as input to planning applications and environmental impact assessment. The other important features of this research were, firstly, the size of the data-set. With over 1200 plots, data reduction through numerical classification was essential and TWINSPAN successfully partitioned the samples into interpretable groups. Secondly, the methods of field survey represent another good example of how field survey needs to be adaptable to cater for more unusual habitat types.

Two-Way Indicator Species Analysis (TWINSPAN), similarity analysis and the National Vegetation Classification (NVC) of the British Isles, and the TABLEFIT and MATCH computer programs (Rodwell, 1991, 1992a,b, 1995a, 2000, 2006; Hill, 1991; Malloch, 1998)

The National Vegetation Classification (NVC) of Great Britain was introduced in case studies in Chapters 1 and 3. In total, some 35,000 samples had been collected by the end of the project, which necessarily required computerised methods for numerical classification and tabular rearrangement. In the early days of the project, back in the 1970s and 1980s, methods of similarity analysis and information analysis were used to assist with sample sorting.

Subsequently, TWINSPAN (Hill, 1979b) was employed extensively and integrated with the *VESPAN III* (Malloch, 1999) and *TURBOVEG* (Hennekens, 1996; Hennekens and Schaminée 2001) computer packages, which facilitated the editing, input and output of sorted tables and the summarisation of tables and results. The analysis involved many successive rounds of analysis and re-analysis of data, corresponding to a process of successive refinement based on informed, yet inevitably still subjective, decision-making on the basis of the results of each round (Rodwell, 2006).

In total, 286 communities were identified in the five volumes of the British National Vegetation Classification (Plate 8.1). They are grouped into the major categories shown in Table 8.8. For each community type, a summary floristic table was produced. One

Table 8.8 The 286 plant communities of the British National Vegetation Classification (NVC). Reproduced with kind permission of the Joint Nature Conservation Committee.

Prefix	Category	Number of communities
W	Woodland and scrub communities (19 classed as woodland, four as scrub and two as 'underscrub')	25
M	Mires	38
H	Heaths	22
MG	Mesotrophic grasslands	13
CG	Calcicolous grasslands	14
U	Calcifugous grasslands and montane communities	21
A	Aquatic communities	24
S	Swamps and tall-herb fens	28
SM	Salt-marsh communities	28
SD	Shingle, strandline and sand-dune communities (one shingle, two strandline and 16 sand-dune communities)	19
MC	Maritime cliff communities	12
OV	Vegetation of open habitats	42
Total		286

example for community type W23 *Ulex europaeus-Rubus fruticosus* scrub is shown in Table 8.9. Three subcommunities are identified in the first three numerical columns of the table: a, *Anthoxanthum odoratum* (sweet vernal grass) subcommunity; b, *Rumex acetosella* (sheep's sorrel) subcommunity; and c, *Teucrium scorodonia* (wood sage) subcommunity with information for the total community in the last column. Only vascular plants, bryophytes and macrolichens occurring with a frequency of 5% or more in any one of the communities or subcommunities is listed. The Roman numerals in the table refer to the frequency of each species on the following scale:

Frequency class		
I	= 1–20% (1 sample in 5)	scarce
II	= 21–40%	occasional
III	= 41–60%	frequent
IV	= 61–80%	constant
V	= 81–100%	constant

Next to the frequency class are values for the range of typical abundances for each species in a community using the Domin scale listed in the NVC case study at the end of Chapter 3. Species with frequencies of V or IV are referred to as constants and are placed towards the top of the table, depending on the nature of any subcommunities. Towards the bottom of the table, species with frequencies of just I or II are referred to as general associates or companions (Rodwell, 2006). Species in the middle of the table are often important in particular subcommunities. In each of the five volumes of *British Plant Communities*, detailed information is provided on the physiognomy and structure of each community and subcommunity, lists of synonyms from other studies and classifications, such as various European vegetation classifications, a general description of typical environmental characteristics, such as climate and soils and local spatial variations, the successional status of the community/subcommunities, and finally the floristic attributes of the community and its position in relation to other NVC types and subcommunities.

Table 8.9 The NVC floristic table for group W23 *Ulex europaeus-Rubus fruticosus* scrub. Reproduced from Rodwell (1991), with kind permission of Cambridge University Press.

	a	b	c	23
Ulex europaeus	V (5–9)	V (2–9)	V (5–10)	V (2–10)
Rubus fruticosus agg.	V (1–6)	IV (1–6)	V (3–9)	V (l–9)
Cytisus scoparius	II (1–10)	II (1–6)		II (1–10)
Rubus idaeus	II (1–3)	I (2–3)		II (1–3)
Agrostis capillaris	V (1–6)	V (2–8)	II (3–4)	IV (1–8)
Holcus lanatus	IV (1–4)	IV (1–3)	I (3)	III (l–4)
Galium saxatile	III (1–3)	III (1–3)		III (1–3)
Rhytidiadelphus squarrosus	III (1–5)	III (1–5)		III (1–5)
Holcus mollis	II (2–8)	II (1–6)		II (1–8)
Eurhynchium praelongum	II (1–7)	II (1–4)		II (1–7)
Cerastium fontanum	II (1)	II (1–2)		I (1–2)
Viola riviniana	II (1–6)	II (3)		I (1–6)
Pseudoscleropodium purum	II (1–5)	II (1–4)		I (1–5)
Anthoxanthum odoratum	IV (1–4)	I (1–5)	I (2–3)	III (1–5)
Potentilla erecta	III (1–2)	I (2–4)		II (1–4)
Poa pratensis	III (2–4)	I (2–4)		II (2–4)
Deschampsia flexuosa	II (2–5)	I (4)		I (2–5)
Calluna vulgaris	II (2)	I (1)		I (1–2)
Rumex acetosella	I (1)	IV (1–3)		II (1–3)
Hypochoeris radicata		IV (1–3)	I (1)	II (1–3)
Senecio jacobaea		III (1–2)		I (1–2)
Plantago lanceolata		III (1–3)		I (1–3)
Crepis capillaris		II (1–2)		I (1–2)
Jasione montana		II (1–2)		I (1–2)
Aira praecox		I (1)		I (1)
Teucrium scorodonia		I (3–5)	V (2–5)	II (2–5)
Hedera helix			II (2–6)	I (2–6)
Brachypodium sylvaticum			II (2–4)	I (2–4)
Pteridium aquilinum	III (2–5)	III (1 –7)	III (1–4)	III (1–7)
Festuca rubra	III (1–4)	II (2–4)	II (3–4)	II (1–4)
Dactylis glomerata	I (1)	II (1–3)	II (2–4)	II (1–4)
Rumex acetosa	II (1–3)	I (1)	I (3)	I (1–3)
Achillea millefolium	I (1)	I (1)	I (2)	I (1–2)
Silene dioica	I (1–3)	I (1–2)	I (3–4)	I (1–4)
Digitalis purpurea	I (1)	I (1)	I (1–3)	I (1–3)
Campanula rotundifolia	I (1–2)	I (1)		I (1–2)
Rhytidiadelphus triquetrus	I (1–8)	I (1)		I (1–8)
Veronica officinalis	I (1)	I (1–2)		I (1–2)
Veronica chamaedrys	1 (1–4)	I (3)		I (3–4)
Arrhenatherum elatius		I (1–2)	I (3–4)	I (1–4)
Number of samples	14	9	9	32
Number of species/sample	18 (15–23)	23 (17–30)	9 (4–14)	16 (4–30)
Vegetation height (cm)	177 (30–250)	140 (80–250)	108 (60–220)	143 (30–250)
Vegetation cover (%)	98 (92–100)	97 (90–100)	100	99 (90–100}
Altitude (m)	34 (6–50)	62 (32–105)	33 (3–60)	41 (3–105)
Slope (*)	8 (0–16)	10 (3–16)	19 (4–45)	13 (0–45)

a *Anthoxanthum odoratum* sub-community

b *Rumex acetosella* sub-community

c *Teucrium scorodonia* sub-community

23 *Ulex europaeus-Rubus fruticosus* scrub (total)

Comparing new samples with the NVC

Two computer programs, *TABLEFIT* (Hill, 1991) and *MATCH* (Malloch, 1998), were written to enable comparison of new samples with the tables produced to characterise each of the NVC plant communities. They both make statistical comparisons between samples and the floristic tables in *British Plant Communities*, using the similarity coefficients described in Chapter 4. *TABLEFIT* is better suited than *MATCH* to the comparison of single samples, and uses quantitative data. *TABLEFIT* lists the top five most similar community types, each of which comes with a 'goodness-of-fit' rating, while *MATCH* lists the top 10.

As an example, of how the NVC has been used, reference can be made to the *Lobelia urens* L. (heath lobelia) case study in Chapter 1. Once the 95 quadrats of this survey had been collected, they were classified into the seven groups of Table 1.2, using TWINSPAN. The quadrats in each group were then compared with standard reference tables for the 286 community types using *TABLEFIT* and the predominant best-fit community types for the samples within each group were taken to characterise that TWINSPAN group with its most similar NVC category. As always, the ecological knowledge of the researcher is paramount in making the interpretations. The best-fit *TABLEFIT* groups for each of the seven TWINSPAN groups are shown in Table 1.2.

The British NVC is extremely important as an example of contemporary phytosociology. It now acts as a yardstick against which virtually all vegetation in Britain is compared and is used in a large amount of description for biological conservation planning and management. As a major exercise in phytosociology, it demonstrates important links between the original European continental approaches to phytosociology described in Chapter 7, and the numerical approaches to classification described in this chapter. The ongoing development of the National Vegetation Classification for the US, presented as a case study in Chapter 7, highlights the continuing relevance and importance of phytosociology in vegetation description and analysis.

Chapter 9

Computer software for the analysis of vegetation and environmental/biotic data

A very wide choice of computer software is now available for the analysis of vegetation and associated environmental/biotic data. There is still a considerable amount of free software that can be obtained over the Internet, but an important consideration for a majority of users is 'user friendliness', and most of the more 'user-friendly' computer packages are only available at cost, although often there are discounts available for students on individual licenses. Many packages can be purchased by individuals for a one-off charge, although there is usually a further one-off charge for any future upgrades. Other packages require the purchase of a time-limited product-key, which tends to make them rather more expensive. A major factor in favour of many commercially available packages is the incorporation of an online help system that reduces the need for book-style manuals and also enables instant access to assistance while actually running an analysis.

Virtually all software has been designed to run under the various versions of the Microsoft Windows® operating system, and users should check individual websites to determine whether Apple versions are also available. All packages are now designed to work with Microsoft Excel® spreadsheets, which makes for ease of data preparation and checking.

Although the many excellent packages described below are more than adequate for many users, particularly undergraduate students and newcomers to vegetation description and data analysis, all serious researchers, academics and postgraduates should now consider learning the R programming language and using the R environment to perform their data analyses. The wide range of techniques described in this book and explained in detail in Legendre and Legendre (1998) have recently been represented very effectively within an R format by Borcard et al. (2011). Information on R and related software is presented later in this chapter.

SOFTWARE FOR BASIC STATISTICAL ANALYSIS

The *MINITAB* statistical package

This is only available under license and at very considerable cost. However, it is an excellent package that is easy to use and has a very good online help facility. Many universities and colleges in higher

Vegetation Description and Data Analysis: A Practical Approach, Second Edition. Martin Kent.
© 2012 John Wiley & Sons, Ltd. Published 2012 by John Wiley & Sons, Ltd.

education will have a site license, renewable annually, which makes it accessible to lecturers, researchers and students. Unfortunately an individual license is very expensive, even though there are some concessions for academic users. All the methods for both exploratory data analysis (EDA) and classical (confirmatory) data analysis (CDA) described in Chapter 5 of this book are included, as well as several multivariate methods such as principal component analysis (PCA) and cluster analysis (similarity analysis). The methods for EDA are particularly good, and there are also very effective methods for summarising, transforming, editing and manipulating data. Graphics are also of high quality. Information on the program is available at: http://www.minitab.com/en-US/products/minitab/default.aspx?WT.srch=1&WT.mc_id=SE004815. There is the opportunity to try the software for free over a limited time period.

Given its popularity and widespread adoption within many institutions of higher education and research, a number of useful introductory textbooks of statistics incorporate the use of *MINITAB*. Good examples are Dytham (2010), Eddison (2000), and Wheeler *et al.* (2004).

QED Statistics (Pisces Conservation Ltd, 2008e)

This very reasonably priced package again covers virtually all of the basic statistical methods mentioned in Chapter 5 and comes with a clearly written guide, demonstration data-sets and a dedicated website at http://www.qedstatistics.com/. A considerable advantage is that it is available for a one-off payment, rather than requiring purchase of a time-limited user-key like the other competing packages in this section.

Statistical Package for the Social Sciences (SPSS)

Although clearly aimed at those involved in social science research, the *SPSS* statistics software incorporates virtually every form of analysis included on *MINITAB* and a great deal more in terms of multivariate analysis, although many more specialist ecological applications are absent. The reason for including it here is that many institutions of higher education and research will have a site license. Individual licenses are very expensive and since the software was taken over by IBM, the software has become incorporated into a much more business-orientated framework that is much less user-friendly for ecologists. The software nevertheless remains very powerful for many standard types of statistical analysis and particularly EDA. There is a good help system and graphics. Information on the program is available at http://www.spss.com/uk/software/statistics/.

Stats Direct Software

Stats Direct is another package that covers virtually the whole range of basic statistics. The main reason for recommending it is its comparative cheapness for individual licenses, especially for students, when compared with *MINITAB* and *SPSS,* and the license key is available for three years rather than just one. Information on the package is available at: http://www.statsdirect.com/.

SOFTWARE FOR THE ANALYSIS OF ECOLOGICAL DATA

PC-ORD (MjM Software; McCune and Mefford, 1999, 2010)

PC-ORD is recommended strongly for many undergraduate and postgraduate students and particularly for those attempting to use multivariate analysis for the analysis of vegetation data for the first time. This spreadsheet-based software (Excel® or Lotus 1-2-3®), now in its 6th version (2010), includes methods for virtually all the methods of multivariate analysis covered in this text. In addition to utilities for editing and transforming data and managing files, *PC-ORD* offers many ordination and classification techniques including non-metric multi-dimensional scaling (NMS), canonical correspondence analysis (CCA), detrended correspondence analysis (DCA), correspondence analysis/reciprocal averaging (CA/RA), principal component analysis (PCA), redundancy analysis (RDA), principal coordinates analysis (PCoA), Bray-Curtis ordination, similarity analysis and two-way clustering, two-way indicator species analysis (TWINSPAN), indicator species analysis, Mantel tests and partial Mantel tests, and multiple response permutation procedures (MRPP). Also included are programs for the computation of diversity indices (Simpson and Shannon). Graphics are particularly strong, with excellent facilities for the production of ordination overlays (quantitative, symbol-coding, colour-coding, grid, joint plot, biplot, successional vector/trajectory analysis and 3D ordination graphics) and dendrograms from classification and cluster analysis. Two particularly important features are the ability to produce publication-quality graphics and the availability of a 3-level autopilot mode for non-metric multidimensional scaling (NMS). The latter is a particularly strong asset. The program can handle very large data-sets.

Most important of all, however, is the availability of a step-by-step introductory guide for students (Peck, 2010) and a comprehensive manual (McCune and Grace, 2002), linked to an online help system within the software itself. The program has evolved and has been refined and optimised over a 30-year period. Given its range of applications and excellent supporting materials, it is very reasonably priced and there are generous discounts for student users. Purchasers can also obtain the *Hyperniche* multiplicative habitat modelling package for non-parametric regression (McCune, 2004, 2006; McCune and Mefford, 2009) at a significant discount.

Further information on *PC-ORD* is available at http://home.centurytel.net/~mjm/pcordwin. htm.

Gilliam and Saunders (2003) have provided a comparative review of an earlier version of *PC-ORD* (v. 4) with *CANOCO* for Windows 4.5 and *SYN-TAX*, and Grandin (2006) evaluated *PC-ORD* (v. 5).

Software from Pisces Conservation Ltd (*CAP* version 4.0 and *ECOM* version 2.0; *SDR* version 4.0 and Fuzzy Grouping version 2.0)

Pisces Conservation Ltd offer a range of software applications for ecologists. The Community Analysis Package *CAP* (v. 4.0) (Pisces Conservation, 2008b) includes programs for most methods of ordination and classification (NMS, DCA, PCA, similarity analysis and TWINSPAN). It also includes some other applications, such as ANOSIM (analysis of similarities) and SIMPER (discriminating between sample groups) and multiple discriminant analysis. A second package *ECOM* (v. 2.0) (Pisces Conservation, 2008a) is centred on multiple regression and performs CCA, RDA and multiple/stepwise regression. There is a very good handbook (Henderson and Seaby, 2008) that provides an excellent simplified introduction for students and new researchers, and examples are taken from across the whole of biology and (very interestingly!) archaeology.

A third package is focused on Species Diversity and Richness (*SDR*) (Pisces Conservation, 2008c) with a wide range of diversity indices and approaches to visualising diversity and species richness characteristics of samples. A fourth package enables users to explore fuzzy grouping or classification (*Fuzzy Grouping* v. 2) (Pisces Conservation, 2008d). Packages are available for a one-off charge with an additional charge for upgrades. The separation of programs into different packages may be seen as an advantage by some users. Again, there are excellent self-help manuals available. The Pisces Conservation Ltd website can be found at http://www.pisces-conservation.com/.

Primer-E (version 6) and *PERMANOVA*+ (Clarke and Warwick, 2001; Clarke and Gorley, 2006)

The *Primer-E* package is another extremely wide-ranging and well-supported set of software that has been developed by K. Robert Clarke and his colleagues at the Plymouth Marine Science Laboratory, UK, and which has been extensively used by marine ecologists. Many of the applications described in this book are available, and an important point is that analyses are based primarily on NMS as a method of ordination and on similarity analysis for numerical classification. In addition, Clarke has been responsible for the development of the ANOSIM (analysis of similarities) and SIMPER (discriminating between sample groups) software that has now also been incorporated into a number of other packages.

The greatest strength of the *Primer-E* package is its graphics, which are some of the best available in the programs described in this section. Some vegetation scientists may find the focus on NMS and similarity analysis limiting, but Clarke is a very strong supporter of these methods, and as described in Chapters 6 and 8, the consensus of many researchers is now that these are the most effective and least distorting methods for ordination and classification. An additional add-on package to *Primer-E* called *PERMANOVA*+ is now available, which enables the analysis of multivariate (or univariate) data within more complex sampling structures, experimental designs and models, using various forms of analysis of variance, and more importantly, adds other ordination methods that are considered to reduce distortion effects in data analysis – principal coordinates analysis (PCoA), distance-based redundancy analysis (dbRDA) and canonical analysis of principal coordinates (CAP) with distance-based canonical correlation.

Further information on *Primer-E* and *PERMANOVA*+ are available at http://www.primer-e.com/.

Multi-Variate Statistical Package *MVSP* (version 3.0) (Kovach Computing Services)

MVSP is an easy-to-use set of programs that performs a range of the multivariate numerical analyses used by ecologists, including principal component analysis (PCA), principal coordinates analysis (PCoA), correspondence/detrended correspondence analyses (RA/CA; DCA) and canonical correspondence analysis (CCA). Diversity indices (Simpson, Shannon and Brillouin) can be calculated. Similarity and cluster analysis are also available in both single and two-way forms, with 23 different distance and similarity measures and seven clustering strategies. There is a wide range of clear graphical output and the software has good facilities for spreadsheet input/output in varying formats, editing and data transformation. The main advantage of *MVSP* is that it is inexpensive when compared with some other similar programs.

Program details are available at http://www.kovcomp.co.uk/mvsp/mvspwbro.html.

MULVA-5

MULVA-5 was written by Otto Wildi and László Orlóci and is now distributed as freeware from the Swiss Federal Research Institute WSL, Switzerland, and is another package that has been progressively developed and upgraded over many years. The software comprises a suite of programs and subprograms for most methods of ordination (PCA, PCoA, CA/RA, DCA, NMS) and classification using similarity analysis, together with extensive editing and data manipulation facilities. The manual for the program was originally published as Wildi and Orlóci (1996) but has recently been augmented by a new text by Wildi (2010).

SYN-TAX (version 5.2)

The *SYN-TAX* software was developed by János Podani from the Department of Plant Taxonomy and Ecology, Eotvos University, Budapest, Hungary (Podani, 2000, 2005), and contains a suite of programs for ordination (PCA, PCoA, CA/RA, DCA, CCA, RDA, NMS and CVA) and classification (similarity analysis, k-means and fuzzy clustering). Programs are also available for dendrogram comparison. There is a printed manual and an accompanying text by Podani (2000). The program is comparatively expensive when compared with others in this section, although there are significant reductions for students.

Further information is available at http://www.exetersoftware.com/cat/syntax/syntax.html.

CANOCO for Windows 4.5

The *CANOCO* package, developed by Cajo J.F. ter Braak and Petr Šmilauer, was originally primarily of interest to ecologists because it was the software that first brought canonical correspondence analysis (CCA) and variants to vegetation scientists. The program has been progressively refined and now contains a wide range or ordination methods, including PCA, PCoA, RDA, dbRDA, CA/RA, DCA, CCA and the detrended version of CCA-DCCA, as well as canonical variates analysis (CVA). Permutation tests are widely used for significance testing where methods are used deductively. The greatest strengths of the *CANOCO* package lie in its facilities for partial ordination analyses of both indirect methods (PCA, CA/RA and DCA) and direct/canonical methods (CCA, DCCA, RDA and CVA), and hypothesis testing from experimental designs linked to various forms of analysis of variance (ANOVA). It is unfortunate that non-metric multidimensional scaling (NMS) has never been added, given its popularity. A full range of graphics is available in the accompanying *CANODRAW for Windows* software. There is a manual, and users are also able to refer to the text by ter Braak and Šmilauer (2003) and Lepš and Šmilauer (2003), which contains a range of interesting case studies. The program is suited to more advanced researchers and is probably less user-friendly than others for newcomers to multivariate analysis.

Further information is available at http://www.pri.wur.nl/UK/products/Canoco/.

JUICE (version 7) software for vegetation classification (Tichý, 2002; Tichý and Holt, 2006) and the *TURBOVEG* database program for phytosociological data (Hennekens, 1996; Hennekens and Schaminée, 2001)

The *JUICE* software is a Microsoft Windows® application for the editing, classification and analysis of large phytosociological tables and databases, and represents one of the best applications for the subjective classification of relevé samples (Tichý, 2002; Tichý and Holt, 2006). This software, with a current maximum capacity of 1 million relevés in one table, includes many functions for easy manipulation of table and header data, calculation of interspecific associations, fidelity measures, average Ellenberg indicator values, preparation of synoptic tables, automatic sorting of relevé tables, and export of table data into other applications. Various options include classification using both *COCKTAIL* (Bruelheide, 1997, 2000) and TWINSPAN (Hill, 1979b) methods.

The *COCKTAIL* software closely follows the procedure followed in the traditional Braun-Blanquet approach, but is improved by statistical measuring of fidelity. Species groupings are formed and continuously optimised using tests of interspecific associations within large databases.

The combination of different species groups using logical operators allows the user to re-define commonly known vegetation units in a formal way. Further detail and information is available in Bruelheide (1997, 2000) and Bruelheide and Jandt (1997). The process of defining phytosociological groups and vegetation units can include subjective decisions, but all these decisions are arrived at in a statistically sound manner.

JUICE is optimised for use in association with the *TURBOVEG* program, which is the most widely used database program for storing phytosociological data in Europe (Hennekens, 1996; Hennekens and Schaminée, 2001). *TURBOVEG* has been used as the standard database package within the European Vegetation Survey (Mucina, 1997a,b; Ewald, 2003). The latest version 7.0 of *JUICE* has well-developed interfaces to other vegetation analysis software, such as *CANOCO*, *PC-ORD*, *MULVA* and *SYN-TAX*, as well as to *R* script, and also has a version of TWINSPAN that enables the full hierarchy of clusters to be displayed.

A freeware version can be downloaded from the website of the Department of Botany, Masaryk University, at http://www.sci.muni.cz/botany/juice.htm.

VegAna (*Ginkgo, Quercus, Fagus* and *Yucca*) software

The *VegAna* (Vegetation Analysis) package contains four programs named *Ginkgo, Quercus, Fagus* and *Yucca*. *Ginkgo* is a suite of programs for multivariate analysis, oriented mainly towards ordination and classification of ecological data. *Quercus* performs phytosociological analysis and sorting of relevé tables, while *Fagus* is a floristic citation editor, handling data from field surveys, bibliographic sources or collections. *Yucca* is a cartographic tool; it allows plotting of distributions of taxa or syntaxa. *VegAna* has been developed by the Department of Vegetal Biology at the University of Barcelona under the direction of Xavier Font I Castell. The multivariate *Ginkgo* module has been the responsibility of Francesc Oliva I Cuyàs, together with programmers Miquel De Cáceres and Richard Garcia.

Ginkgo includes a number of ordination programs – PCA, PCoA, CA/RA and NMS – and various forms of similarity analysis for numerical classification. There are some good graphics, particularly three-dimensional ordination plots. Although it offers a more restricted choice of methods than other packages and its manual could be improved, its main advantage is that it can be obtained as freeware from http://java.sun.com/j2se/1.4.2/download.html. A review of the package is published in Bouxin (2005).

TWINSPAN program (Two-Way Indicator Species Analysis – Hill, 1979b) and the British National Vegetation Classification *TABLEFIT* program (Hill, 1991)

For those who may wish to use it, a freeware version of this program for numerical classification of vegetation data is available at: http://www.ceh.ac.uk/products/software/CEHSoftware-DECORANATWINSPAN.htm. Also the *TABLEFIT* program for the British National Vegetation Classification can be downloaded free from http://www.ceh.ac.uk/products/software/tablefit/download.asp.

The R software and related packages

The increasing importance of R and its application to statistical data analysis, graphical presentation and modelling in vegetation science was introduced at the start of this chapter. R is a high-level programming language that is available as freeware from a site called CRAN (Comprehensive R Archive Network) and which can be downloaded from http://cran.r-project.org/.

Petchey *et al.* (2009) have presented clearly the arguments in favour of ecologists moving to R for all statistical and particularly multivariate data analysis. Various textbooks are now available that enable students and researchers to learn to apply the R language within an ecological and biological context (Crawley, 2005; Braun and Murdoch, 2007; Zuur *et al.*, 2009; Beckerman and Petchey, 2011). For more sophisticated tuition, there are also the books by Crawley (2007) and Stevens (2009).

However, the most important text for more advanced vegetation scientists wishing to use R for data analysis is the book by Borcard *et al.* (2011), which takes all the methods described in this text from exploratory data analysis up to the multivariate methods of ordination and classification and demonstrates how they can be programmed in the R language with appropriate case studies and examples.

The value of R to vegetation scientists has also been greatly increased through the availability of user-created packages, which have been written for many specialised statistical techniques and particularly multivariate analysis and associated graphics. These packages have been developed primarily in R, but sometimes also in Java, C and Fortran. A core set of packages are included with the installation of R, and across all scientific and engineering applications, there were more than 4300 of these (March 2011) available at the Comprehensive R Archive Network (CRAN). All packages are available as freeware. Manuals are available as .pdf files, and there are user support groups and an R journal describing recent developments and updates and new contributions and packages.

Within the CRAN, there are various specialised packages that are relevant to the analysis of vegetation data. These can be downloaded from a large number of 'mirror' sites around the world that can be seen at http://cran.r-project.org/ and the complete list of packages is listed on the same website.

The following three packages available under CRAN are specifically for vegetation and environmental/biotic data analysis.

ADE-4 - Analyse de Données destinée d'abord à la manipulation des données Écologiques et Environnementales avec des procédures Exploratoires d'essence Euclidienne (Ecological Data Analysis: Exploratory and Euclidean methods in Environmental Sciences)

This package comes from researchers at the Biometry and Evolutionary Biology Laboratory at the University of Lyon, France: Daniel Chessel, Anne Dufour and Jean Thioulouse. Some standard ordination analyses are offered, such as PCA and CA/RA, but the package also includes a number of differing approaches, including: multiple correspondence analysis (Tenenhaus and Young, 1985); fuzzy correspondence analysis (Chevenet *et al.*, 1994); analysis of a mixture of numeric variables and factors (Hill and Smith, 1976; Kiers, 1994); non-symmetric correspondence analysis (Kroonenberg and Lombardo, 1999); and decentred correspondence analysis (Dolédec *et al.*, 1995).

There are excellent graphical facilities. The programs are nevertheless more suitable for advanced users.

The program is available under the CRAN scheme at http://cran.r-project.org/, but can also be downloaded directly from the University of Lyon at http://pbil.univ-lyon1.fr/ADE-4/ home.php?lang=eng. An early (pre-R) version of the package is also described in Thioulouse *et al.* (1997).

The Vegan *Community Ecology Package*

The *Vegan* package has been authored by Jari Oksanen at the University of Oulu in Finland, together with a number of other colleagues (Oksanen *et al.*, 2006). The *Vegan* software includes most of the ordination methods described in this text, including CA/RA, DCA, CCA, PCA, RDA and NMS in both full and partial analyses with permutation tests. It also offers methods to analyse for diversity and species richness, dissimilarity measures, ANOVA using dissimilarities, ANOSIM, MRPP, Mantel and partial Mantel tests. Data standardisation and transformations are available and there is an excellent graphical interface. There are both printed manuals, available as .pdf files, and an online help system.

The *Vegan* homepage is at http://vegan.r-forge.r-project.org/. An earlier version of the program has been reviewed by Dixon (2003).

LabDSV: Ordination and Multivariate Analysis for Ecology Programs

David Roberts from the University of Montana, USA, is the author of this package with the full title of Laboratory for Dynamic Synthetic Vegephenomenology (LabDSV). The full range of ordination methods are once again programmed here, but perhaps of greater interest are the tutorials in R and ordination/multivariate analysis available in another related CRAN application called *R Labs for Vegetation Ecologists*. As with the other programs in this section, these are not really suitable for beginners but are excellent for those wishing to develop their analytical skills further and branch out into the R environment using freeware. Fuzzy set ordination is a particularly interesting development (Roberts, 1986, 1989, 2009) and this is available here.

The program and tutorials can be downloaded under the CRAN scheme at http://cran.r-project.org/ and a copy of the manual can be accessed at http://cran.r-project.org/web/packages/labdsv/labdsv.pdf.

Brodgar (Version 2.6.4)

Brodgar is an R-based software package written by Alain Zuur and Elena Ieno, Directors of Highland Statistics Ltd based in Scotland, for exploratory data analysis, univariate analysis, multivariate analysis and time series analysis. It is supplied independently of CRAN from the website http://www.brodgar.com/. Like most other packages, it will perform the following ordination analyses: CA/RA, DCA, CCA, PCA, RDA and NMS in both full and partial analyses with permutation tests. It will also compute all methods of numerical classification/cluster analysis and classification and regression trees. Graphics are very good and there is good online help and tutorial advice linked to the text by Zuur *et al.* (2007). Unlike the other R-related packages above, it comes at a price, but there are good discounts for students and those from developing countries.

CONCLUSION

The aim of this chapter has been to present the wide range of software that is now available to vegetation scientists. To the student beginning their studies, this wealth of software and enormous choice is extremely daunting. For beginners, obtaining access to packages such as *PC-ORD* or the Pisces Conservation software is probably as good a starting point as any, because of their user-friendly nature, but they come at a cost. The paradox seems to be that the best freeware now available is mostly R-related. Thus those researchers intending to develop their skills in vegetation data analysis significantly are definitely encouraged to follow the R route, even though this requires considerable time and effort in the learning process.

A final point to note in this respect is the availability of advice from *The Ordination Website* (Palmer, 2011). This is run by Michael Palmer of Oregon State University and offers a large amount of sensible advice in the use and application of all the methods described in this book, as well as evaluations of some software. Users can also sign up to *ORDNEWS Listserv*, which is a discussion group that provides a forum for topics involving the analysis of multivariate and spatial data. The website is found at http://ordination.okstate.edu/.

Chapter 10

Future developments in vegetation science and quantitative plant ecology

As outlined in the preface and first edition, this text represents an attempt to make vegetation description and data analysis more accessible to teachers, researchers and especially students who have an interest in this field. In particular, the aim has been to provide simple but clear explanations of approaches and techniques, keeping the amount of difficult mathematics and statistics to the minimum necessary for understanding. Most certainly, to many mathematical ecologists, this 'grey box' approach will seem unsatisfactory. However, as before, it is necessary to ask them to demonstrate how they would make the subject more attractive to the majority of students and teachers who have a fascination for vegetation, plants and ecology but whose mathematical background is often very limited.

The book has also tried to adopt both an ecological and practical approach, stressing that it is the ecological context in which vegetation description and data analysis is carried out which is most important, rather than just the methods of data collection and analysis themselves. The methods are only a means to an end, and the case studies at the end of each chapter are intended to emphasise this point. The practical approach is also crucial. It is when students identify their own problem, develop and design their own field programme, collect their own data, select and apply appropriate methods of description and data analysis, and make their own interpretations of results, that the subject really comes alive.

As the reference list demonstrates, the literature on this subject is now very substantial, but it is more essential than ever that students read and are familiar with the key literature in any discipline. There is an increasing tendency for students to assume that they only need to look at core texts or the Internet in order to gain understanding of a subject. As the list of common errors in ecological research provided by Belovsky *et al.* (2004), which is presented at the end of Chapter 1, illustrates, inevitably, research foci move in and out of fashion in ecology and there are grave dangers in assuming that only the most recent research papers have anything meaningful to say. Citation statistics for many ecological journals show that they have a comparatively 'long' half-life. Methods for vegetation description and data analysis remain an extremely important sub-area of ecology which, as suggested in the Preface, over the past 20 years, has tended to be neglected in those parts of the world where they were originally developed (Europe, Scandinavia,

North America and Australia) and have become of greater significance in most other regions of the globe.

In the conclusion to the previous edition, criticisms of the inductive approach behind a great deal of vegetation description and data analysis were highlighted and it was emphasised that methods and analysis should more often be seen as a prelude to more detailed deductive research. To some extent, these criticisms have been taken on board by many plant ecologists and vegetation scientists, but they still remain valid, and both students and researchers need to look carefully at and evaluate the whole philosophy and methodology of their subject.

As ever, there are exciting new developments in prospect. Despite Austin's work on the nature of the continuum, tolerance curves and environmental gradients (Austin and Smith, 1989; Austin, 2005), a fully developed model of plant/vegetation response to both the environment and biotic controls and interactions has yet to appear. In terms of methodology, a pluralistic approach to the application of techniques of ordination and classification now applies (von Wehrden *et al.*, 2009). Perhaps one of the most surprising features of the past 20 years has been the comparatively limited further development of both methods of ordination and numerical classification. Most of the methods widely used at the present time were already extensively used when the first edition of this text was published in 1992.

One of the most significant developments for the future almost certainly lies in Bayesian statistics, which, for obvious reasons, have only been given limited acknowledgement here in this introductory text. Bayesian methods have not, as yet, become widely used in vegetation science, partly because of the often inductive nature of the subject, but indications of future possibilities are there (ter Braak *et al.*, 2003; Witte *et al.*, 2007; Gotelli and Ulrich, 2010; Kuhnert *et al.*, 2010).

Another key area in recent years has been the incorporation of spatial analysis and geostatistics into vegetation science. Approaches to the description of spatial patterns in plant community and environmental/biotic data and how they vary at different spatial scales, together with the problems of the quantification and removal of spatial autocorrelation in bivariate and multivariate analysis, represent extremely important topics for the future (Cushman and McGarigal, 2002; Wagner, 2003, 2004; Couteron and Ollier, 2005; Fortin and Dale, 2005; Kent *et al.*, 2006; Schlup and Wagner, 2008; Guenard *et al.*, 2010). The use of partial CCA ordination to remove the spatial component of variability in the case study in Chapter 6 (Cushman and Wallin, 2002) is a good example of this. Recent advances in the combination of geostatistics and ordination are significant, notably the use of redundancy analysis (RDA) with a principal coordinates of neighbour matrices (PCNM) spatial matrix and Moran's eigenvector maps (Laliberté, 2008; Legendre *et al.*, 2005, 2008; Dray *et al.*, 2006, 2008; Pélissier *et al.*, 2008), and represent an important development, although Gilbert and Bennett (2010) show that many problems still remain with this whole approach. These methods are extremely complex and demanding for beginners, but interestingly, Gilbert and Bennett (2010) suggest that, until some of the more difficult statistical issues are resolved, methods for variance partitioning, for example, should only be used within a descriptive exploratory framework. Clearly, these topics represent the research frontier in vegetation science and quantitative plant ecology, although they are extremely daunting for student novices.

A significant part of teaching and education is about simplifying concepts and methods with the purpose of improving understanding. That has indeed been my core aim in writing this book. I hope that once more, many student ecologists and vegetation scientists will find this text rewarding. I certainly have no doubt that the whole subject of vegetation description and data analysis will continue to prosper and flourish in the future.

References

Aho, K., Roberts, D.W. and Weaver, T. (2008) Using geometric and non-geometric internal evaluators to compare eight vegetation classification methods. *Journal of Vegetation Science* **19**, 549–562.

Alexander, R. and Millington, A. (eds) (2000) *Vegetation Mapping: Patch to Planet*, John Wiley and Sons Ltd, Chichester.

Allen, S.E., Grimshaw, H.M. and Rowland, A.P. (1986) Chemical analysis, in *Methods in Plant Ecology* (2nd edn) (eds P.D. Moore and S.E. Chapman), Blackwell Scientific, London, pp. 285–344.

Allen, T.F.H. and Wileyto, E.P. (1983) A hierarchical model for the complexity of plant communities. *Journal of Theoretical Biology* **101**, 529–540.

Allen, T.F.H. and Hoekstra, T.W. (1990) The confusion between scale-defined levels and conventional levels of organization in ecology. *Journal of Vegetation Science* **1**, 5–12.

Alvey, N.G. *et al.* (1980) *GENSTAT: A General Statistical Program*, Lawes Agricultural Trust, Rothampstead Experimental Station, Numerical Algorithms Group Ltd, Oxford.

Anderberg, M.R. (1973) *Cluster Analysis for Applications*, Academic Press, New York.

Anderson, A.J.B. (1971) Ordination methods in ecology. *Journal of Ecology* **59**, 713–726.

Anderson, D.E., Goudie, A.S. and Parker, A.G. (2007) *Global Environments through the Quaternary: Exploring Environmental Change*, Oxford University Press, Oxford.

Anderson, D.J. and Kikkawa, J. (1986) Development of concepts, in *Community Ecology – Pattern and Process* (eds J. Kikkawa and D.J. Anderson), Blackwell Scientific Publications, Oxford, pp. 3–16.

Anderson, D.R. and Burnham, K.R. (2002) Avoiding pitfalls when using information-theoretic methods. *Journal of Wildlife Management* **66**, 912–918.

Anderson, D.R., Burnham, K.P. and Thompson, W.L. (2000) Null hypothesis testing: problems, prevalence and an alternative. *Journal of Wildlife Management* **64**, 912–923.

Anderson, M.J. and Gribble, N.A. (1998) Partitioning out the variation among spatial, temporal and environmental components in a multivariate dataset. *Australian Journal of Ecology* **23**, 158–167.

Anderson, M.J. and Willis, T.J. (2003) Canonical analysis of principal coordinates: a useful method of constrained ordination for ecology. *Ecology* **84**, 511–525.

Anderson, S. and Marcus, L.F. (1993) Effects of quadrat size on measurements of species diversity. *Journal of Biogeography* **20**, 421–428.

Anderson, T.M., Metzger, K. and McNaughton, S.J. (2007) Rainfall and soils modify plant community response to grazing in Serengeti National Park. *Ecology* **88**, 1191–1201.

Aplin, P. (2005) Remote sensing: ecology. *Progress in Physical Geography* **29**, 104–113.

Araújo, M.B., Densham, P.J., Lampinen, R., *et al.* (2001) Would environmental diversity be a good surrogate for species diversity? *Ecography* **24**, 103–110.

Archaux, F. (2009) Could we obtain better estimates of plot species richness from multiple observer plant censuses? *Journal of Vegetation Science* **20**, 603–611.

Archaux, F., Bergès, L. and Chevalier, R. (2007) Are plant censuses carried out on small quadrats more reliable than on larger ones? *Plant Ecology* **188**, 179–190.

Archaux, F., Gosselin, F., Bergès, L. and Chevalier, R. (2006) Effects of sampling time, quadrat richness and observer on exhaustiveness of plant censuses. *Journal of Vegetation Science* **17**, 299–306.

Archibold, O.W. (1995) *Ecology of World Vegetation*, Chapman and Hall, London.

Armitage, R.P., Weaver, R.E. and Kent, M. (2000) Remote sensing of semi-natural upland vegetation: the relationship between species composition and spectral response, in *Vegetation Mapping: Patch to Planet* (eds R. Alexander and A. Millington), John Wiley and Sons Ltd, Chichester, pp. 83–102.

Armitage, R.P., Kent, M. and Weaver, R.E. (2004) Identification of the spectral characteristics of British semi-natural upland vegetation using direct ordination: a case study from Dartmoor, UK. *International Journal of Remote Sensing* **25**, 3369–3388.

Artigras, F.J. and Yang, J. (2004) Hyperspectral remote sensing of habitat heterogeneity between tide-restricted and tide-open areas in the New Jersey Meadowlands. *Urban Habitats* **2**, 112–129.

Ashman, M.R. and Puri, G. (2002) *Essential Soil Science: a Clear and Concise Introduction to Soil Science*, Blackwell Science, Oxford.

Auerbach, M. and Shmida, A. (1987) Spatial scale and the determination of plant species richness. *Trends in Ecology and Evolution* **2**, 238–242.

Augustine, D.J. and McNaughton, S.J. (2004) Regulation of shrub dynamics by native browsing ungulates on East African rangeland. *Journal of Applied Ecology* **41**, 45–58.

Austin, M.P. (1968) An ordination study of a chalk-grassland community. *Journal of Ecology* **56**, 739–757.

Austin, M.P. (1971) Role of regression analysis in ecology. *Proceedings of the Ecological Society of Australia* **6**, 63–75.

Austin, M.P. (1976) Performance of four ordination techniques assuming three different non-linear response models. *Vegetatio* **33**, 43–49.

Austin, M.P. (1980) Searching for a model for use in vegetation analysis. *Vegetatio* **42**, 11–21.

Austin, M.P. (1981) Permanent quadrats: an interface for theory and practice. *Vegetatio* **46**, 1–10.

Austin, M.P. (1985) Continuum concept, ordination methods and niche theory. *Annual Review of Ecology and Systematics* **16**, 39–61.

Austin, M.P. (1986) The theoretical basis of vegetation science. *Trends in Ecology and Evolution* **1**, 161–164.

Austin, M.P. (1987) Models for the analysis of species response to environmental gradients. *Vegetatio* **69**, 35–45.

Austin, M.P. (1990) Community theory and competition in vegetation, in *Perspectives on Plant Competition* (eds J.B. Grace and D. Tilman), Academic Press, San Diego, CA.

Austin, M.P. (1991a) Vegetation theory in relation to cost-efficient surveys, in *Nature Conservation: Cost Effective Ecological Surveys and Data Analysis* (eds C.R. Margules and M.P. Austin), CSIRO, Canberra, pp. 17–22.

Austin, M.P. (1991b) Vegetation data collection and analysis, in *Nature Conservation: Cost Effective Ecological Surveys and Data Analysis* (eds C.R. Margules and M.P. Austin), CSIRO, Canberra, pp. 37–41.

Austin, M.P. (1999a) A silent clash of paradigms: some inconsistencies in community ecology. *Oikos* **86**, 170–178.

Austin, M.P. (1999b) The potential contribution of vegetation ecology to biodiversity research. *Ecography* **22**, 465–484.

Austin, M.P. (2002) Spatial prediction of species distribution: an interface between ecological theory and statistical modelling. *Ecological Modelling* **157**, 101–118.

Austin, M.P. (2005) Vegetation and environment: discontinuities and continuities, in *Vegetation Ecology* (ed. E. van der Maarel), Blackwell Publishing, Oxford, pp. 52–84.

Austin, M.P. and Orlóci, L. (1966) Geometric models in ecology. II. An evaluation of some ordination techniques. *Journal of Ecology* **54**, 217–227.

Austin, M.P. and Austin, B.O. (1980) Behaviour of experimental plant communities along a nutrient gradient. *Journal of Ecology* **68**, 891–918.

Austin, M.P., Cunningham, R.B. and Fleming, P.M. (1984) New approaches to direct gradient analysis using environmental scalars and statistical curve-fitting procedures. *Vegetatio* **55**, 11–27.

Austin, M.P. and Heyligers, P.C. (1989) Vegetation survey design for conservation: Gradsect sampling of forests in North-eastern New South Wales. *Biological Conservation* **50**, 13–32.

Austin, M.P. and Smith, T.M. (1989) A new model for the continuum concept. *Vegetatio* **83**, 35–47.

Austin, M.P. and Gaywood, M.J. (1994) Current problems of environmental gradients and species response curves in relation to continuum theory. *Journal of Vegetation Science* **5**, 473–482.

Australian National Vegetation Information System (2003) http://www.environment.gov.au/erin/nvis/index.html

Bacaro, G. and Ricotta, C. (2007) A spatially explicit measure of beta diversity. *Community Ecology* **8**, 41–46.

Bakker, J.D. (2008) Increasing the utility of indicator species analysis. *Journal of Applied Ecology* **45**, 1829–1835.

Bakker, J.P., Olff, H., Willems, J.H. and Zobel, M. (1996) Why do we need permanent plots in the study of long-term vegetation dynamics? *Journal of Vegetation Science* **7**, 147–156.

Bakker, P.A. (1979) Vegetation science and nature conservation, in *The Study of Vegetation* (ed. M.J.A. Werger), Junk, The Hague, pp. 249–288.

Ball, D.F. (1986) Site and soils, in *Methods in Plant Ecology* (2nd edn) (eds P.D. Moore and S.E. Chapman), Blackwell Scientific, London, pp. 215–284.

Ball, M.E. (1974) Floristic changes on grasslands and heaths on the Isle of Rhum after a reduction or exclusion of grazing. *Journal of Environmental Management* **2**, 299–318.

Banyikwa, F.F., Feoli, E. and Zuccarello, V. (1990) Fuzzy set ordination and classification of Serengeti short grasslands, Tanzania. *Journal of Vegetation Science* **1**, 97–104.

Barabesi, L. and Fattorinin, L. (1998) The use of replicated plot, line and point sampling for estimating species abundance and ecological diversity. *Environmental and Ecological Statistics* **5**, 353–370.

Barbour, M.G. and Billings, W.D. (1988) *North American Terrestrial Vegetation*, Cambridge University Press, New York.

Barbour, M.G. and Billings, W.D. (2000) *North American Terrestrial Vegetation* (2nd edn), Cambridge University Press, New York.

Barbour, M.G., Burk, J.H., Pitts, W.D., *et al.* (1999) *Terrestrial Plant Ecology* (3rd edn), Addison Wesley Longman Inc., Menlo Park, CA.

Barker, P. (2001) *A Technical Manual for Vegetation Monitoring*. Department of Primary Industries and Water, Hobart, Tasmania, Australia. Available at http://www.dpiw.tas.gov.au/inter.nsf/Publications/Attachments/LBUN-5MJ2NY/$FILE/Manual_screen.pdf.

Barkman, J.J. (1989) A critical evaluation of minimum area concepts. *Vegetatio* **85**, 89–104.

Barkman, J.J. (1990) Controversies and perspectives in plant ecology and vegetation science. *Phytocoenologia* **18**, 565–589.

Barr, C.J. *et al.* (1993) *Countryside Survey 1990. Main Report*. Department of the Environment, London.

Barr, J., Howard, D.C., Bunce, R.G.H., *et al.* (1991) *Changes in Hedgerows in Britain between 1984 and 1990*. Report to Department of the Environment. Institute of Terrestrial Ecology, Grange-over-Sands.

Bartlett, M.S. (1949) Fitting a straight line when both variables are subject to error. *Biometrics* **5**, 207–212.

Bastian, O. (2001) Landscape ecology: towards a unified discipline. *Landscape Ecology* **16**, 757–766.

Batalha, M.A. and Martins, F.R. (2002) Life-form spectra of Brazilian cerrado sites. *Flora* **197**, 452–460.

Batschelet, E. (1981) *Circular Statistics in Biology*, Academic Press, London.

Beals, E.W. (1973) Ordination: mathematical elegance and ecological naïveté. *Journal of Ecology* **61**, 23–35.

Beals, E.W. (1984) Bray-Curtis ordination: an effective strategy for analysis of multivariate ecological data. *Advances in Ecological Research* **14**, 1–55.

Bebbington, A. (2005) The ability of A-level students to name plants (ITI). *Journal of Biological Education* **39**, 62–67.

Beckerman, A.P. and Petchey, O.L. (2011) *Introduction to R for Biological Data Analysis*, Oxford University Press, Oxford.

Becking, R.W. (1957) The Zurich-Montpellier school of phytosociology. *Botanical Review* **23**, 412–488.

Beever, E.A., Swihart, R.K. and Bestelmeyer, B.T. (2006) Linking the concept of scale to studies of biological diversity: evolving approaches and tools. *Journal of Biogeography* **12**, 229–235.

Begon, M., Harper, J.L. and Townsend, C.R. (1990) *Ecology: Individuals, Populations and Communities* (2nd edn), Blackwell Science, Oxford.

Begon, M., Townsend, C.R. and Harper, J.L. (2006) *Ecology: from Individuals to Ecosystems* (4th edn), Blackwell Science, Oxford.

Beilsel, J.-N. and Moreteau, J.-C. (1997) A simple formula for calculating the lower limit of Shannon's diversity index. *Ecological Modelling* **99**, 289–292.

Bekker, R.M., van der Maarel, E., Bruelheide, H. and Woods, K. (2007) Long-term datasets: from descriptive to predictive data using ecoinformatics. *Journal of Vegetation Science* **18**, 458–462.

Belbin, L. and McDonald, C. (1993) Comparing three classification strategies for use in ecology. *Journal of Vegetation Science* **4**, 341–348.

Belovsky, G., Bôtkin, D., Crowl, T., *et al.* (2004) Ten suggestions to strengthen the science of ecology. *BioScience* **54**, 345–351.

Belyea, L.R. and Lancaster, J. (1999) Assembly rules within a contingent ecology. *Oikos* **86**, 402–416.

Benninghoff, W.S. and Southworth, W.C. (1964) Ordering of tabular arrays of phytosociological data by digital computer. *Abstract of the 10th International Botanical Congress*, 331–332.

Benzécri, J.P. (1969) Statistical analysis as a tool to make patterns emerge from data, in *Methodologies of Pattern Recognition* (ed. S. Watanabe), Academic Press, New York, pp. 35–60.

Benzécri, J.P. (1973) *L'analyse des données. Vol. 2. L'analyse des correspondances*, Dunod, Paris.

Bergstedt, J., Westerberg, L. and Milberg, P. (2009) In the eye of the beholder: bias and stochastic variation in cover estimates. *Plant Ecology* **204**, 271–283.

Berryman, S. and McCune, B. (2006) Estimating epiphytic macrolichen biomass from topography, stand structure and lichen community data. *Journal of Vegetation Science* **17**, 157–170.

Bertness, M.D. and Callaway, R. (1994) Positive interactions in communities. *Trends in Ecology and Evolution* **9**, 191–193.

Besag, J. (1981) On resistant techniques and statistical analysis. *Biometrika* **68**, 463–469.

Bezdek, J.C. (1974) Numerical taxonomy with fuzzy sets. *Journal of Mathematical Biology* **1**, 57–71.

Bezdek, J.C. (1981) *Pattern Recognition with Fuzzy Objective Function Algorithms*, Plenum Press, New York.

Bio, A.M.F. (2000) *Does vegetation suit our models? Data and model assumptions and the assessment of species distribution in space*. Nederlandsche Geografische Studies 265. The Royal Dutch Geographical Society/Faculty of Geographical Sciences Utrecht University.

Bio, A.M.F., Alkemade, R. and Barendregt, A. (1998) Determining alternative models for vegetation analysis: a non-parametric approach. *Journal of Vegetation Science* **9**, 5–16.

Birkeland, P.W. (1999) *Soils and Geomorphology* (3rd edn), Oxford University Press, Oxford.

Birks, H.J.B. and Birks, H.H. (1980) *Quaternary Palaeoecology*, Arnold, London.

Birks, H.J.B., Pelgar, S.M. and Austin, H.A. (1996) An annotated bibliography of canonical correspondence analysis and related constrained ordination methods 1986–1993. *Abstracta Botanica* **20**, 17–36.

Bishop, O.N. (1983) *Statistics for Biology* (4th edn), Longman, London.

Bissonette, J.A. (ed.) (1997) *Wildlife and Landscape Ecology: Effects of Pattern and Scale*, Springer, New York.

Blackburn, T.M. and Gaston, K.J. (2004) Macroecology. *Basic and Applied Ecology* **5**, 385–387.

Blackburn, T.M. and Gaston, K.J. (2006) There's more to macroecology than meets the eye. *Journal of Biogeography* **33**, 537–540.

Blake, S.F. (1961) *Geographical Guide to Floras of the World, Vol. 2*, Ashington, Government Printing Office.

Blake, S.F. and Atwood, A.C. (1942; republished 1963) *Geographical Guide to Floras of the World. Vol. 1.* US Department of Agriculture Miscellaneous Publication 401 (1942); Hafner Publishing Co., New York (1963).

Blanchet, F.G., Legendre, P. and Borcard, D. (2008) Forward selection of explanatory variables. *Ecology* **89**, 2623–2632.

Blockeel, T.L. and Long, D.G. (1998) *A Checklist and Census Catalogue of British and Irish Bryophytes*, British Bryological Society, Cardiff.

Bonham, C.D. and Reich, R.M. (2009) Influences of transect relocation errors on line-point estimates of plant cover. *Plant Ecology* **204**, 173–178.

Boose, E.R., Boose, E.F. and Lezberg, A.L. (1998) A practical method for mapping trees using distance measurements. *Ecology* **79**, 819–827.

Borcard, D. and Legendre, P. (1994) Environmental control and spatial structure in ecological communities: an example using orbatid mites (Acari, Oribatei). *Environmental and Ecological Statistics* **1**, 37–53.

Borcard, D. and Legendre, P. (2002) All-scale spatial analysis of ecological data by means of principal coordinates on neighbour matrices. *Ecological Modelling* **153**, 51–68.

Borcard, D., Legendre, P. and Drapeau, P. (1992) Partialing out the spatial component of ecological variation. *Ecology* **73**, 1045–1055.

Borcard, D., Legendre, P., Avois-Jacquet, C. and Tuomisto, H. (2004) Dissecting the spatial structure of ecological data at multiple scales. *Ecology* **85**, 1826–1832.

Borcard, D., Gillet, F. and Legendre, P. (2011) *Numerical Ecology with R*, Springer, New York.

Botta-Dukát, Z., Kovács-Láng, E., Rédei, T., *et al.* (2007) Statistical and biological consequences of preferential sampling in phytosociology: theoretical considerations and a case study. *Folia Geobotanica* **42**, 141–152.

Boulière, F. (ed.) (1983) *Tropical Savannas*, Ecosystems of the World 13, Elsevier, Amsterdam.

Boulière, F. and Hadley, M. (1983) Present-day savannas: an overview, in *Tropical Savannas* (ed. F. Boulière), Ecosystems of the World 13, Elsevier, Amsterdam, 1–15.

Boutin, C. and Keddy, P.A. (1993) A functional classification of wetland plants. *Journal of Vegetation Science* **4**, 591–600.

Bouxin, G. (2005) Ginkgo, a multivariate analysis package. *Journal of Vegetation Science* **16**, 355–359.

Box, E.O. (1996) Plant functional types and climate at the global scale. *Journal of Vegetation Science* **7**, 309–320.

Box, E.O. and Fujiwara, K. (2005) Vegetation types and their broad-scale distribution, in *Vegetation Ecology* (ed. E. van der Maarel), Blackwell Publishing, Oxford, pp. 106–128.

Bradbury, I. (1998) *The Biosphere* (2nd edn), Chichester, John Wiley and Sons Ltd.

Bradu, D. and Gabriel, K.R. (1978) The biplot as a diagnostic tool for models of two-way tables. *Technometrics* **20**, 47–68.

Brady, N.C. and Weil, R.R. (2004) *Elements of the Nature and Properties of Soils*, Prentice Hall, Upper Saddle River, NJ.

Brady, N.C. and Weil, R.R. (2007) *The Nature and Properties of Soils* (14th edn), Prentice Hall, Upper Saddle River, NJ.

Bråkenhielm, S. and Liu, Q. (1995) Comparison of field methods in vegetation monitoring. *Water Air and Soil Pollution* **79**, 75–87.

Branko, K. and Ranka, P. (1994) A generalized standardization procedure in ecological ordination. Tests with principal components analysis. *Journal of Vegetation Science* **5**, 259–262.

Braun, W.J. and Murdoch, D.J. (2007) *A First Course in Statistical Programming with R*, Cambridge University Press, Cambridge.

Braun-Blanquet, J. (1928) *Pflanzensoziologie. Grundzüge der Vegetationskunde*, Springer, Berlin.

Braun-Blanquet, J. (1932/1951) *Plant Sociology: the Study of Plant Communities* (English translation), McGraw-Hill, New York.

Bray, R.J. and Curtis, J.T. (1957) An ordination of the upland forest communities of southern Wisconsin. *Ecological Monographs* **27**, 325–349.

Bredenkamp, G., Chytrý, M., Fischer, H.S., *et al.* (eds) (1998) Vegetation mapping: theory methods and case studies: Introduction. *Applied Vegetation Science* **1**, 161–266.

Breiman, L., Friedman, J.H., Olshen, R.A. and Stone, C.J. (1984) *Classification and Regression Trees*, Chapman and Hall, New York.

Bridge, P.D. (1993) Classification, in *Biological Data Analysis: a Practical Approach* (ed. J.C. Fry), IRL Press/Oxford University Press, Oxford, pp. 219–242.

Briggs, D.J. (1977) *Soils*, Sources and Methods in Geography, Butterworth, London.

Briggs, D.J., Smithson, P., Addison, K. and Atkinson, K. (1997) *Fundamentals of the Physical Environment* (2nd edn), Routledge, London.

British Standards Institute (2007, 2009) *BS8848:2007 + Amendment 1:2009 Specification for the Provision of Visits, Fieldwork, Expeditions, and Adventurous Activities, outside the United Kingdom*. British Standards Institute, London.

Brooker, R.W. and Callaghan, T.V. (1998) The balance between positive and negative plant interactions and its relationship to environmental gradients: a model. *Oikos* **81**, 196–207.

Brooker, R.W. and Callaway, R.M. (2009) Facilitation in the conceptual melting pot. *Journal of Ecology* **97**, 1117–1120.

Brooks, S.P. (2003) Bayesian computation: a statistical revolution. *Philosophical Transactions of the Royal Society of London, Series A*, **361**, 2681–2697.

Brown, G.W. and Mood, A.M. (1951) On median tests for linear hypotheses, in *Proceedings of 2nd Berkeley Symposium on Mathematical Statistics and Probability* (ed. J. Neyman), California University Press, Berkeley and Los Angeles, pp. 159–166.

Brown, J.H. (1995) *Macroecology*, University of Chicago Press, Chicago.

Brown, M.J., Ratkowsky, D.A. and Minchin, P.R. (1984) A comparison of detrended correspondence analysis and principal coordinates analysis using four sets of Tasmanian vegetation data. *Australian Journal of Ecology* **9**, 273–279.

Brown, R.T. and Curtis, J.T. (1952) The upland conifer-hardwood forest of Northern Wisconsin. *Ecological Monographs* **22**, 217–234.

Bruelheide, H. (1997) Using formal logic to classify vegetation. *Folia Geobotanica Phytotaxonomica* **32**, 41–46.

Bruelheide, H. (2000) A new measure of fidelity and its application to defining species groups. *Journal of Vegetation Science* **11**, 167–178.

Bruelheide, H. and Chytrý, M. (2000) Towards unification of national vegetation classifications: a comparison of two methods for the analysis of large data sets. *Journal of Vegetation Science* **11**, 295–306.

Bruelheide, H. and Jandt, U. (1997) Demarcation of communities in large data bases. *Phytocoenologia* **27**, 141–159.

Bryant, D.M., Ducey, M.J., Innes, J.C., *et al.* (2004) Forest community analysis and the point-centred quarter method. *Plant Ecology* **175**, 193–203.

Buckland, S.T., Magurran, A.E., Green, R.E. and Fewster, R.M. (2005) Monitoring change in biodiversity through composite indices. *Philosophical Transactions of the Royal Society B* **360**, 246–254.

Bulla, L. (1994) An index of evenness and its associated diversity measure. *Oikos* **70**, 167–171.

Bullock, J.M. (2006) Plants, in *Ecological Census Techniques: a Handbook* (2nd edn) (ed. W.J. Sutherland), Cambridge University Press, Cambridge, pp. 186–213.

Bunce, R.G.H. and Shaw, M.W. (1973) A standardised procedure for ecological survey. *Journal of Environmental Management* **1**, 129–158.

Bunce, R.G.H., Barr, C.J., Gillespie, M.K. and Howard, D.C. (1996) The ITE land classification: providing an environmental stratification of Great Britain. *Environmental Monitoring and Assessment* **39**, 39–46.

Bunce, R.G.H., Barr, C.J., Gillespie, M.K., *et al.* (1999a) *Vegetation of the British Countryside: the Countryside Vegetation System*. ECOFACT Volume 1, Department of the Environment Transport and the Regions, London.

Bunce, R.G.H., Smart, S.M., van de Poll, H.M., *et al.* (1999b) *Measuring Change in British Vegetation*. ECOFACT Volume 2, Department of the Environment Transport and the Regions, London.

Burel, F. and Baudry, J. (2003) *Landscape Ecology: Concepts, Methods and Applications*, Science Publishers Inc., Enfield, NH, USA.

Burnett, J.D. (ed.) (1964) *The Vegetation of Scotland*, Oliver and Boyd, Edinburgh.

Burnham, K.P. and Anderson, D.R. (2002) *Model Selection and Multimodel Inference: A Practical Information-Theoretic Approach* (2nd edn), Springer-Verlag, New York.

Burrough, P.A. (1995) Spatial aspects of ecological data, in *Data Analysis in Community and Landscape Ecology* (2nd edn) (eds R.H.G. Jongman, C.J.F. ter Braak and O.F.R. van Tongeren), Cambridge University Press, Cambridge, pp. 213–251.

Burrough, P.A. and McDonnell, R.A. (1998) *Principles of Geographical Information Systems* (2nd edn), Oxford University Press, Oxford.

Burt, J.E. and Barber, G.M. (1996) *Elementary Statistics for Geographers* (2nd edn), Guilford Press, New York.

Burrows, C.J. (1990) *Processes of Vegetation Change*, Unwin Hyman, London.

Cain, S.A. (1932) Concerning certain phytosociological concepts. *Ecological Monographs* **2**, 475–508.

Cain, S.A. (1934a) The climax and its complexities. *American Midland Naturalist* **21**, 146–181.

Cain, S.A. (1934b) A comparison of quadrat sizes in a quantitative phytosociology study of Nash's Woods, Posey County, Indiana. *American Midland Naturalist* **15**, 529–566.

Cain, S.A. (1938) The species-area curve. *American Midland Naturalist* **19**, 573–581.

Callan, R. (1999) *The Essence of Neural Networks*, Prentice-Hall, London.

Callaway, R.M. (1995) Positive interactions among plants. *Botanical Review* **61**, 306–349.

Callaway, R.M. (1997) Positive interactions in plant communities and the individualistic-continuum concept. *Oecologia* **112**, 255–279.

Callaway, R.M. (1998) Are positive interactions species-specific? *Oikos* **82**, 202–207.

Callaway, R.M. and Walker, L.R. (1997) Competition and facilitation: a synthetic approach to interactions in plant communities. *Ecology* **78**, 1958–1965.

Cao, Y., Bark, A.W. and Williams, P. (1997) A comparison of clustering methods for river benthic community analysis. *Hydrobiologia* **347**, 25–40.

Carleton, T.J., Stitt, R.H. and Nieppola, J. (1996) Constrained indicator species analysis (COINSPAN): an extension of TWINSPAN. *Journal of Vegetation Science* **7**, 125–130.

Carroll, J.D. (1987) Some multidimensional scaling and related procedures devised at Bell Laboratories, with ecological applications, in *Developments in Numerical Ecology* (eds P. Legendre and L. Legendre), NATO ASI Series G14, Springer-Verlag, Berlin, pp. 65–138.

Causton, D.R. (1988) *Introduction to Vegetation Analysis*, Unwin Hyman, London.

Cavieres, L.A. and Badano, E.I. (2009) Do facilitative interactions increase species richness at the entire community level? *Journal of Ecology* **97**, 1181–1191.

Černá, L. and Chytrý, M. (2005) Supervised classification of plant communities with artificial neural networks. *Journal of Vegetation Science* **16**, 407–414.

Chamberlin, T.C. (1890, 1965) The method of multiple working hypotheses. *Science* **XV**, 92–96, reprinted in *Science* **148**, 754–759.

Champely, S. and Chessel, D. (2002) Measuring biological diversity using Euclidean metrics. *Environmental and Ecological Statistics* **9**, 167–177.

Chapman, J.L. and Reiss, M.J. (1992) *Ecology: Principles and Applications*, Cambridge University Press, Cambridge.

Chatfield, C. (1995) *Problem Solving – a Statistician's Guide* (2nd edn), CRC Press, Boca Raton, FL.

Cherrett, J.M. (ed.) (1989) *Ecological Concepts: the Contribution of Ecology to an Understanding of the Natural World*, British Ecological Society/Blackwell Scientific Publications, Oxford.

Cherrill, A. and McClean, C. (1999) Between-observer variation in the application of a standard method of habitat mapping by environmental consultants in the UK. *Journal of Applied Ecology* **36**, 989–1008.

Chevenet, F., Dolédec, S. and Chessel, D. (1994) A fuzzy coding approach for the analysis of long-term ecological data. *Freshwater Biology* **31**, 295–309.

Chiarucci, A. (2007) To sample or not to sample? That is the question … for the vegetation scientist. *Folia Geobotanica* **42**, 209–216.

Chiarucci, A., Wilson, J.B., Anderson, B.J. and De Dominicis, V. (1999) Cover *versus* biomass as an estimate of species abundance: does it make a difference to the conclusions? *Journal of Vegetation Science* **10**, 35–42.

Chow, T.J. (1970) Lead accumulation in roadside soils and grass. *Nature* **225**, 295–296.

Chytrý, M., Hejcman, M., Hennekens, S.M. and Schellberg, J. (2009) Changes in vegetation types and Ellenberg indicator values after 65 years of fertilizer application in the Rengen Grassland Experiment, Germany. *Applied Vegetation Science* **12**, 167–176.

Chytrý, M. and Otýpkova, Z. (2003) Plot sizes used for phytosociological sampling of European vegetation. *Journal of Vegetation Science* **14**, 563–570.

Clapham, A.R., Tutin, T.G. and Warburg, E.F. (1981) *Excursion Flora of the British Isles* (3rd edn), Cambridge University Press, Cambridge.

Clapham, A.R., Tutin, T.G. and Moore, D.N. (1987) *Flora of the British Isles* (3rd edn), Cambridge University Press, Cambridge.

Clapham, A.R., Tutin, T.G., Warburg, E.F. and Roles, S.J. (2010) *Flora of the British Isles: Illustrations* (illustrations by S.J. Roles), Cambridge University Press, Cambridge.

Clark, J.S. (2005) Why environmental scientists are becoming Bayesians. *Ecology Letters* **8**, 2–15.

Clarke, K.R. (1993) Non-parametric multivariate analysis of changes in community structure. *Australian Journal of Ecology* **18**, 117–143.

Clarke, K.R. and Green, R.H. (1988) Statistical design and analysis for a 'biological effects' study. *Marine Ecology Progress Series* **46**, 213–226.

Clarke, K.R. and Warwick, R.M. (1994) Similarity-based testing for community pattern: the 2-way layout with no replication. *Marine Biology* **118**, 167–176.

Clarke, K.R. and Warwick, R.M. (2001) *Change in Marine Communities: An Approach to Statistical Analysis and Interpretation*, (2nd edn), PRIMER-E Ltd, Plymouth, UK.

Clarke, K.R. and Gorley, R.N. (2006) *PRIMER v.6: User Manual/Tutorial*, PRIMER-E Ltd, Plymouth, UK.

Clements, F.E. (1916) *Plant Succession. An Analysis of the Development of Vegetation*, Carnegie Institute, Publication 242, Washington, DC.

Clements, F.E. (1928) *Plant Succession and Indicators*, H.W. Wilson, New York.

Clements, F.E. (1936) Nature and structure of the climax. *Journal of Ecology* **24**, 254–282.

Cleveland, W.S. (1993) *Visualising Data*, Hobart Press, Summit, NJ.

Cliff, A.D. and Ord, J.K. (1973) *Spatial Autocorrelation*, Pion, London.

Cliff, A.D. and Ord, J.K. (1981) *Spatial Processes: Models and Applications*, Pion, London.

Clifford, H.T. and Stephenson, W. (1975) *An Introduction to Numerical Classification*, Academic Press, New York.

Cochrane, M.A. (2000) Using vegetation reflectance variability for species level classification of hyperspectral data. *International Journal of Remote Sensing* **21**, 2075–2087.

Coker, P.D. (1988) *Some Aspects of the Biogeography of the Hoyfjellet with Special Reference to Hoyrokampen, Boverdal, Southern Norway*. Unpublished PhD. thesis, University College, London.

Cole, M.M. (1986) *The Savannas: Biogeography and Geobotany*, Academic Press, London.

Cole, M.M. (1987) The savannas. *Progress in Physical Geography* **11**, 334–355.

Collins, S.L., Glenn, S.M. and Roberts, D.W. (1993) The hierarchical continuum concept. *Journal of Vegetation Science* **4**, 149–156.

Connell, J.H. and Slatyer, R.O. (1977) Mechanisms of succession in natural communities and their role in community structure and organisation. *American Naturalist* **111**, 1119–1144.

Cook, C.N., Wardell-Johnson, G., Keatley, M., *et al.* (2010) Is what you see what you get? Visual vs. measured assessments of vegetation condition. *Journal of Applied Ecology* **47**, 650–661.

Corley, M.F.V. and Hill, M.O. (1981) *Distribution of Bryophytes in the British Isles*, British Bryological Society, Cardiff.

Cormack, R.M. (1971) A review of classification. *Journal of the Royal Statistical Society, Series A* **134**, 321–367.

Cormack, R.M. (1988) Statistical challenges in the environmental sciences: a personal view. *Journal of the Royal Statistical Society, Series A* **151**, 201–210.

Cornelissen, J.H.C., Lavorel, S., Garnier, E., *et al.* (2003) A handbook of protocols for standardised and easy measurement of plant functional traits worldwide. *Australian Journal of Botany* **51**, 335–380.

Cottam, G., Glenn Goff, F. and Whittaker, R.H. (1978) Wisconsin comparative ordination, in *Ordination of Plant Communities* (ed. R.H. Whittaker), Junk, The Hague, pp. 185–214.

Cousins, S.H. (1991) Species diversity measurement: choosing the right index. *Trends in Ecology and Evolution* **6**, 190–192.

Couteron, P. and Ollier, S. (2005) A generalized, variogram-based framework for multi-scale ordination. *Ecology* **86**, 828–834.

Cox, B. and Moore, P.D. (2005) *Biogeography: an Ecological and Evolutionary Approach* (7th edn), Blackwell, Oxford.

Cox, N.J. (1989) Teaching and learning spatial autocorrelation: a review. *Journal of Geography in Higher Education* **13**, 185–190.

Cox, T.F. and Cox, M.A.A. (1996) *Multidimensional Scaling*, Chapman and Hall, London.

Craine, J.M. (2005) Reconciling plant strategy theories of Grime and Tilman. *Journal of Ecology* **93**, 1041–1052.

Craine, J.M. (2007) Plant strategy theories: replies to Grime and Tilman. *Journal of Ecology* **95**, 235–240

Crawford, R.M.M. and Wishart, D. (1967) A rapid multivariate method for the detection and classification of ecologically related species. *Journal of Ecology* **55**, 505–524.

Crawley, M.J. (ed.) (1997) *Plant Ecology* (2nd edn), Blackwell Science, Oxford.

Crawley, M.J. (2005) *Statistics: an Introduction using R*, John Wiley and Sons Ltd, Chichester.

Crawley, M.J. (2007) *The R Book*, John Wiley and Sons Ltd, Chichester.

Critchley, C.N.R. and Poulton, S.M.C. (1998) A method to optimize precision and scale in grassland monitoring. *Journal of Vegetation Science* **9**, 837–846.

Crowe, T.M. (1979) Lots of weeds: insular phytogeography of vacant urban lots. *Journal of Biogeography* **6**, 169–181.

Curran, P.J. (1983) The problems of remote sensing of vegetation canopies for biomass estimates, in *Ecological Mapping from Ground, Air and Space* (ed. R.M. Fuller), Institute of Terrestrial Ecology, Cambridge, pp. 83–100.

Curtis, J.T. and McIntosh, R.P. (1950) The inter-relations of certain analytic and synthetic phytosociological characters. *Ecology* **31**, 434–455.

Curtis, J.T. and McIntosh, R.P. (1951) An upland forest continuum in the prairie-forest border region of Wisconsin. *Ecology* **32**, 476–496.

Cushman, S.A. and McGarigal, K. (2002) Hierarchical, multi-scale decomposition of species-environment relationships. *Landscape Ecology* **17**, 637–646.

Cushman, S.A. and Wallin, D.O. (2002) Separating the effects of environmental, spatial and disturbance factors on forest community structure in the Russian Far East. *Forest Ecology and Management* **168**, 201–215.

Czekanowski, J. (1909) Zur Differentialdiagnose der Neandertalgruppe. *Korrespondenz-Blatt der Deutschen Gesellschaft für Anthropologie Ethnologie und Urgeschichte* **40**, 44–47.

Czekanowski, J. (1913) *Zarys Metod Statystycznyck w zastosowaniu do antropologii*. Towarzystwo Naukowe Warszawskie, Warszawa. Travaux de la Société des Sciences de Varsovie. III. Classe de sciences mathématiques et naturelles, No. 5.

Dagnelie, P. (1960) Contribution à l'étude des communautés végétales par l'analyse factorielle. *Bull. Serv. Carte. Phytogéogr., Serie B* **5**, 7–71; 93–195.

Dagnelie, P. (1978) Factor analysis, in *Ordination of Plant Communities* (ed. R.H. Whittaker), Junk, The Hague, pp. 215–238.

Dahl, E. (1956) Rondane. Mountain vegetation in South Norway and its relation to the environment. *Skr. Norske. Vid.-Akad.* **I**. Mat.-Naturvid. 3.

Dahl, E. (1968) *Analytical Keys to British Macrolichens* (2nd edn), British Lichen Society, London.

Dahl, E. (1985) *A Survey of the Plant Communities at Finse, Handangervidda, Norway.* Agricultural University of Norway, As-NLH.

Dale, P.E.R. (1979) A pragmatic study of Danserau's universal system for recording vegetation: application in South-East Queensland, Australia. *Vegetatio* **40**, 129–133.

Daniel, W.W. (1978) *Applied Nonparametric Statistics*, Houghton Mifflin, Boston.

Danin, A. and Orshan, G. (1990) The distribution of Raunkaier life-forms in Israel in relation to the environment. *Journal of Vegetation Science* **1**, 41–48.

Danserau, P. (1951) Description and recording of vegetation upon a structural basis. *Ecology* **32**, 172–229.

Danserau, P. (1957) *Biogeography: an Ecological Perspective*, Ronald Press, New York.

Danserau, P., Buell, P.F. and Dagon, R. (1966) A universal system for recording vegetation. II. A methodological critique and an experiment. *Sarracenia* **10**, 1–64.

Dargie, T.C.D. (1984) On the integrated interpretation of indirect site ordinations: a case study using semi-arid vegetation in south-eastern Spain. *Vegetatio* **55**, 37–55.

Dargie, T.C.D. (1986) Species richness and distortion in reciprocal averaging and detrended correspondence analysis. *Vegetatio* **65**, 95–98.

Darlington, A. (1981) *Ecology of Walls*, Heinemann Education, London.

Davies, A., Baker, R.D., Grant, S.A. and Laidlaw, A.S. (eds) (1993) *Sward Measurement Handbook* (2nd edn), British Grassland Society, University of Reading, Reading.

Davis, B.N.K. (1976) Wildlife, urbanisation and industry. *Biological Conservation* **10**, 249–291.

Davison, A.W. (1971) The effects of de-icing salt on roadside verges. I. Soil and plant analysis. *Journal of Applied Ecology* **8**, 555–561.

De'ath, G. (1999) Principal curves: a new technique for indirect and direct gradient analysis. *Ecology* **80**, 2237–2253.

De'ath, G. and Fabricus, K.E. (2000) Classification and regression trees: a powerful yet simple technique for ecological data analysis. *Ecology* **81**, 3178–3192.

de Blois, S., Domon, G. and Bouchard, A. (2002) Landscape issues in plant ecology. *Ecography* **25**, 244–256.

de Cáceres, M., Font, X. and Oliva, F. (2010) The management of vegetation classifications with fuzzy clustering. *Journal of Vegetation Science* **21**, 1138–1151.

del Moral, R. (2007) Limits to convergence of vegetation during early succession. *Journal of Vegetation Science* **18**, 479–488.

del Moral, R. (2009) Increasing deterministic control of primary succession on Mount St.Helens, Washington. *Journal of Vegetation Science* **20**, 1145–1154.

Dengler, J., Löbel, S. and Dolnik, C. (2009) Species constancy depends on plot size – a problem for vegetation classification and how it can be solved. *Journal of Vegetation Science* **20**, 754–766.

Dennis, B. (1996) Discussion: should ecologists become Bayesians? *Ecological Applications* **6**, 1095–1103.

Deshmukh, I. (1986) *Ecology and Tropical Biology*, Blackwell Scientific, Oxford.

de Valpine, P. (2009) Shared challenges and common ground for Bayesian and classical analysis of hierarchical statistical models. *Ecological Applications* **19**, 584–588.

de Wit, C.T. (1960) On competition. *Verslagen Landbouwkundige Onderzoekingen* **66**, 1–82.

Díaz, S. and Cabido, M. (1997) Plant functional types and ecosystem function in relation to global change. *Journal of Vegetation Science* **8**, 463–474.

Díaz, S., Hodgson, J.G., Thompson, K., *et al.* (2004) The plant traits that drive ecosystems: Evidence from three continents. *Journal of Vegetation Science* **15**, 295–304.

Dickinson, G. and Murphy, K. (1998) *Ecosystems*, Routledge Introductions to Environment Series, Routledge, London.

Diekmann, M., Kühne, A. and Isermann, M. (2007) Random vs non-random sampling: effects on patterns of species abundance, species richness and vegetation-environment relationships. *Folia Geobotanica* **42**, 179–190.

Dierschke, H. (1994) *Pflanze soziologie. Grundlagen und Methoden*, Ulmer, Stuttgart.

Dietvorst, E., van der Maarel, E. and van der Putten, H. (1982) A new approach to the minimal area of a plant community. *Vegetatio* **50**, 77–91.

Dietz, E.J. (1983) Permutation tests for association between two distance matrices. *Systematic Zoology* **32**, 21–26.

Dietz, H. and Steinlein, T. (1996) Determination of plant species cover by means of image analysis. *Journal of Vegetation Science* **7**, 131–136.

Digby, P.G.N. and Kempton, R.A. (1987) *Multivariate Analysis of Ecological Communities*, Chapman and Hall, London.

Dinsdale, J., Dale, M.P. and Kent, M. (1997) The biogeography and historical ecology of *Lobelia urens* L. (the heath lobelia) in southern England. *Journal of Biogeography* **24**, 153–175.

Dinsdale, J.M., Dale, M.P. and Kent, M. (2000) Microhabitat availability and seedling recruitment of *Lobelia urens* L. *Seed Science Research* **10**, 471–487.

Dixon, P.M. (2003) VEGAN, a package of R functions for community ecology. *Journal of Vegetation Science* **14**, 927–930.

Doak, D.F., Bigger, D. and Thomson, D. (1998) The statistical inevitability of stability-diversity relationships in community ecology. *The American Naturalist* **151**, 264–276.

Dolédec, S. and Chessel, D. (1994) Co-inertia analysis: an alternative method for studying species-environmental relationships. *Freshwater Biology* **31**, 227–294.

Dolédec, S., Chessel, D. and Olivier, J. (1995) L'analyse des correspondances décentrée: application aux peuplements ichtyologiques du hautrhône. *Bulletin Français de la Pêche et de la Pisciculture* **336**, 29–40.

Down, C.G. (1973) Life-form succession in plant communities on colliery waste. *Environmental Pollution* **5**, 19–22.

Dray, S., Chessel, D. and Thioulouse, J. (2004) Co-inertia analysis and the linking of ecological data tables. *Ecology* **84**, 3078–3089.

Dray, S., Legendre, P. and Peres-Neto, P.R. (2006) Spatial modelling: a comprehensive framework for principal coordinate analysis of neighbour matrices (PCNM). *Environmental Modelling* **196**, 483–493.

Dray, S., Saïd, S. and Déblas, F. (2008) Spatial ordination of vegetation data using a generalization of Wartenberg's multivariate spatial correlation. *Journal of Vegetation Science* **19**, 45–56.

Drury, W.H. and Nisbet, I.C. (1973) Succession. *Journal of the Arnold Arboretum* **54**, 331–368.

Duckworth, J.C., Kent, M. and Ramsay, P.M. (2000) Plant functional types: an alternative to taxonomic plant community description in biogeography? *Progress in Physical Geography* **24**, 515–542.

Dufrêne, M. and Legendre, P. (1997) Species assemblages and indicator species: the need for a flexible asymmetrical approach. *Ecological Monographs* **67**, 345–366.

Dunn, G. and Everitt, B.S. (1982) *An Introduction to Mathematical Taxonomy*, Cambridge University Press, Cambridge.

Du Rietz, G.E. (1921) *Zur Methodologischen Grundlage der Modernen Pflanzensoziologie*, Holzhausen, Vienna.

Du Rietz, G.E. (1942a) Rishedsförband i Tarneträskomradets lagfjällbalte. *Svensk Botanisk Tidskrift* **36**: Uppsala.

Du Rietz, G.E. (1942b) De svenska fjallens vantvärld. Norrland. Natur, befolkning och naringar. *Ymer* **62**, 3–4.

Dytham, C. (2010) *Choosing and Using Statistics: a Biologist's Guide* (3rd edn), Wiley-Blackwell, Chichester.

Eberhardt, L.L. (2003) What should we do about hypothesis testing? *Journal of Wildlife Management* **67**, 265–282.

Eberhardt, L.L. and Thomas, J.M. (1991) Designing environmental field studies. *Ecological Monographs* **61**, 53–73.

Eckblad, J.W. (1991) How many samples should be taken? *BioScience* **41**, 346–348.

Eddison, J. (2000) *Quantitative Investigations in the Biosciences using Minitab*. Chapman & Hall, Boca Raton, FLA.

Egler, F.E. (1954) Philosophical and practical considerations of the Braun-Blanquet system of phytosociology. *Castanea* **19**, 45–60.

Ehrenfeld, J.G., Han, X., Parsons, W.F.J. and Zhus, W. (1997) On the nature of environmental gradients: temporal and spatial variability of soils and vegetation in the New Jersey pinelands. *Journal of Ecology* **85**, 785–798.

Ejrnæs, R. (2000) Can we trust gradients extracted by Detrended Correspondence Analysis? *Journal of Vegetation Science* **11**, 565–572.

Ellenberg, H. (1956) Aufgaben und Methoden der Vegetationskunde, in *Grundlagen der Vegetationsgliederung* (ed. H. Walter), E. Ulmer, Stuttgart, pp. 1–135.

Ellenberg, H. (1979) Zeigerwerte von Gefässpflanzen Mitteleuropas. *Scripta Geobotanica* **9**, 1–122.

Ellenberg, H. (1988) *Vegetation Ecology of Central Europe* (4th edn), Cambridge University Press, Cambridge.

Ellenberg, H., Weber, H.E., Düll, R., *et al.* (1992) Zeigerwerte von Pflanzen in Mitteleuropa. *Scripta Geobotanica* **18**, 1–248.

Elphick, C.S. (2008) How you count counts: the importance of methods research in applied ecology. *Journal of Applied Ecology* **45**, 1313–1320.

Elton, C.S. (1966) *The Pattern of Animal Communities*, Methuen, London.

Elton, C.S. and Miller, R.S. (1954) The ecological survey of animal communities with a practical system of classifying habitats by structural characters. *Journal of Ecology* **42**, 460–496.

Emery, M, (1986) *Promoting Nature in Cities and Towns – a Practical Guide*, Croom Helm, London.

Equihua, M. (1991) Fuzzy clustering of ecological data. *Journal of Ecology* **78**, 519–534.

Erickson, B.H. and Nosanchuk, T.A. (1992) *Understanding Data* (2nd edn), Open University Press, Milton Keynes.

Ertsen, A.C.D., Alkemade, J.R.M. and Wassen, M.J. (1998) Calibrating Ellenberg indicator values for moisture, acidity, nutrient availability and salinity in the Netherlands. *Plant Ecology* **135**, 113–124.

Everitt, B.S., Landau, S. and Leese, M. (2001) *Cluster Analysis* (4th edn), Arnold, London.

Ewald, J. (2003) A critique for phytosociology. *Journal of Vegetation Science* **14**, 291–296.

Eyre, S.R. (1968) *Vegetation and Soils – a World Picture*, Arnold, London.

Ezcurra, E. (1987) A comparison of reciprocal averaging and non-centred principal components analysis. *Vegetatio* **71**, 41–47.

Faber-Langendoen, D., Aaseng, N., Hop, K., *et al.* (2007) Vegetation classification, mapping and monitoring at Voyageurs National Park, Minnesota: a application of the U.S. National Vegetation Classification. *Applied Vegetation Science* **10**, 361–374.

Faber-Langendoen, D., Tart, D.L. and Crawford, R.H. (2009) Contours of the revised U.S. National Vegetation Standard. *Bulletin of the Ecological Society of America* **90**, 87–93.

Farina, A. (2000) *Landscape Ecology in Action*, Kluwer, Dordrecht, The Netherlands.

Farina, A. (2006) *Principles and Methods in Landscape Ecology* (2nd edn), Kluwer, Dordrecht, The Netherlands.

Fasham, M.J.R. (1977) A comparison of non-metric multidimensional scaling, principal components analysis and reciprocal averaging for the ordination of simulated coenoclines and coenoplanes. *Ecology* **58**, 551–561.

Federal Geographic Data Committee (FGDC) (1997/2008) *Vegetation Classification Standard. FGDC-STD-005. Versions 1 and 2*. Vegetation Subcommittee, Federal Geographic Data Committee, FGDC Secretariat, U.S. Geological Survey, Reston, Virginia, USA. Available at http://www.fgdc.gov/standards/projects/FGDC-standards-projects/vegetation/NVCS_V2_FINAL_2008-02.pdf)

Fehmi, J.S. (2009) Confusion among three common plant cover definitions may result in data unsuited for comparison. *Journal of Vegetation Science* **20**, 1–7.

Fenner, M. (1997) Evaluation of methods for estimating vegetation cover in a simulated grassland sward. *Journal of Biological Education* **31**, 49–54.

Feyerabend, P. (1975) *Against Method*, Verso, London.

Feyerabend, P. (1978) *Science in a Free Society*, Verso, London.

Fielding, A.H. (2007) *Cluster Analysis and Classification Techniques for the Biosciences*, Cambridge University Press, Cambridge.

Fischer, H.S. and Bemmerlein, F.A. (1989) An outline for data analysis in phytosociology: past and present. *Vegetatio* **81**, 17–28.

Fisher, R.A. (1940) The precision of discriminant functions. *Annals of Eugenics* **10**, 422–429.

Fitter, A.H. (1987) Spatial and temporal separation of activity in plant communities: prerequisite or consequence of coexistence? in *Organization of Communities Past and Present* (eds J.H.R. Gee and P.S. Giller), British Ecological Society/Blackwell Scientific Publications, Oxford, pp. 119–139.

Floyd, D.A. and Anderson, J.E. (1987) A comparison of three methods for estimating plant cover. *Journal of Ecology* **75**, 221–228.

Ford, E.D. (2000) *Scientific Method for Ecological Research*, Cambridge University Press, Cambridge.

Forman, R.T.T. (1995) *Land Mosaics*, Cambridge University Press, Cambridge.

Forman, R.T.T. and Godron, M. (1986) *Landscape Ecology*, Wiley, New York.

Fortin, M.-J. (1999) Spatial statistics in landscape ecology, in *Landscape Ecological Analysis: Issues and Applications* (eds J.M. Klopatek and R.H. Gardner), Springer-Verlag, New York, pp. 253–279.

Fortin, M.-J, Drapeau, P. and Legendre, P. (1989) Spatial autocorrelation and sampling design in plant ecology. *Vegetatio* **83**, 209–222.

Fortin, M.-J. and Jacquez, G.M. (2000) Randomisation tests and spatially autocorrelated data. *Bulletin of the Ecological Society of America* **81**, 201–206.

Fortin, M.-J., Olson, R.J., Ferson, S., *et al.* (2000) Issues related to the detection of boundaries. *Landscape Ecology* **15**, 453–466.

Fortin, M.-J. and Gurevitch, J. (2001) Mantel tests: spatial structure in field experiments, in *Design and Analysis of Ecological Experiments* (2nd edn) (eds S.M. Scheiner and J. Gurevitch), Oxford University Press, Oxford, pp. 308–326.

Fortin, M.-J and Dale, M. (2005) *Spatial Analysis: a Guide for Ecologists*, Cambridge University Press, Cambridge.

Fosberg, F.R. (1961) A classification of vegetation for general purposes. *Tropical Ecology* **2**, 1–28.

Foster, B.L. and Tilman, D. (2000) Dynamic and static views of succession: Testing the descriptive power of the chronosequence approach. *Plant Ecology* **146**, 1–10.

Franklin, S.E. (2009) *Remote Sensing for Biodiversity and Wildlife Management: Synthesis and Applications*, McGraw-Hill, New York.

Franquet, E., Dolédec, S. and Chessel, D. (1995) Using multivariate analyses for separating spatial and temporal effects within species-environment relationships. *Hydrobiologia* 300–301, 425–531.

French, D.D. and Cummins, R.P. (2001) Classification, composition, richness and diversity of British hedgerows. *Applied Vegetation Science* **4**, 213–228.

Fridriksson, S. (1975) *Surtsey – Evolution of Life on a Volcanic Island*, Butterworth, London.

Fridriksson, S. (1987) Plant colonisation of a volcanic island, Surtsey, Iceland. *Arctic and Alpine Research* **19**, 425–431.

Fridriksson, S. (1989) The volcanic island of Surtsey, Iceland, a quarter century after it 'rose from the sea'. *Environmental Conservation* **16**, 157–162.

Frodin, D.G. (2001) *Guide to Standard Floras of the World* (2nd edn), Cambridge University Press, Cambridge.

Fry, J.C. (1993) Bivariate regression, in *Biological Data Analysis: a Practical Approach* (ed. J.C. Fry), IRL Press/Oxford University Press, Oxford, 81–125.

Gabriel, K.R. (1971) The biplot graphic display of matrices with application to principal component analysis. *Biometrika* **58**, 453–467.

Gabriel, K.R. (1981) Biplot display of multivariate matrices for inspection of data and diagrams, in *Interpreting Multivariate Data* (ed. V. Barnett), John Wiley & Sons Ltd, Chichester, pp. 147–173.

Galán de Mera, A., Hagen, M.A. and Vicente Orellana, J.A. (1999) Aerophyte: a new life form in Raunkaier's classification? *Journal of Vegetation Science* **10**, 65–68.

Gallegos Torell, A. and Glimskär, A. (2009) Computer-aided calibration for visual estimation of vegetation cover. *Journal of Vegetation Science* **20**, 973–983.

Gaston, K.J. (2003) *The Structure and Dynamics of Geographic Ranges*, Oxford University Press, Oxford.

Gaston, K.J. and Spicer, J.I. (2004) *Biodiversity: an Introduction*, Blackwell Science, Oxford.

Gauch, H.G. (1979) *COMPCLUS: a FORTRAN Program for the Initial Clustering of Large Data Sets*. Cornell University, Department of Ecology and Systematics, Ithaca, New York.

Gauch, H.G. (1980) Rapid initial clustering of large data sets. *Vegetatio* **42**, 103–111.

Gauch, H.G. (1982a) Noise reduction by eigenvector ordination. *Ecology* **63**, 1643–1649.

Gauch, H.G. (1982b) *Multivariate Analysis in Community Ecology*, Cambridge Studies in Ecology, Cambridge University Press, Cambridge.

Gauch, H.G. and Whittaker, R.H. (1972a) Coenocline simulation. *Ecology* **53**, 446–451.

Gauch, H.G. and Whittaker, R.H. (1972b) Comparison of ordination techniques. *Ecology* **53**, 868–875.

Gauch, H.G. and Whittaker, R.H. (1976) Simulation of community patterns. *Vegetatio* **33**, 13–16.

Gauch, H.G., Whittaker, R.H. and Wentworth, R.T. (1977) A comparative study of reciprocal averaging and other ordination techniques. *Journal of Ecology* **65**, 157–174.

Gauch, H.G. and Scruggs, W.M. (1979) Variants of polar ordination. *Vegetatio* **40**, 147–153.

Gauch, H.G. and Whittaker, R.H. (1981) Hierarchical classification of community data. *Journal of Ecology* **69**, 135–152.

Gauch, H.G., Whittaker, R.H. and Singer, S.B. (1981) A comparative study of non-metric ordinations. *Journal of Ecology* **69**, 135–152.

Gerrard, J. (2000) *Fundamentals of Soils*, Routledge, London.

Ghilarov, A. (1996) What does 'biodiversity' mean? – scientific problem or convenient myth? *Trends in Ecology and Evolution* **11**, 304–306.

Gibbons, P. and Freudenberger, D. (2006) An overview of methods used to assess vegetation condition at the scale of the site. *Ecological Management and Restoration* **7**, S10–S17.

Gibson, C.W.D. and Brown, V.K. (1986) Plant succession: theory and application. *Progress in Physical Geography* **10**, 473–493.

Gilbert, B. and Bennett, J.R. (2010) Partitioning variation in ecological communities: do the numbers add up? *Journal of Applied Ecology* **47**, 1071–1082.

Gilbert, O.L. (1989) *The Ecology of Urban Habitats*, Chapman and Hall, London.

Gilbertson, D.D., Kent, M. and Pyatt, F.B. (1985) *Practical Ecology for Geography and Biology*, Hutchinson/Unwin Hyman, London.

Giller, P.S. and Gee, J.H.R. (1987) The analysis of community organization: the influence of equilibrium, scale and terminology, in *Organization of Communities Past and Present* (eds J.H.R Gee and P.S. Giller), British Ecological Society/Blackwell Scientific Publications, Oxford, pp. 519–542.

Gilliam, F.S. and Saunders, N.E. (2003) Making more sense of the order: a review of CANOCO for Windows 4.5, PC-ORD Version 4 and SYN-TAX 2000. *Journal of Vegetation Science* **14**, 297–304.

Gillison, A.N. and Anderson, D.J. (eds) (1981) *Vegetation Classification in Australia*, CSIRO, Canberra, Australia.

Giraudel, J.L. and Lek, S. (2001) A comparison of self-organizing map algorithm and some comventional statistical methods for ecological community ordination. *Ecological Modelling* **146**, 329–339.

Gitay, H. and Noble, I.R. (1997) What are functional types and how should we seek them? in *Plant Functional Types* (eds T.M. Smith, H.H. Shugart and F.I. Woodward), Cambridge University Press, Cambridge, pp. 3–19.

Gittins, R. (1969) The application of ordination techniques, in *Ecological Aspects of Mineral Nutrition in Plants* (ed. I.H. Rorison), Blackwell Scientific, Oxford, pp. 37–66.

Gittins, R. (1979) Ecological applications of canonical analysis, in *Multivariate Methods in Ecological Work* (eds L. Orloci, C.R. Rao and W.M. Stiteler), Statistical Ecology, No. 7., Md. Int. Coop., Fairland, pp. 309–535.

Gittins, R., Amir, S., Dupouey, J.-L., *et al.* (1987) Numerical methods in terrestrial plant ecology, in *Developments in Numerical Ecology* (eds P. Legendre and L. Legendre) NATO ASI Series G14, Springer-Verlag, Berlin, pp. 529–558.

Gleason, H.A. (1917) The structure and development of the plant association. *Bulletin of the Torrey Botanical Club* **43**, 463–481.

Gleason, H.A. (1926) The individualistic concept of the plant association. *Bulletin of the Torrey Botanical Club* **53**, 1–20.

Gleason, H.A. (1927) Further views on the succession concept. *Ecology* **8**, 299–326.

Gleason, H.A. (1939) The individualistic concept of the plant association. *American Midland Naturalist* **21**, 92–110.

Glenn-Lewin, D.C. and van der Maarel, E. (1992) Patterns and processes of vegetation dynamics, in *Plant Succession: Theory and Prediction* (eds D.C. Glenn-Lewin, R.K. Peet and T.T. Veblen), Chapman and Hall, New York, pp. 11–59.

Glenn-Lewin, D.C., Peet, R.K. and Veblen, T.T. (eds) (1992) *Plant Succession: Theory and Prediction*, Chapman and Hall, New York.

Godsoe, W. (2010) I can't define the niche but I know it when I see it: a formal link between statistical theory and the ecological niche. *Oikos* **119**, 53-60.

Goff, F.G. and Cottam, G. (1967) Gradient analysis: the use of species and synthetic indices. *Ecology* **48**, 793–806.

Goldsmith, F.B. (1973a) The vegetation of exposed sea cliffs at South Stack, Anglesey. I. The multivariate approach. *Journal of Ecology* **61**, 787–818.

Goldsmith, F.B. (1973b) The vegetation of exposed sea cliffs at South Stack, Anglesey. II. Experimental studies. *Journal of Ecology* **61**, 819–830.

Goldsmith, F.B. (1974) An assessment of the Fosberg and Ellenberg methods of classifying vegetation for conservation purposes. *Biological Conservation* **6**, 3–6.

Goldsmith, F.B. (1975) An evaluation of ecological resources in the countryside for conservation purposes. *Biological Conservation* **8**, 89–96.

Goldsmith, F.B. (1983) Evaluating nature, in *Conservation in Perspective* (eds A. Warren and F.B. Goldsmith), John Wiley and Sons Ltd, Chichester, pp. 233–246.

Goldsmith, F.B. (1991) Vegetation monitoring, in *Monitoring for Conservation and Ecology* (ed. F.B. Goldsmith), Chapman and Hall, London, pp. 77–86.

Gondard, H., Jauffret, S., Aronson, J. and Lavorel, S. (2003) Plant functional types: a promising tool for management and restoration of degraded lands. *Applied Vegetation Science* **6**, 223–234.

Goodall, D.W. (1954) Objective methods for the comparison of vegetation. III. An essay in the use of factor analysis. *Australian Journal of Botany* **1**, 39–63.

Goodall, D.W. (1978) Numerical classification, in *Classification of Plant Communities* (ed. R.H. Whittaker), Junk, The Hague, pp. 247–286.

Goodman, D. (1975) The theory of diversity-stability relationships in ecology. *Quarterly Review of Biology* **50**, 237–266.

Gordon, A.D. (1987) A review of hierarchical classification. *Journal of the Royal Statistical Society Series A* **150**, 119–137.

Gordon, A.D. (1996) Hierarchical classification, in *Clustering and Classification* (eds P. Arabie, L.J. Hubert and G. De Soete), World Scientific Publications, River Edge, NJ, 65–121.

Gorlick, R. (2006) Combining richness and abundance into a single diversity index using matrix analogues of Shannon's and Simpson's indices. *Ecography* **29**, 525–530.

Gorrod, E. and Keith, D. (2009) Observer variation in field assessments of vegetation condition: implications for biodiversity conservation. *Ecological Management and Restoration* **10**, 31–40.

Gotelli, N.J. (2001) *A Primer of Ecology* (3rd edn), Sinauer Associates Inc., Sunderland, MA.

Gotelli, N.J. and Colwell, R.K. (2001) Quantifying biodiversity: procedures and pitfalls in the measurement and comparison of species richness. *Ecology Letters* **4**, 379–391.

Gotelli, N.J. and Ellison, A.M. (2004) *A Primer of Ecological Statistics*, Sinauer Associates Inc., Sunderland, MA.

Gotelli, N.J. and Ulrich, W. (2010) The empirical Bayes approach as a tool to identify non-random species associations. *Oecologia* **162**, 463–477.

Gould, P. and White, R. (1974) *Mental Maps*, Penguin, Harmondsworth.

Gould, P. and White, R. (1986) *Mental Maps* (2nd edn), Allen and Unwin, London.

Gove, J.H., Banapati, P.P. and Taillie, C. (1996) Diversity measurement and comparison with examples, in *Biodiversity in Managed Landscapes* (eds R.C. Szaro and D.W. Johnson), Oxford University Press, New York, pp. 157–175.

Gower, J.C. (1966) Some distance properties of latent root and vector methods used in multivariate analysis. *Biometrika* **53**, 325–338.

Gower, J.C. (1967) A comparison of some methods of cluster analysis. *Biometrics* **23**, 623–637.

Gower, J.C. (1971) A general coefficient of similarity and some of its properties. *Biometrics* **27**, 857–871.

Gower, J.C. (1974) Maximal predictive classification. *Biometrics* **30**, 643–654.

Gower, J.C. (1985) Measures of similarity, dissimilarity and distance, in *Encyclopedia of Statistics, Vol. 5* (eds N.L. Johnson, S. Kotz and C.B. Read), John Wiley and Sons Ltd., New York, pp. 397–405.

Gower, J.C. (1987) Introduction to ordination techniques, in *Developments in Numerical Ecology* (eds P. Legendre and L. Legendre), NATO ASI Series G14, Springer-Verlag, Berlin, 5–64.

Gower, J.C. (1988) Classification, geometry and data analysis, in *Classification and Related Methods of Data Analysis* (ed. H.H. Bock), Elsevier, North-Holland, pp. 3–14.

Grabmeier, J. and Rudolph, A. (2002) Techniques of cluster algorithms in data mining. *Data Mining and Knowledge Discovery* **6**, 303–360.

Graham, M.H. (2003) Confronting multicollinearity in ecological multiple regression. *Ecology* **84**, 2809–2815.

Grandin, U. (2006) PC-ORD version 5: A user-friendly toolbox for ecologists. *Journal of Vegetation Science* **17**, 843–844.

Grant, S.A. (1993) Resource description: vegetation and sward components, in *Sward Measurement Handbook* (2nd edn) (eds A. Davies, R.D. Baker, S.A. Grant and A.S. Laidlaw), British Grassland Society, University of Reading, Reading, pp. 67–97.

Gray, A.J., Crawley, M.J. and Edwards, P.J. (eds) (1987) *Colonisation, Succession and Stability*, 26th Symposium of the British Ecological Society, Blackwell Scientific, Oxford.

Greenacre, M.J. (1981) Practical correspondence analysis, in *Interpreting Multivariate Data* (ed. M.J. Barnett), John Wiley & Sons Ltd, New York, pp. 119–146.

Greenacre, M.J. (1984) *Theory and Applications of Correspondence Analysis*, Academic Press, London.

Greenwood, J.J.D. and Robinson, R.A. (2006) Principles of sampling, in *Ecological Census Techniques: a Handbook* (2nd edn) (ed. W.J. Sutherland), Cambridge University Press, Cambridge, pp. 11–86.

Gregorius, H.R. and Gillet, E.M. (2008) Generalised Simpson diversity. *Ecological Modelling* **211**, 90–96.

Greig-Smith, P. (1980) The development of numerical classification and ordination. *Vegetatio* **42**, 1–9.

Greig-Smith, P. (1983) *Quantitative Plant Ecology* (3rd edn), Studies in Ecology Vol. 9, Blackwell Scientific, Oxford.

Grime, J.P. (1974) Vegetation classification by reference to strategies. *Nature* **250**, 26–31.

Grime, J.P. (1977) Evidence for the existence of three primary strategies in relation to plants and its relevance to ecological and evolutionary theory. *American Naturalist* **11**, 1169–1194.

Grime, J.P. (1979) *Plant Strategies and Vegetation Processes*, John Wiley and Sons Ltd, Chichester.

Grime, J.P. (1993) Ecology sans frontières. *Oikos* **68**, 385–392.

Grime, J.P. (2001) *Plant Strategies, Vegetation Processes and Ecosystem Properties* (2nd edn), John Wiley and Sons Ltd, Chichester.

Grime, J.P. (2007) Plant strategy theories: a comment on Craine (2005). *Journal of Ecology* **95**, 227–230.

Grime, J.P., Thompson, K., Hunt, R., *et al.* (1997) Integrated screening validates primary axes of specialisation in plants. *Oikos* **79**, 259–281.

Grossman, D.H., Faber-Langendoen, D., Weakley, A.S., *et al.* (1998) *Terrestrial Vegetation of the United States – Volume 1. The National Vegetation Classification System: Development, Status, and Applications.* The Nature Conservancy, Washington, USA. Available at http://www.natureserve.org/library/vol1.pdf [accessed 10 February 2010].

Grubb, P.J. (1987) Global trends in species-richness in terrestrial vegetation: a view from the Northern Hemisphere, in *Organization of Communities Past and Present* (eds J.H.R. Gee and P.S. Giller), British Ecological Society/Blackwell Scientific Publications, Oxford, 99–118.

Guenard, G., Legendre, P., Boisclair, D. and Bilodeau, M. (2010) Multiscale codependence analysis: an integrated approach to analyze relationships across scales. *Ecology* **91**, 2952–2964.

Guinochet, M. (1973) *Phytosociologie*, Masson, Paris.

Gunnarsson, T.G., Appleton, G.F., Gíslason, H., *et al.* (2006) Large-scale habitat associations of birds in lowland Iceland: implications for conservation. *Biological Conservation* **128**, 265–275.

Guo, Q. and Rundel, P.W. (1997) Measuring dominance and diversity in ecological communities: choosing the right variables. *Journal of Vegetation Science* **8**, 405–408.

Gurevitch, J., Scheiner, S.M. and Fox, G.A. (2002) *The Ecology of Plants*, Sinauer Associates, Sunderland.

Gustafson, E.J. (1998) Quantifying landscape spatial pattern: what is the state of the art? *Ecosystems* **1**, 143–156.

Haigh, M.J. (1980) Ruderal communities in English cities. *Urban Ecology* **4**, 329–338.

Haines-Young, R. and Petch, J. (1986) *Physical Geography: its Nature and Methods*, Harper and Row, London.

Haines-Young, R., Green, D.R. and Cousins, S. (eds) (1993a) *Landscape Ecology and Geographic Information Systems*, Taylor and Francis, London.

Haines-Young, R., Green, D.R. and Cousins, S. (1993b) Landscape ecology and spatial information systems, in *Landscape Ecology and Geographic Information Systems* (eds R. Haines-Young, D.R. Green and S. Cousins), Taylor and Francis, London, pp. 3–8.

Hairston, N.G. Sr. (1989) *Ecological Experiments. Purpose, Design and Execution*, Cambridge Studies in Ecology, Cambridge University Press, Cambridge.

Halloy, S. (1990) A morphological classification of plants, with special reference to the New Zealand alpine flora. *Journal of Vegetation Science* **1**, 291–304.

Hamilton, A.J. (2005) Species diversity or biodiversity? *Journal of Environmental Management* **75**, 89–92.

Hammond, R. and McCullagh, P. (1978) *Quantitative Techniques in Geography – an Introduction* (2nd edn), Oxford University Press, Oxford.

Hansjörg, D. and Steinlein, T. (1996) Determination of plant species cover by means of image analysis. *Journal of Vegetation Science* **7**, 131–136.

Hanski, I. (1982) Dynamic of regional distribution: the core and satellite species hypothesis. *Oikos* **38**, 210–221.

Hanski, I. (1991) Single-species metapopulation dynamics: concepts, models and observations. *Biological Journal of the Linnean Society* **42**, 17–38.

Hanski, I. (1998) Connecting the parameters of local extinction and metapopulation dynamics. *Oikos* **83**, 390–396.

Hanski, I. (1999) *Metapopulation Ecology*, Oxford University Press, Oxford.

Hanski, I. and Gyllenberg, M. (1993) Two general metapopulation models and the core-satellite species hypothesis. *American Naturalist* **142**, 17–41.

Hanski, I. and Gyllenberg, M. (1997) Uniting two general patterns in the distributions of species. *Science* **275**, 397–400.

Hanski, I., Kouki, J. and Halkka, A. (1993) Three explanations of the positive relationship between distribution and abundance of species, in *Community Diversity: Historical and Geographical Perspectives* (eds R.E. Ricklefs and D. Schliuter), Chicago University Press, Chicago, pp. 108–116.

Hanwell, J.D. and Newson, M.D. (1973) *Techniques in Physical Geography*, Macmillan, London.

Harper, J.L. (1977) *Population Biology of Plants*, Academic Press, London.

Harper, J.L. and Hawkesworth, D.L. (1994) Biodiversity: measurement and estimation. *Philosophical Transactions of the Royal Society of London Series B* **345**, 5–12.

Harris, J.A., Birch, P. and Palmer, J.P. (1996) *Land Restoration and Reclamation: Principles and Practice*, Longman, Harlow.

Harrison, C., Limb, M. and Burgess, J. (1987) Nature in the city – popular values for a living world. *Journal of Environmental Management* **25**, 347–362.

Harte, J., Conlisk, E., Ostling, A., *et al.* (2005) A theory of spatial structure in ecological communities at multiple spatial scales. *Ecological Monographs* **75**, 179–197.

Hartigan, J.A. and Wong, M.A. (1979) A k-means clustering algorithm. *Applied Statistics – Journal of the Royal Statistical Society Series C* **28**, 100–108.

Hastie, T.J. and Stuetzle, W. (1989) Principal curves. *Journal of the American Statistical Association* **84**, 502–516.

Hatheway, W.H. (1971) Contingency-table analysis of rain forest vegetation, in *Statistical Ecology, Vol. 3.* (eds G.P. Patil, E.C. Pielou and W.E. Waters), Pennsylvania State University Press, pp. 271–313.

Hawkins, B.A. and Agrawal, A.A. (2005) Latitudinal gradients. *Ecology* **86**, 2261–2262.

Haykin, S. (1999) *Neural Networks* (2nd edn), Prentice-Hall, Upper Saddle River, NJ.

Hédl, R. (2007) Is sampling subjectivity a distorting factor in surveys for vegetation diversity? *Folia Geobotanica* **42**, 191–198.

Heikkinen, J. and Mäkipää, R. (2010) Testing hypotheses on shape and distribution of ecological response curves. *Ecological Modelling* **221**, 388–399.

Heltshe, J.F. and Forrester, N.E. (1983a) Estimating diversity using quadrat sampling. *Biometrics* **39**, 1073–1076.

Heltshe, J.F. and Forrester, N.E. (1983b) Estimating diversity using the jacknife procedure. *Biometrics* **39**, 1–11.

Heltshe, J.F. and Forrester, N.E. (1985) Statistical evaluation of the jacknife estimate of diversity when using quadrat samples. *Ecology* **66**, 107–111.

Henderson, P.A. (2003) *Practical Methods in Ecology*, Blackwell Science, Oxford.

Henderson, P. and Seaby, R. (2008) *A Practical Handbook for Multivariate Methods*, Pisces Conservation Ltd., Lymington, UK.

Hennekens, S.M. (1996) *TURBO(VEG): Software Package for Input, Processing and Presentation of Phytosociological Data.* IBN-DLO Wageningen, NL and University of Lancaster, UK.

Hennekens, S.M. and Schaminée, J.H.L. (2001) TURBOVEG, a comprehensive data base management system for vegetation data. *Journal of Vegetation Science* **12**, 589–591.

Herben, T. (1996) Permanent plots as tools for plant community ecology. *Journal of Vegetation Science* **7**, 195–202.

Hilbert, D.W. and Ostendorf, B. (2001) The utility of artificial neural networks for modelling the distribution of vegetation in past, present and future climates. *Ecological Modelling* **146**, 311–327.

Hilborn, R. and Mangel, M. (1997) *The Ecological Detective: Confronting Models with Data*, Princeton University Press, Princeton, NJ.

Hill, J.L. and Hill, R.A. (2001) Why are tropical rain forests so species-rich? Classifying, reviewing and evaluating theories. *Progress in Physical Geography* **25**, 326–354.

Hill, M.O. (1973a) Diversity and evenness: a unifying notation and its consequences. *Ecology* **54**, 427–432.

Hill, M.O. (1973b) Reciprocal averaging: an eigenvector method of ordination. *Journal of Ecology* **61**, 237–250.

Hill, M.O. (1974) Correspondence analysis. A neglected multivariate method. *Journal of the Royal Statistical Society, Series C* **23**, 340–354.

Hill, M.O. (1977) Use of simple discriminant functions to classify quantitative data, in *First International Symposium on Data Analysis and Informatics* Vol. 1 (eds E. Diday, L. Lebart, J.P. Pages and R. Tomassone), Institute de Recherche d'Informatique et d'Automatique, Le Chesnay, pp. 181–199.

Hill, M.O. (1979a) *DECORANA – a FORTRAN Program for Detrended Correspondence Analysis and Reciprocal Averaging.* Cornell University, Department of Ecology and Systematics, Ithaca, NY.

Hill, M.O. (1979b) *TWINSPAN – a FORTRAN Program for arranging Multivariate Data in an Ordered Two Way Table by Classification of the Individuals and the Attributes.* Cornell University, Department of Ecology and Systematics, Ithaca, NY.

Hill, M.O. (1989) Computerized matching of relevés and association tables, with an application to the British National Vegetation Classification. *Vegetatio* **83**, 187–194.

Hill, M.O. (1991) *TABLEFIT. Program Manual (Version 1).* Institute of Terrestrial Ecology, Huntingdon.

Hill, M.O., Bunce, R.G.H. and Shaw, M.W. (1975) Indicator species analysis, a divisive polythetic method of classification and its application to a survey of native pinewoods in Scotland. *Journal of Ecology* **63**, 597–613.

Hill, M. and Smith, A. (1976) Principal component analysis of taxonomic data with multi-state discrete characters. *Taxon* **25**, 249–255.

Hill, M.O. and Gauch, H.G. (1980) Detrended correspondence analysis, an improved ordination technique. *Vegetatio* **42**, 47–58.

Hill, M.O. and Carey, P.D. (1997) Prediction of yield in the Rothamsted Park Grass Experiment by Ellenberg indicator values. *Journal of Vegetation Science* **8**, 579–586.

Hill, M.O., Mountford, J.O., Roy, D.B. and Bunce, R.G.H. (1999) *Ellenberg's Indicator Values for British Plants.* ECOFACT Vol. 2, Technical Annex, Institute of Terrestrial Ecology, Huntingdon.

Hill, M.O., Roy, D.B., Mountford, J.O. and Bunce, R.G.H. (2000) Extending Ellenberg's indicator values to a new area: an algorithmic approach. *Journal of Applied Ecology* **37**, 3–15.

Hill, M.O. and Šmilauer, P. (2005) *TWINSPAN for Windows Version 2.3.* Huntingdon and České Budějovice, Centre for Ecology and Hydrology and University of South Bohemia.

Hinch, S.G. and Somers, K.M. (1987) An experimental evaluation of the effect of data centering, data standardization and outlying observations on principal component analysis. *Coenoses* **2**, 19–23.

Hirschfeld, H.O. (1935) A connection between correlation and contingency. *Proceedings of the Cambridge Philosophical Society* **31**, 520–527.

Hirst, C.N. and Jackson, D.A. (2007) Reconstructing community relationships: the impact of sampling error, ordination approach and gradient length. *Diversity and Distributions* **13**, 361–371.

Hoagland, B.W. and Collins, S.L. (1997) Gradient models, gradient analysis and hierarchical structure in plant communities. *Oikos* **78**, 23–30.

Hoaglin, J.D., Mosteller, F. and Tukey, J.W. (1983) *Understanding Robust and Exploratory Data Analysis*, John Wiley and Sons Ltd, New York.

Hodgson, J.G., Wilson, P.J., Hunt, R., *et al.* (1999) Allocating C-S-R plant functional types: a soft approach to a hard problem. *Oikos* **83**, 282–294.

Hoekstra, T.W., Allen, T.F.H. and Flather, C.H. (1991) Implicit scaling in ecological research. *BioScience* **41**, 148–159.

Hooper, M.D. (1970) Dating hedges. *Area* **2**, 63–65.

Hopkins, B. (1955) The species-area relations of plant communities. *Journal of Ecology* **43**, 209–226.

Hopkins, B. (1957) Pattern in the plant community. *Journal of Ecology* **45**, 451–463.

Hopkins, B. (1965) *Forest and Savanna*, Heinemann, London.

Horning, N., Robinson, J.A., Sterling, E.J., *et al.* (2010) *Remote Sensing for Ecology and Conservation*, Oxford University Press, Oxford.

Hotelling, H. (1933) Analysis of a complex of statistical variables into principal components. *Journal of Educational Psychology* **24**, 417–441; 498–520.

Huerta-Martínez, F.M., Vázquez-García, J.A., García-Moya, E., *et al.* (2004) Vegetation ordination at the southern Chihuahuan Desert (San Luis Potosi, Mexico). *Plant Ecology* **174**, 79–87.

Huisman, J., Olff, H. and Fresco, L.F.M. (1993) A hierarchical set of models for species response analysis. *Journal of Vegetation Science* **4**, 37–46.

Hunt, R., Hodgson, J.G., Thompson, K., *et al.* (2004) A new practical tool for deriving a functional signature for herbaceous vegetation. *Applied Vegetation Science* **7**, 163–170.

Huntley, B.J. and Walker, B.H. (eds) (1982) *Ecology of Tropical Savannas.* Ecological Studies 42, Springer-Verlag, Berlin.

Hurlbert, S.H. (1971) The nonconcept of species diversity: a critique and alternative parameters. *Ecology* **52**, 577–586.

Hurlbert, S.H. (1984) Pseudoreplication and the design of ecological field experiments. *Ecological Monographs* **54**, 187–211.

Hurlbert, S.H. (2004) On misinterpretation of pseudoreplication and related matters: a reply to Oksanen. *Oikos* **104**, 591–597.

Huston, M.A. (1979) A general hypothesis of species diversity. *American Naturalist* **113**, 81–101.

Huston, M.A. (2003) *Biological Diversity – the Coexistence of Species in Changing Landscapes* (2nd edn), Cambridge University Press, Cambridge.

Huston, M.A. and Smith, T. (1987) Plant succession: life history and competition. *American Naturalist* **130**, 168–198.

Hutcheson, K. (1970) A test for comparing diversities based on the Shannon formula. *Journal of Theoretical Biology* **29**, 151–154.

Hutchings, M.J. (1983) Plant diversity in four chalk grassland sites with different aspects. *Vegetatio* **53**, 179–189.

Hutchinson, G.E. (1957) Concluding remarks. *Cold Spring Harbour Symposium on Quantitative Biology* **22**, 415–427.

Inchausti, P. (1994) Reductionist approaches in community ecology. *American Naturalist* **143**, 201–221.

Isaac, N.J.B., Mallet, J. and Mace, G.M. (2004) Taxonomic inflation: Its influence on macroecology and conservation. *Trends in Ecology and Evolution* **19**, 464–469.

Jaccard, P. (1901) Étude comparative de la distribution florale dans une portion des Alpes et du Jura. Bulletin *Soc. Vaud. Sc. Nat.* **37**, 547–579.

Jaccard, P. (1912) The distribution of the flora of the alpine zone. *New Phytologist* **11**, 37–50.

Jaccard, P. (1928) Die statistisch-floristische method als grundlage derpflanzensoziologie. *Abderhalden, Handbuch Biologisch Arbeitsmethod* **11**, 165–202.

Jackson, D.A. and Somers, K.M. (1991) Putting things in order: the ups and downs of detrended correspondence analysis. *American Naturalist* **137**, 704–712.

James, F.C. and McCulloch, C.E. (1990) Multivariate analysis in ecology and systematics: panacea or Pandora's box? *Annual Review of Ecology and Systematics* **21**, 129–166.

Janssen, J.G.M. (1975) A simple clustering procedure for preliminary classification of very large sets of phytosociological results. *Vegetatio* **30**, 67–71.

Janzen, D. and CBL Plant Working Group (2009) A DNA barcode for land plants. *Proceedings of the National Academy of Sciences USA* **106**, 12794–12797.

Jennings, M.D., Loucks, O., Glenn-Lewin, D., *et al.* (2002) *Standards for Associations and Alliances of the U.S. National Vegetation Classification. Version 1.0.* Ecological Society of America Vegetation Classification Panel. Available at www.esa.org/vegweb/vegstds_v1.htm

Jennings, M.D., Faber-Langendoen, D., Loucks, O.L., *et al.* (2008) *Description, Documentation and Evaluation of Associations and Alliances within the U.S. National Vegetation Classification. Version 5.2.* Ecological Society of America, Vegetation Classification Panel. Available at http:/www.esa.org/vegweb/docFiles?ESA_Guidelines_Version_5.2.pdf

Jennings, M.D., Faber-Langendoen, D., Loucks, O.L., *et al.* (2009) Standards for associations and alliances of the US National Vegetation Classification. *Ecological Monographs* **79**, 173–199.

Jerram, R. and Drewitt, A. (1998) *Assessing Vegetation Condition in the English Uplands.* English Nature Research Report No. 264, English Nature, Peterborough.

Jetz, W., Rahbek, C. and Lichstein, J.W. (2005) Local and global approaches to spatial data analysis in ecology. *Global Ecology and Biogeography* **14**, 97–98.

Johnson, D.H. (1995) Statistical sirens: the allure of nonparametrics. *Ecology* **76**, 1998.

Johnson, D.H. (1999) The insignificance of statistical significance testing. *Journal of Wildlife Management* **63**, 763–772.

Johnson, D.H. (2002) The role of hypothesis testing in wildlide science. *Journal of Wildlife Management* **66**, 272–276.

Johnson, E.A. and Miyanishi, K. (2008) Testing the assumptions of chronosequences in succession. *Ecology Letters* **11**, 419–431.

Johnston, C.A. (1998) *Geographic Information Systems in Ecology*, Blackwell Science, Oxford.

Johnston, R.J. (1978) *Multivariate Statistical Analysis in Geography*, Longman, London.

Joliffe, I.T. (2002) *Principal Component Analysis* (2nd edn), Springer-Verlag, New York/London.

Jones, C.G. and Callaway, R.M. (2007) The third party. *Journal of Vegetation Science* **18**, 771–776.

Jongman, R.H.G., ter Braak, C.J.F. and van Tongeren, O.F.R. (1995) *Data Analysis in Community and Landscape Ecology* (2nd edn), Cambridge University Press, Cambridge.

Jonsson, B.G. and Moen, J. (1998) Patterns in species associations in plant communities: the importance of scale. *Journal of Vegetation Science* **9**, 327–332.

Kaufman, L. and Rousseeuw, P.J. (1990) *Finding Groups in Data, an Introduction to Cluster Analysis*, John Wiley and Sons Ltd, New York.

Keddy, P.A. (2001) *Competition* (2nd edn), Kluwer Academic Publishers, Dordrecht.

Keddy, P.A. (2007) *Plants and Vegetation. Origins, Processes, Consequences*, Cambridge University Press.

Keith, D. and Gorrod, E. (2006) The meanings of vegetation condition. *Ecological Management and Restoration* **7**, S7–S9.

Kenkel, N.G. and Orlóci, L. (1986) Applying metric and non-metric multidimensional scaling to ecological studies: some new results. *Ecology* **67**, 919–928.

Kennedy, K.A. and Addison, P.A. (1987) Some consideration for the use of visual estimates of plant cover in biomonitoring. *Journal of Ecology* **75**, 151–157.

Kent, M. (1972) *A Method for the Survey and Classification of Marginal Land in Agricultural Landscapes.* Discussion Papers in Conservation 1, University College, London.

Kent, M. (1977) BRAYCURT and RECIPRO – two programs for ordination in ecology and biogeography. *Computer Applications* **4**, 589–647.

Kent, M. (1982) Plant growth in colliery spoil reclamation. *Applied Geography* **2**, 83–107.

Kent, M. (1987a) Reclamation of deep coal mining wastes with particular reference to Britain and Western Europe, in *Environmental Consequences of Energy Production* (eds S.K. Majumdar, F.J. Brenner and E.W. Miller), Pennsylvania Academy of Science, 61–77.

Kent, M. (1987b) Island biogeography and habitat conservation. *Progress in Physical Geography* **11**, 91–102.

Kent, M. (2005) Biogeography and macroecology. *Progress in Physical Geography* **29**, 256–264.

Kent, M. (2006) Numerical classification and ordination methods in biogeography. *Progress in Physical Geography* **30**, 399–408.

Kent, M. (2007a) Biogeography and landscape ecology. *Progress in Physical Geography* **31**, 345–355.

Kent, M. (2007b) Biogeography and macroecology: now a significant component of physical geography. *Progress in Physical Geography* **31**, 643–657.

Kent, M. (2009a) Space: making room for space in physical geography, in *Key Concepts in Geography* (2nd edn) (eds N. Clifford, S. Holloway, S. Rice and G. Valentine), Sage Publications, London, pp. 97–118.

Kent, M. (2009b) Biogeography and landscape ecology: the way forward – gradients and graph theory. *Progress in Physical Geography* **33**, 424–436.

Kent, M. and Wathern, P. (1980) The vegetation of a Dartmoor catchment. *Vegetatio* **43**, 163–172.

Kent, M. and Smart, N. (1981) A method for habitat assessment in agricultural landscapes. *Applied Geography* **1**, 9–30.

Kent, M. and Ballard, J. (1988) Trends and problems in the application of classification and ordination methods in plant ecology. *Vegetatio* **78**, 109–124.

Kent, M. and Coker, P. (1992) *Vegetation Description and Analysis a Practical Approach*, John Wiley and Sons Ltd, Chichester.

Kent, M., Weaver, R.E., Gilbertson, D., *et al.* (1996) The present-day machair vegetation of the southern Outer Hebrides, in *The Outer Hebrides: The Last 14,000 Years* (eds D. Gilbertson, M. Kent and J. Grattan), Sheffield Academic Press, Sheffield, pp. 133–145.

Kent, M., Gill, W.J., Weaver, R.E. and Armitage, R.P. (1997) Landscape and plant community boundaries in biogeography. *Progress in Physical Geography* **21**, 315–353.

Kent, M., Stevens, A. and Zhang, L. (1999) Urban plant ecology patterns and processes: a case study of the flora of the City of Plymouth, Devon, England. *Journal of Biogeography* **26**, 1281–1298.

Kent, M., Moyeed, R.A., Reid, C.L., *et al.* (2006) Geostatistics, spatial rate of change analysis and boundary detection in plant ecology and biogeography. *Progress in Physical Geography* **30**, 201–231.

Kercher, S.M., Frieswyk, C.B. and Zedler, J.B. (2003) Effects of sampling teams and estimation methods on the assessment of plant cover. *Journal of Vegetation Science* **14**, 899–906.

Kershaw, K.A. (1968) Classification and ordination of Nigerian savanna vegetation. *Journal of Ecology* **56**, 467–482.

Kershaw, K.A. and Looney, J.H.H. (1985) *Quantitative and Dynamic Plant Ecology* (3rd edn), Arnold, London.

Kéry, M. (2010) *Introduction to WinBUGS for Ecologists*, Academic Press, London.

Keylock, C.J. (2005) Simpson diversity and Shannon-Wiener index as special cases of a generalised entropy. *Oikos* **109**, 203–204.

Kiers, H. (1994) Simple structure in component analysis techniques for mixtures of qualitative and quantitative variables. *Psychometrika* **56**, 197–212.

Kikkawa, J. (1986) Complexity and stability, in *Community Ecology: Pattern and Process* (eds J. Kikkawa and D.J. Anderson), Blackwell Scientific, Oxford, pp. 41–62.

Kirby, K.J., Blines, T., Burn, A., *et al.* (1986) Seasonal and observer differences in vascular plant records from British woodlands. *Journal of Ecology* **74**, 123–132.

Klimeš, L. (2003) Scale-dependent variation in visual estimates of grassland plant cover. *Journal of Vegetation Science* **14**, 815–821.

Klimeš, L., Dančák, M., Hájek, M., *et al.* (2001) Scale-dependent biases in species counts in a grassland. *Journal of Vegetation Science* **12**, 699–704.

Klopatek, J.M. and Gardner, R.H. (eds) (1999) *Landscape Ecological Analysis: Issues and Applications*, Springer, New York.

Knox, R.G. (1989) Effects of detrending and rescaling on correspondence analysis: solution stability and accuracy. *Vegetatio* **83**, 129–136.

Koenig, W.D. (1999) Spatial autocorrelation of ecological phenomena. *Trends in Ecology and Evolution* **14**, 22–26.

Koenig, W.D. and Knops, J.M.H. (1998) Testing for spatial autocorrelation in ecological studies. *Ecography* **21**, 423–429.

Kohonen, T. (1982) Self-organized formation of topologically correct feature maps. *Biological Cybernetics* **43**, 59–69.

Kohonen, T. (1997) *Self-Organizing Maps* (2nd edn), Springer, Berlin.

Kolasa, J. (1989) Ecological systems in hierarchical perspective: breaks in community structure and other consequences. *Ecology* **70**, 36–47.

Koleff, P., Gaston, K.J. and Lennon, J.J. (2003) Measuring beta diversity for presence-absence data. *Journal of Animal Ecology* **72**, 367–382.

Krebs, C.J. (1999) *Ecological Methodology* (2nd edn), Addison Wesley Longman, Menlo Park, CA.

Krebs, C.J. (2001) *Ecology: the Experimental Analysis of Distribution and Abundance* (5th edn), Longman/Harper and Row, New York.

Kroonenberg, P. and Lombardo, R. (1999) Non-symmetric correspondence analysis: a tool for analysing contingency tables with a dependence structure. *Multivariate Behavioral Research* **34**, 367–396.

Kruskal, J.B. (1964a) Multidimensional scaling by optimizing goodness of fit to a nonmetric hypothesis. *Psychometrika* **29**, 1–27.

Kruskal, J.B. (1964b) Nonmetric multidimensional scaling: a numerical method. *Psychometrika* **29**, 115–129.

Kruskal, J.B. and Landwehr, J.M. (1983) Icicle plots: better displays for hierarchical clustering. *American Statistician* **37**, 162–168.

Küchler, A.W. (1967) *Vegetation Mapping*, Ronald Press, New York.

Kuhn, T.S. (1962) *The Structure of Scientific Revolutions*, University of Chicago Press, Chicago.

Kuhn, T.S. (1970) Logic of discovery or the psychology of research? in *Criticism and the Growth of Knowledge* (eds I. Lakatos and A. Musgrave), Cambridge University Press, Cambridge, 1–24.

Kuhnert, P.M., Martin, T.G. and Griffiths, S.P. (2010) A guide to eliciting and using expert knowledge in Bayesian ecological models. *Ecology Letters* **13**, 900–914.

Kuiters, A.T., Kramer, K., van der Hagen, H.G.J.M. and Schaminée, J.H.J. (2009) Plant species diversity, species turnover and shifts in functional traits in coastal dune vegetation: results from permanent plots over a 52-year period. *Journal of Vegetation Science* **20**, 1053–1063.

Kunick, W. (1982) Comparison of the flora of some cities of the central European lowlands, in *Urban Ecology* (eds R. Bornkamm, J.A. Lee and M.R.D. Seaward), 2nd European Ecological Symposium, Blackwell Scientific, Oxford, pp. 13–22.

Kupfer, J.A. (1995) Landscape ecology and biogeography. *Progress in Physical Geography* **19**, 18–34.

Kusch, J., Weber, C., Idelberger, S. and Koob, T. (2004) Foraging habitat preferences of bats in relation to food supply and spatial vegetation structures in a western European low mountain range forest. *Folia Zoologica* **53**, 113–128.

Lájer, K. (2007) Statistical tests as inappropriate tools for data analysis performed on non-random samples of plant communities. *Folia Geobotanica* **42**, 115–122.

Laliberté, E. (2008) Analyzing or explaining beta diversity? Comment. *Ecology* **89**, 3232–3237.

Laliberté, E. and Legendre, P. (2010) A distance-based framework for measuring functional diversity from multiple traits. *Ecology* **91**, 299–305.

Lambert, J.M., Meacock, S.E., Barrs, J. and Smartt, P.F.M. (1973) AXOR and MONIT: Two new polythetic–divisive strategies for hierarchical classification. *Taxon* **22**, 173–176.

Lance, G.N. and Williams, W.T. (1966) A generalised sorting strategy for computer classification. *Nature* **211**, 218.

Lance, G.N. and Williams, W.T. (1967) A general theory of classification sorting strategies. I. Hierarchical systems. *Computer Journal* **9**, 373–380.

Lance, G.N. and Williams, W.T. (1975) REMUL: a new divisive polythetic classificatory program. *Australian Computer Journal* **7**, 109–112.

Laska, G. (2001) The disturbance and vegetation dynamics: a review and an alternative framework. *Plant Ecology* **157**, 77–99.

Latour, J.B., Reiling, R. and Sloof, W. (1994) Ecological standards for eutrophication and desiccation: perspectives for a risk assessment. *Water Air and Soil Pollution* **78**, 265–277.

Laurance, W.F., Albernaz, A.K.M., Schroth, G., *et al.* (2002) Predictors of deforestation in the Brazilian Amazon. *Journal of Biogeography* **29**, 737–748.

Lavorel, S., Grigulis, K., McIntyre, S., *et al.* (2008) Assessing functional diversity in the field – methodology matters! *Functional Ecology* **22**, 134–147.

Law, R. and Watkinson, A.R. (1989) Competition, in *Ecological Concepts: the Contribution of Ecology to an Understanding of the Natural World* (ed. J.M. Cherrett), British Ecological Society/Blackwell Scientific Publications, Oxford, pp. 243–284.

Lawesson, J.E. (2003) pH optima for Danish forest species compared with Ellenberg reaction values. *Folia Geobotanica* **38**, 403–418.

Lawesson, J.E., Diekmann, M., Eilertsen, O., *et al.* (1997) The Nordic Vegetation Survey – concepts and perspectives. *Journal of Vegetation Science* **8**, 455–458.

Lawesson, J.E., Fosaa, A.M. and Olsen, E. (2003) Calibration of Ellenberg indicator values for the Faroe Islands. *Applied Vegetation Science* **6**, 53–62.

Le Duc, M.G., Yang, L. and Marrs, R.H. (2007) A database application for long-term ecological field experiments. *Journal of Vegetation Science* **18**, 509–516.

Lee, J.D. and Lee, T.D. (1982) *Statistics and Numerical Methods in BASIC for Biologists*, Van Nostrand Reinhold, New York.

Legendre, P. (1993) Spatial autocorrelation: trouble or a new paradigm? *Ecology* **74**, 1659–1673.

Legendre, P. and Legendre, L. (eds) (1987) *Developments in Numerical Ecology*, NATO ASI Series G14, Springer-Verlag, Berlin.

Legendre, P. and Fortin, M.-J. (1989) Spatial pattern and ecological analysis. *Vegetatio* **80**, 107–138.

Legendre, P. and Legendre, L. (eds) (1998) *Numerical Ecology* (2nd English edn), *Development and Environmental Modelling* **20**, 1–853 (Elsevier, Amsterdam).

Legendre, P. and Anderson, M.J. (1999) Distance-based redundancy analysis: testing multi-species responses in multifactorial ecological experiments. *Ecological Monographs* **69**, 1–24.

Legendre, P. and Gallagher, E.D. (2001) Ecologically meaningful transformations for ordination of species data. *Oecologia* **129**, 271–280.

Legendre, P., Dale, M.R.T., Fortin, M.-J., *et al.* (2002) The consequences of spatial structure for the design and analysis of ecological field surveys. *Ecography* **25**, 601–615.

Legendre, P., Borcard, D. and Peres-Neto, P.R. (2005) Analyzing beta diversity: partitioning the spatial variation of community composition data. *Ecology* **75**, 435–450.

Legendre, P., Borcard, D. and Peres-Neto, P.R. (2008) Analyzing or explaining beta diversity? Comment. *Ecology* **89**, 3238–3244.

Legg, C.J. (1992) Putting concrete vegetation into abstract boxes. *Bulletin of the British Ecological Society* **xxiii** (1), 28–30.

Leibold, M.A. (1995) The niche concept revisited: mechanistic models and community context. *Ecology* **76**, 1371–1382.

Lek, S. and Guégan, J.F. (1999) Artificial neural networks as a tool in ecological modelling, an introduction. *Ecological Modelling* **120**, 65–73.

Lele, S.R. and Dennis, B. (2009) Bayesian methods for hierarchical models: are ecologists making a Faustian bargain? *Ecological Applications* **19**, 581–584.

Lennon, J.J. (2000) Red-shifts and red herrings in geographical ecology. *Ecography* **23**, 101–113.

Lennon, J.J., Koleff, P., Greenwood, J.J.D. and Gaston, K.J. (2004) Contribution of rarity and commonness to patterns of species richness. *Ecology Letters* **7**, 81–87.

Lennon, J.J., Beale, C.M., Reid, C.L., *et al.* (2011) Are richness patterns of common and rare species equally well explained by environmental variables? *Ecography.* **34**, 529–539.

Lepš, J. (2005) Diversity and ecosystem function, in *Vegetation Ecology* (ed. E. van der Maarel), Blackwell Publishing, Oxford, pp. 199–237.

Lepš, J. and Hadincova, V. (1992) How reliable are our vegetation analyses? *Journal of Vegetation Science* **3**, 119–124.

Lepš, J. and Šmilauer, P. (2003) *Multivariate Analysis of Ecological Data using CANOCO*, Cambridge University Press, Cambridge.

Lepš, J. and Šmilauer, P. (2007) Subjectively sampled vegetation data: don't throw the baby out with the bath water. *Folia Geobotanica* **42**, 169–178.

Leuschner, C. (2005) Vegetation and ecosystems, in *Vegetation Ecology* (ed. E. van der Maarel), Blackwell Publishing, Oxford, pp. 85–105.

Levin, S.A. (1992) The problem of pattern and scale in ecology. *Ecology* **73**, 1949–1967.

Link, W.A. and Barker, R.J. (2010) *Bayesian Inference with Ecological Examples*, Academic Press, San Diego, CA.

Lintz, H.E., McCune, B., Gray, A.N. and McCulloh, K.A. (2011) Quantifying ecological thresholds from response surfaces. *Ecological Modelling* **222**, 427–436.

Lockwood, J.L., Powell, R.D., Nott, P. and Pimm, S.L. (1997) Assembling ecological communities in time and space. *Oikos* **80**, 549–553.

Longley, P.A., Goodchild, M., Maguire, D.J. and Rhind, D.W. (2010) *Geographic Information Systems and Science* (3rd edn), Wiley-Blackwell, Chichester.

Lookingbill, T.R. and Urban, D. (2005) Gradient analysis, the next generation: towards more plant-relevant explanatory variables. *Canadian Journal of Forest Research* **35**, 1744–1753.

Loram, A., Thompson, K., Warren, P.H. and Gaston, K.J. (2008) Urban domestic gardens (XII): the richness and composition of the flora in five UK cities. *Journal of Vegetation Science* **19**, 321–330.

Loreau, M., Naeem, S. and Inchausti, P. (eds) (2002) *Biodiversity and Ecosystem Functioning. Synthesis and Perspectives*, Oxford University Press, Oxford.

Lortie, C.J., Brooker, R.W., Choler, P., *et al.* (2004) Rethinking plant community theory. *Oikos* **107**, 433–438.

Louppen, J.M.W. and van der Maarel, E. (1979) CLUSLA: a computer program for the clustering of large phytosociological data sets. *Vegetatio* **40**, 107–114.

Lowe, J.J. and Walker, M.J.C. (1997) *Reconstructing Quaternary Environments* (2nd edn), Longman, Harlow.

Ludwig, J.A. and Reynolds, J.F. (1988) *Statistical Ecology: a Primer on Methods and Computing*, John Wiley & Sons Ltd, New York.

Lukacs, P.M., Thompson, W.L., Kendall, W.L., *et al.* (2007) Concerns regarding a call for pluralism of information theory and hypothesis testing. *Journal of Applied Ecology* **44**, 456–460.

Luken, T.O. (ed.) (1990) *Directing Ecological Succession*, Chapman and Hall, London.

Mabberley, D.J. (1991) *Tropical Rain Forest Ecology* (2nd edn), Blackie, Glasgow.

MacArthur, R.H. and Wilson, E.O. (1967) *The Theory of Island Biogeography*, Princeton University Press, Princeton, NJ.

MacDonald, A., Stevens, P., Armstrong, H., *et al.* (1998) *A Guide to Upland Habitats: Surveying Land Management Impacts*, Scottish Natural Heritage, Battleby.

MacMahon, J.A. (1981) Successional processes: comparisons among biomes with special reference to the probable role and influence on animals, in *Forest Succession: Concepts and Applications* (eds D.C. West, H.H. Shugart and D.B. Botkin), Springer-Verlag, New York, pp. 277–304.

MacNally, R. (2002) Multiple regression and inference in ecology and conservation biology: further comments on identifying important predictor variables. *Biodiversity and Conservation* **11**, 1397–1401.

MacNeil, A. (2008) Making empirical progress in observational ecology. *Environmental Conservation* **35**, 193–196.

MacQueen, J. (1967) Some methods for classification and analysis of multivariate observations, in *Proceedings of the Fifth Berkeley Symposium on Mathematical Statistics and Probability* (eds L.M. Le Cam and J. Neyman), University of California Press, Berkeley, CA, pp. 281–297.

Magurran, A.E. (1988) *Ecological Diversity and its Measurement*, Princeton University Press/Croom Helm, London.

Magurran, A.E. (2004) *Measuring Biological Diversity* (2nd edn), Blackwell, Oxford.

Maher, W.A., Cullen, P.W. and Norris, R.H. (1994) Framework for designing sampling programs. *Environmental Monitoring and Assessment* **30**, 139–162.

Malloch, A.J.C. (1998) *MATCH II. A computer program to Aid the Assignment of Vegetation Data to the Communities and Subcommunities of the National Vegetation Classification*. Institute of Environmental and Biological Sciences, University of Lancaster.

Manly, B.F.J. (1997) *Randomization, Bootstrap and Monte Carlo Methods in Biology* (2nd edn), Chapman and Hall, London.

Manly, B.F.J. (2005) *Multivariate Statistical Methods: a Primer* (3rd edn), Chapman and Hall, London.

Mantel, N. (1967) The detection of disease clustering and a generalised regression approach. *Cancer Research* **27**, 209–220.

Marsh, C. and Elliott, J. (2008) *Exploring Data: an Introduction to Data Analysis for Social Scientists* (2nd edn), Polity Press, Cambridge.

Marsili-Libelli, S. (1989) Fuzzy clustering of ecological data. *Coenoses* **2**, 95–106.

Mason, N.W.H., Mouillet, D., Lee, W.G. and Wilson, J.B. (2005) Functional richness, functional evenness and functional divergence: the primary components of functional diversity. *Oikos* **111**, 112–118.

Matthews, J.A. (1992) *The Ecology of Recently Deglaciated Terrain*, Cambridge University Press, Cambridge.

May, R.M. (1989) Levels of organization in ecology, in *Ecological Concepts: the Contribution of Ecology to an Understanding of the Natural World* (ed. J.M. Cherrett), British Ecological Society/Blackwell Scientific Publications, Oxford, pp. 339–363.

McArdle, B.H. and Anderson, M.J. (2001) Fitting multivariate models to community data: a comment on distance-based redundancy analysis. *Ecology* **82**, 290–297.

McCann, K.S. (2000) The diversity-stability debate. *Nature* **405**, 228–233.

McCarthy, M.A. (2007) *Bayesian Methods for Ecology*, Cambridge University Press, Cambridge.

McCarthy, M.A. and Masters, P. (2005) Profiting from prior information in Bayesian analyses of ecological data. *Journal of Applied Ecology* **42**, 1012–1019.

McCollin, D., Jackson, J.J., Bunce, R.G.H., *et al.* (2000) Hedgerows as habitat for woodland plants. *Journal of Environmental Management* **60**, 77–90.

McCook, L.J. (1994) Understanding ecological community succession: causal models and theories: a review. *Vegetatio* **110**, 115–148.

McCune, B. (1994) Improving community analysis with the Beals smoothing function. *Ecoscience* **1**, 82–86.

McCune, B. (1997) Influence of noisy environmental data on canonical correspondence analysis. *Ecology* **78**, 2617–2623.

McCune, B. (2004) *Nonparametric multiplicative regression for habitat modeling*. Available at http://www.pcord.com/NPMRintro.pdf

McCune, B. (2006) Nonparametric habitat models with automatic interactions. *Journal of Vegetation Science* **17**, 819–830.

McCune, B. and Beals, E.W. (1993) History of the development of Bray-Curtis ordination, in *John T. Curtis. Fifty Years of Wisconsin Plant Ecology* (eds J.S. Fralish, R.P. McIntosh and O.L. Loucks), Wisconsin Academy of Science, Art and Letters, Madison, Wisconsin, pp. 67–79.

McCune, B. and Grace, J.B. (2002) *Analysis of Ecological Communities*, MjM Software Design, Gleneden Beach, OR.

McCune, B. and Mefford, M.J. (1999) *PC-ORD. Multivariate Analysis of Ecological Data*. Version 4. MjM Software Design, Gleneden Beach, OR.

McCune, B. and Mefford, M.J. (2009) *Hyperniche: Nonparametric Multiplicative Habitat Modelling*. Version 2.0. MjM Software Design, Gleneden Beach, OR.

McCune, B. and M. J. Mefford. (2010) *PC-ORD. Multivariate Analysis of Ecological Data*. Version 6. MjM Software, Gleneden Beach, OR.

McDermid, G.J., Franklin, S.E. and LeDrew, E.F. (2005) Remote sensing for large-area habitat mapping. *Progress in Physical Geography* **29**, 449–474.

McGarigal, K. and Cushman, S.A. (2005) The gradient concept of landscape structure, in *Issues and Perspectives in Landscape Ecology* (eds J. Wiens and M. Moss), Cambridge University Press, Cambridge, pp. 112–119.

McGarigal, K., Cushman, S. and Stafford, S. (2000) *Multivariate Statistics for Wildlife and Ecology Research*, Springer-Verlag, Heidelberg.

McIntosh, R.P. (1967a) An index of diversity and the relation of certain concepts of diversity. *Ecology* **48**, 392–404.

McIntosh, R.P. (1967b) The continuum concept of vegetation. *Botanical Review* **33**, 130–187.

McIntosh, R.P. (1986) *The Background of Ecology – Concept and Theory*, Cambridge University Press, Cambridge.

McNaughton, S.J. (1983) Serengeti grassland ecology: the role of composite environmental factors and contingency in community organisation. *Ecological Monographs* **53**, 291–320.

McNaughton, S.J. (1985) Ecology of a grazing ecosystem: the Serengeti. *Ecological Monographs* **55**, 259–294.

McNaughton, S.J., Stronach, N.R.H. and Georgiadis, N.J. (1998) Combustion in natural fires and global emissions budgets. *Ecological Applications* **8**, 464–468.

McPherson, G. (1989) The scientist's view of statistics – a neglected area. *Journal of the Royal Statistical Society Series A* **152**, 221–240.

McQuitty, L.L. (1960) Hierarchical linkage analysis for the isolation of types. *Educational and Psychological Measurement* **20**, 55–67.

McVean, D.N. and Ratcliffe, D.A. (1962) *Plant Communities of the Scottish Highlands. A Study of Scottish Mountain, Moorland and Forest Vegetation*, Monographs of the Nature Conservancy, No. 1, HMSO, London.

Meades, W.J. (1983) Heathlands, in *Biogeography and Ecology of the Island of Newfoundland* (ed. G.R. South), Junk, The Hague, pp. 267–318.

Meentenmeer, V. and Box, E.O. (1987) Scale effects in landscape studies, in *Landscape Heterogeneity and Disturbance* (ed. M.G. Turner), Springer, New York, pp. 15–33.

Méot, A., Legendre, P. and Borcard, D. (1998) Partialling out the spatial component of ecological variations: questions and propositions in the linear modelling framework. *Environmental and Ecological Statistics* **5**, 1–27.

Michaelsen, J., Schimel, D.S., Friedl, M.A., *et al.* (1994) Regression tree analysis of satellite and terrain data to guide vegetation sampling and surveys. *Journal of Vegetation Science* **5**, 673–686.

Mielke, P.W. Jr. and Berry, K.J. (2001) *Permutation Methods: a Distance Function Approach*, Springer, New York.

Mielke, P.W. Jr., Berry, K.J. and Brier, G.W. (1981) Application of multi-response permutation procedures for examining seasonal changes in monthly sea-level pressure patterns. *Monthly Weather Review* **109**, 120–126.

Mielke, P.W. Jr., Berry, K.J. and Johnson, E.S. (1976) Multiresponse permutation procedures for *a priori* classifications. *Communications in Statistics* **A5**, 1409–1424.

Milberg, P., Bergstedt, J., Fridman, J., *et al.* (2008) Observer bias and random variation in vegetation monitoring data. *Journal of Vegetation Science* **19**, 633–644.

Millar, R.B., Anderson, M.J. and Zunun, G. (2005) Fitting nonlinear environmental gradients to community data: a general distance-based approach. *Ecology* **86**, 2245–2251.

Milligan, G.W. (1996) Clustering validation: results and implications for applied analysis, in *Clustering and Classification* (eds P. Arabie, L.J. Hubert and G. De Soete), World Scientific, Singapore, pp. 341–375.

Milner, C. (1978) Shetland ecology surveyed. *Geographical Magazine* **50**, 730–753.

Minchin, P.R. (1987a) An evaluation of the relative robustness of techniques for ecological ordination. *Vegetatio* **69**, 89–107.

Minchin, P.R. (1987b) Simulation of multidimensional community patterns: towards a comprehensive model. *Vegetatio* **71**, 145–156.

Minchin, P.R. and Rennie, L.D. (2010) Does the Hellinger transformation make PCA a viable method for community ordination? Paper presented at 95th Ecological Society of America Annual Meeting, August 2010. Available at http://eco.confex.com/eco/2010/techprogram/P24489.HTM [accessed 17 February 2011].

Mistral, M., Buck, O., Meier-Behrmann, D.C., *et al.* (2000) Direct measurement of spatial autocorrelation at the community level in four plant communities. *Journal of Vegetation Science* **11**, 911–916.

Mistry, J. (2000) *World Savannas*, Prentice-Hall, New York.

Mitchell, J. (1977) *The effect of bracken distribution on moorland vegetation and soils*. Unpublished PhD thesis, University of Glasgow.

Moerman, D.E. and Estabrook, G.F. (2006) The botanist effect: counties with maximal species richness tend to be home to universities and botanists. *Journal of Biogeography* **33**, 1969–1974.

Molinari, J. (1989) A calibrated index for the measurement of evenness. *Oikos* **56**, 319–326.

Molinari, J. (1996) A critique of Bulla's paper on diversity indices. *Oikos* **76**, 577–582.

Moore, G.W., Benninghoff, W.S. and Dwyer, P.S. (1967) A computer method for the arrangement of phytosociological tables. *Proceedings of the Association for Computer Machinery* **20**, 297–299.

Moore, J.J. (1962) The Braun-Blanquet system – a reassessment. *Journal of Ecology* **50**, 701–709.

Moore, J.J., Fitzsimons, P., Lambe, E. and White, J. (1970) A comparison and evaluation of some phytosociological techniques. *Vegetatio* **20**, 1–20.

Moraczewski, I.R. (1993a) Fuzzy logic for phytosociology. I. Syntaxa as vague concepts. *Vegetatio* **106**, 1–11.

Moraczewski, I.R. (1993b) Fuzzy logic for phytosociology. II. Generalizations and predictions. *Vegetatio* **106**, 13–20.

Moraczewski, I.R. (1996) Fuzzy sets as a tool for ecological data analysis. *Coenoses* **11**, 55–68.

Moravec, J. (1971) A simple method for estimating homogeneity of sets of phytosociological relevés. *Folia Geobotanica Phytotaxonomie* **6**, 147–170.

Moravec, J. (1992) Is the Zurich-Montpellier approach still unknown in vegetation science of the English-speaking countries? *Journal of Vegetation Science* **3**, 277–278.

Morin, P.J. (1999) *Community Ecology*, Blackwell Science, Oxford.

Mortimer, A.M. (1974) *Studies of Germination and Establishment of Selected Species with Reference to the Fates of Seeds*. Unpublished PhD thesis, University of Wales.

Motyka, J. (1947) O zadaniach I metodach badan geobotanicznych. Sur les buts et les methods des recherché géobotaniques. *Annales Universitatis Mariae Curie-Sklodowska (Lubin, Polonia), Sectio C, Supplementum I*.

Mouchet, M., Guilhaumon, F., Villéger, S., *et al.* (2008) Towards a consensus for calculating dendrogram-based functional doiversity indices. *Oikos* **117**, 794–800.

Mouchet, M.A., Villéger, S., Mason, N.W.H. and Mouillot, D. (2010) Functional diversity measures: an overview of their redundancy and their ability to discriminate community assembly rules. *Functional Ecology* **24**, 867–876.

Mouillot, D., Mason, N.W.H. and Wilson, J.B. (2007) Is the abundance of species determined by their functional traits? A new method with a test using plant communities. *Oecologia* **152**, 729–737.

Mucina, L. (1997a) Conspectus of classes of European vegetation. *Folia Geobotanica et Phytotaxa* **32**, 117–172.

Mucina, L. (1997b) Classification of vegetation: past, present and future. *Journal of Vegetation Science* **8**, 751–760.

Mucina, L. and van der Maarel, E. (1989) Twenty years of numerical syntaxonomy. *Vegetatio* **81**, 1–15.

Mucina, L., Rodwell, J.S., Schaminee, J.H.J. and Dierschke, H. (1993) European vegetation survey: current state of some national programmes. *Journal of Vegetation Science* **4**, 429–438.

Mucina, L., Schaminée, J.H.J. and Rodwell, J.S. (2000) Common data standards for recording relevés in field survey for vegetation classification. *Journal of Vegetation Science* **11**, 769–772.

Mueller-Dombois, D. and Ellenberg, H. (1974) *Aims and Methods of Vegetation Ecology*, John Wiley and Sons Ltd, New York.

Muttlak, H.A. and Sabooghi-Alvandi, S.M. (1993) A note on the line intercept sampling method. *Biometrics* **49**, 1209–1215.

Nature Conservancy Council (1979) *Nature Conservation in Urban Areas: Challenge and Opportunity*. Nature Conservancy Council, London.

Navas, M.L. and Violle, C. (2009) Plant traits related to competition: how do they shape the functional diversity of communities? *Community Ecology* **10**, 131–137.

Naveh, Z. and Lieberman, A.S. (1994) *Landscape Ecology: Theory and Application* (2nd edn), Springer-Verlag, New York.

Nee, S., Gregory, R.D. and May, R.M. (1991) Core and satellite species: theory and artefacts. *Oikos* **62**, 83–87.

Neitlich, P. and McCune, B. (1997) Hotspots of epiphytic lichen diversity in two young managed forests. *Conservation Biology* **11**, 172–182.

Nelder, J.A. (1986) Statistics, science and technology: the address of the President (with proceedings). *Journal of the Royal Statistical Society Series A* **149**, 109–121.

Nesje, A., Bakke, J., Dahl, S.O., *et al.* (2008) Norwegian mountain glaciers in the past, present and future. *Global and Planetary Change* **60**, 10–28.

Newbould, P. (1965) Production ecology and the International Biological Programme. *Geography* **49**, 98–104.

Nicholson, M. and McIntosh, R.P. (2002) H.A. Gleason and the individualistic hypothesis revisited. *Bulletin of the Ecological Society of America* **83**, 133–142.

Nilsson, C. (1992) Increasing the reliability of vegetation analyses by using a team of two investigators. *Journal of Vegetation Science* **3**, 565.

Nilsson, I.N. and Nilsson, S.G. (1985) Experimental estimates of census efficiency and pseudoturnover on islands: error trend and between-observer variation when recording vascular plants. *Journal of Ecology* **73**, 65–70.

Noble, I.R. and Slatyer, R.O. (1980) The use of vital attributes to predict successional changes in plant communities subject to recurrent disturbances. *Vegetatio* **43**, 5–21.

Noest, V. and van der Maarel, E. (1989) A new dissimilarity measure and a new optimality criterion in phytosociological classification. *Vegetatio* **83**, 157–165.

Noy-Meir, I. (1973) Data transformations in ecological ordination. I. Some advantages of non-centring. *Journal of Ecology* **61**, 329–341.

Noy-Meir, I. and van der Maarel, E. (1987) Relations between community theory and community analysis in vegetation science: some historical perspectives. *Vegetatio* **69**, 5–15.

Noy-Meir, I., Walker, D. and Williams, W.T. (1975) Data transformation in ecological ordination. II. On the meaning of standardization. *Journal of Ecology* **63**, 779–800.

Oden, N.L. (1984) Assessing the significance of a spatial correlogram. *Geographical Analysis* **16**, 1–16.

Oden, N.L. and Sokal, R.R. (1986) Directional autocorrelation: an extension of spatial correlograms in two dimensions. *Systematic Zoology* **35**, 608–617.

Odum, E.P. (1997) *Ecology: a Bridge Between Science and Society*, Sinauer Associates, Sunderland, MA.

Økland, R.H. (1996) Are ordination and constrained ordination alternative or complementary strategies in general ecological studies? *Journal of Vegetation Science* **7**, 289–292.

Økland, R.H. (1999) On the variation explained by ordination and constrained ordination axes. *Journal of Vegetation Science* **10**, 131–136.

Økland, R.H. (2003) Partitioning the variation in a plot-by-species data matrix that is related to *n* sets of exploratory variables. *Journal of Vegetation Science* **14**, 693–700.

Økland, R.H. (2007) Wise use of statistical tools in ecological field studies. *Folia Geobotanica* **42**, 123–140.

Økland, R.H. and Bendiksen, E. (1985) The vegetation of the forest-alpine transition in the Grunningdalen area, Telemark, Southern Norway. *Sommerfeltia* **2**, 1–224.

Økland, R.H. and Eilertsen, O. (1994) Canonical correspondence analysis with variance partitioning: some comments and an application. *Journal of Vegetation Science* **5**, 117–126.

Oksanen, J. (1983) Ordination of boreal heath-like vegetation with principal components analysis, correspondence analysis and multidimensional scaling. *Vegetatio* **52**, 181–189.

Oksanen, J. (1988) A note on the occasional instability of detrending in correspondence analysis. *Vegetatio* **74**, 29–32.

Oksanen, J. and Minchin, P.R. (1997) Instability of ordination results under changes in input data order: explanation and remedies. *Journal of Vegetation Science* **8**, 447–454.

Oksanen, J. and Minchin, P.R. (2002) Continuum theory revisited: what shape are species responses along ecological gradients? *Ecological Modelling* **157**, 119–129.

Oksanen, J., Kindt, R., Legendre, P. and O'Hara, B. (2006) *Vegan: Community Ecology Package*. Available at: http://cc.oulu.fi/~jarioksa/.

Oksanen, L. (2001) Logic of experiments in ecology: is pseudoreplication a pseudoissue? *Oikos* **94**, 27–38.

Olano, J.M., Loidi, J.J., González, A. and Escudero, A. (1998) Improving the interpretation of fuzzy partitions in vegetation science with constrained ordinations. *Plant Ecology* **134**, 113–118.

Olff, H. and Bakker, J.P. (1998) Do intrinsically dominant and subordinate species exist? A test statistic for field data. *Applied Vegetation Science* **1**, 15–20.

Olmstead, M.A., Wample, R., Greene, S. and Tarara, J. (2004) Nondestructive measurement of vegetative cover using digital image analysis. *Hortscience* **39**, 55–59.

O'Neill, R.V. (1989) Perspectives in hierarchy and scale, in *Perspectives in Ecological Theory* (eds J. Roughgarden, R.M. May and S.A. Levin), Princeton University Press, Princeton, NJ, pp. 140–156.

Orlóci, L. (1966) Geometric models in ecology. I. The theory and application of some ordination methods. *Journal of Ecology* **54**, 193–215.

Orlóci, L. (1967a) Data centering: a review and evaluation with reference to component analysis. *Systematic Zoology* **16**, 208–212.

Orlóci, L. (1967b) An agglomerative method for the classification of plant communities. *Journal of Ecology* **55**, 193–206.

Orlóci, L. (1972) On information analysis in phytosociology, in *Grundfragen und Methoden in der Pflanzensoziologie* (eds E. van der Maarel and R. Tüxen), Junk, The Hague, pp. 75–88.

Orlóci, L. (1974) Revision for the Bray and Curtis ordination. *Canadian Journal of Botany* **52**, 1773–1776.

Orlóci, L. (1978) *Multivariate Analysis in Vegetation Research*, Junk, The Hague.

Osborne, P.L. (2000) *Tropical Ecosystems and Ecological Concepts*, Cambridge University Press, Cambridge.

Pakeman, R.J., Lepš, J., Kleyer, M., *et al.* (2009) Relative climatic, edaphic and management controls of plant functional trait signatures. *Journal of Vegetation Science* **20**, 148–159.

Pakeman, R.J. and Quested, H.M. (2007) Sampling plant functional traits: what proportion of the species need to be measured? *Applied Vegetation Science* **10**, 91–96.

Pakeman, R.J., Reid, C.L., Lennon, J.J. and Kent, M. (2008) Possible interactions between environmental factors in determining species optima. *Journal of Vegetation Science* **19**, 201–208.

Palmer, M.W. (1993) Putting things in even better order – the advantages of CCA. *Ecology* **74**, 2215–2230.

Palmer, M.W. (2011) *The Ordination Web Page*. www.okstate.edu/artsci/botany/ordinate.htm

Parker, V.T. (2001) Conceptual problems and scale limitations of defining ecological communities: a critique of the CI concept (Community of Individuals). *Perspectives in Plant Ecology, Evolution and Systematics* **4**, 80–96.

Pawlowski, B. (1966) Review of terrestrial plant communities. A. Composition and structure of plant communities and methods of their study, in *The Vegetation of Poland* (ed. W. Szafer), Pergamon Press/PWN Polish Scientific Publishers, Warsaw, pp. 241–281.

Pearson, K. (1901) On lines and planes of closest fit to systems of points in space. *Philosophical Magazine, Sixth Series* **2**, 559–572.

Peck, J.E. (2010) *Multivariate Analysis for Community Ecologists: Step-by-Step using PC-ORD*. MjM Software Design, Gleneden Beach, OR.

Peet, R.K. (2008) A decade of effort by the ESA Vegetation Panel leads to a new federal standard. *Ecological Society of America Bulletin* **89**, 210–211.

Peet, R.K., Knox, R.G., Case, J.S. and Allen, R.B. (1988) Putting things in order: the advantages of detrended correspondence analysis. *American Naturalist* **131**, 924–934.

Pélissier, R., Couteron, P. and Dray, S. (2008) Analyzing or explaining beta diversity? Comment. *Ecology* **89**, 3227–3232.

Petchey, O.L. and Gaston, K.J. (2002) Functional diversity, species richness and community composition. *Ecology Letters* **5**, 402–411.

Petchey, O.L., Hector, A. and Gaston, K.J. (2004) How do different measures of functional diversity perform? *Ecology* **85**, 847–857.

Petchey, O.L. and Gaston, K.J. (2006) Functional diversity: back to basics and looking forward. *Ecology Letters* **9**, 741–758.

Petchey, O.L. and Gaston, K.J. (2007) Dendrograms and measuring functional diversity. *Oikos* **116**, 1422–1426.

Petchey, O.L. and Gaston, K.J. (2009) Dendrograms and measures of functional diversity: a second instalment. *Oikos* **118**, 1118–1120.

Petchey, O.L., Beckerman, A.P. and Childs, D.Z. (2009) Shock and awe by statistical software – Why R? *Bulletin of the British Ecological Society* **40** (4), 55–58.

Peterken, G.P. (1967) *Guide to the Check Sheet for IBP Areas*, IBP Handbook No. 4, Blackwell Scientific, Oxford.

Peters, D.P.C., Gosz, J.R., Pockman, W.T., *et al.* (2006) Integrating patch and boundary dynamics to understand and predict biotic transitions at multiple scales. *Landscape Ecology* **21**, 19–33.

Peterson, E.B. and McCune, B. (2001) Diversity and succession of epiphytic macrolichen communities in low-elevation managed conifer forests in western Oregon. *Journal of Vegetation Science* **12**, 511–524.

Pickett, S.T.A. and White, P.S. (eds) (1985) *The Ecology of Natural Disturbance and Patch Dynamics*, Academic Press, Orlando, FL.

Pickett, S.T.A. and Kolasa, J. (1989) Structure of theory in vegetation science. *Vegetatio* **83**, 7–15.

Pickett, S.T.A. and Cadenasso, M.L. (1995) Landscape ecology: spatial heterogeneity in ecological systems. *Science* **269**, 331–334.

Pickett, S.T.A. and Cadenasso, M.L. (2005) Vegetation dynamics, in *Vegetation Ecology* (ed. E. van der Maarel), Blackwell Publishing, Oxford, pp. 172–198.

Pickett, S.T.A., Cadenasso, M.L. and Meiners, S.J. (2008) Ever since Clements: from succession to vegetation dynamics and understanding to intervention. *Applied Vegetation Science* **12**, 9–21.

Piclou, E.C. (1969) *An Introduction to Mathematical Ecology*, John Wiley and Sons Ltd, New York.

Pielou, E.C. (1975) *Ecological Diversity*, John Wiley and Sons Ltd, New York.

Pignatti, S. (1980) Reflections on the phytosociological approach and the epistemological basis of vegetation science. *Vegetatio* **42**, 181–185.

Pillar, V.D. (1998) Sampling sufficiency in ecological surveys. *Abstracta Botanica* **22**, 37–48.

Pillar, V.D. (1999) On the identification of optimal plant functional types. *Journal of Vegetation Science* **10**, 631–640.

Pillar, V.D. & Sosinski, E.E. (2003) An improved method for searching plant functional types by numerical analysis. *Journal of Vegetation Science* **14**, 323–332.

Pinel-Alloul, B., Niyonsenga, T. and Legendre, P. (1995) Spatial and environmental components of freshwater zooplankton structure. *Ecoscience* **2**, 1–19.

Pisces Conservation Ltd (2008a) *Ecom 2*. Pisces Conservation Ltd., Lymington, UK. www.pisces-conservation.com/software.html.

Pisces Conservation Ltd (2008b) *Community Analysis Package 4*. Pisces Conservation Ltd., Lymington, UK. www.pisces-conservation.com/software.html.

Pisces Conservation Ltd (2008c) *Species Diversity and Richness*. Pisces Conservation Ltd., Lymington, UK. www.pisces-conservation.com/software.html.

Pisces Conservation Ltd (2008d) *Fuzzy Grouping*. Pisces Conservation Ltd., Lymington, UK. www.pisces-conservation.com/software.html.

Pisces Conservation Ltd (2008e) *QED Statistics*. Pisces Conservation Ltd., Lymington, UK. www.pisces-conservation.com/software.html.

Podani, J. (1990) Comparison of fuzzy classifications. *Coenoses* **5**, 17–21.

Podani, J. (1994) *Multivariate Analysis in Ecology and Systematics*, SPB Publishing, The Hague.

Podani, J. (1996) On the sensitivity of ordination and classification methods to variation in the input order of data. *Journal of Vegetation Science* **8**, 153–156.

Podani, J. (2000) *Introduction to the Exploration of Multivariate Biological Data*, Backhuys, Leiden.

Podani, J. (2005) Multivariate exploratory analysis of ordinal data in ecology: pitfalls, problems and solutions. *Journal of Vegetation Science* **16**, 497–510.

Podani, J. (2006) Braun-Blanquet's legacy and data analysis in vegetation science. *Journal of Vegetation Science* **17**, 113–117.

Poland, J. and Clement, E. (2009) *The Vegetative Key to the British Flora*, John Poland, Southampton, in association with the Botanical Society of the British Isles.

Ponomarenko, S. and Alvo, R. (2000) *Perspectives on Developing a Canadian Classification of Ecological Communities*. Canadian Forest Service, Science Branch, Information report ST-X-18E, Natural Resources Canada, Ottawa.

Poole, R.W. (1974) *An Introduction to Quantitative Ecology*, McGraw-Hill, Tokyo.

Poore, M.E.D. (1955a,b,c) The use of phytosociological methods in ecological investigations. I. The Braun-Blanquet system. *Journal of Ecology* **43**, 226–244. II. Practical issues involved in trying to apply the Braun-Blanquet system. *Journal of Ecology* **43**, 245–269. III. Practical applications. *Journal of Ecology* **43**, 606–651.

Poore, M.E.D. (1956) The use of phytosociological methods in ecological investigations. IV. General discussion of phytosociological problems. *Journal of Ecology* **44**, 28–50.

Poore, M.E.D. and McVean, D.N. (1957) A new approach to Scottish mountain vegetation. *Journal of Ecology* **45**, 401–439.

Popper, K.R. (1972a) *The Logic of Scientific Discovery* (6th revised impression), Hutchinson, London.

Popper, K.R. (1972b) *Objective Knowledge*, Oxford University Press, Oxford.

Popper, K.R. (1976) *Unended Quest: an Intellectual Autobiography*, Fontana, London.

Prance, G.T. (1977) The phytogeographical subdivisions of Amazonia and their influence on the selection of biological reserves, in *Extinction is for Ever* (eds G.T. Prance and T.S. Elias,) New York Botanic Garden, New York, pp. 193–213.

Prance, G.T. (1978) Conservation problems in the Amazon Basin, in *Earthcare: Global Protection of Natural Areas* (ed. E.A. Schofield), 14th Proceedings of the Biennial Wilderness Conference, Westview Press, Colorado, pp. 191–207.

Prance, G.T. (1996) Islands in Amazonia. *Philosophical Transactions of the Royal Society of London Series B* **351**, 823–833.

Prentice, I.C. (1977) Non-metric ordination models in ecology. *Journal of Ecology* **65**, 85–94.

Prentice, I.C. (1980) Vegetation analysis and order invariant gradient models. *Vegetatio* **42**, 27–34.

Primack, R. and Corlett, R. (2011) *Tropical Rain Forests: an Ecological and Biogeographical Comparison* (2nd edn), Wiley/Blackwell Science, Chichester.

Pugnaire, F.I. and Valladares, F. (eds) (1999) *Handbook of Functional Plant Ecology*, Marcel Dekker, Inc., New York.

Purvis, A. and Hector, A. (2000) Getting the measure of diversity. *Nature* **405**, 212–219.

Purvis, O.W., Coppins, B.J., Hawksworth, D.L., *et al.* (1992) *The Lichen Flora of Great Britain and Ireland*, Natural History Museum Publications and British Lichen Society, London.

Purvis, O.W., Coppins, B.J. and James, P.W. (1994) *Checklist of Lichens of Great Britain and Ireland*, British Lichen Society, London.

Putman, R.J. (1994) *Community Ecology*, Chapman and Hall, London.

Pyšek, P. and Leps, J. (1991) Response of a weed community to nitrogen fertilization: a multivariate analysis. *Journal of Vegetation Science* **2**, 237–244.

Pyšek, P., Chocholoušková, Z., Pyšek, A., *et al.* (2004) Trends in species diversity and composition of urban vegetation over three decades. *Journal of Vegetation Science* **15**, 781–788.

Qinghong, L. and Bråkenhielm, S. (1995) A statistical approach to decompose ecological variation. *Water, Air and Soil Pollution* **85**, 1587–1592.

Quenouille, M.H. (1959) *Rapid Statistical Calculations*, Griffin, London.

Quinn, G.P. and Keough, M.J. (2002) *Experimental Design and Data Analysis for Biologists*, Cambridge University Press, Cambridge.

Rackham, O. (1977) Hedgerow trees: their history, conservation and renewal. *Arboricultural Journal* **3**, 169–177.

Rahbek, C. (2005) The role of spatial scale and perception of large-scale species richness patterns. *Ecology Letters* **8**, 224–239.

Rahel, F.J. (1990) The hierarchical nature of community persistence: a problem of scale. *American Naturalist* **136**, 328–344.

Ramenskii, L.G. (1930) Zur Methodik der vergleichenden Bearbeitung und Ordnung von Pflanzenlisten und anderen Objekten, die durch mehrere, verschiedenartig wirkende Factoren bestimmt werden. *Beiträge zur Biologie der Pflanzen* **18**, 269–304.

Ramenskii, L.G. (1938) *Introduction to the Geobotanical Study of Complex Vegetation*, Sflkhozgiz, Moscow.

Ramsay, P.M., Kent, M., Reid, C.L. and Duckworth, J.C. (2006) Taxonomic, morphological and structural surrogates for the rapid assessment of vegetation. *Journal of Vegetation Science* **17**, 747–754.

Randall, R.E. (1978) *Theories and Techniques in Vegetation Analysis*. Oxford University Press, Oxford.

Rao, C.R. (1964) The use and interpretation of principal component analysis in applied research. *Sankhya A* **26**, 329–358.

Raunkaier, C. (1928) Dominansareal artstaethed of formationsdominanter. *Kgl. Danske Vidensk Selsk. Biol. Meddel.* **7**, 1.

Raunkaier, C. (1934) *The Life Forms of Plants and Statistical Plant Geography*, Clarendon Press, Oxford.

Raunkaier, C. (1937) *Plant Life Forms*, Clarendon Press, Oxford.

Ravan, S.A., Roy, P.S. and Sharma, C.M. (1995) Space remote sensing for spatial vegetation characterisation. *Journal of Bioscience* **20**, 427–438.

Rich, T.C.G. and Woodruff, E.R. (1992) Recording bias in botanical surveys. *Watsonia* **19**, 73–95.

Richards, P.W. (1996) *The Tropical Rain Forest: an Ecological Study* (2nd edn), Cambridge University Press, Cambridge.

Ricotta, C. (2004) A parametric diversity measure combining the relative abundances and taxonomic distinctiveness of species. *Diversity and Distributions* **10**, 143–146.

Ricotta, C. (2007) Random sampling does not exclude spatial dependence: the importance of neutral models for ecological hypothesis testing. *Folia Geobotanica* **42**, 153–160.

Ricotta, C. and Avena, G. (2000) The remote sensing approach in broad-scale phenological studies. *Applied Vegetation Science* **3**, 117–122.

Ricotta, C. and Avena, G. (2006) On the evaluation of ordinal data with conventional multivariate procedures. *Journal of Vegetation Science* **17**, 839–842.

Rieley, J.O. and Page, S.E. (1990) *Ecology of Plant Communities – a Phytosociological Account of the British Vegetation*, Longman, London.

Ringvall, A., Petersson, H., Stahl, G. and Lamas, T. (2005) Surveyor consistency in presence/absence sampling for monitoring vegetation in a boreal forest. *Forest Ecology and Management* **212**, 109–117.

Ripley, B.D. (1996) *Pattern Recognition and Neural Networks*, Cambridge University Press, Cambridge.

Robbins, J.A. and Matthews, J.A. (2009) Pioneer vegetation on glacier forelands in southern Norway: emerging communities? *Journal of Vegetation Science* **20**, 889–902.

Robbins, J.A. and Matthews, J.A. (2010) Regional variation in successional trajectories and rates of vegetation change on glacier forelands in South-Central Norway. *Arctic, Antarctic and Alpine Research* **42**, 351–361.

Roberts, C.N. (1998) *The Holocene: an Environmental History* (2nd edn), Blackwell Publishers, Oxford.

Roberts, D.W. (1986) Ordination on the basis of fuzzy set theory. *Vegetatio* **66**, 123–131.

Roberts, D.W. (1989) Fuzzy systems vegetation theory. *Vegetatio* **83**, 71–80.

Roberts, D.W. (2009) Comparison of multidimensional fuzzy set ordination with CCA and db-RDA. *Ecology* **90**, 2622–2634.

Roberts, M.J. and Russo, R. (1999) *A Student's Guide to Analysis of Variance*, Routledge, London.

Robotnov, T.A. (1979) Concepts of ecological individuality of plant species and of the continuum in the works of L.G. Ramenskii. *Soviet Journal of Ecology* **9**, 417–422.

Rodwell, J.S. (ed.) (1991) *British Plant Communities. Vol. 1. Woodlands and Scrub*, Cambridge University Press, Cambridge.

Rodwell, J.S. (ed.) (1992a) *British Plant Communities. Vol. 2. Mires and Heaths*, Cambridge University Press, Cambridge.

Rodwell, J.S. (ed.) (1992b) *British Plant Communities. Vol. 3. Grasslands and Montane Communities*, Cambridge University Press, Cambridge.

Rodwell, J.S. (ed.) (1995a) *British Plant Communities. Vol. 4. Aquatic Communities, Swamps and Tall-herb Fens*, Cambridge University Press, Cambridge.

Rodwell, J.S. (ed.) (1995b) The European Vegetation Survey questionnaire: an overview of phytosociological data, vegetation survey programmes and data bases in Europe. *Ann. Bot. (Roma)* **53**, 87–98.

Rodwell, J.S. (ed.) (2000) *British Plant Communities. Vol. 5. Maritime Communities and Vegetation of Open Habitats*, Cambridge University Press, Cambridge.

Rodwell, J.S. (2006) *National Vegetation Classification: User's Handbook*. Joint Nature Conservation Committee (JNCC), Peterborough. Available at http://www.jncc.gov.uk/pdf/pub06_NVCusershandbook2006.pdf

Rodwell, J.S., Mucina, L., Pignatti, S., *et al.* (1997) European vegetation survey: The context of the case studies. *Folia Geobotanica* **32**, 113–115.

Rodwell, J.S., Pignatti, S., Mucina, L. and Schaminée, J.H.J. (1995) European Vegetation Survey: Update on progress. *Journal of Vegetation Science* **6**, 759–762.

Rohlf, F.J. (1974) Methods of comparing classifications. *Annual Review of Ecology and Systematics* **5**, 101–113.

Roleček, J., Chytrý, M., Hájek, M., *et al.* (2007) Sampling design in large-scale vegetation studies: do not sacrifice ecological thinking to statistical purism! *Folia Geobotanica* **42**, 199–208.

Roleček, J., Lubomir, T., Zelený, D. and Chytrý, M. (2009) Modified TWINSPAN classification in which the hierarchy represents cluster heterogeneity. *Journal of Vegetation Science* **20**, 596–602.

Rorison, I.H. (1969) Ecological inferences from laboratory experiments on mineral nutrition, in *Ecological Aspects of the Mineral Nutrition of Plants* (ed. I.H. Rorison), British Ecological Society Symposium, Blackwell Scientific Publications, Oxford, pp. 155–175.

Rosenzweig, M.L. (1995) *Species Diversity in Space and Time*, Cambridge University Press, Cambridge.

Ross, S.M. (1986) Vegetation change on highway verges in south-east Scotland. *Journal of Biogeography* **13**, 109–117.

Roughgarden, J., Running, S. and Matson, P. (1991) What does remote sensing do for ecology? *Ecology* **72**, 1918–1922.

Roux, G. and Roux, M. (1967) A propos de quelques méthodes de classification en phytosociologie. *Revue de Statistique Appliquée* **15**, 59–72.

Rowe, J.S. (1961) The level of integration concept and ecology. *Ecology* **42**, 420–427.

Salisbury, E.J. (1943) The flora of bombed areas. *Nature* **151**, 462–466.

Sánchez-Mata, D. (2003) The phytosociological approach in North American studies: Some considerations on phytosociological nomenclature. *Folia Geobotanica* **32**, 415–418.

Sandel, B. and Smith, A.B. (2009) Scale as a lurking factor: incorporating scale-dependence in experimental ecology. *Oikos* **118**, 1284–1291.

Sanderson, J. and Harris, L.D. (eds) (2000) *Landscape Ecology – a Top-down Approach*, Lewis Publishers, Boca Raton.

Schaffers, A.P. and Sýkora, K.V. (2000) Reliability of Ellenberg indicator values for moisture, nitrogen and soil reaction: a comparison with field measurements. *Journal of Vegetation Science* **11**, 225–244.

Scheiner, S.M. (2001) Theories, hypotheses and statistics, in *Design and Analysis of Ecological Experiments* (2nd edn) (eds S.M. Scheiner and J. Gurevitch), Oxford University Press, Oxford, pp. 3–13.

Scheiner, S.M. and Gurevitch, J. (eds) (2001) *Design and Analysis of Ecological Experiments* (2nd edn), Oxford University Press, Oxford.

Schlup, B.M. and Wagner, H.H. (2008) Effects of study design and analysis on the spatial community structure detected by multiscale ordination. *Journal of Vegetation Science* **19**, 621–632.

Schluter, D. and Ricklefs, R.E. (1993) Species diversity: an introduction to the problem, in *Species in Ecological Communities – Historical and Ecological Perspectives* (eds R.E. Ricklefs and D. Schluter), University of Chicago Press, Chicago, pp. 1–10.

Schmera, D., Podani, J. and Ers, T. (2009) Measuring the contribution of community members to functional diversity. *Oikos* **118**, 961–971.

Schmidtlein, S. (2005) Imaging spectroscopy as a tool for mapping Ellenberg indicator values. *Journal of Applied Ecology* **42**, 966–974.

Schmidtlein, S. and Sassin, J. (2004) Mapping of continuous floristic gradients in grasslands using hyperspectral imagery. *Remote Sensing and Environment* **92**, 126–138.

Schneider, D.C. (2001) The rise of the concept of scale in ecology. *BioScience* **51**, 545–553.

Schoener, T.W. (1989) The ecological niche, in *Ecological Concepts: the Contribution of Ecology to an Understanding of the Natural World* (ed. J.M. Cherrett), British Ecological Society/Blackwell Scientific Publications, Oxford, pp. 79–113.

Schrader-Frechette, K.S. and McCoy, E.D. (1993) *Method in Ecology*, Cambridge University Press, Cambridge.

Scott, W.A. and Hallam, C.J. (2003) Assessing species misidentification rates through quality assurance of vegetation monitoring. *Plant Ecology* **165**, 101–115.

Scull, P., Franklin, J. and Chadwick, O.A. (2004) The application of classification tree analysis to soil type prediction in a desert landscape. *Ecological Modelling* **181**, 1–15.

Seefeldt, S.S. and Booth, D.T. (2006) Measuring plant cover in sagebrush steppe rangelands: a comparison of methods. *Environmental Management* **37**, 703–711.

Semenova, G.V. and van der Maarel, E. (2000) Plant functional types – a strategic perspective. *Journal of Vegetation Science* **11**, 917–922.

Shenstone, J.C. (1912) The flora of London building sites. *Journal of Botany* **50**, 117–124.

Shepard, R.N. (1962) The analysis of proximities: multidimensional scaling with an unknown distance function. *Psychometrika* **27**, 125–139; 219–246.

Shimwell, D.W. (1971) *Description and Classification of Vegetation*, Sidgewick and Jackson, London.

Shmida, A. and Wilson, M.V. (1985) Biological determinants of species diversity. *Journal of Biogeography* **12**, 1–20.

Shugart, H.H. (1997) Plant and ecosystem functional types, in *Plant Functional Types* (eds T.M. Smith, H.H. Shugart and F.I. Wodward), Cambridge University Press, Cambridge, pp. 20–43.

Shurin, J.B., Gergel, S.E., Kaufman, D.M., *et al.* (2001) Letter: In defense of ecology. *The Scientist* **15**, 6–7.

Sibley, D. (1987) *Spatial Applications of Exploratory Data Analysis*, Concepts and Techniques in Modern Geography No. 49, Geo Books, Norwich.

Siegel, S. (1956) *Non-parametric Statistics for the Social Sciences*, McGraw-Hill/Kogakusha, New York and Tokyo.

Silk, J. (1979) *Statistical Concepts in Geography*, George Allen and Unwin, London.

Silvertown, J. (2004) Plant coexistence and the niche. *Trends in Ecology and Evolution* **19**, 605–611.

Silvertown, J. and Charlesworth, D. (2001) *Introduction to Plant Population Biology* (4th edn), Blackwell Science, Oxford.

Simpson, E.H. (1949) Measurement of diversity. *Nature* **163**, 688.

Smart, S.M., Clarke, R.T., van de Poll, H.M., *et al.* (2003a) National-scale vegetation change across Britain: an analysis of sample-based surveillance data from the Countryside Surveys of 1990 and 1998. *Journal of Environmental Management* **67**, 239–254.

Smart, S.M., Robertson, J.C., Shield, E.J. and van de Poll, H.M. (2003b) Locating eutrophication effects across British vegetation between 1990 and 1998. *Global Change Biology* **9**, 1763–1774.

Smartt, P.F.M. (1978) Sampling for vegetation survey: a flexible systematic model for sample location. *Journal of Biogeography* **5**, 43–56.

Smartt, P.F.M., Meacock, S.E. and Lambert, J.M. (1974) Investigations into the properties of quantitative vegetational data. I. Pilot study. *Journal of Ecology* **62**, 735–759.

Smartt, P.F.M., Meacock, S.E. and Lambert, J.M. (1976) Investigations into the properties of quantitative vegetational data. II. Further data type comparisons. *Journal of Ecology* **64**, 41–78.

Smith, B. and Wilson, J.B. (1996) A consumer's guide to evenness indices. *Oikos* **76**, 70–82.

Smith, E.P. (1998) Randomisation methods and the analysis of multivariate ecological data. *Environmetrics* **9**, 37–51.

Smith, R.J., and Atkinson, K. (1975) *Techniques in Pedology: a Handbook for Environmental and Resource Studies*, Elek Science, London.

Smith, S.M. (1995) Distribution-free and robust statistical methods: viable alternatives to parametric statistics. *Ecology* **76**, 1997–1998.

Smith, T. and Huston, M. (1989) A theory of the spatial and temporal dynamics of plant communities. *Vegetatio* **83**, 49–69.

Smith, T.M., Shugart, H.H., Woodward, F.I. and Burton, P.J. (1993) Plant functional types, in *Vegetation Dynamics and Global Change* (eds A.M. Solomon and H.H. Shugart), Chapman and Hall, London, pp. 272–292.

Smits, N.A.C., Schaminée, J.H.J. and van Duuren, L. (2002) 70 years of permanent plot research in the Netherlands. *Applied Vegetation Science* **5**, 121–126.

Sneath, P.H.A. and Sokal, R.R. (1973) *Numerical Taxonomy*, Freeman, San Francisco.

Sobolev, L.N. and Utekhin, V.D. (1978) Russian (Ramensky) approaches to community systematization, in *Ordination of Plant Communities* (ed. R.H. Whittaker), Junk, The Hague, pp. 71–98.

Sodhi, N.S. and Brook, B.W. (2006) *South-east Asian Biodiversity in Crisis*, Cambridge University Press, Cambridge.

Sokal, R.R. (1974) Classification: purposes, principles, progress and prospects. *Science* **185**, 1115–1123.

Sokal, R.R. (1979) Testing statistical significance of geographic variation patterns. *Systematic Zoology* **28**, 627–632.

Sokal, R.R. and Michener, C.D. (1958) A statistical method for evaluating systematic relationships. *University of Kansas Science Bulletin* **38**, 1409–1438.

Sokal, R.R. and Oden, N.L. (1978a) Spatial autocorrelation in biology. 1. Methodology. *Biological Journal of the Linnean Society* **10**, 199–228.

Sokal, R.R. and Oden, N.L. (1978b) Spatial autocorrelation in biology. 2. Some biological implications and four applications of ecological and evolutionary interest. *Biological Journal of the Linnean Society* **10**, 229–249.

Sokal, R.R., Oden, N.L. and Thomson, B.A. (1998a) Local spatial autocorrelation in biological variables. *Biological Journal of the Linnean Society* **65**, 41–62.

Sokal, R.R., Oden, N.L. and Thomson, B.A. (1998b) Local spatial autocorrelation in a biological model. *Geografiska Annaler* **65**, 41–62.

Song, Y.C. (1988) The essential characteristics and main types of the broad-leaved evergreen forest in China. *Phytocoenologia* **16**, 105–123.

Song, Y.C. (1995) On the global position of the evergreen broad-leaved forests of China, in *Vegetation Science in Forestry: Global Perspective Based on Forest Ecosystems of East and South-east Asia* (eds E.O. Box, R.K. Peet, T. Masuzawa, *et al.*), Springer, Berlin, pp. 69–84.

Song, Y.C. and Wang, X.R. (1995) *Vegetation and Flora of Tiantong National Forest Park, Zheijiang Province, China* (in Chinese with English summary), Shanghai Science and Technology Literature Press, Shanghai.

Sørensen, T. (1948) A method of establishing groups of equal amplitude in plant sociology based on similarity of species content. *Det Kongelige Danske Videnskabernes Selskab, Biologiske Skrifter, Bind V, Nr. 4*, Copenhagen.

Sousa, W.P. (1984) The role of disturbance in natural communities. *Annual Review of Ecology and Systematics* **15**, 353–391.

South, G.R. (ed.) (1983) *Biogeography and Ecology of Newfoundland*, Junk, The Hague.

Southall, E.J., Dale, M.P. and Kent, M. (2003) Spatial and temporal analysis of vegetation mosaics for conservation: poor fen communities in a Cornish valley mire. *Journal of Biogeography* **30**, 1427–1443.

Southwood, T.R.E. and Henderson, P.A. (2000) *Ecological Methods* (3rd edn), Oxford, Blackwell Science.

Spellerberg, I.F. (1991) *Monitoring Ecological Change*, Cambridge University Press, Cambridge.

Spellerberg, I.F. (1992) *Evaluation and Assessment for Conservation*, Chapman and Hall, London.

Spellerberg, I.F. and Fedor, P.J. (2003) A tribute to Claude Shannon (1916–2001) and a plea for more rigorous use of species richness, species diversity and the 'Shannon-Wiener' Index. *Ecography* **12**, 177–180.

Spencer, H.J. and Port, G.R. (1988) Effects of roadside conditions on plants and insects. II. Soil conditions. *Journal of Applied Ecology* **25**, 709–715.

Spencer, H.J., Scott, N.E., Port, G.R. and Davison, A.W. (1988) Effects of roadside conditions on plants and insects. I. Atmospheric conditions. *Journal of Applied Ecology* **25**, 699–707.

Stace, C. (2010) *New Flora of the British Isles* (3rd edn), Cambridge University Press, Cambridge.

Stephens, P.A., Buskirk, S.W., Hayward, G.D. and Martínez del Rio, C. (2005) Information theory and hypothesis testing: a call for pluralism. *Journal of Applied Ecology* **42**, 4–12.

Stephens, P.A., Buskirk, S.W., Hayward, G.D. and Martínez del Rio, C. (2007) A call for statistical pluralism answered. *Journal of Applied Ecology* **44**, 461–463.

Stevens, M.H. (2009) *A Primer of Ecology with R*, Springer, New York.

Stohlgren, T.J., Coughenour, M.B., Chong, G.W., *et al.* (1997) Landscape analysis of plant diversity. *Landscape Ecology* **12**, 155–170.

Storch, D. and Gaston, K.J. (2004) Untangling ecological complexity on different scales of space and time. *Basic and Applied Ecology* **5**, 389–400.

Sukopp, H., Hejny, S. and Kowarik, I. (eds) (1990) *Urban Ecology: Plants and Plant Communities in Urban Environments*, SPB Academic Publishing, The Hague, The Netherlands.

Sun, D., Hnatiuk, R.J. and Neldner, V.J. (1997) Review of vegetation classification and mapping systems undertaken by major forested land management agencies in Australia. *Australian Journal of Botany* **45**, 929–948.

Sutherland, W.J. (ed.) (2006a) *Ecological Census Techniques: a Handbook* (2nd edn), Cambridge University Press, Cambridge.

Sutherland, W.J. (2006b) Planning a research project, in *Ecological Census Techniques: a Handbook* (2nd edn) (ed. W.J. Sutherland), Cambridge University Press, Cambridge, pp. 1–10.

Sykes, J.M., Horrill, A.D. and Mountford, M.D. (1983) Use of visual cover estimates as quantitative estimates of some British woodland taxa. *Journal of Ecology* **51**, 141–186.

Tabeni, S., Mastrantonio, L. and Ojeda, R.A. (2007) Linking small desert mammal distribution to habitat structure in a protected and grazed landscape of the Monte, Argentina. *Acta Oecologia* **31**, 259–269.

Tansley, A.G. (1911) *Types of British Vegetation*, Cambridge University Press, Cambridge.

Tansley, A.G. (1920) The classification of vegetation and the concept of development. *Journal of Ecology* **8**, 118–149.

Tansley, A.G. (1935) The use and abuse of vegetational concepts and terms. *Ecology* **16**, 284–307.

Tansley, A.G. (1939) *The British Islands and their Vegetation*, Cambridge University Press, Cambridge.

Tansley, A.G. (1949) *Britain's Green Mantle: Past, Present and Future*, Allen and Unwin, London.

Tausch, R.J., Charlet, D.A., Weixelman, D.A. and Zamudio, D.C. (1995) Patterns of ordination and classification instability resulting from changes in input data order. *Journal of Vegetation Science* **6**, 897–902.

Tenenhaus, M. and Young, F. (1985) An analysis and synthesis of multiple correspondence analysis, optimal scaling, dual scaling, homogeneity analysis and other methods for quantifying categorical multivariate data. *Psychometrika* **50**, 91–119.

ter Braak, C.J.F. (1982) *DISCRIM – a modification to TWINSPAN (Hill, 1979) to construct simple discriminant functions and to classify attributes, given a hierarchical classification of samples*. Report C82 ST10756. TNO Institute of Mathematics, Information Processing and Statistics, Wageningen.

ter Braak, C.J.F. (1983) Principal components biplots and alpha and beta diversity. *Ecology* **64**, 454–462.

ter Braak, C.J.F. (1985) Correspondence analysis of incidence and abundance data: properties in terms of a unimodal response model. *Biometrics* **41**, 859–873.

ter Braak, C.J.F. (1986a) Canonical correspondence analysis: a new eigenvector technique for multivariate direct gradient analysis. *Ecology* **67**, 1167–1179.

ter Braak, C.J.F. (1986b) Interpreting a hierarchical classification with simple discriminant function: an ecological example, in *Data Analysis and Informatics* 4 (eds E. Diday *et al.*), North-Holland, Amsterdam, pp. 11–21.

ter Braak, C.J.F. (1987) The analysis of vegetation-environment relationships by canonical correspondence analysis. *Vegetatio* **64**, 69–77.

ter Braak, C.J.F. (1988a) *CANOCO – a FORTRAN Program for Canonical Community Ordination by [Partial] [Detrended] [Canonical] Correspondence Analysis (Version 2.0)*. TNO Institute of Applied Computer Science, Wageningen.

ter Braak, C.J.F. (1988b) CANOCO – an extension of DECORANA to analyse species-environment relationships. *Vegetatio* **75**, 159–160.

ter Braak, C.J.F. (1988c) Partial canonical correspondence analysis, in *Classification and Related Methods of Data Analysis* (ed. H.H. Bock), Elsevier, North Holland, pp. 551–558.

ter Braak, C.J.F. (1994) Canonical community ordination: 1. Basic theory and linear methods. *Ecoscience* **1**, 127–140.

ter Braak, C.J.F. (1995) Ordination, in *Data Analysis in Community and Landscape Ecology* (eds R.H.G. Jongman, C.J.F. ter Braak and O.F.R van Tongeren), Cambridge University Press, Cambridge, pp. 91–173.

ter Braak, C.J.F., Hoijtink, H., Akkermans, W. and Verdonschot, P.F. (2003) Bayesian model-based cluster analysis for predicting macrofaunal communities. *Ecological Modelling* **160**, 235–248.

ter Braak, C.J.F. and Prentice, I.C. (1988) A theory of gradient analysis. *Advances in Ecological Research* **18**, 271–317.

ter Braak, C.J.F. and Šmilauer, P. (2003) *CANOCO Reference Manual and User's Guide to Canoco for Windows: Software for Canonical Community Ordination (Version 4.5)*. Microcomputer Power, Ithaca, New York.

Thioulouse, J., Chessell, D., Dolédec, S. and Olivier, J.-M. (1997) ADE-4: a multivariate analysis and display software. *Statistics and Computing* **7**, 75–83.

Thompson, K., Hodgson, J.G., Grime, J.P., *et al.* (1993) Ellenberg numbers revisited. *Phytocoenologia* **23**, 277–289.

Thompson, K., Hodgson, J.G., Smith, R.M., *et al.* (2004) Urban domestic gardens (III): Composition and diversity of lawn floras. *Journal of Vegetation Science* **15**, 373–378.

Tibshirani, R. (1992) Principal curves revisited. *Statistics and Computing* **2**, 183–190.

Tichý, L. (2002) JUICE, software for vegetation classification. *Journal of Vegetation Science* **13**, 451–453.

Tichý, L. and Holt, J. (2006) *JUICE, program for management, analysis and classification of ecological data*. Program manual, available at http://www.sci.muni.cz/botany/juice/.

Tilman, D. (1982) *Resource Competition and Community Structure*. Princeton University Press, Princeton, NJ.

Tilman, D. (1988) *Plant Strategies and the Dynamics and Structure of Plant Communities*. Princeton University Press, Princeton, NJ.

Tilman, D. (1996) Biodiversity: population versus ecosystem stability. *Ecology* **77**, 350–363.

Tilman, D. (1999) The ecological consequences of changes in biodiversity: a search for general principles. *Ecology* **80**, 1455–1474.

Tilman, D. (2001) Functional diversity, in *Encyclopedia of Biodiversity* (ed. A. Levin), Academic Press, San Diego, CA, pp. 109–120.

Tilman, D. (2007) Resource competition and plant traits: a response to Craine *et al.* (2005). *Journal of Ecology* **95**, 231–234.

Tilman, D., Lehman, C.L. and Bristow, C.E. (1998) Diversity-stability relationships: statistical inevitability or ecological consequence. *American Naturalist* **151**, 277–282.

Tilman, D. and Pacala, S. (1993) The maintenance of species richness in plant communities, in *Species in Ecological Communities – Historical and Ecological Perspectives* (eds R.E. Ricklefs and D. Schluter), University of Chicago Press, Chicago, pp. 13–25.

Tivy, J. (1982) *Biogeography: a Study of Plants in the Ecosphere* (1st edn), Oliver and Boyd, Edinburgh.

Tivy, J. (1993) *Biogeography: a Study of Plants in the Ecosphere* (3rd edn), Longman, London.

Tobler, W.R. (1970) A computer movie simulating urban growth in the Detroit region. *Economic Geography* **46**, 234–240.

Todd, J.A. and White, P.S. (2009) A new cost-distance model for human accessibility and an evaluation of accessibility bias in permanent vegetation plots in Great Smoky Mountains National Park, USA. *Journal of Vegetation Science* **20**, 1099–1109.

Tomlinson, R. (1981) A rapid sampling technique suitable for expedition use, with reference to the vegetation of the Faroe Islands. *Biological Conservation* **20**, 69–81.

Tong, S.T.Y. (1989) On non-metric multidimensional scaling ordination and interpretation of the Mattoral vegetation in lowland Murcia. *Vegetatio* **79**, 65–74.

Townend, J. (2002) *Practical Statistics for Environmental and Biological Scientists*, John Wiley and Sons, Chichester.

Townsend, C. and Begon, M. (2002) *Essentials of Ecology* (2nd edn), Blackwell Publishing, Oxford.

Treitz, P.M., Howerth, P.J., Shuffling, R.C. and Smith, P. (1992) Application of detailed ground information to vegetation mapping with high spatial resolution digital imagery. *Remote Sensing of Environment* **42**, 65–82.

Trodd, N.M. (1996) Analysis and representation of heathland vegetation from near-ground level remote sensing. *Global Ecology and Biogeography Letters* **5**, 206–216.

Tuittila, E.-S., Väliranta, M., Laine, J. and Korhola, A. (2007) Quantifying patterns and controls of mire vegetation succession in a southern boreal bog in Finland using partial ordinations. *Journal of Vegetation Science* **18**, 891–902.

Tukey, J.W. (1969) Analysing data: sanctification or detective work. *American Psychologist* **24**, 83–91.

Tukey, J.W. (1977) *Exploratory Data Analysis*, Addison-Wesley, Reading, MA.

Turner, M.G. (2005a) Landscape ecology in North America: past, present and future. *Ecology* **86**, 1967–1974.

Turner, M.G. (2005b) Landscape ecology: what is the state of the science? *Annual Review of Ecology Evolution and Systematics* **36**, 319–344.

Turner, M.G. (2010) Disturbance and landscape dynamics in a changing world. *Ecology* **91**, 2833–2849.

Turner, M.G., Gardner, R.H. and O'Neill, R.V. (2001) *Landscape Ecology in Theory and Practice*, Springer, New York.

Tutin, T.G., Heywood, V.H., Burges, N.A., *et al.* (1964–93) *Flora Europaea* (5 vols), Cambridge University Press, Cambridge.

Underwood, A.J. (1997) *Experiments in Ecology: their Logical Design and Interpretation Using Analysis of Variance*. Cambridge University Press, Cambridge.

Urban, D.L. (2002) Classification and regression trees, in *Analysis of Ecological Communities* (eds B. McCune and J.B. Grace), MjM Software, Gleneden Beach, OR.

Urban, D., Goslee, S., Pierce, K. and Lookingbill, T. (2002) Extending community ecology to landscapes. *Ecoscience* **9**, 200–212

Usher, M.B. (ed.) (1986) *Wildlife Conservation Evaluation*, Chapman and Hall, London.

van Andel, J. (2005) Species interactions structuring plant communities, in *Vegetation Ecology* (ed. E. Van der Maarel), Blackwell Publishing, Oxford, pp. 238–264.

van Andel, J., Bakker J.P. and Grootjans, A.P. (1993) Mechanisms of vegetation succession: a review of concepts and perspectives. *Acta Botanica Neerlandica* **42**, 413–433.

van den Brink, P.J. and ter Braak, C.J.F. (1997) Principal response curves: analysis of time-dependent multivariate responses of biological community to stress. *Environmental Toxicology and Chemistry* **18**, 138–148.

van den Brink, P.J. and ter Braak, C.J.F. (1998) Multivariate analysis of stress in experimental ecosystems by principal response curves and similarity analysis. *Aquatic Ecology* **32**, 163–178.

van den Wollenberg, A.L. (1977) Redundancy analysis – an alternative to canonical correlation analysis. *Psychometrika* **42**, 207–219.

van der Maarel, E. (1975) The Braun-Blanquet approach in perspective. *Vegetatio* **30**, 13–19.

van der Maarel, E. (1979) Transformation of cover-abundance values in phytosociology and its effects on community similarity. *Vegetatio* **39**, 97–114.

van der Maarel, E. (1984a) Vegetation science in the 1980s, in *Trends in Ecological Research for the 1980s* (eds J.H. Cooley and F.B. Golley), Plenum Press, New York, 89–110.

van der Maarel, E. (1984b) Dynamics of plant populations from a synecological viewpoint, in *Perspectives on Plant Ecology* (eds R. Dirzo and J. Sarukhan), Sinauer Associates Inc., Sunderland, MA.

van der Maarel, E. (1988) Vegetation dynamics: patterns in time and space. *Vegetatio* **77**, 7–19.

van der Maarel, E. (1989) Theoretical vegetation science on the way. *Vegetatio* **83**, 1–6.

van der Maarel, E. (1993) Relations between sociological-ecological species groups and Ellenberg indicator values. *Phyto-coenologia* **23**, 343–362.

van der Maarel, E. (ed.) (2005a) *Vegetation Ecology*, Blackwell Publishing, Oxford.

van der Maarel, E. (2005b) Vegetation ecology – an overview, in *Vegetation Ecology* (ed. E. van der Maarel), Blackwell Publishing, Oxford, pp. 1–51.

van der Maarel, E. (2007) Transformation of cover-abundance values for appropriate numerical treatment – alternatives to the proposals by Podani. *Journal of Vegetation Science* **18**, 767–770.

van der Maarel, E., Janssen, J.G.M. and Louppen, J.M.W. (1978) TABORD – a program for structuring phytosociological tables. *Vegetatio* **38**, 143–156.

van der Maarel, E., Boot, R., van Dorp, D. and Rijntjes, J. (1985) Vegetation succession on the dunes near Oostvoorne, the Netherlands, a comparison of the vegetation in 1959 and 1980. *Vegetatio* **58**, 137–187.

van der Maarel, E., Espejel, I. and Moreno-Casasola, P. (1987) Two-step vegetation analysis based on very large data sets. *Vegetatio* **68**, 139–143.

van Emden, H. (2008) *Statistics for Terrified Biologists*, Blackwell Publishing, Oxford.

van Groenewoud, H. (1992) The robustness of correspondence, detrended correspondence and TWINSPAN analysis. *Journal of Vegetation Science* **3**, 239–246.

Vanha-Majamaa, I., Salemaa, M., Tuominen, S. and Mikkola, K. (2000) Digitized photographs in vegetation analysis – a comparison of cover estimates. *Applied Vegetation Science* **3**, 89–94.

van Tongeren, O.F.R. (1986) FLEXCLUS: an interactive program for classification and tabulation of ecological data. *Acta Botanica Nederlandia* **35**, 137–142.

van Tongeren, O.F.R. (1995) Cluster analysis, in *Data Analysis in Community and Landscape Ecology* (eds R.H.G. Jongman, C.F.J. ter Braak and O.F.R. van Tongeren), Cambridge University Press, Cambridge, pp. 174–212.

Vayssieres, M.P., Plant, R.E. and Allen-Diaz, B.H. (2000) Classification trees: an alternative approach for predicting species distributions. *Journal of Vegetation Science* **11**, 679–694.

Vellend, M. (2001) Do commonly used indices of β-diversity measure species turnover? *Journal of Vegetation Science* **12**, 545–552.

Vellend, M., Lilley, P.L. and Starzomski, B.M. (2008) Using subsets of species in biodiversity surveys. *Journal of Applied Ecology* **45**, 161–169.

ver Hoef, J.M., Cressie, N., Fisher, R.N. and Case, T.J. (2001) Uncertainty and spatial linear models for ecological data, in *Spatial Uncertainty for Ecology: Implications for Remote Sensing and GIS Applications* (eds C.T. Hunsaker, M.F. Goodchild, M.A. Friedl and T.J. Case), Springer, New York, pp. 214–237.

Vevle, O. (1999) *Ellenbergs økologiske faktortal. Liste for moser og lav utarbeidet for norske forhold.* Available at http://www2.hit.no/af/nv/nvlink/flora/hit1mos141299.htm.

Villaseñor, J.L., Ibarra-Manriquez, G., Meave, J.A. and Ortiz, E. (2005) Higher taxa as surrogates of plant biodiversity in a megadiverse country. *Conservation Biology* **19**, 232–238.

Villéger, S., Mason, N.W.H. and Mouillot, D. (2008) New multidimensional functional diversity indices for a multifaceted framework in functional ecology. *Ecology* **89**, 2290–2301.

Vittoz, P. and Guisan, A. (2007) How reliable is the monitoring of permanent vegetation plots? A test with multiple observers. *Journal of Vegetation Science* **18**, 413–422.

von Post, H. (1862) *Försök till en systematik uppställning af vextstallena i mellersta Sverige*, Bonnier, Stockholm.

von Wehrden, H., Hanspach, J., Bruelheide, H. and Wesche, K. (2009) Pluralism and diversity: trends in the use and application of ordination methods 1990–2007. *Journal of Vegetation Science* 1–11.

Wackernagel, H. (2003) *Multivariate Geostatistics: an Introduction with Applications* (3rd edn), Springer, Berlin/London.

Wadsworth, R. and Treweek, J. (1999) *Geographical Information Systems for Ecology: an Introduction*, Longman, Harlow.

Wagner, H.H. (2003) Spatial covariance in plant communities: integrating ordination, geostatistics and variance testing. *Ecology* **84**, 1045–1057.

Wagner, H.H. (2004) Direct multi-scale ordination with canonical correspondence analysis. *Ecology* **85**, 342–351.

Waichler, W.S., Miller, R.F. and Doescher, P.S. (2001) Community characteristics of old-growth western juniper woodlands. *Journal of Range Management* **54**, 518–527.

Waite, S. (2000) *Statistical Ecology in Practice*, Prentice-Hall, London.

Walker, D. (1989) Diversity and stability, in *Ecological Concepts: the Contribution of Ecology to an Understanding of the Natural World* (ed. J.M. Cherrett), British Ecological Society/Blackwell Scientific Publications, Oxford, pp. 115–145.

Walker, J., Sharpe, P.J.H., Penridge, L.K. and Wu, H. (1989) Ecological field theory: the concept and field tests. *Vegetatio* **83**, 81–95.

Walker, L.R. and del Moral, R. (2003) *Primary Succession and Ecosystem Rehabilitation*, Cambridge University Press, Cambridge.

Walker, L.R. and del Moral, R. (2008) Lessons from primary succession for restoration of severely damaged habitats. *Applied Vegetation Science* **12**, 55–67.

Walker, L.R., Walker, J. and Hobbs, R.J. (eds) (2007) *Linking Restoration and Ecological Succession*, Springer, New York.

Walsh, S.J., Davis, F.W. and Peet, R.K. (1994) Application of remote sensing and geographic information systems in vegetation science. *Journal of Vegetation Science* **5**, 657–672.

Wamelink, G.W.W., Goedhart, P.W., van Dobben, H.F. and Berendse, F. (2005) Plant species as predictors of soil pH: replacing expert judgement with measurements. *Journal of Vegetation Science* **16**, 461–470.

Wamelink, G.W.W., Joosten, V., van Dobben, H.F. and Berendse, F. (2002) Validity of Ellenberg indicator values judged from physico-chemical field measurements. *Journal of Vegetation Science* **13**, 269–278.

Wang, K., Franklin, S.E., Guo, X., *et al.* (2009) Problems in remote sensing of landscapes and habitats. *Progress in Physical Geography* **33**, 747–768.

Wang, X.-H., Kent, M. and Fang, X.-F. (2007) Evergreen broad-leaved forest in Eastern China: its ecology and conservation and the importance of resprouting in forest restoration. *Forest Ecology and Management* **245**, 76–87.

Ward, J.H. (1963) Hierarchical grouping to optimize an objective function. *American Statistical Association Journal* **58**, 236–244.

Wardlaw, A.C. (1985) *Practical Statistics for Experimental Biology*, John Wiley and Sons Ltd, Chichester.

Waring, R.H. (1989) Ecosystems: fluxes of matter and energy, in *Ecological Concepts: the Contribution of Ecology to an Understanding of the Natural World* (ed. J.M. Cherrett), British Ecological Society/Blackwell Scientific Publications, Oxford, pp. 17–41.

Wartenberg, D., Ferson, S. and Rohlf, F.J. (1987) Putting things in order: a critique of detrended correspondence analysis. *American Naturalist* **129**, 434–448.

Wassen, M.J., Barendregt, A., Palczynski, A., *et al.* (1990) The relationship between fen vegetation gradients, groundwater flow and flooding in an undrained valley mire at Biebrza, Poland. *Journal of Ecology* **78**, 1106–1122.

Wathern, P. (1976) *The Ecology of Development Sites*. Unpublished PhD. thesis, Department of Landscape Architecture, University of Sheffield.

Watts, S. and Halliwell, L. (eds) (1996) *Essential Environmental Science Methods and Techniques*, Routledge, London.

Way, J.M. (ed.) (1969) *Road Verges: their Function and Management*, Nature Conservancy, Monks Wood Experimental Station, Abbots Ripton, Huntingdon.

Webb, L.J. (1978) A structural comparison of New Zealand and South-East Australian rain forests and their tropical affinities. *Australian Journal of Ecology* **3**, 7–21.

Webb, L.J., Tracey, J.G., Williams, W.T. and Lance, G.N. (1970) Studies in the numerical analysis of complex rain forest communities. V. A comparison of the properties of floristic and physiognomic-structural data. *Journal of Ecology* **58**, 203–232.

Webb, L.J., Tracey, J.G. and Williams, W.T. (1976) The value of structural features in tropical forest typology. *Australian Journal of Ecology* **1**, 3–28.

Weber, H.E., Moravec, J. and Theurillat, J.-P. (2000) International Code of Phytosociological Nomenclature, 3rd edition. *Journal of Vegetation Science* **11**, 739–768.

Weiher, E., Clarke, G.D.P. and Keddy, P.A. (1998) Community assembly rules, morphological dispersion, and the coexistence of plant species. *Oikos* **81**, 309–322.

Weiher, E. and Keddy, P.A. (eds) (1999a) *Ecological Assembly Rules: Perspectives, Advances, Retreats*, Cambridge University Press, Cambridge.

Weiher, E. and Keddy, P.A. (1999b) Assembly rules as general constraints on community composition, in *Ecological; Assembly Rules. Perspectives, Advances, Retreats* (eds E. Weiher and P. Keddy), Cambridge University Press, Cambridge, pp. 251–271.

Weiher, E., van der Werf, A., Thompson, K., *et al.* (1999) Challenging Theophrastus: a common core list of traits for functional plant ecology. *Journal of Vegetation Science* **10**, 609–620.

Werger, M.J.A. (1974a) The place of the Zurich-Montpellier method in vegetation science. *Folia Geobotanica et Phytotax-onomica* **9**, 99–109.

Werger, M.J.A. (1974b) On concepts and techniques applied in the Zurich-Montpellier method of vegetation survey. *Bothalia* **11**, 309–323.

Werger, M.J.A. and Sprangers, J.T.C. (1982) Comparison of floristic and structural classification of vegetation. *Vegetatio* **50**, 175–183.

Wessells, K.J., van Jaarsveld, A.S., Grimbeek, J.D. and van der Linde, M.J. (1998) An evaluation of the gradsect biological survey method. *Biodiversity and Conservation* **7**, 1093–1121.

Westfall, R.H., Dednam, G., van Rooyen, N. and Theron, G.K. (1982) PHYTOTAB – a program package for Braun-Blanquet tables. *Vegetatio* **49**, 35–37.

Westfall, R.H., Theron, G.K. and Rooyen, N. (1997) Objective classification and analysis of vegetation data. *Plant Ecology* **132**, 137–154.

Westhoff, V. and van der Maarel, E. (1978) The Braun-Blanquet approach, in *Classification of Plant Communities* (ed. R.H. Whittaker), Junk, The Hague, pp. 289–374.

Westoby, M. (1999) Generalization in functional plant ecology: the species sampling problem, plant ecology strategy schemes and phylogeny, in *Handbook of Functional Plant Ecology* (eds F.I. Pugnaire and F. Valladares), Marcel Dekker, Inc., New York, pp. 847–872.

Westoby, M., Falster, D.S., Moles, A.T., *et al.* (2002) Plant ecological strategies: Some leading dimensions of variation between species. *Annual Review of Ecology and Systematics* **33**, 125–159.

Westoby, M. and Leishman, M. (1997) Categorizing plant species into functional types, in *Plant Functional Types* (eds T.M. Smith, H.H. Shugart and F.I. Woodward), Cambridge University Press, Cambridge, pp. 104–121.

Wheeler, B.D. (1980) Plant communities of rich-fen systems in England and Wales. I. Introduction. Tall sedge and reed communities. *Journal of Ecology* **68**, 365–396.

Wheeler, B., Shaw, G. and Barr, S. (2004) *Statistical Techniques in Geographical Analysis* (3rd edn), David Fulton Publishers, London.

White, R.E. (2006) *Principles and Practice of Soil Science: the Soil as a Natural Resource* (4th edn), Blackwell, Oxford.

Whitmore, T.C. (1998) *An Introduction to Tropical Rain Forests* (2nd edn), Clarendon Press, Oxford.

Whitney, G.G. (1985) A quantitative analysis of the flora and plant communities of a representative midwestern U.S. town. *Urban Ecology* **9**, 143–160.

Whitney, G.G. and Adams, S.D. (1980) Man as a maker of new plant communities. *Journal of Applied Ecology* **17**, 431–448.

Whittaker, R.H. (1948) *A Vegetation Analysis of the Great Smoky Mountains.* Unpublished PhD thesis, University of Illinois, Urbana.

Whittaker, R.H. (1951) A criticism of the plant association and climatic climax concepts. *NorthWest Science* **25**, 17–31.

Whittaker, R.H. (1953) A consideration of climax theory: the climax as a population and pattern. *Ecological Monographs* **23**, 41–78.

Whittaker, R.H. (1956) Vegetation of the Great Smoky Mountains. *Ecological Monographs* **26**, 1–80.

Whittaker, R.H. (1960) Vegetation of the Siskiyou Mountains, Oregon and California. *Ecological Monographs* **30**, 279–338.

Whittaker, R.H. (1962) Classification of plant communities. *Botanical Review* **28**, 1–239.

Whittaker, R.H. (1965) Dominance and diversity in land plant communities. *Science* **147**, 250–260.

Whittaker, R.H. (1967) Gradient analysis of vegetation. *Biological Review* **42**, 207–264.

Whittaker, R.H. (1972) Evolution and measurement of species diversity. *Taxon* **21**, 213–251.

Whittaker, R.H. (1975) *Communities and Ecosystems* (2nd edn), Macmillan, London.

Whittaker, R.H. (ed.) (1978a) *Ordination of Plant Communities*, Junk, The Hague.

Whittaker, R.H. (ed.) (1978b) *Classification of Plant Communities*, Junk, The Hague.

Whittaker, R.H., Levin, S.A. and Root, R.B. (1973) Niche, habitat and ecotope. *American Naturalist* **107**, 321–338.

Whittaker, R.H. and Levin, S.A. (1977) The role of mosaic phenomena in natural communities. *Theoretical Population Biology* **12**, 117–139.

Whittaker, R.H. and Gauch, H.G. (1978) Evaluation of ordination techniques, in *Ordination of Plant Communities* (ed. R.H. Whittaker), Junk, The Hague, pp. 277–336.

Whittaker, R.J. (1989) The vegetation of the Storbreen Gletschervorfeld, Jotunheimen, Norway. III. Vegetation-environment relationships. *Journal of Biogeography* **16**, 413–433.

Whittaker, R.J. (1991) The vegetation of the Storbreen Gletschervorfeld, Jotunheimen, Norway. IV. Short-term vegetation change. *Journal of Biogeography* **18**, 41–52.

Whittaker, R.J., Bush, M.B. and Richards, K. (1989) Plant recolonization and vegetation succession on the Krakatau Islands, Indonesia. *Ecological Monographs* **59**, 59–123.

Whittaker, R.J. and Fernández-Palacios, J.M. (2007) *Island Biogeography: Ecology, Evolution, and Conservation* (2nd edn), Oxford University Press, Oxford.

Whittaker, R.J., Willis, K.J. and Field, R. (2001) Scale and species-richess: towards a general hierarchical theory of species diversity. *Journal of Biogeography* **28**, 453–470.

Wiegleb, G. (1989) Explanation and prediction in vegetation science. *Vegetatio* **83**, 17–34.

Wiens, J.A. (1989) Spatial scaling in ecology. *Functional Ecology* **3**, 385–397.

Wiens, J.A. and Moss, M. (eds) (2005) *Issues and Perspectives in Landscape Ecology*, Cambridge University Press, Cambridge.

Wildi, O. (2010) *Data Analysis in Vegetation Ecology*, Wiley/Blackwell, Oxford.

Wildi, O. and Orlóci, L. (1996) *Numerical Exploration of Community Patterns* (2nd edn), SPB Academic Publishing, The Hague, The Netherlands.

Williams, G.J., Harris, J.R., Kemp, P.R., *et al.* (1979) Introducing the principles of vegetation sampling in the laboratory. *American Biology Teacher* **41**, 14–17.

Williams, W.T. (1971) Principles of clustering. *Annual Review of Ecology and Systemetics* **2**, 303–326.

Williams, W.T. (ed.) (1976a) *Pattern Analysis in Agricultural Science*, Elsevier, New York.

Williams, W.T. (1976b) Hierarchical divisive strategies, in *Pattern Analysis in Agricultural Science* (ed. W.T. Williams), Elsevier, New York.

Williams, W.T., Lambert, J.M. and Lance, G.N. (1966) Multivariate methods in plant ecology. V. Similarity analysis and information analysis. *Journal of Ecology* **54**, 427–445.

Willis, A.J. (1997) The ecosystem: an evolving concept viewed historically. *Functional Ecology* **11**, 268–275.

Willner, W. (2006) The association concept revisited. *Phytocoenologia* **36**, 67–76.

Wilson, J.B. (1991a) Methods for fitting dominance/diversity curves? *Journal of Vegetation Science* **2**, 35–46.

Wilson, J.B. (1999) Assembly rules in plant communities, in *Ecological Assembly Rules. Perspectives, Advances, Retreats* (eds E. Weiher and P. Keddy), Cambridge University Press, Cambridge, pp. 130–164.

Wilson, J.B. (2003) The deductive method in community ecology. *Oikos* **101**, 216–217.

Wilson, J.B. (2007) Priorities in statistics, the sensitive feet of elephants, and don't transform data. *Folia Geobotanica* **42**, 161–167.

Wilson, J.B., Gitay, H., Steel, J.B. and King, E.M. (1998) Relative abundance distributions in plant communities: effects of species richness and of spatial scale. *Journal of Vegetation Science* **9**, 213–220.

Winston, J.E. (1999) *Describing Species: Practical Taxonomic Procedure for Biologists*, Columbia University Press, New York.

Wishart, D. (1969) An algorithm for hierarchical classifications. *Biometrics* **25**, 165–170.

Witte, J.-P.M., Wójcik, R.B., Torfs, P.J.J.F., *et al.* (2007) Bayesian classification of vegetation types with Gaussian mixture density fitting to indicator values. *Journal of Vegetation Science* **18**, 605–612.

Woodward, F.I., Smith, T.M. and Shugart, H.H. (1997) Defining plant functional types: the end view, in *Plant Functional Types* (eds T.M. Smith, H.H. Shugart and F.I. Woodward), Cambridge University Press, Cambridge, pp. 355–359.

Yeo, M.J.M., Blackstock, T.H. and Stevens, D.P. (1998) The use of phytosociological data in conservation assessment: a case study of lowland grasslands in mid-Wales. *Biological Conservation* **86**, 125–138.

Yoccoz, N.G., Nichols, J.D. and Boulinier, T. (2001) Monitoring of biological diversity in space and time. *Trends in Ecology and Evolution* **16**, 446–453.

Yoccoz, N.G., Nichols, J.D. and Boulinier, T. (2003) Monitoring of biological diversity – a response to Danielsen *et al.* *Oryx* **37**, 410.

Zhang, L., Gao, Z., Armitage, R. and Kent, M. (2008) Spectral characteristics of plant communities from saltmarshes: A case study from Chongming Dongtan, Yangtze estuary, China. *Frontiers in Environmental Science and Engineering in China* **2**, 187–197.

Zitko, V. (1994) Principal component analysis in the evaluation of environmental data. *Marine Pollution Bulletin* **28**, 718–722.

Zobel, M. (1992) Plant species coexistence – the role of historical, evolutionary and ecological factors. *Oikos* **65**, 314–320.

Zuur, A.F., Ieno, E.N. and Elphick, C.S. (2010) A protocol for data exploration to avoid common statistical problems. *Methods in Ecology and Evolution* **1**, 3–14.

Zuur, A.F., Ieno, E.N. and Meesters, E.M. (2009) *A Beginner's Guide to R*, Springer, Berlin/New York.

Zuur, A.F., Ieno, E.N. and Smith, G.M. (2007) *Analysing Ecological Data*, Springer, Berlin/New York.

Index